Y0-AGH-782

NMR Spectroscopy in Pharmaceutical Analysis

NMR Spectroscopy in Pharmaceutical Analysis

Edited by

ULRIKE HOLZGRABE
Institute of Pharmacy and Food Chemistry, University of Würzburg, Germany

IWONA WAWER
Department of Physical Chemistry, Faculty of Pharmacy, Medical University of Warsaw, Poland

BERND DIEHL
Spectral Service, Köln, Germany

ELSEVIER

Amsterdam • Boston • Heidelberg • London • New York • Oxford
Paris • San Diego • San Francisco • Singapore • Sydney • Tokyo

Elsevier
Linacre House, Jordan Hill, Oxford OX2 8DP, UK
Radarweg 29, PO Box 211, 1000 AE Amsterdam, The Netherlands

First edition 2008

British Library Cataloguing in Publication Data
A catalogue record for this book is available from the British Library

Library of Congress Cataloging-in-Publication Data
A catalog record for this book is available from the Library of Congress

ISBN: 978-0-444-53173-5

Printed and bound in Hungary
08 09 10 10 9 8 7 6 5 4 3 2 1

Contents

Preface

Nuclear magnetic resonance (NMR) was discovered in 1945/1946 by two groups of physicists, and it is still a technique with rapid progress with regard to both method development and new fields of application. It has spread from Physics to Chemistry, Biochemistry, Pharmacy, Physiology, Food Science, Veterinary and Medicine. Mostly used by chemists as a tool for structure elucidation and confirmation of synthesized and natural compounds, it has also moved into the emerging field of magnetic resonance imaging (MRI). MRI has proven to be indispensable for clinicians in diagnosis and monitoring of pathology.

Moreover, for more than one decade, people started to use NMR spectroscopy for quantitative purposes. Whereas the determination of the enantiomeric excess of a chiral compound has a long-standing tradition in the organic chemistry area of asymmetric synthesis, the evaluation of drugs and agrochemicals by means of NMR spectroscopy has not yet been established from the point of view of licensing authorities all over the world even though the methods have been proved to be applicable. NMR spectroscopy in medicine also attempts to use the quantitative approaches to the biochemical characterization of disease, investigating biopsies, tissue extracts or body fluids.

The book deals with quantitative NMR spectroscopy by demonstrating the possibilities of the technique and by training the people who have not yet used NMR spectroscopy for assessment of drugs, natural products, plant extracts, agrochemicals or food.

The book is divided into three parts: Part I gives the fundamentals of solution and solid-state NMR spectroscopy and its quantitative application as well as the basics of hyphenated techniques. In Part II, principal application fields are presented and modi operandi discussed: drug evaluation, polymer characterization, natural compound assessment in mixtures, investigation of drug formulations, metabolic profiling, DOSY (diffusion-ordered spectroscopy) for faked drugs and quality control of agrochemicals. Finally, in Part III, special applications from various fields are described, e.g. vaccine evaluation, metabolic studies using ^{19}F and ^{32}P NMR, 2D techniques for glycosaminoglycan analysis, assessment of the inhibitory potency of antibiotics by MRI or hypernation of LC–UV–NMR–MS and its application in industries. This increasing significance of qNMR is currently emphasized by the quality evaluation of heparine contaminated with anaphylactoid oversulfated chrondroitin sulfate. Chapter 2 in Part II and Chapter 4 in Part III give corresponding information.

Taken together, we tried to collect all possible applications of NMR spectroscopy in the field of quantitative pharmaceutical analysis. The book is written for both students in chemistry, pharmacy or related disciplines who are already familiar with NMR spectroscopy and application chemists in pharmaceutical, agrochemical and food industries. We are sure that many stimulating ideas may emerge when reading the book.

Ulrike Holzgrabe, Bernd Diehl and Iwona Wawer
Würzburg, Cologne and Warsaw, 2008

List of Contributors

S. Balayssac
Groupe de RMN Biomédicale, Laboratoire SPCMIB, Université Paul Sabatier, Toulouse cedex, France

M. Bernstein
AstraZeneca R&D Charnwood, Loughborough, Leics., Great Britain

T. Beyer
Institute of Pharmacy and Food Chemistry, University of Würzburg, Germany

J. Cheung
Bob Wells and Associates, Killara, New South Wales, Australia

M.-A. Delsuc
Centre de Biochimie Structurale, CNRS, Université Montpellier I, Montpellier, France

B. Diehl
Spectral Service, Köln, Germany

C. Faber
Institute of Physics, University of Würzburg, Germany

F. Fang
Department of Chemistry, University of California, Physical Sciences I, Riverside, CA, USA

V. Gilard
Groupe de RMN Biomédicale, Laboratoire SPCMIB, Université Paul Sabatier, Toulouse cedex, France

T. Gostan
NMRtec, Illkirch Graffenstaden, France

M. Guerrini
Instituto Ricerche Chimiche e Biochimiche “G. Ronzoni”, Citta Studi, Milano, Italy

H. Hasse
Institut für Technische Thermodynamik und Thermische Verfahrenstechnik, Universität Stuttgart, Germany

K.A. Hamersky
Department of Chemistry, University of California, Physical Sciences I, Riverside, CA, USA

U. Holzgrabe
Institute of Pharmacy and Food Chemistry, University of Würzburg, Germany

J.M. Hook
NMR Facility, Universtity of NSW, Sydney, Australia

E. Humpfer
Bruker BioSpin GmbH, Silberstreifen, Rheinstetten, Germany

C. Jones
Laboratory for Molecular Structure, National Institute for Biological Standards and Control Blanche Lane, South Mimms, United Kingdom

E. Kellenbach
Organon NV, Quality & Regulatory Unit, Analytical Development, Part of Schering-Plough Corporation Oss, The Netherlands

C.K. Larive
Department of Chemistry, University of California, Physical Sciences I, Riverside, CA, USA

M. Maiwald
Merck KGaA, Zentrale Forschungsanalytik, Darmstadt, Germany

M. Malet-Martino
Groupe de RMN Biomédicale, Laboratoire SPCMIB, Université Paul Sabatier, Toulouse cedex, France

F. Malz
Federal Institute for Materials Research and Testing, Berlin, Germany, Dr. Frank Malz, German Institute for Polymers (DKI), Darmstadt, Germany

R. Martino
Groupe de RMN Biomédicale, Laboratoire SPCMIB, Université Paul Sabatier, Toulouse cedex, France

C.E. Merrywell
Department of Chemistry, University of California, Physical Sciences I, Riverside, CA, USA

M. Mörtter
Bruker Biospin GmbH, Silberstreifen, Rheinstetten, Germany

Y. Prigent
Fédération de Chimie, Université Paul Sabatier, Toulouse cedex, France

N. Ravenscroft
Department of Chemistry, University of Cape Town, South Africa

P. Rinke
Bruker Biospin GmbH, Silberstreifen, Rheinstetten, Germany

K. Sanders
Organon NV, Quality & Regulatory Unit, Analytical Development, Part of Schering-Plough Corporation Oss, The Netherlands

H. Schäfer
Bruker Biospin GmbH, Silberstreifen, Rheinstetten, Germany

B. Schütz
Bruker Biospin GmbH, Silberstreifen, Rheinstetten, Germany

C. Sleigh
AstraZeneca R&D Charnwood, Loughborough, Leics., Great Britain

M. Spraul
Bruker BioSpin GmbH, Silberstreifen, Rheinstetten, Germany

O. Steinhof
Institut für Technische Thermodynamik und Thermische Verfahrenstechnik, Universität Stuttgart, Germany

F. Taulelle
Tectospin, Institut Lavoisier, UMR CNRS 8180, Université de Versailles St-Quentin en Yvelines, Versailles cedex, France

G. Torri
Instituto Ricerche Chimiche e Biochimiche "G. Ronzoni", Citta Studi, Milano, Italy

S. Trefi
Groupe de RMN Biomédicale, Laboratoire SPCMIB, Université Paul Sabatier, Toulouse cedex, France

I. Wawer
Department of Physical Chemistry, Faculty of Pharmacy, Medical University of Warsaw, Poland

A.G. Webb
Department of Bioengineering, Director Huck Institute Magnetic Resonance Centre, Penn State University, University Park, PA, USA

R.J. Wells
Bob Wells and Associates, Gordon, New South Wales, Australia

I.D. Wilson
AstraZeneca, Dept. of Drug Metabolism and Pharmacokinetics, Mereside, Alderley Park, Macclesfield, Cheshire, UK

G. Zomer
Netherlands Vaccine Institute, Unit Research and Development, Bilthoven, The Netherlands

List of Editors

Prof. Dr. U. Holzgrabe
Institute of Pharmacy and Food Chemistry
University of Würzburg
Am Hubland
D-97074 Würzburg, Germany

Prof. Dr. I. Wawer
Department of Physical Chemistry
Faculty of Pharmacy
Medical University of Warsaw
ul. Banacha 1
PL-02-097 Warszawa

Dr. B. Diehl
Spectral Service
Emil-Hoffmann-Str. 33
D-50996 Köln, Germany

PART I

FUNDAMENTALS AND TECHNIQUES

CHAPTER 1

Principles in NMR Spectroscopy

B. Diehl

Contents

Abstract

More than any other analytical method the nuclear magnetic resonance (NMR) spectroscopy provides information about the chemical structure and the dynamics of organic molecules. The interpretation of the NMR data depends on a minimum amount of basic information that will be provided in this chapter. For a better appliance, these basics of one- and two-dimensional NMR techniques are demonstrated by means of practical examples.

Keywords: chemical shift, relaxation, spin–spin coupling, molecular dynamics, heteronuclear spectra, stereochemistry, structure elucidation

1. Short History

Nuclear magnetic resonance (NMR) spectroscopy has been developed to be the most powerful analytical method. It allows the visualisation of single atoms and molecules in various media in solution as well as in solid state. NMR is non-destructive and gives molar response that allows structure elucidation and quantification simultaneously. Magnetic interactions between NMR-active nuclei along covalent bindings result in spin–spin (^nJ-) couplings. Through-space interactions can be detected using the nuclear Overhauser effect (NOE). Both interactions enable three-dimensional structure elucidation.

The steady progress of NMR spectroscopy can clearly be seen in the list of Noble Prize winners. In 1944 the first Nobel Prize in physics was awarded to Rabi for the development of a resonance method that enables recording of the magnetic properties of atomic nuclei. Bloch and Purcell received the Prize in 1952 as a tribute to the first practical NMR experiments, which were carried out independently by both of them in 1945 at different places. By then, the NMR spectroscopy started to become more than a physical experiment. By the discovery of the "chemical shift" the method has become a tool for chemists in structure elucidation. The first useful NMR spectrometers were continuous wave (CW) instruments using permanent or electromagnets. Their utility came to an end with the upcoming superconductor magnets in the 1970s. However, only since Ernst developed the basics of the Fourier transformation (FT) method, the foundation of the modern NMR spectroscopy methods was laid. Since NMR spectroscopy was by then a domain of physicians, Ernst was the first chemist in the list of Nobel Prize winners in 1991. A decade later, Wüthrich was the second honoured chemist. He received the Prize in 2002 for the elucidation of three-dimensional structures of macromolecules. The NMR technique has become an important tool in other scientific fields, especially in medicine. It is not surprising that only one year later the NMR technique was honoured again, and the Nobel Prizes were awarded to Lauterbach and Mansfield for their research in magnetic resonance imaging. Rightly, the NMR community expects further Prizes in one of the widespread application areas of NMR spectroscopy in the future.

This book shall give its readers an overview about the NMR techniques used in pharmaceutical applications and help the method to become accepted as the most significant analytical tool in the pharmacopoeia. It is written for pharmacists and chemists and cannot be an NMR textbook. A list of recommended literature is given in references,[1–7] including basic information on NMR imaging.[8] Even in the World Wide Web very useful information is available to understand the basics of NMR spectroscopy.[9] However, a short summary of some basics and important principles of NMR is given in this chapter, using the example of a full structure elucidation on rifamycin and its derivatives by one- and two-dimensional ^{1}H- and ^{13}C NMR techniques. This substance was chosen rather for practical reasons, as it is one of the drugs routinely analysed in our NMR laboratory. However, rifamycin shows a wide variety of interesting substructures and is therefore most suited to illustrate the basics of NMR spectroscopy.

2. THE NMR EXPERIMENT

Within the group of spectroscopic methods the NMR spectroscopy uses the lowest irradiation energy for excitation (Figure 1). Owing to the low-energy level excitation, relaxation and sensitivity of NMR spectroscopy are specifically different from other spectroscopic methods.

2.1. Excitation, relaxation and sensitivity

Molecular spectroscopy is a subject of quantum physics. Excitation of molecular movement or vibration as used with infrared (IR) spectroscopy or the excitation of an electron in a higher π-orbital as used for ultraviolet/visible (UV/VIS) spectroscopy is possible in a normal surrounding. The excitation, in principle, requires two different energy levels: for IR and UV spectroscopy, respectively, these conditions are given at any place without the need of technical resources. However, this is not the case in NMR spectroscopy. The irradiated energy can only interact with the nuclear spin quantum status, if an artificially produced magnetic field affects the generation of two different energy levels.

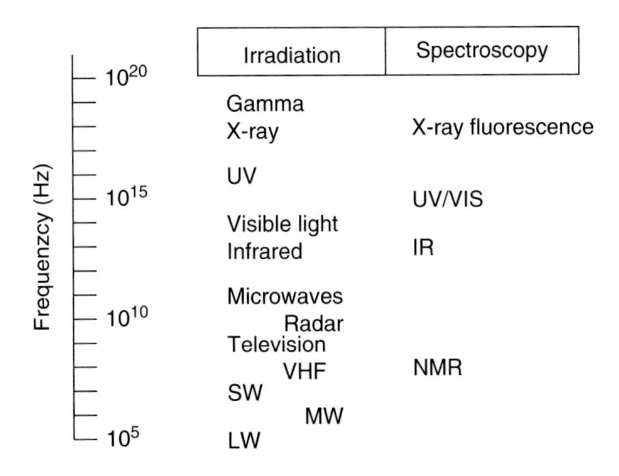

Figure 1 Energy levels of different spectroscopic methods.

A proton is a charged elementary particle. Similar to a spinner, its movement results in the generation of a magnetic field and therefore protons can be described as elementary magnets. The spin quantum number of a proton is 1/2. The possible angular impulse quantum numbers m_I therefore restrict the different energy states to two $m_I = \pm 1/2$. Without a magnetic field ($B_0 = 0$) both the quantum states are equivalent; therefore, no excitation is possible. In case a magnetic field is installed, different energy levels A and A^* are generated (Figure 2). The energy difference $\Delta E = h\nu$ is a linear function of the magnetic field strength. The higher the magnetic field, the higher is ΔE and the resonance frequency ν, which is known as the MHz value for different spectrometer types. For simplification, the ^{1}H NMR resonance frequency is used for defining the magnetic field strength instead of the SI unit tesla.

The large numbers of singular elementary magnets are arranged in both the possible energy levels A and A^*. A major number will stay in the lower energetic A state. The higher the magnetic field strength, the more elementary magnets will occupy this level. This affects the sensitivity of an NMR experiment. The higher the ability of energy absorption, the higher is the sensitivity of the method. Only the excess number of spins in the lower A level can be excited into the higher A^* level. If the population in both the levels is equal, no further excitation is possible. From the Boltzmann distribution, the difference of the population $\Delta n = n_A - n_{A^*}$ can be calculated by

$$\frac{n_A}{n_{A^*}} = e^{\Delta E/kt} \tag{1}$$

Δn is very high for spectroscopic methods, which use high-energy radiation, e.g. UV/VIS but very low for NMR. At 600 MHz, the ratio n_A/n_{A^*} is only 0.999904. The Boltzmann distribution therefore describes not only the strong dependence of the sensitivity of NMR experiments on the magnetic field strength, but also the relative insensitivity compared with other spectroscopic methods.

To understand the data evaluation within an NMR experiment, the rotating coordinate system is a useful tool. In the laboratory coordinate system, the elementary magnetic moments are added to the macroscopic magnetisation, which is in

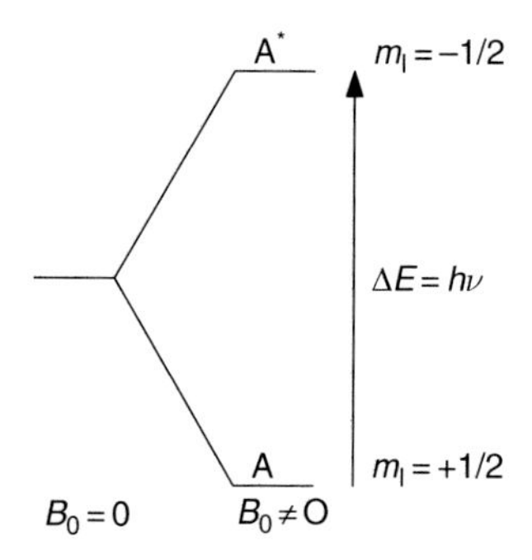

Figure 2 Energy levels of spin 1/2 states split by an external magnetic field.

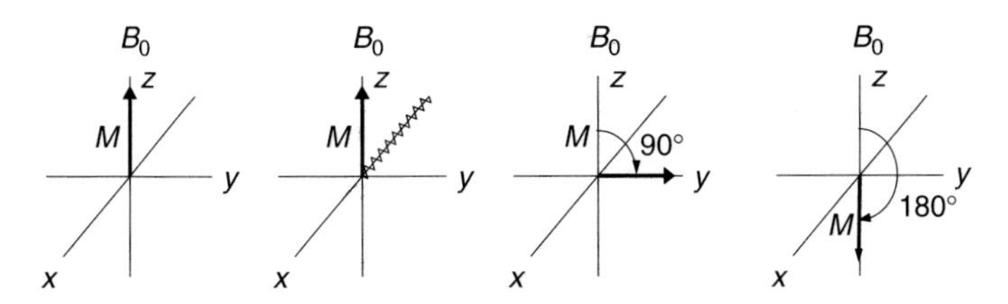

Figure 3 Scheme of a 90 and 180° pulse in a rotating coordinate system.

precession along the z direction of the B_0 field. If the observer position rotates with the same frequency as the precessing spins, the magnetic vectors become static. This is called the rotating coordinate system (Figure 3).

If the system is in an energetic equilibrium, for the NMR experiment the z vector is of most interest. After irradiation over a defined time period coming from the x direction, the z magnetisation sets in a clockwise movement. The 90° pulse – a value normally in μs units – is defined by the time in which the z magnetisation is transferred in the y direction. A doubled irradiation time leads to the 180° pulse.

The NMR data are recorded in the y (and x) direction during the acquisition time (AQ). After a 90° pulse, the y magnetisation is at a maximum. Within the AQ of the NMR experiment, the macroscopic magnetisation will fall back into the equilibrium state. This effect is called spin lattice relaxation. The measurable magnetisation tends to zero in an exponential curve. In case of the 180° pulse, no y magnetisation is detectable. The rotating coordinate system is spinning in a defined frequency that can be defined as the frequency in the middle of an NMR spectrum. Resonances of different protons can be of higher or lower frequency. After the excitation pulse, this results in a splitting of the magnetisation into different vectors placed in the x/y plane. Within the AQ, the different vectors are rotating within the x/y plane and relax to the equilibrium state. This effect can be figuratively described as a spinning and slowly closing umbrella. The row data measured in the NMR experiment is called the free induction decay (FID). Examples are given in Figures 4–6.

At this stadium, the NMR experiment becomes a domain of mathematics. FT allows a number of data manipulations with a positive intention, zero filling, line shape modelling, apodisation and line broadening by different mathematical formulas; these mathematical treatments of the raw data allow an improving of the signal-to-noise ratio (S/N) or an increase in the digital and spectroscopic resolution.

2.2. Relaxation

"What goes up must come down". This is a well-known idiom which is also essential for modern FT/NMR spectroscopy. To overcome the low sensitivity resulting in a low S/N ratio, a higher number of NMR experiments can be accumulated in practice. During data accumulation, the real signals are increased and the statistical noise vanished. The condition for such an approach is that the

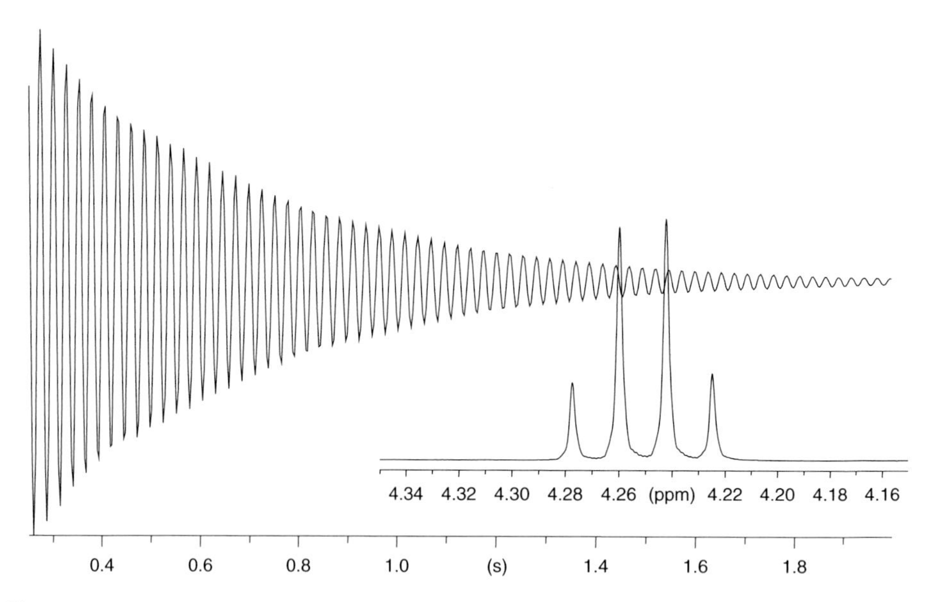

Figure 4 FID and spectrum, no line broadening.

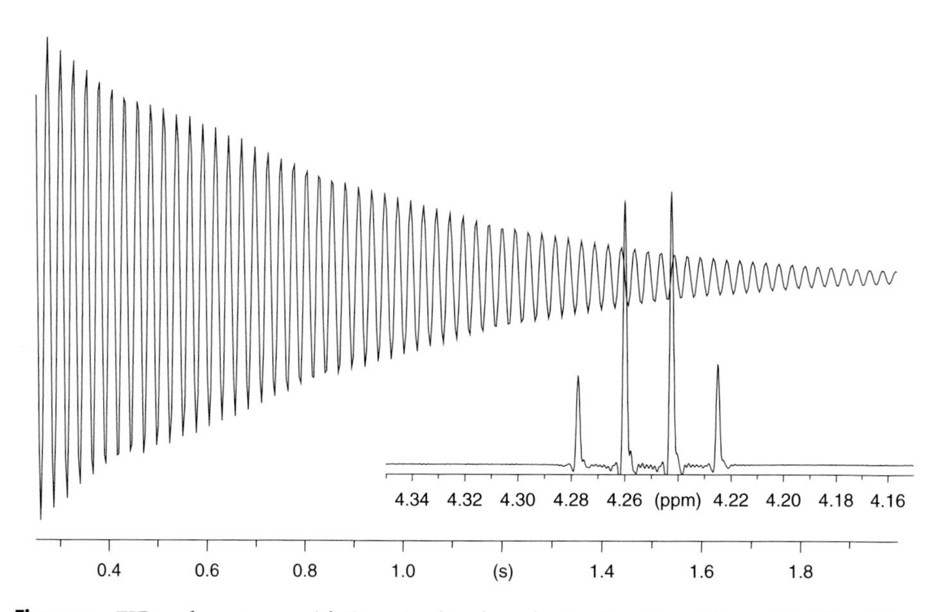

Figure 5 FID and spectrum with Gaussian line broadening function, LB = −0.7, GB = 0.9.

equilibrium state is reached before the next excitation. The relaxation mechanism must therefore be considered. The energy of an excited state has to be returned to the surroundings under the restrictions of quantum mechanics, called transversal relaxation. In other spectroscopic experiments, we observe a spontaneous energy

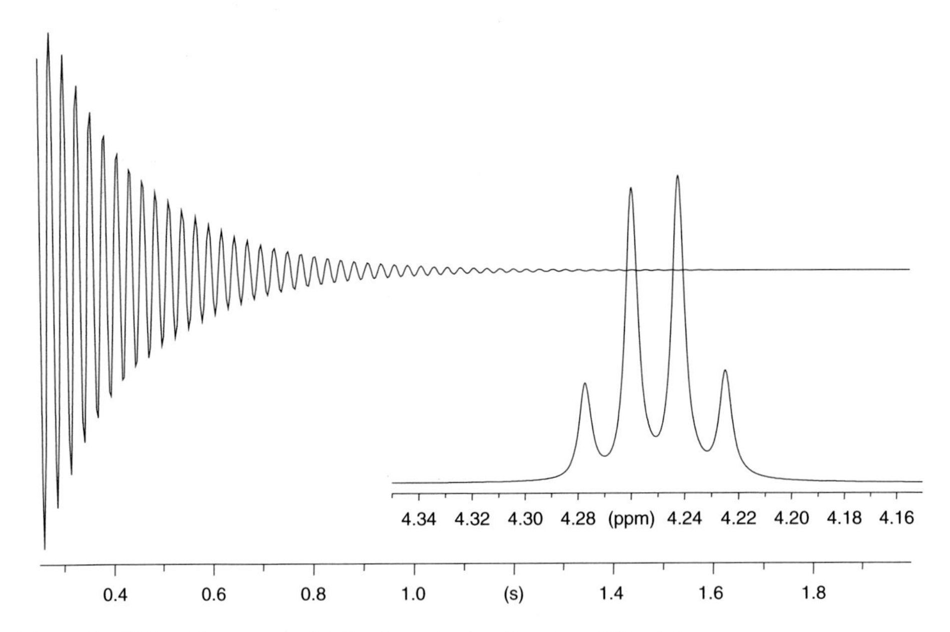

Figure 6 FID with exponential line broadening function, LB = 1.4.

emission. The low ΔE between the levels A and A* in NMR experiments is responsible for the lack of this possibility. Another phenomenon, the induced emission is necessary. The excitation is performed using a defined irradiation, which is nothing but an electromagnetic alternating field. A similar electromagnetic alternating field is necessary for the relaxation but is not available as a new irradiation from outside. However, the reason for NMR experiments to operate is molecular movement. Within the magnetic field, the orientated micromagnets of all NMR-active nuclei change their relative positions and consequently the magnetic influence of its neighbours (Figure 7). Observing the changes in the magnetic surroundings of a

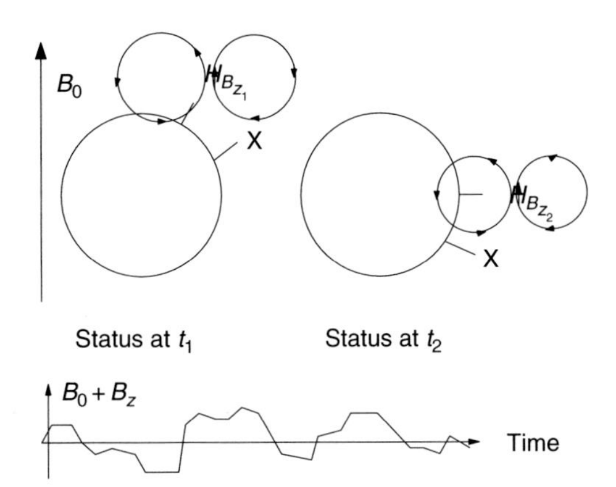

Figure 7 Transversal relaxation induced by the Brauns molecular movement.

Table 1 Important NMR-active nuclei and properties

Isotope	Spin quantum no. I	Natural abundance (%)	NMR frequency (MHz) at 7.05 T
^{1}H	1/2	99.98	300
^{11}B	3 × 1/2	80.42	96.25
^{13}C	1/2	1.11	75.43
^{14}N	1	99.63	21.67
^{15}N	1/2	0.37	30.40
^{17}O	5 × 1/2	0.037	40.67
^{19}F	1/2	100	282.23
^{29}Si	1/2	4.7	59.6
^{31}P	1/2	100	121.44

single nucleus during its molecular movement over the course of time, an alternating field is short term generated, which is responsible for an induced emission. This short and fragmental description shows the complexity of theoretical relaxation phenomenon. The relaxation influences the NMR experiments, causes smaller difference between proton and heteronuclear experiments and is essential for solid-state NMR. A description of solid-state NMR basics is beyond the scope of this chapter.

A new method of S/N enhancement offers the cryoprobe technique. Using a supercooled detector system, the electronically caused noise is reduced to a minimum. At the same magnetic field strength, a sensitivity enhancement of factor 5 is possible.

3. Chemical Shift

The resonance frequency is proportional to the magnetic field, and therefore each active nucleus should give one characteristic signal. Fortunately, for a chemist the magnetic field B_0 is modified by the chemical surroundings of atoms in different positions in a molecule. In fact, a modified field B_{eff} is interacting with different atoms, so the resonance conditions change. B_{eff} can be stronger or weaker than B_0. The following two reasons have to be considered for these effects: the change in the electronic surroundings and the anisotropy.

3.1. Electronic density

Each atom is surrounded by electrons. At higher electronic density, B_{eff} is weaker, the difference between the energy levels A and A*, ΔE, is lower, and the resonance frequency ν is lower. A signal will appear upfield (on the right side) of a spectrum. If the electronic density is lowered, the opposite effect is observed. The corresponding atom is de-shielded and its resonance frequency arises. The signal appears downfield (on the left side of a spectrum) (Figure 8).

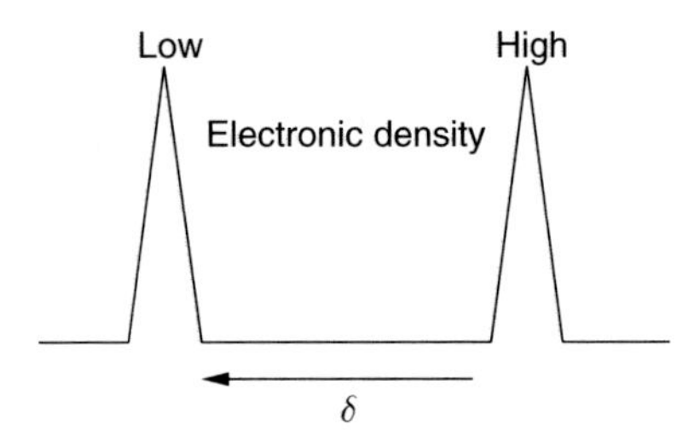

Figure 8 Electronic density and chemical shift.

Table 2 Electronegativity and chemical shift

	δ^{1}H NMR	δ^{13}C NMR
CH_4	0.13	−2.1
CH_3I	1.98	−24.0
CH_3Br	2.45	9.6
CH_3Cl	2.84	25.6
CH_2Cl_2	5.2	52.0
$CHCl_3$	7.25	77.0
CH_3F	4.14	71.6
CH_3OH	3.30	49.5

Table 3 Charge dependency of the chemical shift

Aromatic ring system	δ^{1}H NMR	δ^{13}C NMR
Cyclopentadienyl (negative)	5.35	102.1
Benzene (neutral)	7.26	128.5
Tropyllium (positive)	8.83	155.4

De-shielding is caused by neighbouring atoms with high electronegativity (oxygen, halogens, etc.). Higher electronegativity effects a stronger downfield shift (Table 2). This effect is accumulative. A similar effect is observed in charged molecules. Positive charge shifts downfield and negative charge causes upfield shift. Partial charges show a diminished effect (Table 3).

3.2. Anisotropy

Electrons in double or triple bonds are free to move between the atoms in defined orbits. The magnetic field B_0 induces a movement of these electrons, resulting again in the generation of an additional magnetic field owing to the electron movements. This fact is comparable with the principle of dynamo and electric

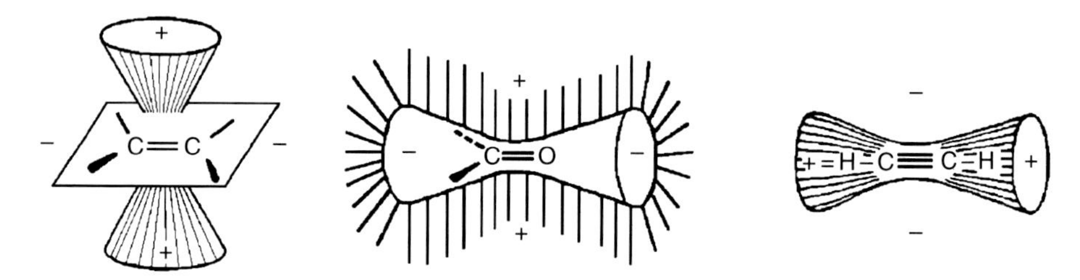

Figure 9 Anisotropy of multiple bonds.

motors. However, the induced magnetic field is not constant-linear in the z direction (the direction of B_0) but depends on the molecular geometry and the type of multiple bond (Figure 9). The resulting additional field is not isotropic, and therefore the effect is defined as anisotropy.

Depending on the relative position of the observed nucleus, the anisotropy of a multiple bond affects the increased or decreased local B_{eff}. The comparison of ^{1}H NMR chemical shifts in cyclohexane ($\delta = 1.4$ ppm), 1,4-cyclohexadiene ($\delta = 5.7$ ppm at the double bond) and benzene ($\delta = 7.1$ ppm) demonstrates the downfield effect, which is very useful in the separation of aromatic (and arylic) protons respectively from aliphatic carbohydrate bonded protons in ^{1}H NMR and ^{13}C NMR, respectively. Carbonyl groups mostly appear downfield (160–220 ppm) in ^{13}C NMR spectra. The so-called ring current effect (Figure 10) in aromatic systems enhances the anisotropy, and therefore aromatic bonded protons appear more downfield compared to aryl protons.

Aldehyde and formyl protons appear in the region of 10 ppm in ^{1}H NMR spectra. The magnitude of the anisotropic effect depends on the distance, and therefore the effect is weaker for alkyl groups located on double bonds (e.g. acetate). Because of the common molecular geometry, most of the NMR signals of protons and carbons are downfield-shifted by anisotropic effects. However, in some cases the upfield effect is observable when a proton is affected by the ring plane of an aromatic system. Negative chemical shifts are the most spectacular and rare (see Figure 38). Interpretation of anisotropic effects is therefore an important tool in elucidation of the three-dimensional structure.

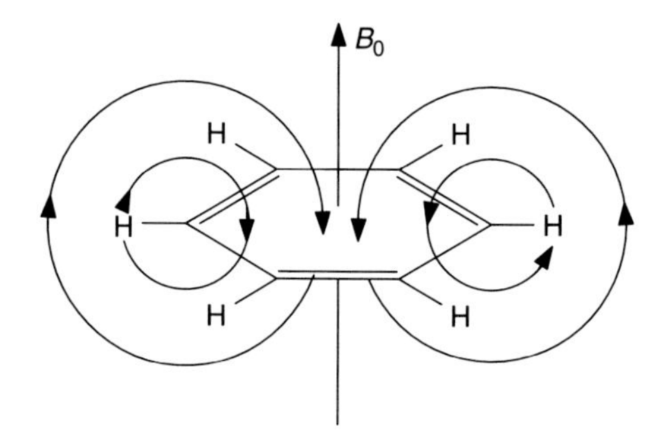

Figure 10 Ring current effect in benzene.

3.3. Mesomerism

Inductive and mesomeric effects in electronic density distribution are well known from the aromatic substitution reaction mechanism. A substitute with a free orbital is able to increase the electronic density in the *ortho* and *para* positions, which leads to a shielding of the corresponding carbon and hydrogen atoms resulting in an upfield shift. Especially in case of oxygen and nitrogen as well as the halogens, the opposite inductive effect due to the electronegativity is exceeded (Figure 11). In contrast, the negative induction is clearly observable at the downfield shift of the responding ipso carbon atoms. The shift of aromatic carbon and hydrogen atoms can be calculated considering all $\pm$ inductive and $\pm$ mesomeric effects, respectively. In ^{1}H NMR spectra the strong anisotropy of substitutes such as nitro, sulphate, carboxyl and aryl must be included in the calculation. Increment tables are available in many standard textbooks.[10] The magnitude of the discussed substituent chemical shift (SCS) effects is expressed in the difference to the chemical shift of the substituted and the ground molecule. An example for nitrobenzene and aniline is given in Figure 12.

3.4. Steric effects

A steric component of the chemical shift is observable, especially in ^{13}C NMR. In a homologous series of methylbenzenes, the chemical shift of methyl groups changes depending on the neighbours. The electronic surroundings of the middle methyl group in 1,2,3-trimethylbenzene is more compact to the outer and the methyl groups of toluene and *o*-xylene (see Table 4). Increased electronic density causes a higher shielding, which results in an upfield shift of approximately 5 ppm. In increment tables, these upfield effects are often called the γ-effect, as a result of an additional substituent over three bonds.

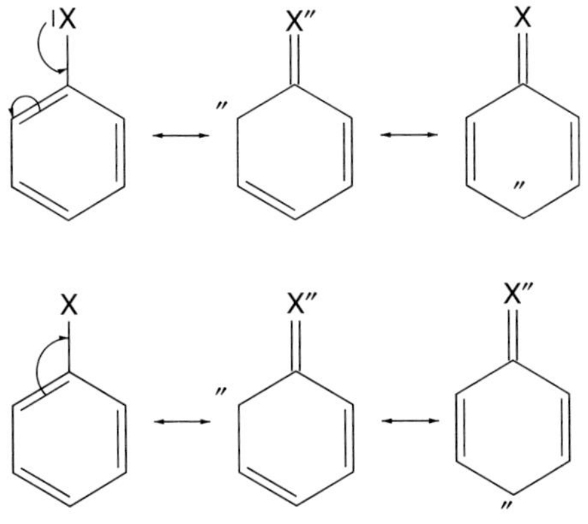

Figure 11 Mechanism of positive mesomerism in benzene.

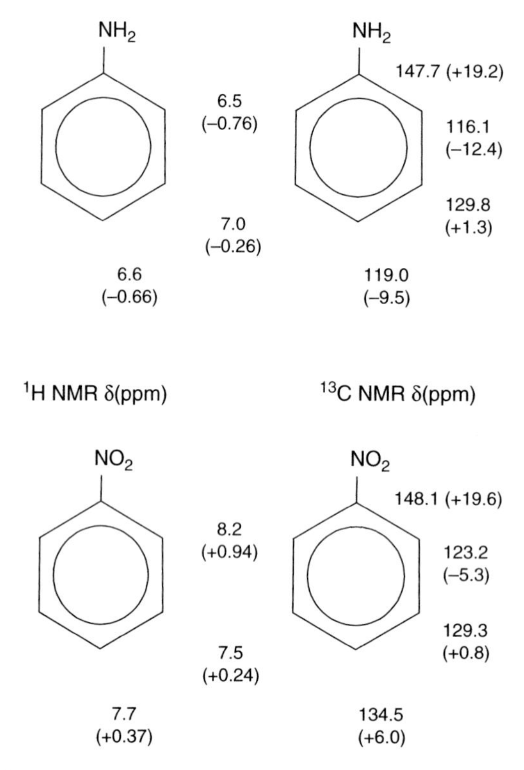

Figure 12 ^{1}H- and ^{13}C NMR chemical shifts of benzene derivatives; SCS values are given in parentheses.

Table 4 Chemical shift and steric interaction of neighboured methyl groups

	δ ^{13}C NMR
Toluene	21.3
o-Xylene	19.7
1,2,3-trimethylbenzene (1/3)	20.5
1,2,3-trimethylbenzene (2)	15.3

4. CALIBRATION AND RELATIVE SCALE PPM

Owing to the relation between the magnetic field strength and the resonance frequency in the NMR experiment, the NMR scale spectra must be normalised. Using the spectrometer frequency as reference, the values of the chemical shifts are given in ppm, not in MHz or Hz. Otherwise, the spectra at different magnetic fields are not comparable. Thus, 1 ppm in a ^{1}H NMR spectrum at 7.3 T is equivalent to 300 Hz and 1 ppm at 14.6 T is equivalent to 600 Hz. However, ^{1}H NMR spectra change their appearance with changing magnetic field strengths because the

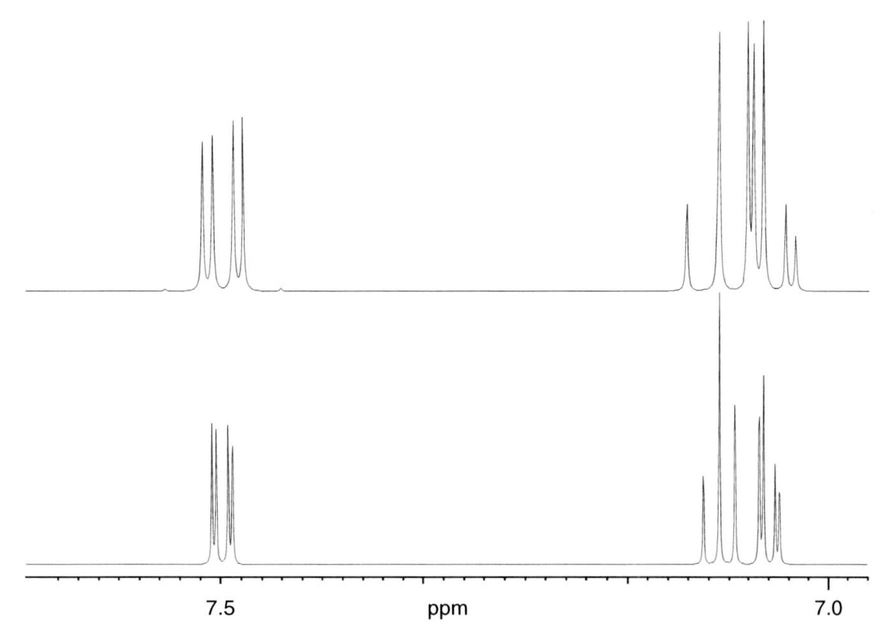

Figure 13 Comparison of a multiplets at 300 MHz (upper) and at 600 MHz (lower).

coupling constants between NMR-active nuclei are field-independent. While the absolute frequency of the resonances increases with the field, the multiplet structures remain in the same magnitude (Figure 13). After normalisation of the NMR scale, the multiplets seem to be compressed. In cases of similar magnitudes of chemical shift, differences and coupling constants of higher order multiplets may become first order and the interpretation of a high field spectrum will be easier.

The common calibration of NMR spectra is done using internal standards, e.g. tetramethylsilan (TMS), as its signal is defined to $\delta = 0$ ppm in both, ^{1}H- and ^{13}C NMR spectra. Lists of conventionally used standards are summarised in standard NMR textbooks. The signal of the solvents can be used as the internal reference, too.

5. Spin–Spin Coupling

5.1. The coupling constant

The magnetic interaction of NMR-active nuclei results in the spin–spin coupling. In practice, homo- and heteronuclear couplings are usually distinguished. The two magnetic moments of the elementary magnets (spin-up and spin-down for nuclei with the spin quantum number $I = 1/2$) increase (resp. decrease) the B_{eff} of a neighboured atom with the same magnitude. According to the Boltzmann distribution, the statistic number for both quantum states is nearly identical. This

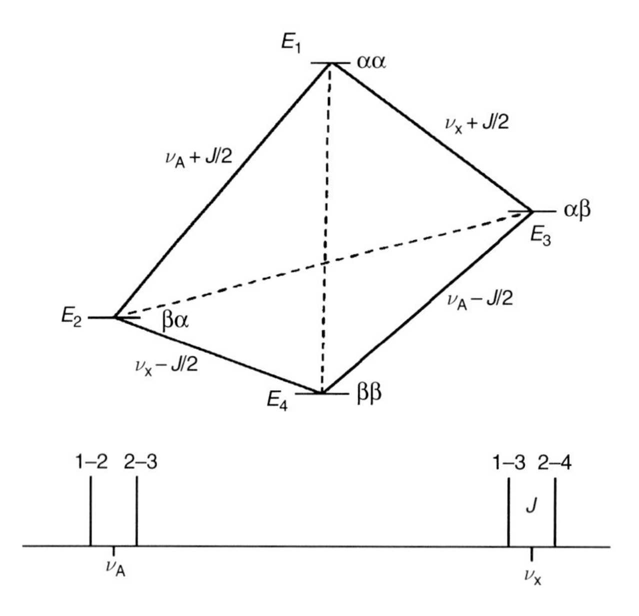

Figure 14 Scheme of the energy levels in an A,X spin system.

fact results in two different B_{eff} for the macroscopic magnetisation. A doublet signal instead of a singlet is the consequence. The spin–spin coupling needs a covalent bonding. The coupling constant is defined as $^{n}J_{A,X}$, where n is the number of chemical bonds between the two coupling atoms A and X. The coupling constant is independent of the field strength, and has a plus or minus prefix and it is mutual to the coupled atoms ($^{n}J_{A,X} = {}^{n}J_{X,A}$). The magnitude decreases with the number of increasing chemical bonds between A and X. In practice, the four bonds are the maximum distance for the observable couplings (Figure 14).

5.2. Multiplicity

With the number n of coupling partners, the distribution of possible spin states increases statistically. One neighboured atom with $I = 1/2$ enables two different states, two chemical equivalent atoms enable three states, and so on (Figure 15). The multiplicity can be calculated from Eq. (2), the intensity distribution according to the binomial distribution visible in Pascal's triangle function:

$$M = 2nI + 1 \qquad (2)$$

Coupling with one single deuterium the most interesting nucleus with $I = 1$ splits the signal of its coupling partners into a 1:1:1 triplet. This is, for example, apparent with the solvent signal of $CDCl_3$ in ^{13}C NMR spectra, resp. the signals remaining by the incomplete deuterated solvents (e.g. CD_2H-OD in CD_3OD, which gives a quintet).

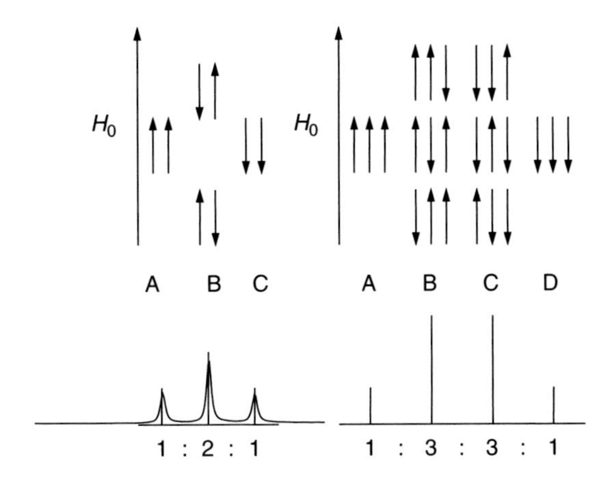

Figure 15 Multiplicity and intensity distribution for the homonuclear coupling $^{3}J_{H,H}$ of an ethyl group.

5.3. Roof effect

In homonuclear coupling systems, the intensity distribution differs from the Pascal's function when the chemical shift between two coupling atoms decreases. Formally, this is a change from an A,X to an A,M or an A,B system. The intensities must now be calculated using the following equations:

$$\delta_A - \delta_B = xy \tag{3}$$

$$I_{1,2} = \frac{1 \pm \mathrm{J}_{\mathrm{A,B}}}{x + J_{\mathrm{A,B}}} = \frac{1 \pm \mathrm{J}_{\mathrm{A,B}}}{y - J_{\mathrm{A,B}}} \tag{4}$$

The resulting figure is called the roof effect (Figure 16). In the special case of high-symmetry molecules the system will become A,A′, which means the same chemical shift for neighbouring atoms. However, even if the resulting signal is visible as a singlet, the coupling still exists but is not observable due to the infinite roof effect.

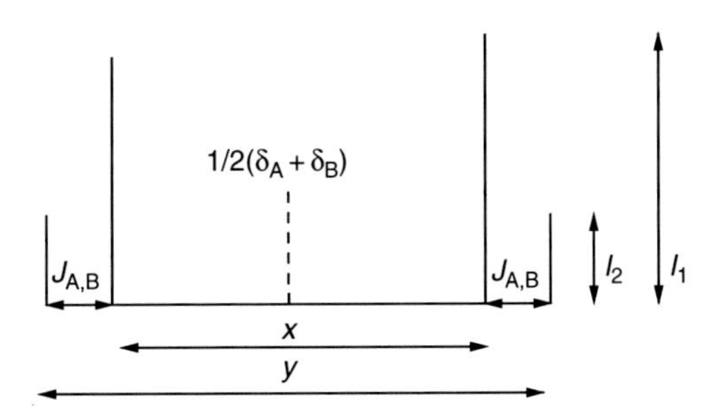

Figure 16 Roof effect of spin–spin couplings.

In each CH_2 group, a germinal coupling $^{2}J_{H,H}$ exists, which is observable if both the protons are chemically non-equivalent for stereochemical reasons.

5.4. Angular dependence of the coupling constant

The value of the vicinal coupling constant is angular dependent. Owing to this fact, the coupling constants are a useful tool for elucidation of the three-dimensional structure. The non-linear Karplus curve (Figure 17) shows the correlation between the angle and the $^{3}J_{A,X}$, and is not restricted to homonuclear couplings. This is demonstrated in Figure 18, which shows the anomeric protons of the α/β mixtures of glucose and mannose. The integral values in addition show the different ratio. Under the given conditions, the α/β ratio is approximately 1:1.5 for glucose and 2:1 for mannose.

The magnitude of geminal coupling constant is also angular dependent. Most of these CH_2 group couplings are not directly detectable because of its magnetic equivalence. Diastereotopic (see Section 9.1) CH_2 groups appear as an A,B system. Electronegative substituents change the hybridisation state and thereby the angle between the protons, resulting in different geminal coupling constants, as demonstrated in Figure 18.

5.5. Heteronuclear coupling

The spin–spin coupling is not restricted to protons (homonuclear couplings). All NMR-active isotopes show spin–spin couplings among each other. The type and

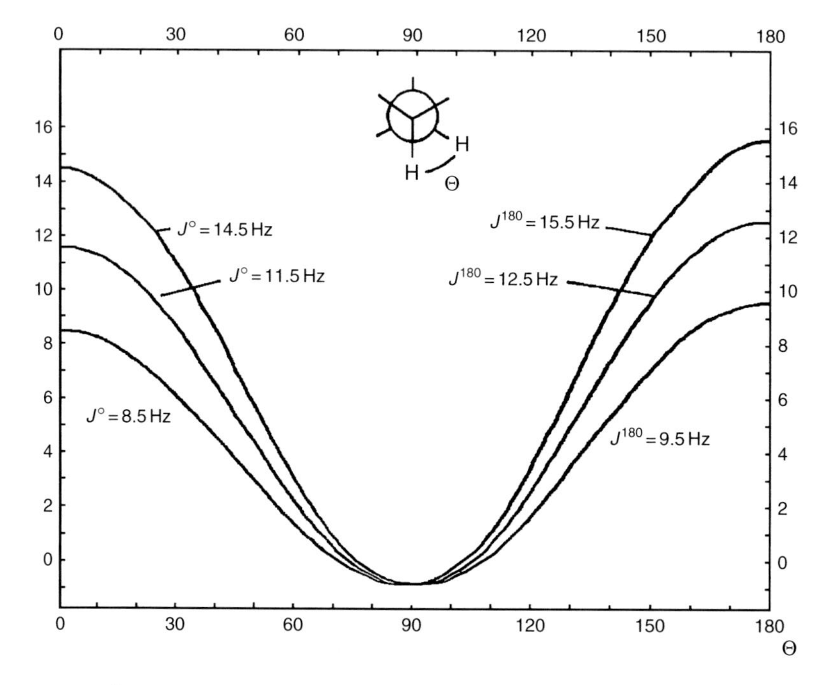

Figure 17 Karplus curve.

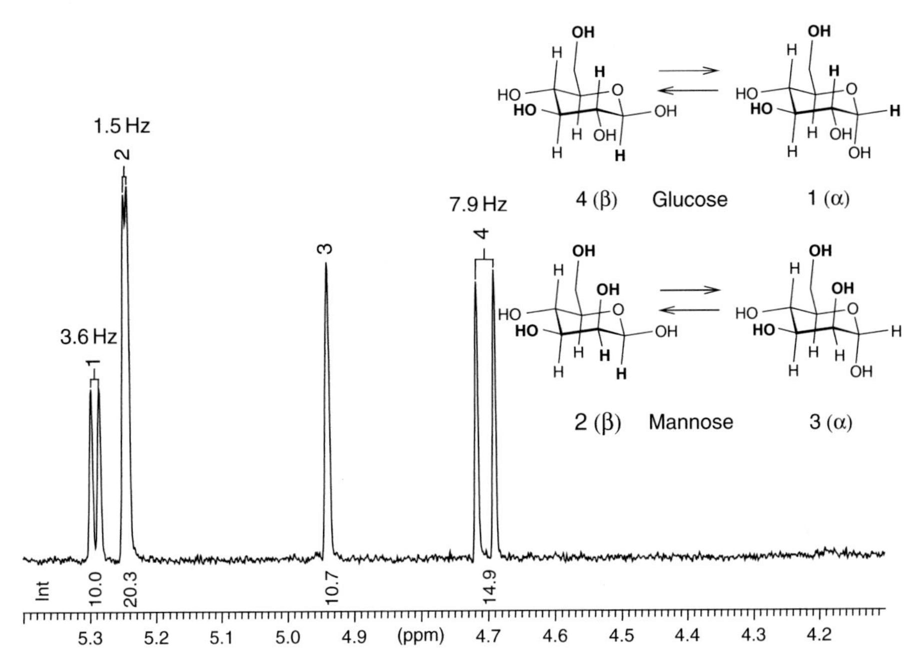

Figure 18 ^{1}H NMR spectrum (detail) of the α, β equilibrium of glucose and mannose in D_2O.

the intensity depend on the natural abundance and the spin quantum number (see Table 1). A single nucleus with $I = 1/2$ couples another signal to a duplet. The 100% natural abundance of ^{31}P and ^{19}F allows the handling of its coupling in equivalence to an additional proton. Each phosphorous or fluorine causes a duplication of the homonuclear multiplets. All other important atoms in organic molecules have lower amounts of NMR-active nuclei, especially carbon, silicon and nitrogen. The main part of the inactive nuclei (e.g. ^{12}C and ^{28}Si) shows no magnetical interaction, or as in the case of ^{14}N the coupling mostly is not observable due to its high quadrupole moment. Exceptions are the highly symmetrical ammonium compounds as shown in the ^{13}C NMR signal of glycerophosphocholine (GPC), which shows two couplings with ^{14}N (1:1:1 $^{2}J_{C,N}$ triplets due to $I = 1$) and four with ^{31}P ($^{2}J_{C,P}$ and $^{3}J_{C,P}$ doublets due to $I = 1/2$) (Figure 19).

The small number of active nuclei leads to a satellite spectrum, where the intensity or integral area of each satellite is half the natural percentage of the isotopes abundance. In any case, coupling constants (in Hz) of a coupling pair of atoms must have identical values. In the homonuclear case of proton–proton coupling, this fact is observable at the multiplet structures in the ^{1}H NMR spectrum. In case of heteronuclear couplings, the coupling constant is observable in the spectra of both nuclei. For example, the heteronuclear $^{1}J_{C,H}$ coupling of methanol is detected in ^{13}C NMR (gated decoupling) and in the ^{1}H NMR spectra as the satellites (Figure 20). Another satellite example is ^{29}Si- and ^{13}C-satellite system is shown in Chapter 2.

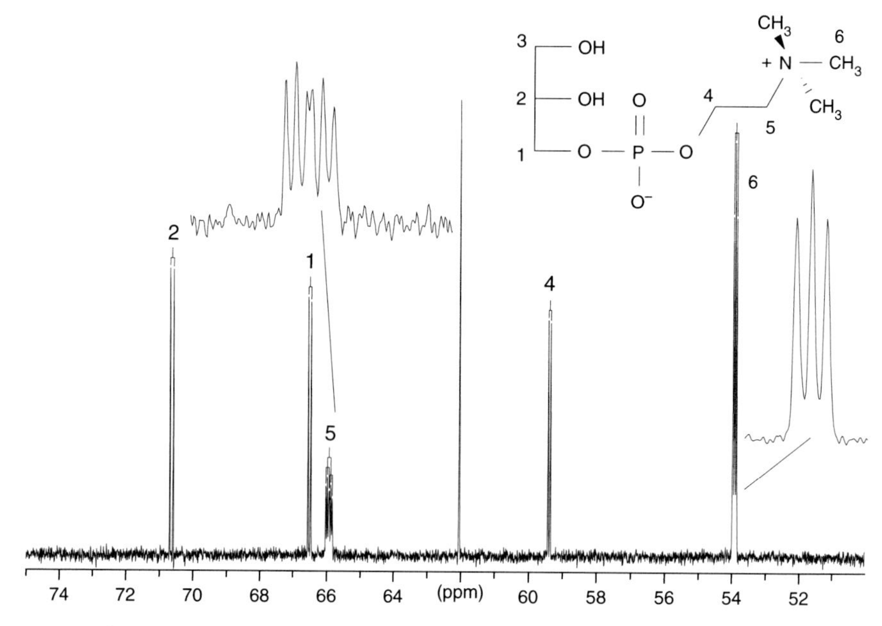

Figure 19 ^{13}C NMR spectrum of glycerolphosphocholine (GPC) in D_2O containing ^{31}P and ^{14}N couplings.

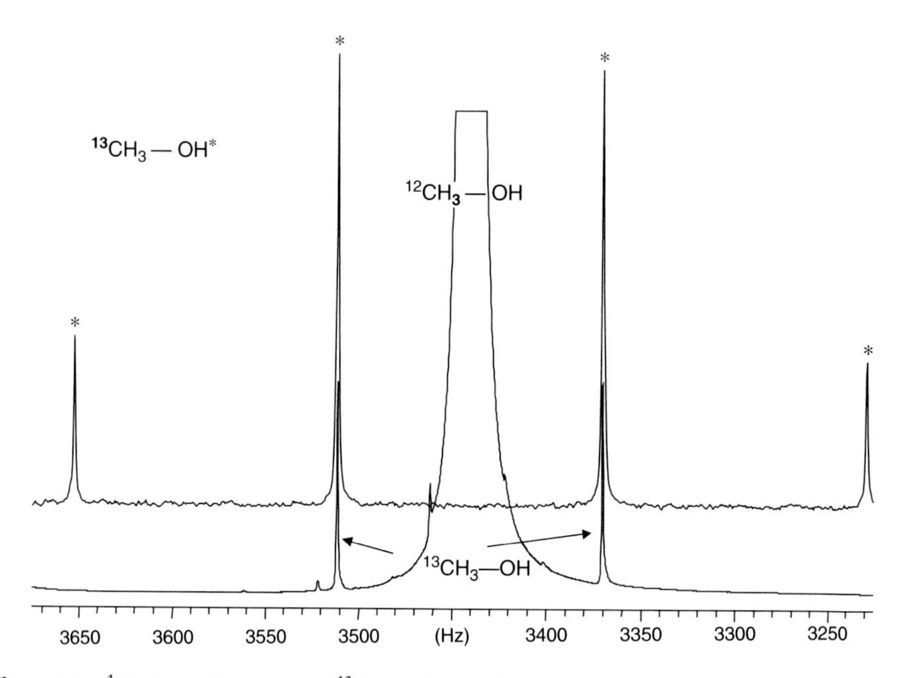

Figure 20 ^{1}H NMR (bottom) and ^{13}C NMR gated decoupling spectrum of methanol.

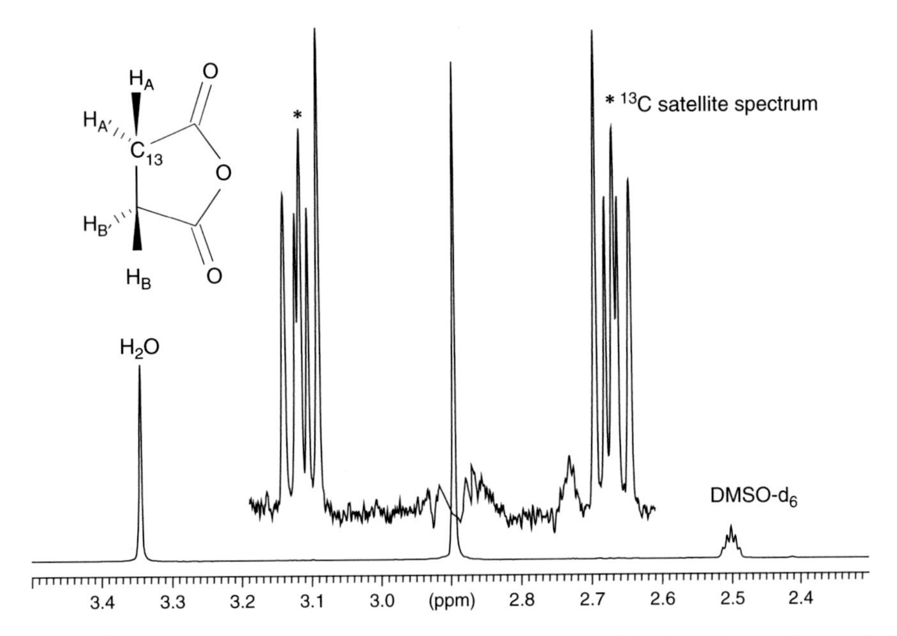

Figure 21 ^{1}H NMR spectrum of succinic anhydride in DMSO-d$_6$, and superimposed the ^{13}C NMR satellite sub-spectrum.

In the spectra of organic and organometallic molecules, different characteristic satellites are observable by coupling with ^{15}N, ^{29}Si, ^{11}B or ^{119}Sn and ^{121}Sn, respectively – the latter results in a double satellite system.

5.6. ^{13}C NMR satellites in ^{1}H NMR spectra

Molecular symmetry often results in singlet signals in ^{1}H NMR spectra. The presence of a ^{13}C-isotope annihilates the symmetry of ethylene and ethyl groups but not of methylene and methyl groups. The invisible but present coupling $^{3}J_{H,H}$ between symmetrical protons is observable in the ^{13}C NMR satellites. The multiplet structure of those asymmetric subsystems and the evaluated coupling constants emphasise the chemical shift information a pure singlet cannot provide, which is demonstrated in the ^{1}H NMR spectrum of succinic anhydride (Figure 21). The superimposed satellite spectrum was performed using signal suppression of the ^{12}C-bonded singlet.

6. HETERONUCLEAR SPECTRA

For historical and practical reasons, a differentiation between proton and heteronuclear NMR was generally undertaken. However, this distinction is not justified regarding chemical shift and spin coupling, as the basics of NMR spectroscopy are identical for all nuclei. In practice, a differentiation is reasonable for a few important principles, i.e. the occurrence of nuclei. Only a few kinds of atoms are basic components of organic molecules – proton, carbon, oxygen and sulphur, nitrogen and phosphorus. In pharmaceuticals fluorine also plays a role that cannot be neglected. The most important reason for the characteristics of the NMR spectra is the natural abundance of the NMR-active nuclei. As a result of physical properties, only the isotopes with $I = 1/2$ are fairly detectable (Table 1). The most common isotopes ^{12}C, ^{16}O and ^{32}S are inactive nuclei because their nuclear spin quantum number is $I=0$. The main isotope of nitrogen is ^{14}N (>99.5%) with $I=1$. The quadrupole moment enables a sufficient spectroscopy.

Proton and carbon show the most important variety while building the backbone of all organic molecules. Therefore, only the combination of the natural abundance of NMR-active isotopes and the occurrence within an organic molecule lead to a homonuclear neighbourhood and consequently to a complex pattern seen in ^{1}H NMR spectra. For carbon, only the second condition is given: its high number and variety in organic molecules. Nevertheless, only 1.1% of all carbon atoms are NMR-active as ^{13}C, resulting in the isolation of ^{13}C atoms in molecules. Only 1.1% of the ^{13}C have a homonuclear neighbour. The resulting coupling appears as a satellite subspectrum and normally is not visible in a ^{13}C NMR spectrum because of its low intensity.

^{31}P and ^{19}F, however, are 100% $I = 1/2$ nuclei, but compared with protons and carbons they are often in isolated positions. Therefore, homonuclear coupling is rare. An overview on the NMR properties of non-metallic elements is published by Kalinowsky et al.[11]

6.1. Decoupling

To summarise the last subsection, nearly all observed couplings in NMR spectra of any active nuclei are couplings with protons. ^{13}C, ^{19}F and ^{31}P appear as singlet signals because of the proton decoupling, which inhibits the heteronuclear coupling. Furthermore, these heteronuclei mostly have no neighbours for homonuclear couplings. For example, a proton decoupled ^{31}P NMR spectrum is shown in Figure 22. Each phosphorous singlet represents a discrete molecule. This type of spectroscopy enables a selective analysis of complex mixtures.

6.2. Quantitative heteronuclear NMR

The selectivity, the singlet signal structure and the 100% natural abundance make ^{31}P- and ^{19}F NMR important quantitative methods for the analysis of complex natural mixtures (see Part II, Chapter 3, Section 5.3) or of by-products in drugs. Detection of amounts lower than 0.1% is mostly not a problem. The NMR spectra can be read easily as chromatograms; the absolute molar response enables the quantification as 100% method when all signals of the drug and its by-products are summarised or absolute in percent (wt/wt) when using a suitable internal standard. Qualitative and quantitative NMR of ^{19}F- and ^{31}P-containing drugs is analytical, and should be implemented in each pharmacopoeia. ^{13}C NMR has few more restrictions in quantitative NMR analysis which, however, can be negotiated (see Chapter 3).

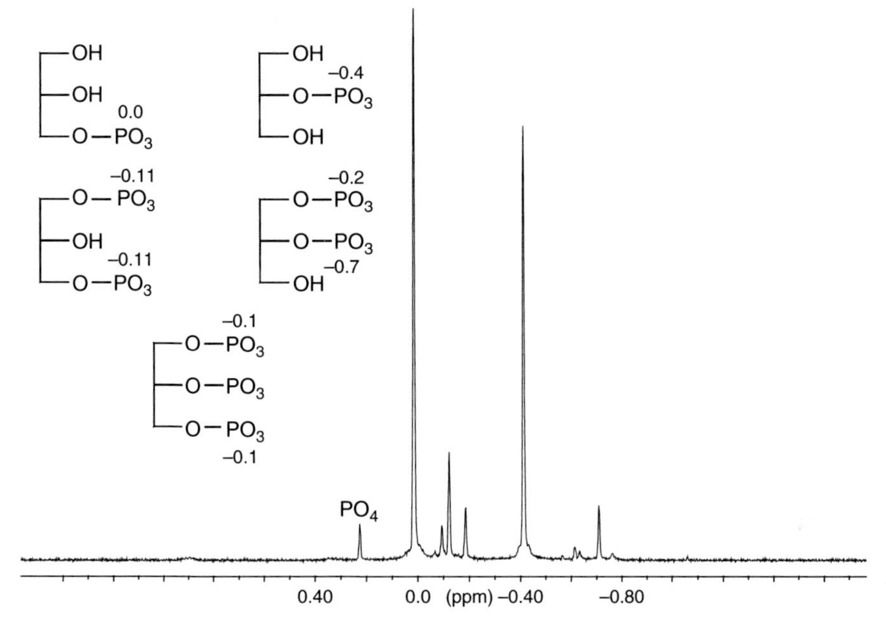

Figure 22 ^{31}P NMR (^{1}H decoupled) of glycerophosphates in methanol-d_4, mixture of isomers and homologues.

7. Molecular Dynamics

NMR is an experiment with time. Molecular dynamics are observable during the excitation and relaxation. Amide structures are the well-known examples. In the NMR spectra of dimethylformamide (DMF), both methyl groups show different chemical shifts due to the hindered rotation (Figure 23). This dynamic effect is temperature dependent. At higher temperatures, the rotation around the C–N bond is faster, the separated signals become broader and disappear at coalescence temperature. On further heating, only an averaged broad signal of both methyl groups appears in the middle of the two singlets, getting sharper with increase in temperature.

A more sophisticated example is the ^{1}H NMR spectrum of *N*-formylmesazalin (Figure 24). Owing to its asymmetric energetic potential, the number of the rotational isomers A and B are different, represented by the ratio of the corresponding NMR signals. In the case of symmetrical DMF the isomer ratio is 1:1 and in the case of asymmetrical *N*-formylmesalazine it is about 4:1.

Molecular dynamic effects are not restricted to amide structures but due to the importance of peptides in pharmaceutical application it becomes a basic tool for NMR interpretation in this field. Many special techniques exist utilising the molecular dynamics, e.g. the two-dimensional DOSY (diffusion-ordered spectroscopy); the detailed demonstration is beyond the scope of this book.

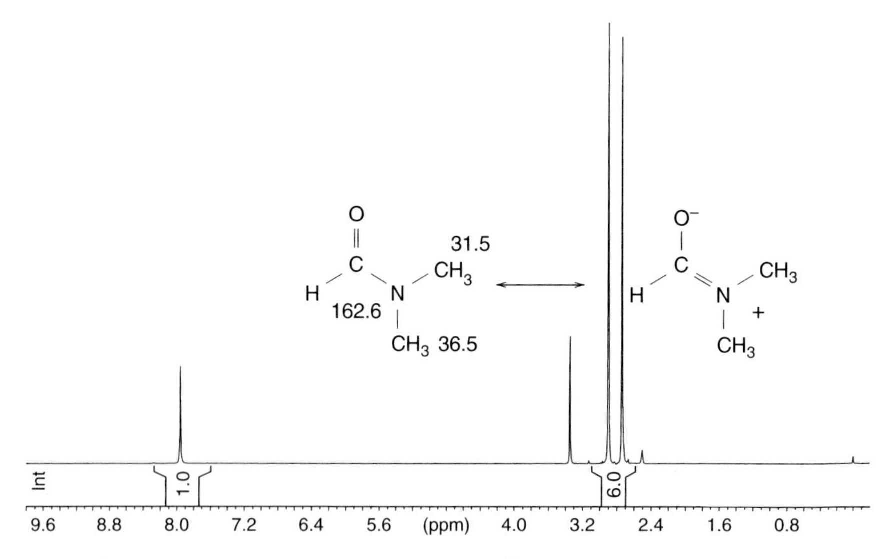

Figure 23 ^{1}H NMR spectrum of DMF in DMSO-d_6. ^{13}C NMR chemical shifts in parentheses.

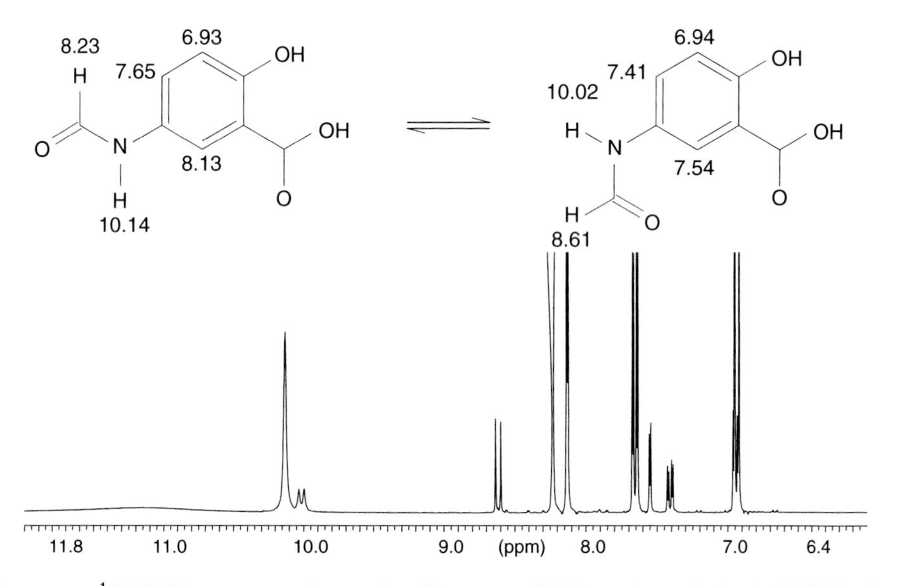

Figure 24 ^{1}H NMR spectrum, the rotational isomers of *N*-formylmesalazine in DMSO-d_6.

7.1. Deuterium exchange and solvent effects

Alcohol and amine protons (OH and NH) often are observable in aprotic and dried solvents only. Because of the fast chemical exchange especially, their couplings with neighboured protons are mostly not detected. Intentionally, the deuterium exchange of OH, NH and similar bonded protons is a common tool in spectrum simplification. By measuring in or by adding D_2O or methanol-d_4 the corresponding protons are substituted by deuterium. The specific resonance signals and in addition the coupling with neighboured protons disappear as demonstrated for *N*-formylmesazalin after adding methanol-d_4 (cf. Figure 24 with Figure 25).

The chemical shift depends on several effects discussed above. In addition to the molecular-immanent electronegativity, anisotropy and mesomerism, the "surrounding" conditions also influence the chemical shifts. These are concentration, temperature, solvent, and especially the pH value in aqueous solutions. Hence spectra interpretation is complicated and a library search similar to gas chromatography/mass spectrometry (GC/MS) is almost impossible; in practice, these effects are useful. In accordance with chromatography, the modification of temperature and solvent (especially mixtures of solvents) enables the separation of interfering signals.

The solvent effect may be demonstrated in an ^{19}F NMR spectrum of flufenoxuron in DMSO-d_6 as shown in Figure 26. The signal of one by-product at $\delta = -60.6$ ppm in an amount of about 1% interferes with the ^{13}C NMR satellite of the main component. By successive additions of benzene, the corresponding signal can be shifted until a baseline separation allows the exact quantification. For ^{19}F NMR or ^{31}P NMR, expensive deuterated solvents need not be used for signal shift.

The change of pH values leads to similar results in the by-product analysis of phosphorous-containing pharmaceutics using ^{31}P NMR. In addition, pH value-dependent measurements of the chemical shifts enable the evaluation of pK_a values even in mixtures as demonstrated by the ^{31}P NMR measurements of a mixture of glyphosate, aminomethylphosporic acid (AMPA) and phosphate (Figures 27 and 28).

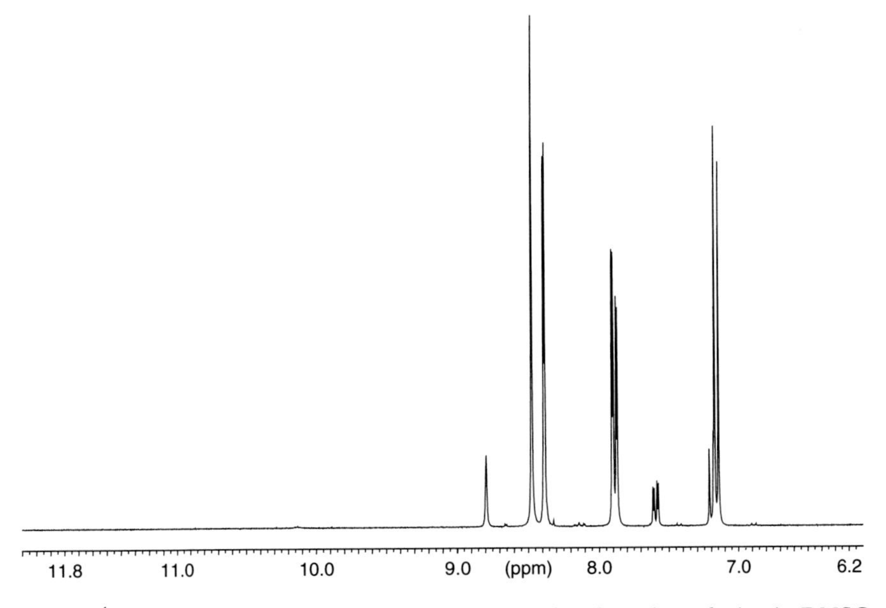

Figure 25 ^{1}H NMR spectrum, the rotational isomers of *N*-formylmesalazine in DMSO-d_6 after adding methanol-d_4.

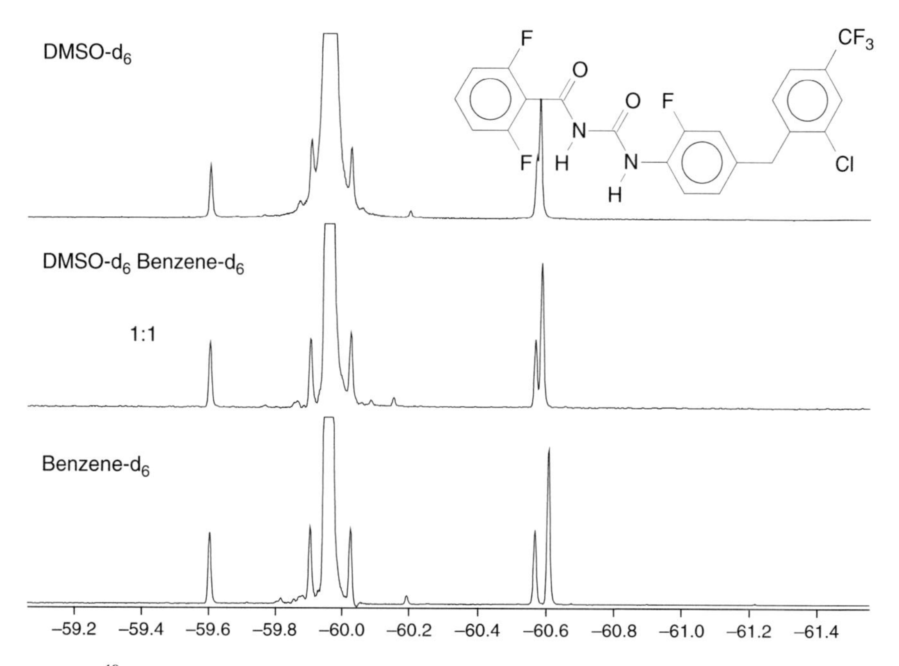

Figure 26 ^{19}F NMR spectrum of flufenoxuron in different solvent systems.

Figure 27 Chemical structures of glyphosate (left) and AMPA (right).

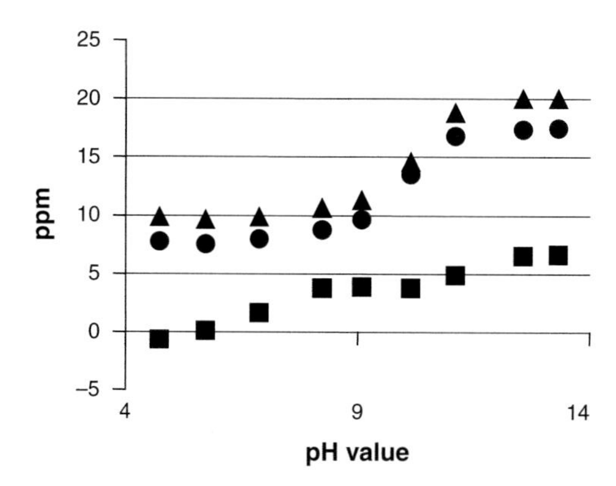

Figure 28 ^{31}P NMR titration curve of glyphosate, •; AMPA, ▲; and phosphate, ■.

To shift the water signal, change in temperature is only another feature of NMR in practical experience. High-temperature measurements at 80°C are common in NMR of hydrocarbon polymers (see Chapter 2).

8. Chemical Derivatisation

A common tool in chromatography is the derivatisation of the target molecules to enable volatility of polar molecules to perform GC or to incorporate a chromophor, which allows UV detection in high-performance liquid chromatography (HPLC). In NMR spectroscopy the *Mosher reagent*[12] derivatisation is well known for the determination of the enantiomeric excess. Enantiomeric pure trifluoromandelic acid chloride is used to create diastereomeric esters of chiral alcohols. The detection is performed using ^{19}F NMR and allows a selective and sensitive quantification. Many other chemical derivatisations are conceivable in principle to solve analytical problems if NMR is more often used as an alternative for chromatographic methods.

In this chapter only one quantitative method will be presented as example: the detection and quantification of hydroxylamine in a hydroxyurea formulation.[13] All chromatographic methods failed and the detection limit in a standard ^{1}H NMR analysis is about 1% due to the broad signals of NH_2 and OH protons of

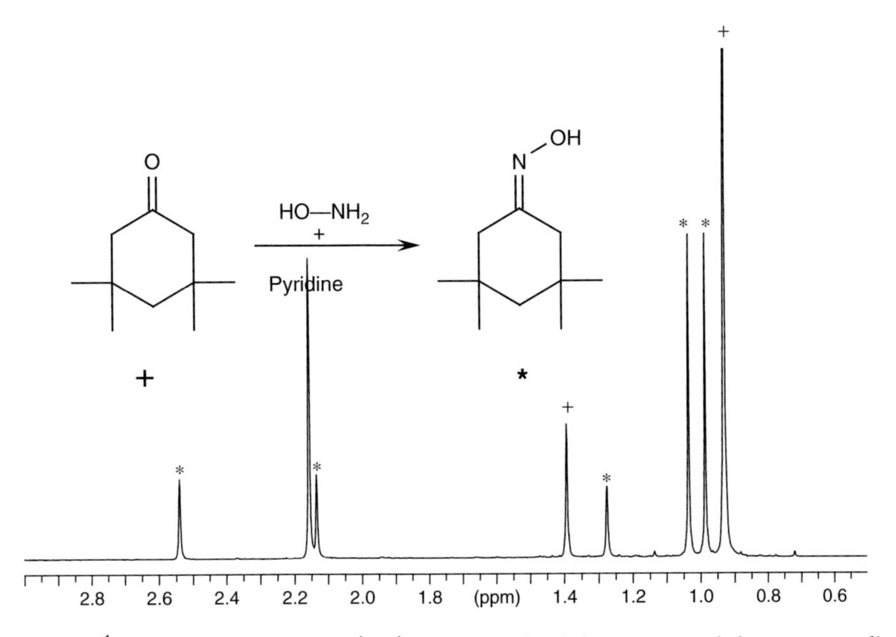

Figure 29 ^{1}H NMR spectrum of 3,3′,5,5′-tetramethylcyclohexanone and the corresponding oxime as a result of the chemical derivatisation of hydroxylamine in pyridine-d_5.

hydroxylamine in the given matrix. The reaction of the amine with a ketone forms a Schiff's base while the urea does not react. The detection of the resulting signals enables the quantification with a limit of detection of 50 ppm or even less depending on the experimental conditions.

Different ketones have been tested. Besides the most simple acetone, 3,3′,5,5′-tetramethylcyclohexanone very well fulfils the criteria of a derivatisation reagent. The reaction is nearly quantitative even in aqueous solution; the resulting singlets of the methyl and methylene groups of the oxime (Figure 29) are intensive and well separated from those of the remaining ketone. By means of the *E*/*Z* isomerism the methyl and the methylene protons appear as five singlets, which can be used independently or in any combination for the quantification of hydroxylamine. If the exact weight of the test item and the ketone is known, the ratio of the integral areas of oxime and the remaining ketone signals correlates directly with the total amount of hydroxylamine. The method was validated by successive addition of hydroxylamine to the hydroxyurea formulation.

9. STEREOCHEMISTRY

9.1. Diastereomerism

Since high-field superconducting magnets have been available, structure elucidation with respect to the stereochemistry of a molecule is a domain of NMR spectroscopy. The chemical shift and the coupling constants are significant for the stereochemistry. Diastereomeric molecules differ in their physical properties and therefore show different NMR spectra. Cyfluthrin includes three centres of chirality – one at the cyanhydrine position and two in the cyclopropane ring – which theoretically lead to 4 diastereomers consisting of 2 enantiomers each (Table 5). In the so-called β-cyfluthrin the ring protons are in *cis*-configuration; therefore, only the enantiomeric pairs *RS* and *SR* are present.

The splitting for both β-cyfluthrin forms is shown in ^{13}C-, ^{19}F- and ^{1}H NMR spectra, and parts of the last splitting are presented in Figure 30. The chemical shifts and the vicinal coupling constants are summarised in Table 6. The ratio of both diastereomers is approximately 2:1. Changing of the stereoisomeric centres in the cyclopropane ring leads to remarkable differences.

Table 5 Possible stereoisomers of cyfluthrin (for structural formula, see Figure 30)

Cyanhydrin	Cyclopropane	Cyanhydrin	Cyclopropane	
R	*Trans (SS)*	*S*	*Trans (RR)*	α-Cyfluthrin
R	*Trans (RR)*	*S*	*Trans (SS)*	α-Cyfluthrin
R	*Cis (RS)*	*S*	*Cis (SR)*	β-Cyfluthrin
R	*Cis (RS)*	*S*	*Cis (SR)*	β-Cyfluthrin

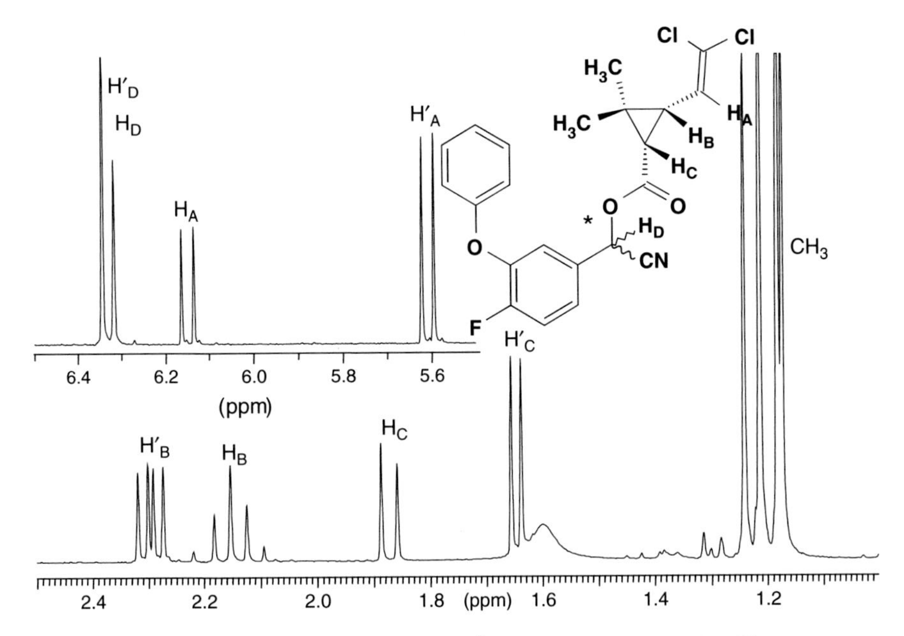

Figure 30 Cyfluthrin, diastereomeric splitting, ^{1}H NMR spectrum in $CDCl_3$.

Table 6 Chemical shift and coupling constants of cyfluthrin isomers

No.	δ (ppm)	$^3J_{H,H}$ (Hz)	No	δ (ppm)	$^3J_{H,H}$ (Hz)	$\Delta\delta$ (ppm)
A	6.15	8.7	A′	5.60	8.1	−0.60
B	2.15	8.5/8.7	B′	2.30	8.1/5.3	+0.15
C	1.88	8.5	C′	1.64	5.3	−0.24
D	6.32	–	D	6.34	–	+0.02

9.2. Atrop or axial chirality

The ^{1}H NMR spectrum of acetochlor (Figure 31) represents two kinds of isomerism effects. The first is the E/Z isomerisation caused by the hindered rotation of the amide bond, and the second is an atrop or axial chirality. The chemical bond between the nitrogen and the benzene ring acts as the stereo centre. Consequently, methylene groups are splitted in the A,B systems. Chemical shifts and corresponding coupling constants are summarised in Table 7.

9.3. Enantiomeric excess

Enantiomeric molecules show identical NMR spectra. The use of chiral shift agents such as cyclodextrines or 2,2,2-trifluor-1(9-anthranyl)-ethanol (TFAE) enables the formation of diastereomeric complexes that results in the splitting of the NMR

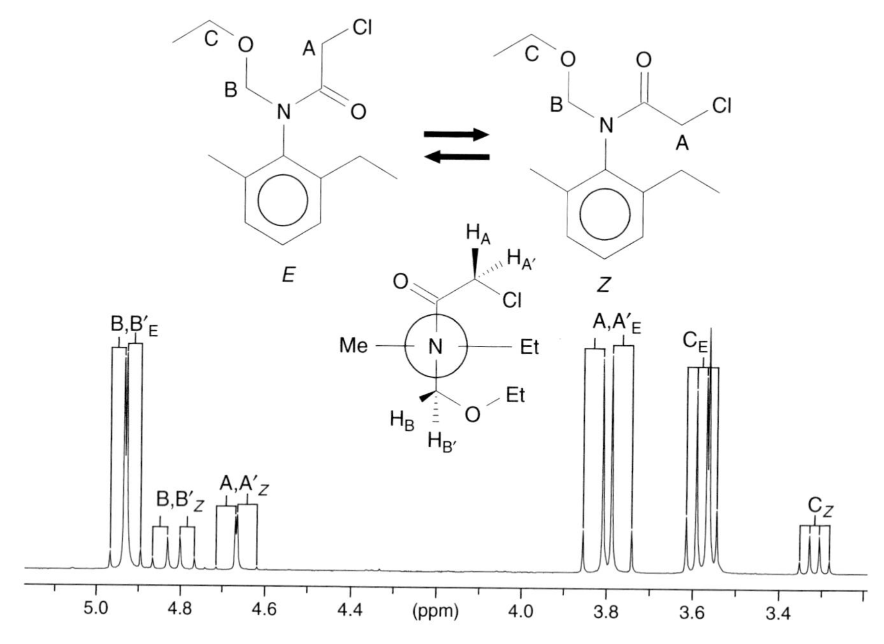

Figure 31 Atrop chirality and E/Z rotamers of acetochlor methylene groups, ^{1}H NMR spectrum in $CDCl_3$ and benzene-d_6 (2:1).

Table 7 Chemical shift and coupling constants of acetochlor methylene groups

Proton	δ (ppm)	Integral	$^2J_{H,H}$ (Hz)	$^3J_{H,H}$ (Hz)
A_E	3.80	10.0	13.9	–
A_Z	4.67	2.2	13.6	–
B_E	4.93	10.0	10.0	–
B_Z	4.81	2.2	10.5	–
C_E	3.58	10.0	–	7.0
C_Z	3.32	2.2	–	7.0

signals of the enantiomeric molecules. The determination of the enantiomeric excess is possible.[14] Figure 32 shows the methyl group of mexiletin before and after the addition of TFAE. The spectrum represents a racemic mixture.

9.4. Diastereotopy

A special effect is the diastereotopy (pro-chirality) often observed at methyl and methylene groups. The chemical non-equivalence results in the splitting of apparently identical methyl groups as demonstrated for valine (Figure 33). The methyl group shows two doublets of equal intensity. The different rotational isomer structures, shown in several Newman projections, demonstrate the non-equivalence

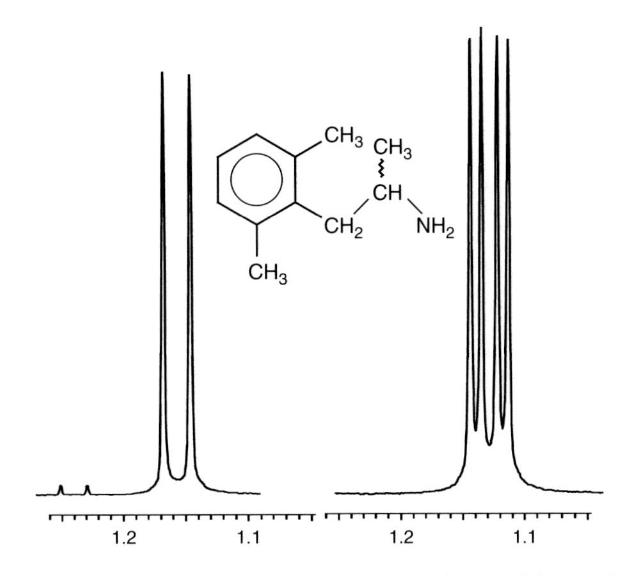

Figure 32 Enantiomeres of mexiletin in $CDCl_3$, before and after addition of TFAE.

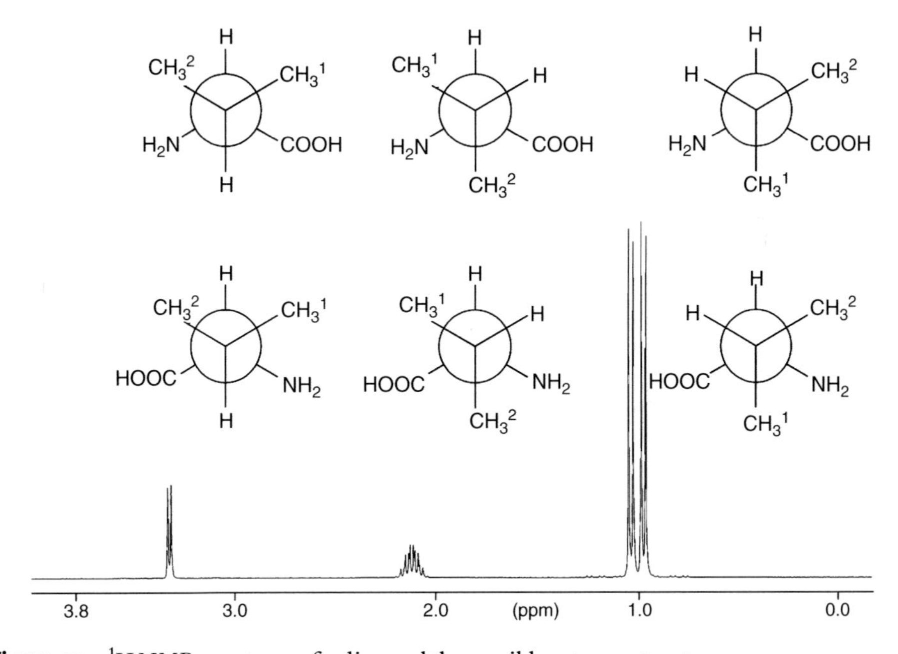

Figure 33 ^{1}H NMR spectrum of valine and the possible rotamer structures.

even if the rotation around the C–C bond is fast. The diastereotopic splitting is not eliminated at high temperatures. The spectra of both enantiomeric pure molecules (and of course a racemic mixture) are identical. It is obvious that the methyl groups have different shifts in the ^{13}C NMR spectrum, too.

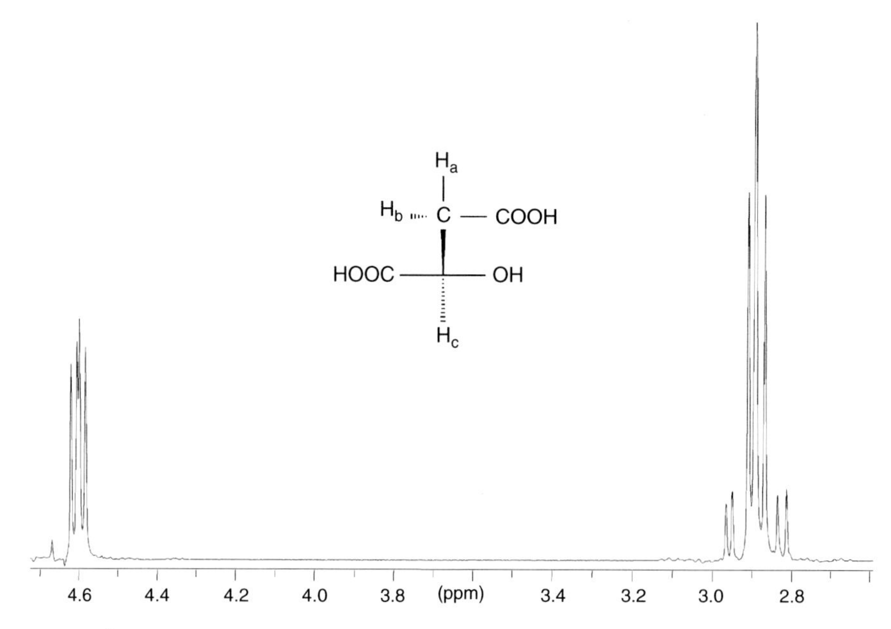

Figure 34 ^{1}H NMR spectrum of malic acid in D_2O.

A higher complexity of the ^{1}H NMR spectra is observed when methylene protons are affected by a chiral centre as can be seen in malic acid (Figure 34). One may expect that the methylene protons H_a and H_b appear as a doublet due to the vicinal coupling with the neighbouring methane, but the methylene protons of

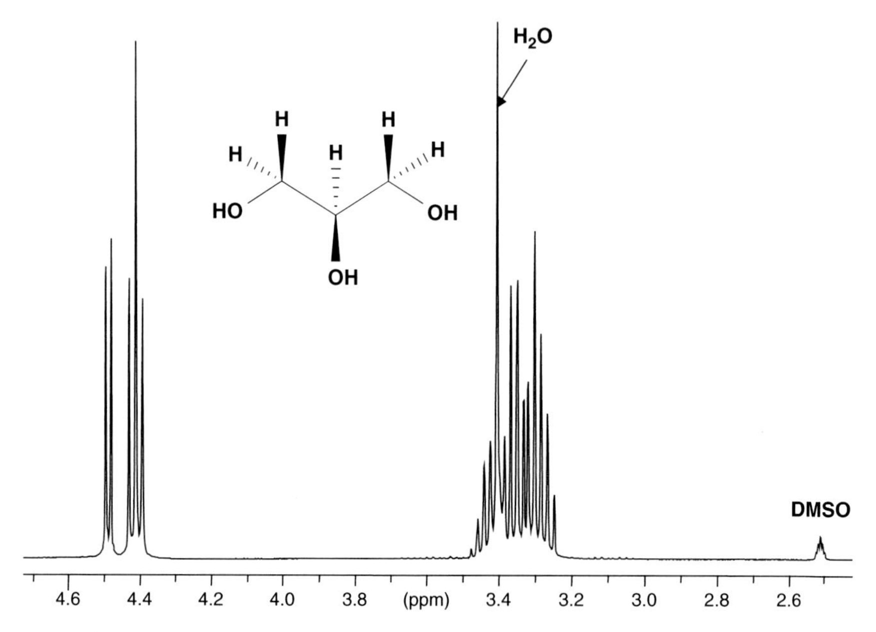

Figure 35 ^{1}H NMR spectrum of glycerol in DMSO-d_6.

malic acid are chemical non-equivalent. Therefore, both the protons (H_a and H_b) have different chemical shifts and the geminal coupling is observable as an A,B system showing a strong roof effect. The mean angle between the methine proton H_c and the methylene protons is not identical and results in a different coupling constant. H_c appears as a doublet of doublets and not as a triplet.

A rule of thumb says that methylene groups in the neighbourhood of an asymmetric centre show diastereomeric splitting of NMR signals. The rule is valid for all NMR-active nuclei. With this knowledge the interpretation of the ^{1}H NMR spectrum of the small molecule glycerol (which is an essential backbone for the structures of lipids) should be easy (Figure 35).

10. Two-Dimensional Methods for Structure Elucidation

The most important two-dimensional NMR techniques are demonstrated in this chapter in combination with the practical use of the NMR techniques and principles. As an example, structure elucidation and signal assignment of rifamycin were chosen (Figure 36). NMR spectroscopic data were published earlier.[10,15,16] The test compound was dissolved in DMSO-d_6.

First, the ^{13}C NMR spectrum recorded in the attached proton test (APT) mode is interpreted (Figure 37). The multiplicity of the carbon atoms is encoded in the positive and negative intensities of the signals: CH_3 and CH groups appear negative, and CH_2 groups and quaternary atoms positive. The number of carbon atoms and the multiplicity are in compliance with the molecular formula $C_{37}H_{46}NO_{12}Na$ and the chemical structure. The signal of the solvent DMSO-d_6 appears as a septet due to the coupling with deuterium at $\delta = 39.5$ ppm. The signal of the solvent was used for calibration (Figure 38).

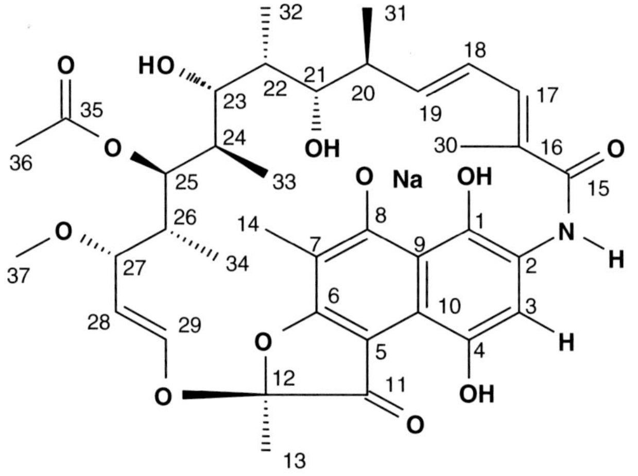

Figure 36 Chemical structure of rifamycin.

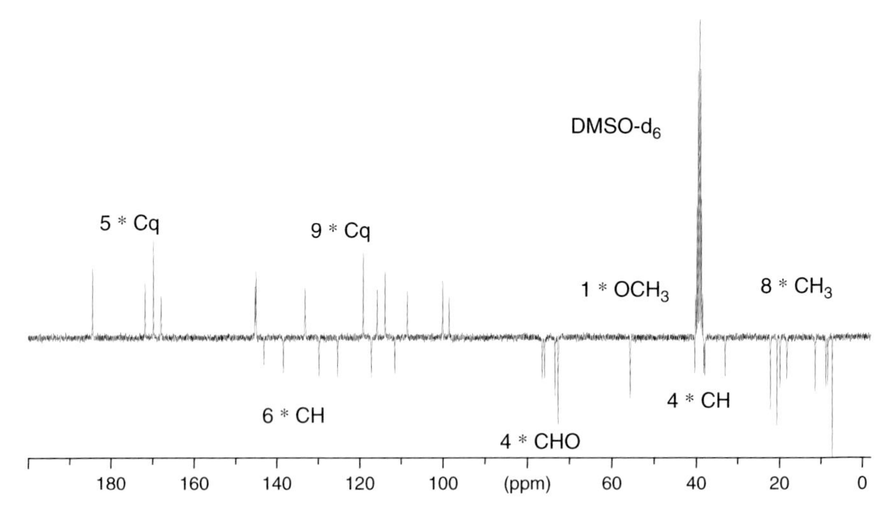

Figure 37 ^{13}C NMR spectrum (APT) of rifamycin.

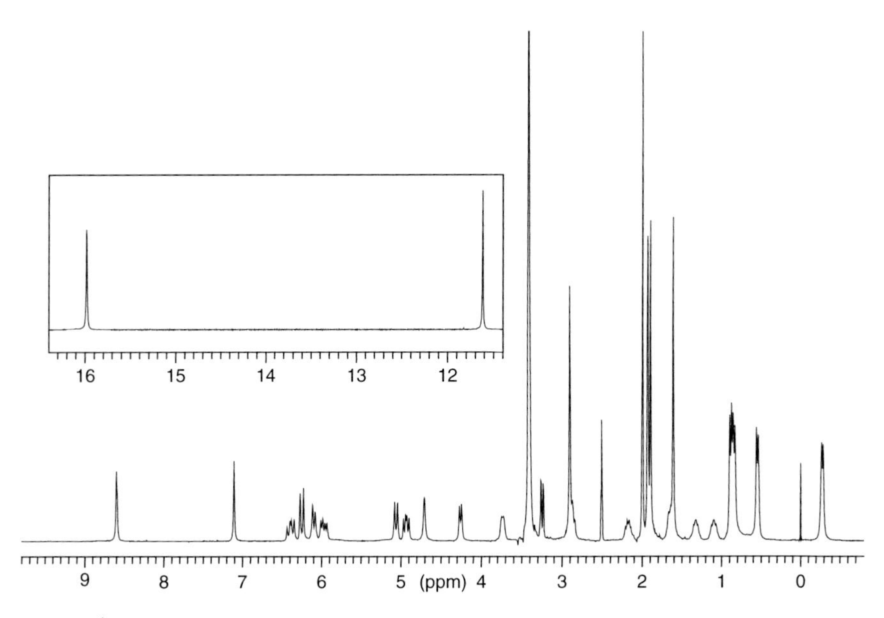

Figure 38 ^{1}H NMR spectrum of Rifamycin in DMSO-d_6.

The ^{1}H NMR spectrum shows signals that are widely distributed in the whole spectrum. The signal of the solvent of DMSO-d_6 appears at $\delta = 2.5$ ppm, and the appropriate water signal at $\delta = 3.4$ in combination with six H_2O from the crystal water. By integrating the singlets and multiplets, the groups fulfil the rule of integer multiples.

Four singlets appear between $\delta = 16$ and 7 ppm. Only the signal at $\delta = 7.1$ shows ^{13}C satellites, and therefore it can be assigned as the aromatic proton. In sum, five OH and one NH protons are expected.

The spectrum further contains nine methyl groups, five as a singlet and four as a doublet having a vicinal coupling with one methine proton. The number of both types is in accordance with the chemical structure. The different chemical shifts of the methyl signals are the result of different physical effects. The group at $\delta = 2.9$ ppm is the methoxy shifted from electronegative oxygen to downfield. The four methyl singlets between $\delta = 1.7$ and 2.1 are connected to an sp^2-hybridised carbon (C=O and C=O double bond). The anisotropy causes a slight downfield shift. The chemical shift of the remaining four methyl groups connected to an sp^3-hybridised carbon atom should show its resonance at approximately $\delta = 1$ ppm. Only two of them fulfil these expectations. Especially the negative chemical shift at $\delta = -0.4$ ppm is a very important indicator for the three-dimensional structure of rifamycin. The methyl group is located in the area of the aromatic ring system, which results in an upfield shift due to the anisotropy of the ring current-induced magnetic field.

The oxygen-connected methine protons appear between $\delta = 2.9$ and 5.2 and the double-bonded protons between $\delta = 4.5$ and 7.5. Not all protons can be assigned by analysing the chemical shifts and coupling constants (Figure 39).

A common tool in spectra simplification is the exchange of OH and NH protons by adding D_2O or CD_3OD or by dissolving the sample in one of the protic solvents. The latter approach was chosen, as a better spectral resolution could be expected. The three very downfield-shifted signals and the doublets at $\delta = 4.7$ and 4.2 disappear. One OH proton is missed. Because of the chemical structure it exchanges fast with the water.

However, a detailed signal assignment of all proton and carbon signals is only possible by the use of two-dimensional methods.

Several modifications of H, H COSY (correlation spectroscopy) methods are available in a modern NMR spectrometer. A COSY 45 should demonstrate the principles for the signal assignment of rifamycin (Figures 40 and 41). Mostly, the

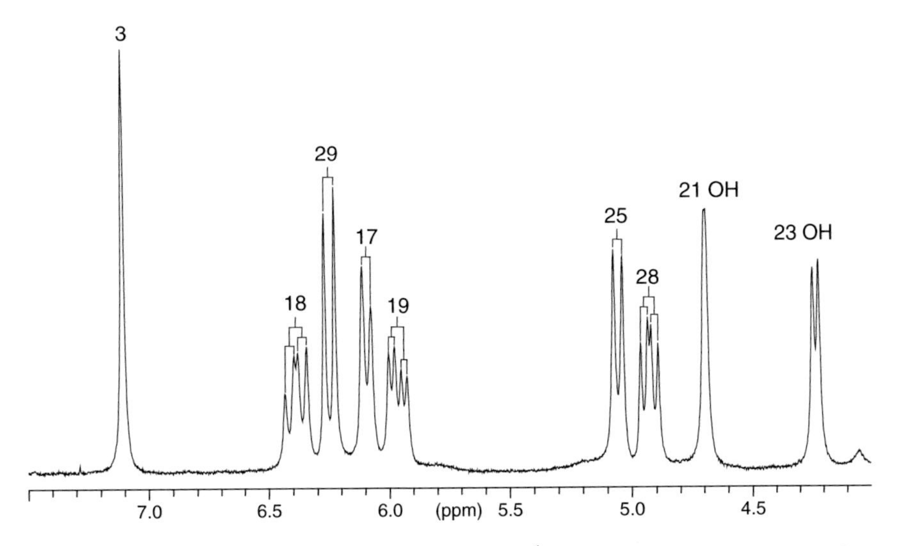

Figure 39 Aryl system of rifamycin with spin analysis, ^{1}H NMR spectrum in DMSO-d_6.

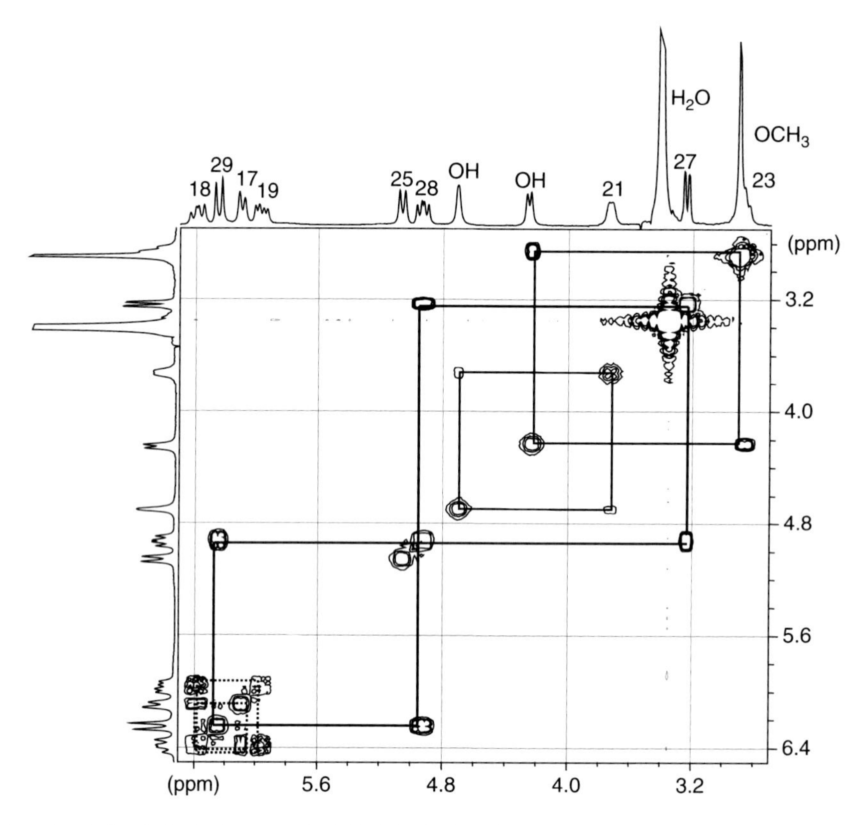

Figure 40 COSY 45 of rifamycin in DMSO-d_6, detail.

vicinal and geminal homonuclear couplings give cross signals that are used to assign neighboured protons. In rifamycin, the two arylic systems are interfered with the signals of two OH and one oxygen-bonded methine proton. The spin analysis given in Table 8 may lead to ambiguous interpretation due to improper evaluation of the coupling constants. The two-dimensional analysis is clear and so the aryl systems including the C29 and C28 (resp. 17, 18 and 19) can be distinguished (see Figure 40). Jumping from one cross signal to another, the substructures of the molecule become observable.

Furthermore, many other connections between protons are visible, e.g. the pairs of CH_3–CH groups. Even very small coupling constants that are not resolved in the one-dimensional ^{1}H NMR spectrum like the connection between H17 and the methyl group H30 – a $^4J_{H,H}$ long-range coupling – show a cross signal in Figure 41.

The most popular heteronuclear COSY variant is the HMQC (heteronuclear multiple quantum correlation). The connection of protons and carbons is detected using the proton signals in a so-called inverse mode. The focus is laid on the large $^1J_{C,H}$ coupling constant and only direct bonded atoms show a cross signal in the two-dimensional spectrum. The pairs of carbon atoms and protons are separated in different groups and circled in Figure 42.

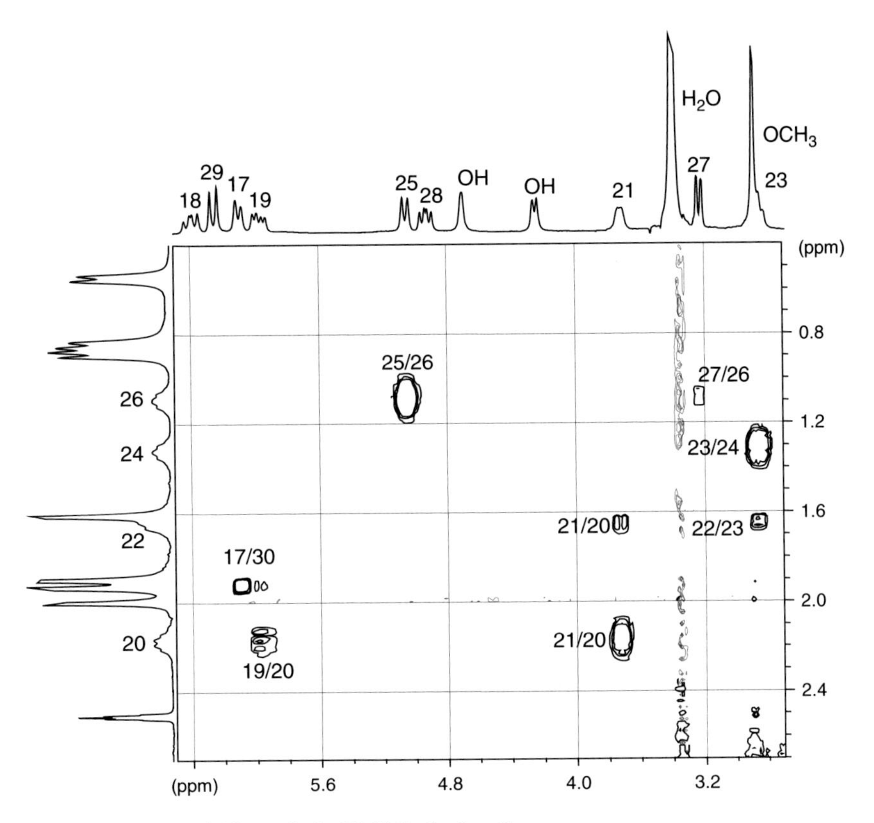

Figure 41 COSY 45 of rifamycin in DMSO-d_6, detail.

Table 8 Aryl protons of rifamycin with coupling constants

No.	Chemical Shift (ppm)	*J* (Hz)
18	6.38	15.7/10.9
29	6.25	12.6
17	6.09	10.8
19	5.96	15.7/7.3
25	5.05	10.9
28	4.93	12.6/8.6
21 (OH)	4.69	1.9
23 (OH)	4.23	7.7

A second heteronuclear method is based on the long-range couplings, mostly the $^2J_{H,C}$ and $^3J_{H,C}$. This heteronuclear multiple bond correlation (HMBC) spectrum (Figure 43) enables the connection of single pairs and substructures to form a molecular backbone. The long-range correlation is not restricted to protons bonded directly on carbon atoms. It also allows the connection of OH and NH protons, a fact which is important for the signal assignment of ^{13}C NMR signals of the aromatic ring in rifamycin.

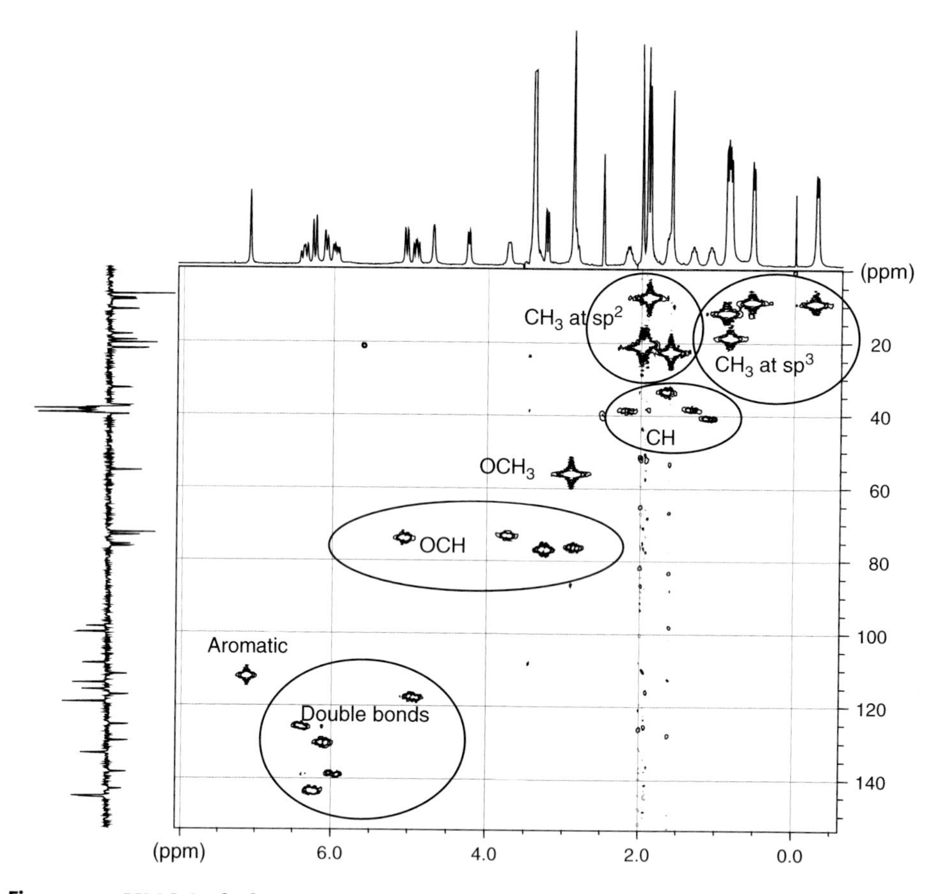

Figure 42 HMQC of rifamycin in DMSO-d_6.

Figure 43 HMBC of rifamycin in DMSO-d_6.

The magnitude of $^3J_{H,C}$ is angular dependent and sometimes not observable. For practical reasons, a connection table should be created, which includes the expected (grey fields) and the detected cross signals (xx for long range and o for ^1J). In Table 9 this is done for rifamycin. Only in ^3J the connection of OH at $\delta = 15.9$ ppm and the carbon atom 9 at $\delta = 114.1$ is missed. In addition, a long-range connection appears with low intensity between H3 and C5.

Table 9 Long-range connections by HMBC analysis of the aromatic part of rifamycin

No.	δ (ppm)	OH 15.9	OH 11.6	NH 8.6	CH 7.1	CH3 1.9
1	145.2	*xx*		*xx*	*xx*	
2	119.4	*xx*		*xx*	*xx*	
3	111.5		*xx*	*xx*	*o*	
4	145.0		*xx*			
5	98.7				*x*	
6	184.5					*xx*
7	100.0					*xx*
8	172.0					*xx*
9	114.1					
10	115.8		*xx*		*xx*	

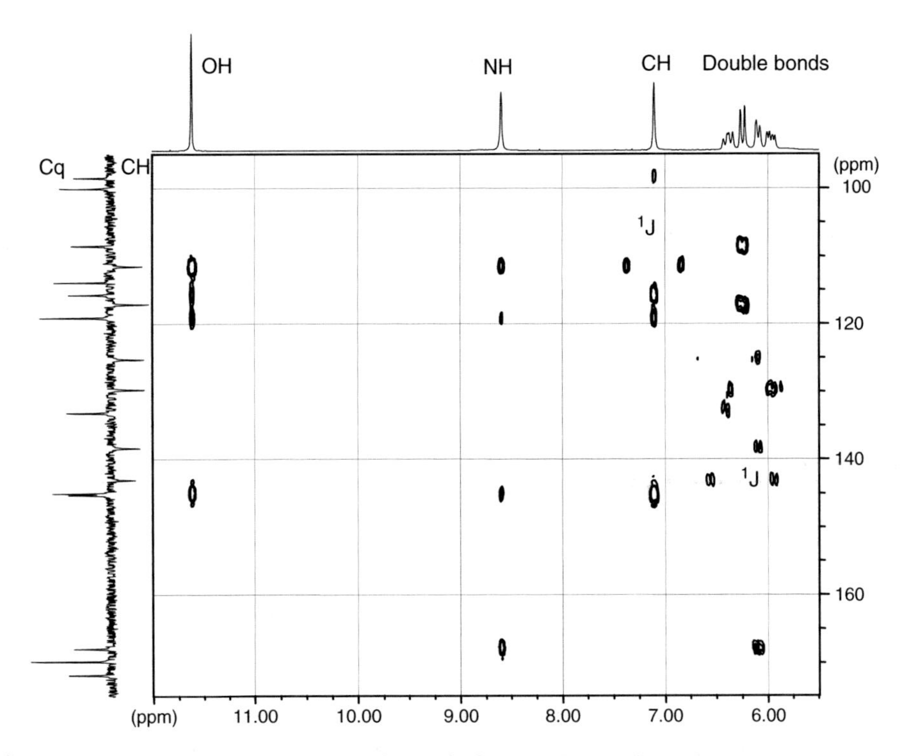

Figure 44 Substructure, a mesomeric form of rifamycin (aromatic part).

Table 10 ^1H NMR and ^{13}C NMR chemical shifts of rifamycin, δ (ppm)

No.	^1H NMR	^{13}C NMR	No.	^1H NMR	^{13}C NMR
1	(OH) 15.9	145.2	20	2.1	40.5
2		119.4	21	3.7 (OH) 4.7	73.6
3	7.1	111.5	22	1.6	33.2
4	(OH) 11.6	145.0	23	2.8(OH) 4.2	76.1
5		98.7	24	1.3	38.4
6		184.5	25	5.1	72.8
7		100.0	26	1.1	38.1
8	(OH) -	172.0	27	3.3	76.7
9		114.1	28	4.9	117.4
10		115.8	29	6.3	143.1
11		184.5	30	1.95	133.2
12		108.7	31	0.84	18.5
13	1.61	22.4	32	0.88	11.6
14	1.90	7.5	33	0.55	8.6
15		168.0	34	−0.3	9.0
16		133.2	35		170.0
17	6.1	129.7	36	2.0	20.8
18	6.4	125.3	37	2.9	55.8
19	6.0	138.4	NH	8.6	

The unusual downfield shift of the phenol protons is caused by hydrogen-bridged bonds. However, these protons are located at the C1 and C4 positions, where, in contrast, no OH is located at the C8. Rifamycin is a sodium salt, and the assignment of the chemical shifts indicates a quinoide structure in two mesomeric forms, anyway in dried DMSO-d_6 as solvent (Figure 44). The given examples do not completely explain the use of NMR techniques but give a helpful assistance in the understanding of modern NMR in pharmaceutical analysis (Table 10).

11. Experimental Data

All NMR spectra given in this chapter were performed using a Bruker 300 MHz Avance 1 spectrometer equipped with a broad band inverse (BBI) probe. Data evaluation was performed using WIN-NMR.

REFERENCES

1. R.R. Ernst, G. Bodenhausen, A. Wokaun, Principles of Nuclear Magnetic Resonance in One and Two Dimensions, Clarendon Press, Oxford, 1987.
2. J.K.M. Sanders, B.K. Hunter, Modern NMR Spectroscopy, a Guide for Chemists, Oxford University Press, New York, 1987.
3. D.M. Grant and R.K. Harris, (Eds), Encyclopaedia of Nuclear Magnetic Resonance, J. Wiley & Sons, Chichester, Vol. 1–8, 1996.

4. T.D.W. Claridge, High-Resolution NMR Techniques in Organic Chemistry, Pergamon Press, Oxford, 1999.
5. W.R. Croasmun, R.M.K. Carlson, Two Dimensional NMR Spectroscopy, Wiley-VCH, New York, 1994.
6. H. Kessler, M. Gehrke, C. Griesinger, Angew. Chem. 100 (1988) 507–554.
7. R.R. Ernst, G. Bodenhausen, A. Wokaun, Principles of Nuclear Magnetic Resonance in One and Two Dimensions, Clarendon Press, Oxford, 1987.
8. P.A. Rinck, Magnetic Resonance in Medicine. The Basic Textbook of the European Magnetic Resonance Forum, ABW-Wissenschaftsverlag, Berlin, 2003.
9. http://www.org.chemie.tu-muenchen.de/lehre/lehredateien/scrips/ociv.pdf
10. H.-O. Kalinowsky, S. Berger, S. Braun, 13C-NMR-Spectroscopie, Thieme, 1984.
11. H.-O. Kalinowsky, S. Berger, S. Braun, NMR Spectroscopy of the Non-Metallic Elements, Wiley-VCH, Weinheim, 1997.
12. H.S. Mosher, J.A. Dale, J. Am. Chem. Soc. 90 (1968), 2581–2598
J.A. Dale, H.S. Mosher, J. Am. Chem. Soc. 95 (1973), 512–519.
13. Spectral Service, not published.
14. U. Holzgrabe, I. Wawer, B. W .K. Diehl, NMR Spectroscopy in Drug Developement and Analysis, Wiley-VCH, Weinheim, 1999.
15. L. Santos, M.A. Madeiros, S. Santos, M.C. Costa, R. Tavares, M.J.M. Curto, J. Mol. Struct. 28 (2001), 563–564.
16. S. Martini, C. Bonechi, G. Corbini, A. Donati, C. Rossi, Bioorg. Med. Chem.12(9) (2004), 2163–2172.

CHAPTER 2

Quantitative NMR in the Solution State NMR

F. Malz

Contents

Abstract

Nuclear magnetic resonance (NMR) spectroscopy is by definition a quantitative spectroscopic tool because the intensity of a resonance line is directly proportional to the number of resonant nuclei (spins). This fact enables accurate and precise determinations of the amount of substance. With the increase of sensitivity due to stronger and stronger static magnetic fields, quantitative NMR can be widely used in various fields of applications.

Keywords: relative method, absolute method, linearity, robustness, specificity, selectivity, accuracy, precision, uncertainty, sensitivity

1. Introduction

Nuclear magnetic resonance (NMR) spectroscopy is one of the most important and widespread analytical methods in academic and industrial research. It enables a unique and, in principle, quantitative determination of the relative amount of molecular groups, thus offering a tool to quantify entire substances even in mixtures. The first quantitative measurements (qNMR) have been described in the literature in 1963 by Jungnickel and Forbes[1] and Hollis.[2] In the first case the intramolecular proton ratios in 26 pure organic substances have been determined, whereas Hollis et al. have analyzed the amount fractions of three analytes, aspirin, phenacetine, and caffeine, in respective mixtures. For a long time there was a lack of acceptance of qNMR as a precise tool even though there were a number of reports regarding the achievable high precision of quantization by NMR. Textbooks on NMR often do not emphasize the quantitative aspects. However, in recent years, quantitative NMR spectroscopic applications received major attention, supported by the technical progress of modern NMR techniques that overcame the problems of comparatively low sensitivity. This leads to a reduction of the previously widespread scepticism toward qNMR results. Griffith and Irving[3] discussed that an exact sample preparation, spectra acquisition, and signal integration can provide assays for simple model compounds with expanded uncertainties of about 1% relative. Maniara et al.[4] performed an extensive validation study of ^{1}H- and ^{31}P NMR measurements, focussing on assays for model and agrochemical compounds, which resulted in standard uncertainties of about 0.5% relative. In a series of papers, Wells et al.[5–7] demonstrated the great power of modern qNMR when performed as a validated method. They assessed the assay of some technical grade agrochemicals with a detailed uncertainty budget[8] on the basis of the guide to the expression of uncertainty measurement (GUM).[9] The great importance of quantitative NMR spectroscopic measurements using modern NMR techniques and validated procedures was recently demonstrated in three theses.[10–12]

Today, quantitative ^{1}H-, ^{13}C-, ^{31}P-, and ^{19}F NMR in liquids is used routinely, e.g., in pharmacy,[13–17] in agriculture,[5,6,18] in material science,[19] and for military purposes,[20] where the assay or content determinations of substances (and impurities) are the key issues. Furthermore, in 1998, Jancke[21] and later the committee for chemical measurements (Comité Consultatif pour la Quantité de Matière, CCQM)[22,23] have discussed the potential of qNMR as a primary method according to the definition of the CCQM[24] based on the fact that the NMR signal response (more precisely, the integrated signal area) is directly proportional to the number of nuclei contributing to the signal.

Few NMR intercomparisons have been published so far.[25] It has been shown that experienced laboratories may reach levels of accuracy and precision better than 99% relative, if carefully validated measurement and data-handling parameters are used.[26]

2. Basics

The basics of high-resolution NMR can be found in Chapter 1, Part I, of this book and in many textbooks,[27–31] with particular focus on quantitative measurements.[32–34] They will not be presented here in detail.

The most important fundamental relation of qNMR is that the signal response (integrated signal area) I_x in a spectrum is directly proportional to the number nuclei N_x generating the corresponding resonance line

$$I_x = K_S N_x \tag{1}$$

where K_s is a spectrometer constant. Usually, the NMR signal of a single substance consists of several resonance lines. However, it is sufficient to select a single resonance line specific for this sample composition. In this case, N_x represents the relative number of spins (i.e., protons) that cause this resonance.

In NMR experiments with correct parameter settings (for structure analysis of a single substance and for quantitative mixture analysis) all molecules in the observation volume underlie the same experimental conditions. In this case, K_s cancels and the intensity ratios in the spectrum are directly related to the ratios of numbers of nuclei at different sites 1, 2, ... (having different chemical shifts) according to

$$\frac{I_1}{I_2} = \frac{N_1}{N_2} \tag{2}$$

However, a number of criteria (acquisition parameters) have to be fulfilled for ensuring this (see below).

3. Quantitative NMR Spectroscopy

3.1. Relative method

Determining ratios is the easiest way to obtain quantitative results by NMR. The molar ratio n_X/n_Y of two compounds X and Y can be calculated straightforwardly using

$$\frac{n_X}{n_Y} = \frac{I_X}{I_Y}\frac{N_Y}{N_X} \tag{3}$$

Consequently, the amount fraction of a compound X in a mixture of m components is given by

$$\frac{n_X}{\sum_{i=1}^{m} n_i} = \frac{I_X/N_X}{\sum_{i=1}^{m} I_i/N_i} \times 100\% \tag{4}$$

irrespective of the solvent signal in which the mixture is dissolved.

It should be noted that qNMR is the most important method for quantifying the ratios of isomers, diastereomers, and enantiomers,[35] because of its excellent selectivity for structural analysis combined with the easy and fast quantification procedure. Even the knowledge of the molar masses of the components is not required. Enantiomers, although having identical spectra, can also be differentiated by NMR, when chiral solvents or complexing agents are used.[36,37]

3.2. Absolute method

As described above, NMR spectra include only information about the ratio of intensities. However, adding a standard to a mixture, absolute results of analysis such as assays, contents, or concentrations can be obtained.

The most popular method for determination of absolutes values by qNMR is the main component analysis using an internal standard. A one-point calibration has to be carried out by adding a standard (of known assay) gravimetrically to a known amount (mass) of mixture such that the intensity ratio of the analyte signals of interest and the standard are nearly equally strong. The following special requirements for those standards have been described elsewhere:[3,38] (i) inexpensive, (ii) stable and chemical inert, (iii) available in pure form, (iv) nonhygroscopic, (v) soluble in most NMR solvents, and (vi) show an simple singlet spectrum. It should be noted that it is impossible to find a universal reference substance for qNMR assessments. A number of possible reference compounds were published in the previous years. Pauli et al.[39] listed them for the ^{1}H qNMR in a review article, and Martino et al.[40] for ^{19}F- and ^{31}P qNMR (Table 1).

Table 1 List of reported reference materials as internal standard for ^{1}H, ^{19}F, and ^{31}P qNMR assessments

^{1}H	^{19}F	^{31}P
1,3,5-benzenetricarboxylic acid	*Para*-fluorobenzoic acid sodium salt (FBEN)	Dimethyl-methylphosphonate
1,3,5-trimethoxybenzene	Sodium fluoroacetate	Methylphosphonic acid (MPA)
1,3,5-trioxane	Trifluoroacetic acid	Phenylphosphinic acid (PPA)
1,4-bis(TMS)-benzene	*Para*-fluoro-D-phenylalanine	Triphenylmethylphosphonium bromide
1,4-dinitrobenzene		glyphosate
1,4-dioxane		3-Aminopropylphosphonic acid
Anthracene		
Benzyl benzoate		
Biphenyl		
Dimethyl isophthalate		
Dimethylformamide		
Dimethylsulfone		
Formic acid		
Hexamethylcyclotrisiloxane		
Maleic acid		
Methenamine		
Phloroglucinol		
Sodium acetate		
Tert-butanol		
Tetramethylpyrazine		
TSP-d_4		
Etacrynic acid		
2,5-dimethylfuran		

The assay of the analyte P_X can be calculated directly from the NMR using a standard of known assay P_{Std}:

$$P_X = \frac{I_X}{I_{Std}} \frac{N_{Std}}{N_X} \frac{M_X}{M_{Std}} \frac{m_{Std}}{m} P_{Std} \tag{5}$$

with M_X and M_{Std} being the molar masses of analyte and standard, m and m_{Std} the weights of the sample and standard, and P_X and P_{Std} the assays of analyte and standard, respectively.

If a contamination of the analyte by the chosen standard must be avoided, external standards have to be used. This can be achieved by employing the external standard either in a concentric tube (putting a capillary, filled with the dissolved standard, inside the analyte NMR tube)[20,41] or in a separate precision tube (using two NMR tubes filled with the analyte and standard solutions, respectively)[42,43]. In both cases, analyte and standard should be dissolved in the same solvent and the volumes of the tubes (capillary) have to be calibrated before. Capillaries are commonly used for determining deuterated assays or the water content of solvents.

The standard addition method is a further possibility of absolute value determinations. If known amounts of the active compound are added to the solution in several steps, the content can be calculated without knowing the molar mass of the analyte.

In 1998, Akoka et al.[44,45] published a new technique called ERETIC (electronic reference to access in vivo concentrations). Here, an electronic reference signal is fed in the resonance circuit of the probe during the acquisition time using a free coil in the probe (heteronuclear channel). Amplitude, FWHH (full-width at half-height), and frequency (chemical shift position) can be set by the operator. The advantage of this technique is that the electronic signal can be shifted to free spectral range avoiding superposition with the analyte signals. Since then a couple of applications of ERETIC in ^{1}H qNMR, ^{13}C qNMR,[46] two-dimensional (2D) NMR,[47] solid-state NMR,[48] and MRI (magnetic resonance imaging)[49] were published.

In contrast to these direct methods of quantification, it is also possible to quantify the assay of an analyte by the indirect method. If all impurities show up in the NMR spectrum, if they can be assigned structurally and if they can be measured quantitatively, the assay is simply the difference from the 100% value (the so-called 100% method). This approach is limited for partially overlapping signals, and it is impossible for impurities not containing the observed nucleus (e.g., inorganic impurities (NaCl) in case of ^{1}H NMR).

4. VALIDATION

Often the terms validation, certification, characterization, and verification are used in the same context causing confusion because of improper use. In 1970s, the term validation appeared for the first time (1975 in Europe, 1978 in the United States) in the literature. Several different definitions of validation have been described.[50] The combination of the ISO Guide 25[51] and the EN 45001[52] to the new international norm DIN EN ISO/IEC 17025[53] yields a unique definition for the validation of analytical methods as the "confirmation by examination and the provision of objective evidence that the particular requirements for a specific intended use are fulfilled." As a straightforward conclusion, nonstandard methods like NMR must be validated. The validation process requires the testing of linearity, robustness, parameters of accuracy (repeatability, comparability, and measurement uncertainty), specificity, and selectivity.[54] Intercomparisons or round robin tests complete the validation.

4.1. Linearity

The linearity of an analytical procedure or method is defined as its ability to obtain test results that are directly proportional to the amount of substance in the sample.

qNMR as a method itself is linear because the intensity of the response signal is directly proportional to the amount of nuclei contributing to this signal (Eq. (1)) and no calibration is necessary for the determination of molar ratios of mixtures. The evaluation of 13 model mixtures containing octamethylcyclotetrasiloxane (D_4) and 1,2,4,5-tetramethyl benzene (Dur) in different molar ratios (from 2.3 to 99.0 mol/mol for D_4) dissolved in $CDCl_3$ confirmed the linearity of NMR spectroscopy. Figure 1 shows the experimentally determined molar ratios for D_4 versus the gravimetric reference values. Linear regression yielded a correlation coefficient of 0.99992 and a regression line of $y = 1.004x$.

For assays or content determinations the linearity will be evaluated by five various contents of the substance in a range 70–130%, according to the content of analyte in test sample. Figure 2 shows exemplarily the linearity test for the determination of ginsenosid using dimethylsulfon as internal standard.

In general, a regression $r > 0.999$ is achievable.

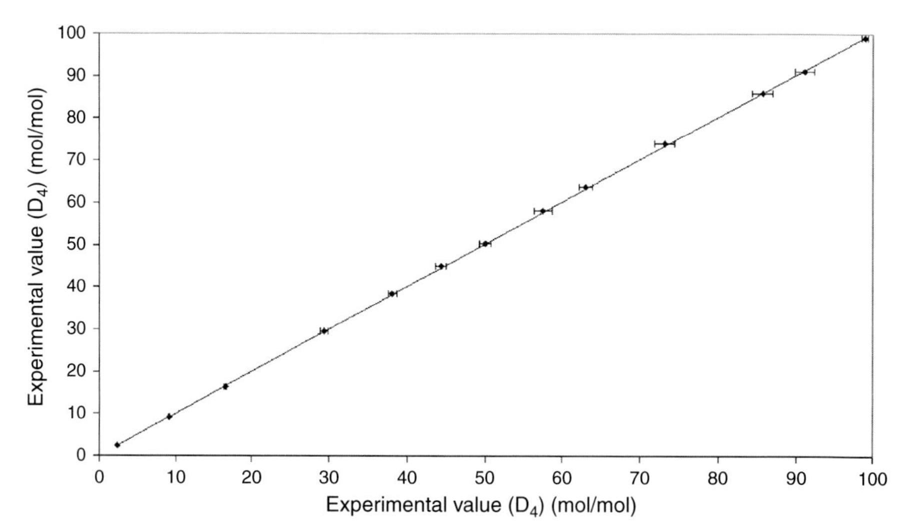

Figure 1 Test of linearity. Theoretical and experimental molar ratios of 13 model mixtures, calculated for Dur correlation coefficient $r^2 = 0.99992$.

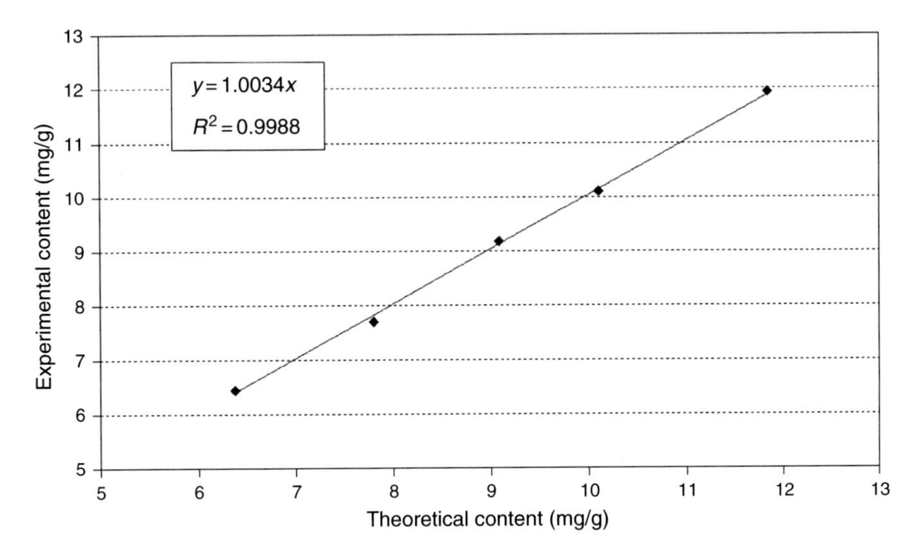

Figure 2 Test of linearity. Linearity is evaluated by five various contents in a range from 70% to 130%, according to the content of ginsenosid in test sample.

4.2. Robustness

For the evaluation of the robustness of the method all important parameters of the data acquisition, processing, and evaluation of the NMR must be varied stepwise in wide ranges. These parameter ranges must be determined, which will definitely yield wrong measuring results; vice versa, the parameter sets required for accurate and precise measurements must be found and fixed. By varying the spectrometer parameters sets, three different effects can be distinguished: (i) no significant influence (robust), (ii) significant influence on the signal-to-noise ratio (S/N), and (iii) systematic change of correct signal intensity. However, most of the parameters do not have significant effects on the accuracy or precision of the method within their evaluated ranges.

4.2.1. Acquisition parameters

The pulse excitation must be uniform for the entire spectral width (SW) of interest, which requires short pulses (typically 10 μs). In this case, ^{1}H NMR spectra can be acquired quantitatively. However, spectra of heavier nuclei (^{19}F, ^{31}P, etc.) with larger chemical shift ranges may suffer intensity distortion, particularly if measured at very high magnetic fields.[31]

The repetition time τ (or recycling time) depends on the longest longitudinal relaxation time T_1 of all signals of interest. The T_1 relaxation is described by

$$M_z = M_0\left(1 - e^{-(\tau/T_1)}\right) \tag{6}$$

with M_z and M_0 being the magnetization along the z-axis (response factor) after waiting time τ and at thermal equilibrium, respectively. Routinely, measurements are done by choosing $\tau = 5T_1$. Hence, 99.3% of the equilibrium magnetization (signal) is measured.[31] For nuclei with very long T_1 values, e.g., ^{13}C, ^{29}Si, and ^{31}P, paramagnetic relaxation reagents such as chromium(III) acetylacetonate ($Cr(acac)_3$) are added to shorten the relaxation times.

Heteronuclear NMR experiments of X nuclei (^{13}C, ^{19}F, etc.) with simultaneous ^1H broadband decoupling cause inherent intensity distortions by the nuclear Overhauser effect (NOE). This NOE effect on the spectrometer constant K_S can be described[55] by

$$K_s = K_0(1+\eta)\left(\frac{1-e^{-(\tau/T_1)}}{1-\cos\alpha\, e^{-(\tau/T_1)}}\right)\sin\alpha \qquad (7)$$

where K_0 is a constant instrumental factor, η the nuclear Overhauser enhancement factor, and α the flip angle of the excitation pulse. To suppress possible intensity distortions below 1%, the following rules must be fulfilled: (i) ^1H decoupling is applied only during the signal acquisition time (inverse gated technique) to minimize η, (ii) the repetition must be set to (5–7) $\times$ T_1,[56] and (iii) 90° pulse should be used for the excitation.

The required acquisition time t_{aq} depends on the smallest line width in the spectrum, and truncation of the NMR signal in the time domain (free induction decay (FID)) must be avoided. If truncation occurs, signal forms with (substantial) "wiggles" appear in the spectra, and wrong intensities will result in combination with FID baseline correction modes.[57] Usually, the signal should decay completely halfway through the acquisition period. It also ensures that enough data points describe the NMR lines (at least five data points above the half-height) such that the integration procedure does not cause artificial distortions (intensity error <1%).[58–60]

However, most of the acquisition parameters are robust. They can be varied within wide ranges. In case that all resonance lines have similar line widths, the values of the preacquisition delay is not important. Very short delay between 5 and ~30 µs are used in general. It should be noted that for those short delay times the first data point is modified by the pulse ringdown due to the high Q factor of the probes. As only the first data point is affected, the Fourier transform causes minor distortions of the baseline; but these are not significant because the NMR lines are by orders of magnitude narrower and the baseline has to be corrected carefully around the resonance position anyway. The reason for choosing a short preacquisition delay is that if NMR lines with different transverse relaxation times occur (e.g., NH protons), a longer preacquisition delay will change the relative peak areas and lead to systematic errors for the intensity ratios.[34] Furthermore, an increase in sample temperature did not change the results for this sample. Even drastic changes of the receiver gain (however, without overload of the ADC) do not change the results because of the large word length of modern ADCs.

Variations of the pulse excitation angle and the number of scans influence the S/N. A smaller flip angle decreases the S/N for a given experiment time and repetition delay. For short pulse angles (~30° and less) the repetition time between the experiments can be shortened so that a better S/N can be obtained in the same experimental time (the so-called Ernst angle).

If spectrometers that do not have digital filtering capability are used, it must be ensured that the analog filter width is set properly. Otherwise, intensity distortion may occur for resonance lines close to the ends of the filter range. It was found by the author (using a 600 MHz spectrometer) that if the frequency offset of a resonance line from the RF is 3 kHz, a filter width of 9.6 kHz causes an unacceptable signal damping of about 2.1%.

4.2.2. Processing parameter

Generally windowing functions were used to enhance the S/N at the cost of the resolution of the spectrum (broader resonances) prior to the Fourier transformation. For qNMR measurements, exponential multiplication (em) is typical used. A larger line broadening factor (lb) for the em function improves the S/N, but the simultaneous line broadening of the signal may complicate the integration routine if an adjacent resonance is close. Following the suggestion of Günther,[61] the best compromise is to use a small line broadening (e.g., lb = ~0.3 Hz) but a larger number of scans to increase the S/N.

Another parameter prior to the Fourier transformation is the zero filling, i.e., the adding of data points with the value zero to increase the digital resolution. Half of the collected data points in the time domain (FID) will be trashed by the Fourier transformation. Usually, a zero filling by a factor of 2 is used. This implies the doubling of data points in the time domain (FID) to maintain the original digital resolution in the frequency domain (spectrum). Higher factors of zero filling are permitted, but will have no effect.

4.2.3. Spectra evaluation (integration)

The operator is commonly the main source of error of qNMR measurements for various reasons. Baseline and phase correction must be performed with very high precision to ensure accurate results.[33,57] Frequency-dependent phase errors of larger than 10° resulting from improper spectra processing cannot be corrected with the BIAS and SLOPE functions of the integration routine and results in significant deviations of the calculated areas of larger than 1%. However, as such phase errors can be seen easily in the spectra, they can be avoided. It was found that automatic routines cannot solve this problem in general; therefore, the operator's experience determines the quality of the results. Furthermore, the integration limits for the NMR lines have to be set according to the FWHH of each signal. Extending the integration over a frequency range of 64 times the FWHH value ensures that 99% of the entire signal intensity (Lorentzian) is obtained.

Instead of the signal integration a line shape fit can be used to determine the signal areas of interest. The problem is that a unique profile for the line shape does not exist, because it depends – among others – on the shim stage (field homogeneity) of the magnets, including the probes.

Table 2 Validated parameter setup for quantitative ^{1}H NMR experiments, listed for a 9.4 T NMR spectrometer (proton frequency of 400 MHz)

Parameter	Nomenclature of Bruker	Value
90° pulse strength	Pl1	Instrument specific
90° pulse length	P1	Instrument specific
Spin rotation		No
Measurement temperature	TE	300 K
Frequency of excitation	o1	Middle of spectrum
Pulse angle		30°
Preacquisition delay	DE	6 μs
Acquisition time	AQ	5.9 s
Relaxation delay	D1	$\geq(7/3) \times$ longest T_1
Sweep width	SW	14 ppm
Filter width	FW	$\geq$20 ppm
Number of FID points	TD	64 k
Number of scans	ns	declined of reached S/N
Signal-to-noise ratio	S/N	$\geq$200:1
Line broadening (em)	lb	0.3 Hz
Number of frequency points	SI	64 k

Detailed validations of qNMR using single pulse excitation of ^{1}H NMR and ^{1}H inverse-gated ^{31}P NMR spectra proved that the qNMR is robust, if lengths of acquisition times, repetition times, and the S/N ratios are set according to the rules mentioned above.[4,10–12] Table 2 shows a validated parameter setup for quantitative ^{1}H NMR measurements.

4.3. Specificity and selectivity

There is one prerequisite for the qNMR procedure outlined above. The NMR signal used for quantitation, the so-called monitor signal, must be assigned definitively to one of the structure groups of the substance to be analyzed. In the easiest case these signals are singlets or simple multiplets. This means that the signals consist of one or a few Lorentzian lines of a typical FWHH of about 1 Hz.

In more complicated cases, the monitor signals may consist of multiplets due to coupling with other nuclei in the molecule and/or of superconductions with a further (impurity) signal within the same spectrum. It must be ensured that the chosen signal belongs exclusively to only one molecular unit of the analyte (e.g., —CH, —CH_2, or —CH_3 for ^{1}H NMR) with no contribution to a possible impurity. Therefore, the specificity and selectivity must be checked for each sample prior to qNMR investigations. Specificity means the ability to assess unequivocally the analyte of interest in the presence of other components (here for NMR: the unambiguous assignment of all NMR lines to the structure of the analyte). The selectivity of a method is given by the ability to determine analytes of interest in a complex mixture without interference from other components in the mixture.

There are different strategies and various NMR experimental tools to check the specificity and selectivity. The most important are listed here:

1. An internal signal integration of several analyte resonances may identify hidden impurity signals due to larger intensity of the respective integral.
2. Higher magnetic fields will increase the spectral resolution.
3. Measurement of spectra of other nuclei, e.g., ^{13}C NMR can yield more details of the structure.
4. Direct comparison of the intensities of the same signal group for two different samples of different preparation route. If there are impurities in one sample, the difference spectrum may show the impurity signals.
5. Consideration of the stereochemistry and the corresponding spectroscopic consequences.
6. Homo- and heteronuclear correlation spectra (Correlation Spectroscopy (cosy), Heteronuclear Multiple Quantum Coherence (HMQC), ^{1}H-deducted Multiple bond correlation (HMBC) may show signal correlations of hidden impurity signals with resolved resonances.
7. Calculation of the theoretical spectrum and comparison with the experimental one. A potential impurity may show up.
8. A change of the temperature can lead to a simpler spectrum (dynamic effects; e.g., breakdown of stereochemistry) with the chance to observe impurity signals.
9. Spectra simplification by homonuclear decoupling can also produce simpler spectra.
10. Analysis of spectra of high-performance liquid chromatography (HPLC) separated (offline or online) fractions.

For the selectivity of the method it must be ensured that this monitor signal belongs exclusively to the analyte without any contributions from an impurity. If it nevertheless happens, one can (i) change the temperature, (ii) change the pH value, (iii) add specific shift reagents, and (iv) use other solvents or different concentration to separate such overlapping signals.

If these attempts fail, one may search for another (resolved) signal of the same impurity. If the specific impurity resonances can be assigned unambiguously and determined quantitatively, the hidden impurity signal can be subtracted from the total signal area.

Another serious problem may arise if the impurity signals cannot be distinguished from the analyte at all. This situation was met in a purity determination of atrazine containing simazine and propazine as impurities.[62] The authors used elevated temperatures and homodecoupling to tackle the problem.

4.4. Accuracy

The accuracy of an analytical procedure or method expresses the closeness of agreement between the value that is accepted either as a conventional true value or an accepted reference value and the value found.

The direct proportionality between the intensity and the amount of nuclei of a resonance line combined with the ability to be a potential primary ratio method,[63–65] as defined by the CCQM[66] warrant the accuracy of the experimental

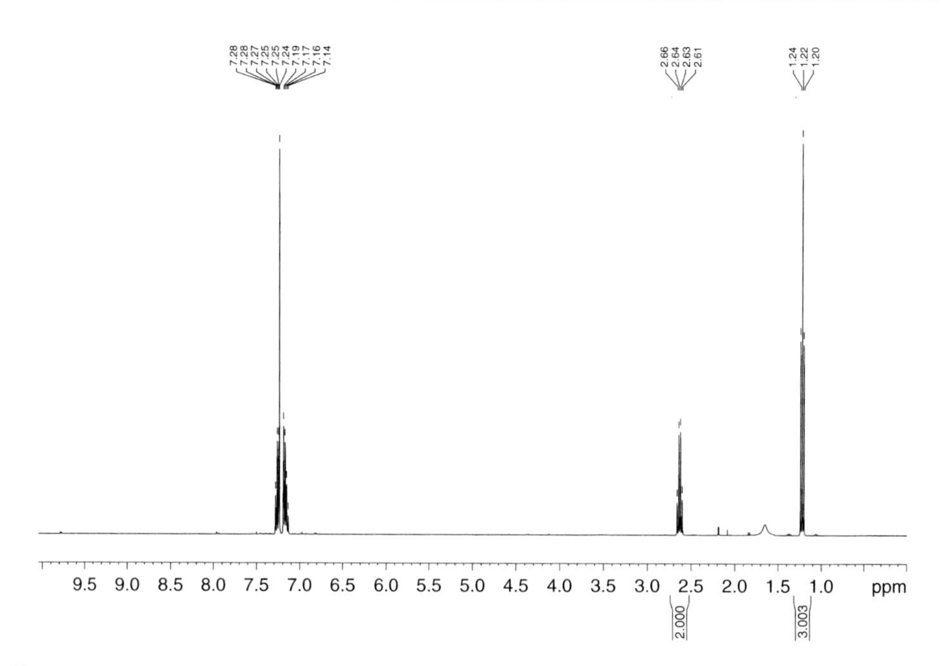

Figure 3 Test of accuracy. Accuracy is evaluated by recovering the natural proton ratio of CH_2 to CH_3 using the reference sample ethylbenzene in $CDCl_3$.

results. Thereby, a certified reference material (CRM, primary standard) must be used as a standard for assay or content determinations, or a standard whose assay ca traced back to a CRM (secondary standard). The accuracy of qNMR might be checked by any substance with at least two measurable signal intensities just by stoichiometry. For example, the intensity ratio of 2:3 for the proton signals of the CH_2 and the CH_3 group of ethylbenzene will be found exactly in the ^{1}H NMR spectrum (Figure 3):

Deviations less than 0.2% relative are achievable.

4.5. Precision

The precision of an analytical procedure or method expressed the closeness of agreement between a series of measurements obtained from multiple sampling of the same homogenous sample. Precision may be considered at two levels: repeatability and comparability.

The precision of the integration procedure of qNMR depends on the S/N (see Figure 4) of the signals of interest. This means for a precise integration a maximum S/N is required:

As shown in Figure 4, an S/N of at least 200:1 is necessary for every resonance line, which should be integrated, for a precision better than 99% (uncertainty less than 1%). Thereby, all influences of the inhomogeneity of the magnetic field as well as by the acquisition, processing, and evaluation steps are included.

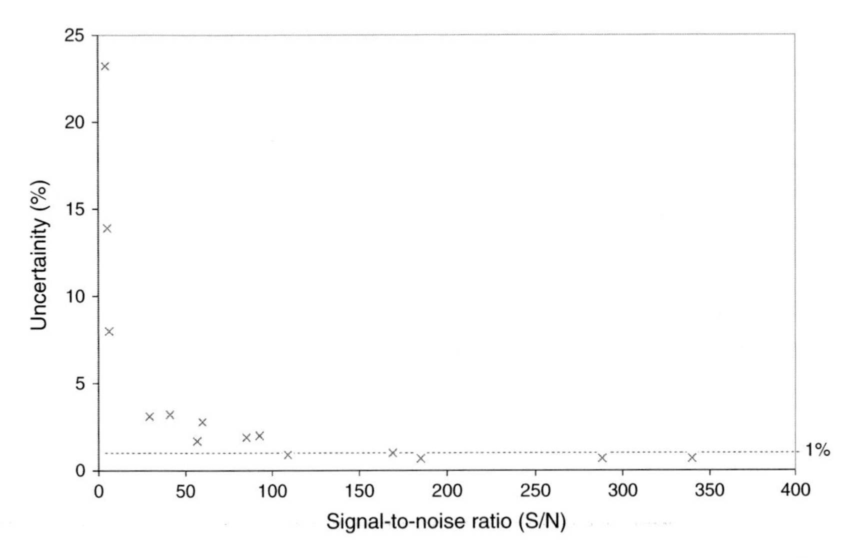

Figure 4 Test of precision. S/N and its influence to the precision/uncertainty of qNMR. Dotted line represents an uncertainty level of 1% relative.

4.5.1. Repeatability and comparability

According to the ICH (International Conference on Harmonization) guidelines the precision of repeatability will be acquired by six repeated determinations ($n = 6$ sample preparations). In contrast, the precision of comparability (or intermediate) will be evaluated by a second analyst and/or a second NMR spectrometer with a different magnetic field strength. For example, coefficients of variations (CV) for repeatability and comparability of 0.7% and 0.6%, respectively were obtained by the validation of the assay determination of kavain using quantitative ^{1}H NMR.

4.6. Measurement uncertainty

In contrast to the ICH guidelines, the DIN EN ISO 17025 requires for each analytical result a value for the uncertainty of the measurements, based on a complete uncertainty budget. The latter must include all uncertainty contributions of the input quantities of the measurement equation. The two international guidelines (GUM[9] and EURACHEM[67]) describe the procedure how the measurement uncertainty for analytical methods has to be determined. According to these two guidelines all uncertainties $u(x_i)$ corresponding to the parameters of the measurement equation must be taken into account. For Eq. (3) the combined uncertainty for the determination of molar ratios $u_c(n_x/n_y)$ is given by

$$u_c\left(\frac{n_x}{n_y}\right) = \frac{n_x}{n_y}\sqrt{\left(\frac{u(I_x/I_y)}{I_x/I_y}\right)^2} \tag{8}$$

with

$$u\left(\frac{I_x}{I_{\mathrm{Std}}}\right)=\sqrt{\frac{\sum_{k=1}^{n}(x_k-\overline{x})^2}{n(n-1)}} \tag{9}$$

where x_k represents the result of a single measurement k (of n measurements) and $\overline{x}$ is the mean value of these n measurements.

This combined uncertainty contains and describes all uncertainties and errors of the whole measurement procedure.

In the case of assay determinations (Eq. (5)), the combined uncertainty $u_c(P_x)$ contains additional quantities and it can calculated using

$$u_c(P_x)=P_x\sqrt{\left(\frac{u(I_x/I_{\mathrm{Std}})}{I_x/I_{\mathrm{Std}}}\right)^2+\left(\frac{u(M_x)}{M_x}\right)^2+\left(\frac{u(M_{\mathrm{Std}})}{M_{\mathrm{Std}}}\right)^2} \tag{10}$$

$$+\sqrt{\left(\frac{u(m_x)}{m_x}\right)^2+\left(\frac{u(m_{\mathrm{Std}})}{m_{\mathrm{Std}}}\right)^2+\left(\frac{u(P_{\mathrm{Std}})}{P_{\mathrm{Std}}}\right)^2}$$

For the uncertainties of molar masses $u(M)$, the number of atoms N_j of the element j (e.g., C, H, ...) and the uncertainties of the atom masses $u(j)$ have to be considered as well:

$$u(M)=\sqrt{\sum_{j=1}^{n}\left(N_j\cdot u(j)\right)^2} \tag{11}$$

Further, the uncertainties of the initial weighings $u(m_i)$ are given by the uncertainty parameters of the used balance in terms of repeatability $u_w(m)$ and nonlinearity $u_{\mathrm{nonlinear}}(m)$ (these parameters are provided by the manufacturer)

$$u(m_i)=\sqrt{u_w^2(m)+2u_{\mathrm{non\text{-}linear}}^2(m)} \tag{12}$$

The uncertainty of the assay of the standard $u(P_{\mathrm{Std}})$ has to be considered accordingly.

The extended measurement uncertainty U is always connected with a confidence interval:

$$U=k\cdot u_c(y) \tag{13}$$

A factor $k=2$ corresponds to a confidence interval of 95% and is popularly used.

A complete uncertainty budget is listed exemplarily for the content determination in Table 3.

4.7. Round robin tests

Round robin tests are practical tools to generalize intra-laboratory validation studies. For example, Figure 5 shows the importance of a validated protocol (SOP) for spectra acquisition, processing, and evaluation by qNMR.

Table 3 Extended uncertainty budget for assay determinations by qNMR according to the GUM

Uncertainty	Value	*u(x)*	Relative *u(x)* (%)
Integration	98.81%	0.14%	0.14
Molar mass of analyte (g/mol)	154.24	0.008	0.005
Molar mass of standard (g/mol)	194.19	0.008	0.004
Weight of analyte (mg)	10.94	0.03	0.27
Weight of standard (mg)	10.06	0.03	0.30
Assay of standard	100.00%	0.15%	0.15
Combined uncertainty			0.45
		0.45% (g/g)	
Measurement uncertainty ($k=2$)		0.9% (g/g)	

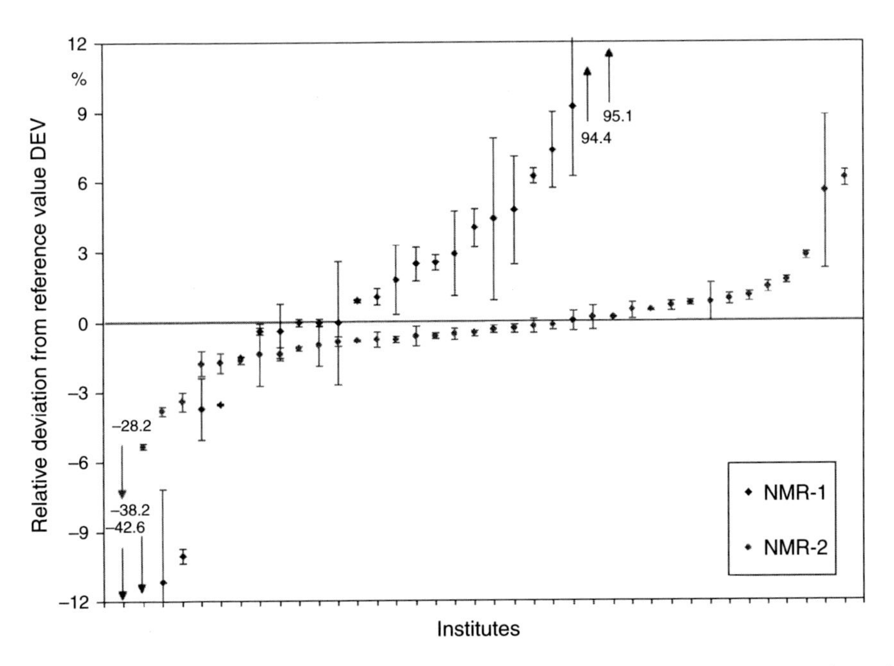

Figure 5 Results of the two round robin tests: NMR-1 (participants ought to analyze the sample by their own routine procedure) and NMR-2 (participants ought to analyze the sample by a validated procedure for acquisition, processing, and evaluation).

Thirty-three laboratories from universities, research institutes, and companies took part. Spectrometers of different manufacturers (Bruker, Varian, and Jeol) and with different B_0 field strengths (^{1}H NMR frequencies between 200 and 600 MHz) were used. All participants were asked to analyze simple model mixtures (five components, Figure 6) during two round robin tests (i) by their own routine (NMR-1) and (ii) using a validated protocol for spectra acquisition, processing, and evaluation (NMR-2).

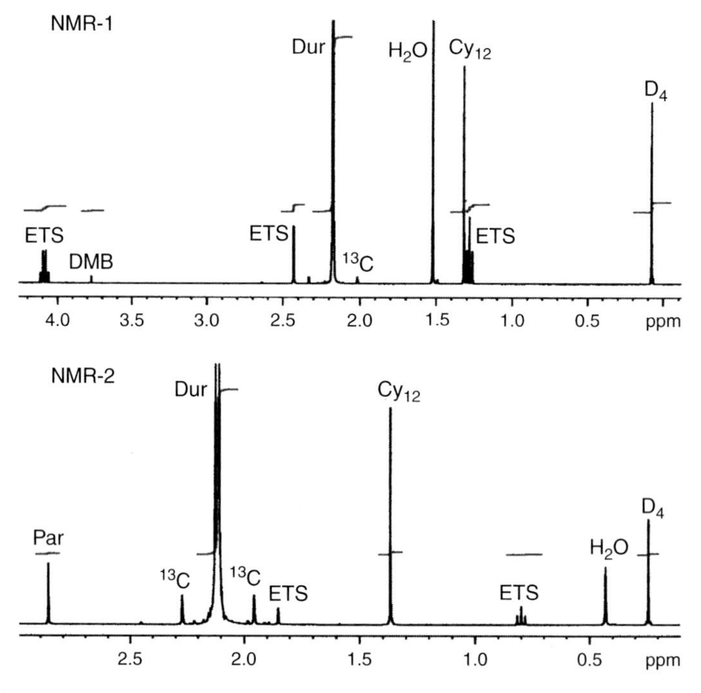

Figure 6 ^{1}H NMR spectra of the two round robin model mixtures NMR-1 and NMR-2 showing only the aliphatic region that ought to be analyzed.

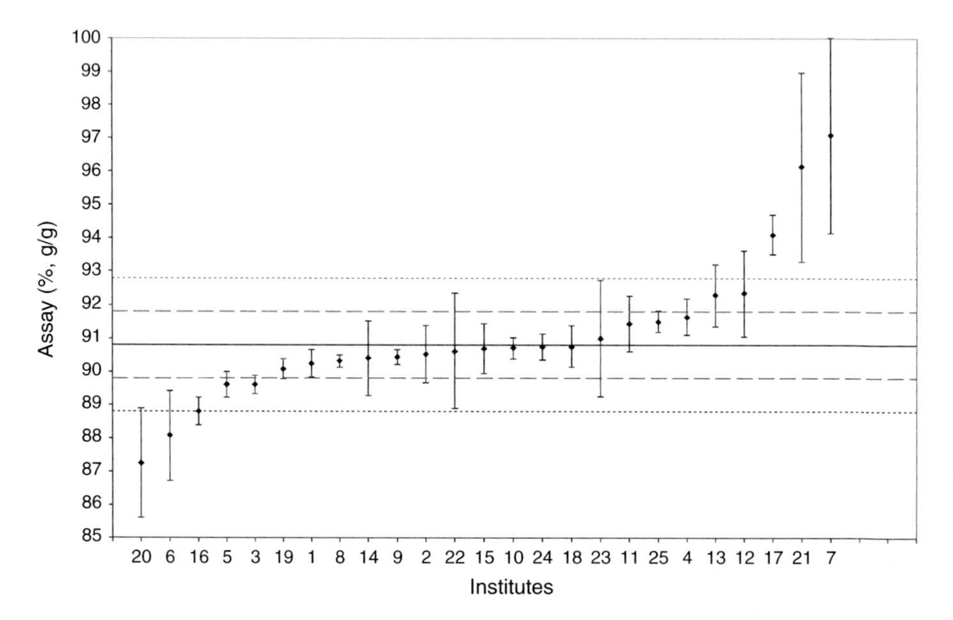

Figure 7 Results of a national round robin test (NMR-4) to determine the assay of rutin with ^{1}H qNMR using a validated procedure (SOP).

In addition, Figure 7 displays the results of a further national round robin test: the content of rutin was determined using 2-hydroxy-3,5-dinitrobenzoicaid as internal standard. Of the participants, 15 of 25 obtained results with a deviation less than $\pm 1\%$. Five other participants were within $\pm 2\%$. Results with deviation $>2\%$ are traced back to problems in the sample preparation by the participants (weighting procedure).

5. Sensitivity

The disadvantage of qNMR is its poor sensitivity in comparison to other spectroscopic and chromatographic methods. As described above, an ensemble of spins exhibits in two energy levels α and β, when put in a static magnetic field B_0. The Boltzmann relationship described the population ratio n_α/n_β between these two levels:

$$\frac{n_\alpha}{n_\beta} = e^{-(\gamma h B_0/2\pi k T)} \tag{14}$$

where γ is the gyromagnetic constant (characteristic constant for each nucleus type). The population difference is inherently small, but it increases with the field strength. For a high-field NMR spectrometer such as a 600 MHz spectrometer (14.1 T), the limit of detection (LOD, S/N = 3) for ^{1}H NMR is in the nanomolecular range (10^{-9}). In contrast, the LOD of IR spectroscopy and MS are around 10^{-14} and 10^{-18} mol, respectively.[68] It should be noted that the natural abundance and the gyromagnetic constant also influence the sensitivity of qNMR. For example, the ^{13}C sensitivity is about 6000 times less as than that for protons. This means that the sensitivity of an NMR depends on the magnetic field strength (hardware configuration), the type of observed nucleus and of the used pulse sequence. Furthermore, a well-tuned homogeneous static magnetic field across the sample maximized the intensity of the resonance lines in the spectra by tuning and matching the probe as well as shimming the magnetic field.

For these reasons, it is important to increase the NMR sensitivity. Szántay et al.[68] discussed four ways in which the sensitivity (or better the S/N) can be maximized:

1. Increasing the field strength: Currently, the state of the art of well-controlled high-end NMR spectrometer is the 600–800 MHz NMR spectrometer with magnetic field strengths of 14.1 and 18.8 T, respectively. They become increasingly standard for a wide variety of liquid applications.
2. Maximizing the sensitivity of detection of the RF coil: The signal intensity S will be increased by using microprobes with small outer-diameter sample tubes. Currently, 3, 1.7, and 1 mm are available in the market with active volumes less than 1 μl. The LOD for ^{1}H NMR using those microprobes is in the picomole range (10^{-12}).
3. Maximizing the number of spins within the active volume: The signal intensity is proportional to the number of spin within the active volume. For this reason, tubes with a smaller volume needed were used to increase the concentration of

the sample. The so-called "Shigemi tubes" optimized the sample solution volume as close as possible to the active volume by using a sample/glass interface. In addition, tubes with a smaller outer-diameter such as 3, 1.7, and 1 mm, which can be placed inside of a 5 mm probe, are becoming more and more popular.

4. Decreasing the noise: Reducing the thermal noise in the receiver coil by cooling the probe enhanced dramatically the S/N. Cryoprobes, in which the transmitter/receiver coils are cooled to about 25 K (sample itself remains at room temperature), are getting more applications in the field of impurity investigations by NMR.

It is interesting to note that all these values presented above for the LODs of qNMR are determined by the definition of the LOD for an S/N = 3. However, this method is only valid for response signals in the form of Gauss lines. In the case of NMR with Lorentzian lines as response signals, the LOD has to be calculated by the standard deviation of the response σ and the slope S of a calibration curve:

$$\mathrm{LOD} = 3.3\frac{\sigma}{S} \tag{15}$$

6. Conclusion

qNMR has found widespread applications, particularly due to specific advantages as the following: (i) the possibility to determine structures at a molecular level, (ii) intensity calibrations are not needed for determination of ratios (signal area is directly proportional to the number of nuclei, method is linear), (iii) relatively short measuring times, (iv) its nondestructive character, (v) no prior isolation of the analyte in a mixture, which means rather easy sample preparation and handling, (vi) quantification by main component analysis (direct method), (vii) the possibility of a simultaneously determining more than one analyte in a mixture, and (viii) the unique combination of being both a powerful method of performing structural analysis and a competitive quantitative analytical tool with the ability to be a potential primary ratio method. The last advantage warranted the accuracy of the experimental results, i.e. if a CRM was used as standard or a standard whose assay was traced back to a CRM. Validations show that measurement uncertainties of 1% and less are achievable for assay determinations by ^{1}H qNMR using an internal standard.

However, poor sensitivity in comparison to other spectroscopic and chromatographic methods is the main disadvantage of qNMR.

Acknowledgment

We thank the Federal Ministry of Economics and Labour (BMWA), the German Federation of Industrial Cooperative Research Associations "Otto von Guericke" (AiF), and the German Research Association of Medicine Manufacturers (FAH) for support of the project AiF-No: 13843 N/1.

REFERENCES

1. J.L. Jungnickel, J.W. Forbes, Anal. Chem. 35 (1963) 938–942.
2. D.P. Hollis, Anal. Chem. 35 (1963) 1682–1684.
3. L. Griffiths, A.M. Irving, Analyst 123 (1998) 1061–1068.
4. G. Maniara, K. Rajamoorthi, R. Srinivasan, G.W. Stockton, Anal. Chem. 70 (1998) 4921–4928.
5. R.J. Wells, J. Cheung, The chemistry preprint server CPS: analchem/0103002 (2001).
6. R.J. Wells, J.M. Hook, T.S. Al-Deen, D.B. Hibbert, J. Agric. Food Chem. 50 (2002) 3366–3374.
7. T.S. Al-Deen, D.B. Hibbert, J.M. Hook, R.J. Wells, Anal. Chim. Acta 474 (2002) 125–135.
8. T.S. Al-Deen, D.B. Hibbert, J.M. Hook, R.J. Wells, Accred. Qual. Assur. 9 (2004) 55–63.
9. BIPM, Guide to the Expression of Uncertainty Measurement, second edition, International Organization for Standardization, Geneva, Switzerland, 1995.
10. T.S. Al Deen, Validation of quantitative nuclear magnetic resonance (QNMR) spectroscopy as a primary ratio analytical method for assessing the purity of organic compounds: a metrological approach, PhD thesis, University of New South Wales, Sydney, Australia, 2002.
11. F. Malz, Quantitative NMR-Spektroskopie als Referenzverfahren in der analytischen Chemie, PhD thesis, Humboldt-Universitaet Berlin, Germany, 2003, http://edoc.huberlin.de/dissertationen/malz-frank-2003-06-30/PDF/Malz.pdf.
12. R. Deubner, Quantitative NMR Spektroskopie zur Reinheitsbestimmung von Arzneistoffen, PhD thesis, University of Würzburg, Germany, 2004, http://opus.bibliothek.uniwuerzburg.de/opus/volltexte/2004/836/pdf/DissertationDeubner.pdf.
13. U. Holzgrabe, B.W. Diehl, I. Wawer, J. Pharm. Biomed. Anal. 17 (1998) 557–616.
14. A. Vailaya, T. Wang, Y. Chen, M. Huffman, J. Pharm. Biomed. Anal. 25 (2001) 577–588.
15. J. Forshed, F.O. Andersson, S.P. Jacobsson, J. Pharm. Biomed. Anal. 29 (2002) 495–505.
16. S. Meshitsuka, Y. Morio, H. Nagashima, R. Teshima, Clin. Chim. Acta 312 (2001) 25–30.
17. U. Sahrbacher, A. Pehlke-Rimpf, G. Rohr, W. Eggert-Kruse, H.R. Kalbitzer, J. Pharm. Biomed. Anal. 28 (2002) 827–840.
18. Z. Xia, L.G. Akim, D.S. Argyropoulos, J. Agric. Food Chem. 49 (2001) 3573–3578.
19. N. Papke, J. Karger-Kocsis, Eur. Polym. J. 37 (2001) 547–557.
20. T.J. Henderson, Anal. Chem. 74 (2002) 191–198.
21. H. Jancke, Nachr. Chem. Tech. Lab. 46 (1998) 720–722.
22. B. King, Accred. Qual. Assur. 5 (2000) 429–436.
23. B. King, Accred. Qual. Assur. 5 (2000) 266–271.
24. M.J. Milton, T.J. Quinn, Metrologia 38 (2001) 289–296.
25. T.G. Alexander, S.A. Koch, J. Assoc. Agric. Chemists 50 (1967) 676–678.
26. F. Malz, H. Jancke, J. Pharm. Biomed. Anal. 38 (2005) 813–823.
27. A.E. Derome, Modern NMR Techniques for Chemistry Research, Pergamon Press, Oxford, 1987.
28. H. Günther, NMR Spectroscopy, Wiley, Chichester, 1995.
29. H. Friebolin, Basic One-Dimensional and Two-Dimensional NMR Spectroscopy, Wiley-VCH, Weinheim, 1998.
30. M.H. Levitt, Spin Dynamics. Basics of Nuclear Magnetic Resonance, Wiley-VCH, Chichester, 2001.
31. R. Freeman, Handbook of Nuclear Magnetic Resonance, second edition, Addison Wesley Longman, Edinburgh, 1997.
32. M.L. Martin, G.J. Martin, J.-J. Delpeuch, Practical NMR Spectroscopy, chapter 9: Practical Aspects of Intensity Measurement and Quantitative Analysis, Heyden & Son, London, 1980.
33. D.L. Rabenstein, D.A. Keire, Pract. Spectrosc. 11 (1991) 323–369.
34. C. Szantay Jr, Trends Anal. Chem. 11 (1992) 332–344.
35. R. Deubner, U. Holzgrabe, Magn. Reson. Chem. 40 (2002) 762–766.
36. G.M. Hanna, Enantiomer 5 (2000) 303–312.
37. J.S. Salsbury, P.K. Isbester, Magn. Reson. Chem. 43 (2005) 910–917.
38. C.K. Larive, D. Jayawickrama, L. Orfi, Appl. Spectrosc. 51 (1997) 1531–1536.
39. G.F. Pauli, B.U. Jaki, D.C. Lankin, J. Nat. Prod. 68 (2005) 133–149.

40. R. Martino, V. Gilard, F. Desmoulin, M. Malet-Martino, J. Pharm. Biomed. Anal. 38 (2005) 871–891.
41. D.B. Fulton, B.G. Sayer, A.D. Bain, H.V. Malle, Anal. Chem. 64 (1992) 349–353.
42. M.V. Laycock, P. Thibault, S.W. Ayer, J.A. Walter, Nat. Toxins 2 (1994) 175–183.
43. U.P. Ramsey, D.J. Douglas, J.A. Walter, J.L.C. Wright, Nat. Toxins 6 (1996) 137–146.
44. S. Akoka, L. Barantin, M. Trierweiler, Anal. Chem. 71 (1999) 2554–2557.
45. G.S. Remaud, V. Silvestrs, S. Akoka, Accred. Qual. Assur. 10 (2005) 415–420.
46. E.G.R. Caytan, E. Tenailleau, S. Akoka, Talanta 71 (2007) 214–218.
47. N.S.A. Michel, J. Magn. Reson. 168 (2004) 118–123.
48. F. Ziarellí, Solid State Nucl. Magn. Reson. 29 (2006) 214–218.
49. F.C.C. Franconi, L. Lemaire, V. Lehmann, L. Barantin, A. Akoka, Magn. Reson. Imaging 20 (2002) 587–592.
50. S. Kromidas, Validierung in der Analytik, Wiley-VCH, Wienheim, 1999.
51. ISO/IEC Guide 25, General requirements for the competence of calibration and testing laboratories, Geneva, 1990.
52. DIN EN 45001, General requirements for the competence of testing and calibration laboratories , Geneva, 1997.
53. EN ISO/IEC 17025, General requirements for the competence of testing and calibration laboratories, Geneva, 1999.
54. S. Kromidas, Nachr. Chem. Tech. Lab. 46 (1998) 28–32.
55. F. El-Shahed, K. Doerffel, R. Radeglia, J. Prakt. Chem. 321 (1979) 859–864.
56. J.R. Mooney, Analytical NMR, Field, L.D. and Sternhell, S. (eds), John Wiley & Sons: Chichester, 1989.
57. D.L. Rabenstein, K.K. Millis, E.J. Strauss, Anal. Chem. 60 (1988) 1380A–1391A.
58. F.G. Herring, P.S. Philips, J. Magn. Reson. 62 (1985) 19.
59. J.-P. Grivet, Signal Treatment and Signal Analysis in NMR, Elsevier, Amsterdam, 1996.
60. K. McLeod, M.B. Comisarow, J. Magn. Reson. 84 (1989) 490–500.
61. H. Günther, NMR Spectroscopy, Wiley, Chichester, 1995.
62. F. Malz, H. Jancke, Anal. Bioanal. Chem. 385 (2006) 760–765.
63. H. Jancke, F. Malz, W. Hässelbarth, Accred. Qual. Assur. 10 (2005) 421–429.
64. B. King, Accred. Qual. Assur. 5 (2000) 429–436.
65. B. King, Accred. Qual. Assur. 5 (2000) 266–271.
66. M.J. Milton, T.J. Quinn, Metrologia 38 (2001) 289–296.
67. EURACHEM, Quantifying Uncertainty in Analytical Measurements, second edition, London, 2000.
68. C. Szántay Jr, Z. Béni, G. Balogh, T. Gáti, Trends Anal. Chem. 25 (2006) 806–820.

CHAPTER 3

qNMR in Solid State

I. Wawer

Contents

Abstract

^{13}C CPMAS NMR (cross-polarization magic angle spinning nuclear magnetic resonance) spectra recorded under restricted, well-defined regime provide quantitative data. Following a protocol which addresses the issues affecting such measurements, a calibration plot has to be produced. Reliable quantitative estimation of polymorphs (or mixed phases) by NMR relies on comparison of integrated intensity of two separate lines in one CPMAS spectrum. The reported limits of detection and quantification under the given probe and acquisition conditions are 0.5–1% and 1–9%, respectively.

Keywords: protocol for quantitation, cross-polarization, ERETIC, Polymorph, drug formulations, pseudoephedrine, sildenafil, suxamethonium

1. Introduction to Solid-state NMR

In the nuclear magnetic resonance (NMR) spectrum recorded for solution the lines are narrow, the width at half-height is usually below 1 Hz. An attempt to record a spectrum of powder sample in standard NMR tube gives a line of several kHz broad and clearly shows that we need another technique for solids. Fundamentals of solid-state NMR are given in the books,[1,2] suitable both for those dealing with NMR technique for the first time and also for those already using NMR in the field of analytical chemistry. A short introduction to solid-state NMR exemplifies basic principles upon which this technique relies.

In solution all the interactions apart from chemical shift and indirect coupling are averaged to zero by thermal motions of molecules. In the solids, the molecules are rigid and kept together by crystal packing forces.

Magnetic interactions in the solid state are described by a Hamiltonian H, which is the sum of several contributions: Zeeman interaction (the same as in solution), direct dipole–dipole interaction, magnetic shielding (giving chemical shifts), scalar spin–spin coupling to other nucleus, and for nuclei with spin > ½, also quadrupolar interactions.

$$H = H_{\text{Zeeman}} + H_{\text{dipolar}} + H_{\text{chemical shift}} + H_{\text{spin coupling}} + H_{\text{quadrupolar}} \quad (1)$$

The most important is the dipole–dipole interaction with neighbor nuclei. Magnetic fields of nuclei I (^{1}H) in the neighborhood of the nucleus S (e.g., ^{13}C, ^{15}N) under observation generate local fields: $B_{\text{loc}} \sim \mu / r^{3}{}_{\text{IS}}(3\cos^{2}\theta_{\text{IS}} - 1)$, which depend on the magnetic moment μ of the nuclei and their mutual orientation $(3\cos^{2}\theta_{\text{IS}} - 1)$, where θ_{IS} is an angle between the internuclear vector and the direction of static magnetic field.

The interaction is significant for ^{1}H and ^{19}F with large magnetic moments, but decreases rapidly with the distance r_{IS}. Such dipolar nuclear interactions are orders of magnitude larger than chemical shifts and, since many mutual interactions are present simultaneously, the result is a broad resonance line.

The dipolar interactions vanish for $\cos^{2}\theta = 1/3$ (because $(3 \times 1/3) - 1 = 0$), thus, if the sample is rotated around an axis inclined at the angle $\theta = 54°44''$ (the magic angle) to the magnetic field, the line broadening is significantly suppressed.

Dipolar ^{13}C–^{13}C interactions practically do not exist because of a scarce amount of ^{13}C (1%, the "dilute nuclei"), ^{1}H–^{13}C dipolar interactions contribute significantly to the line broadening but can be removed through high-power proton decoupling (PD), like in solution NMR. To obtain a high-resolution spectrum for solids, the line-narrowing procedures such as magic angle spinning (MAS) or special pulse sequences have to be applied. Usually, a combination of three techniques is used: MAS, PD, and cross-polarization (CP). The effect is illustrated in Figure 1. Solid-state ^{13}C CPMAS NMR spectrum of antimalarial drug *artemisinin* exhibits narrow "liquid-like" resonances, the static spectrum covers the whole spectral region.

How fast should the sample rotate? The effect of MAS is illustrated in Figure 2 ^{13}C CPMAS NMR spectra of *suxamethonium chloride* (succinylcholine), a drug widely used in emergency medicine and anesthesia to induce muscle relaxation, were recorded at various rotational speeds 4–12 kHz. The compound consists of two acetylcholine molecules linked by their acetyl groups (Scheme 1).

^{13}C MAS spectrum exhibited only five resonances since the molecule is symmetric; the lack of aromatic carbons leaves ~100 ppm space between carbonyl and aliphatic carbons resonances. A set of spinning sidebands (marked with asterisks) appears with a center band at the isotropic chemical shift of C=O at 170 ppm. This signal is easy to identify, it does not move upon increasing rotational speed, and its position is given by isotropic shielding $\sigma_{\text{iso}} = 1/3(\sigma_{11} + \sigma_{22} + \sigma_{33})$. The three components $\sigma_{ii} (i = 1, 2, 3)$ of the shielding tensor (and three angles which specify their orientation with respect to the direction of magnetic field) describe the anisotropy of the shielding. According to the convention, $\sigma_{11} < \sigma_{22} < \sigma_{33}$, the anisotropy is defined as: $\Delta\sigma = \sigma_{33} - 1/2(\sigma_{11} + \sigma_{22})$, and asymmetry parameter

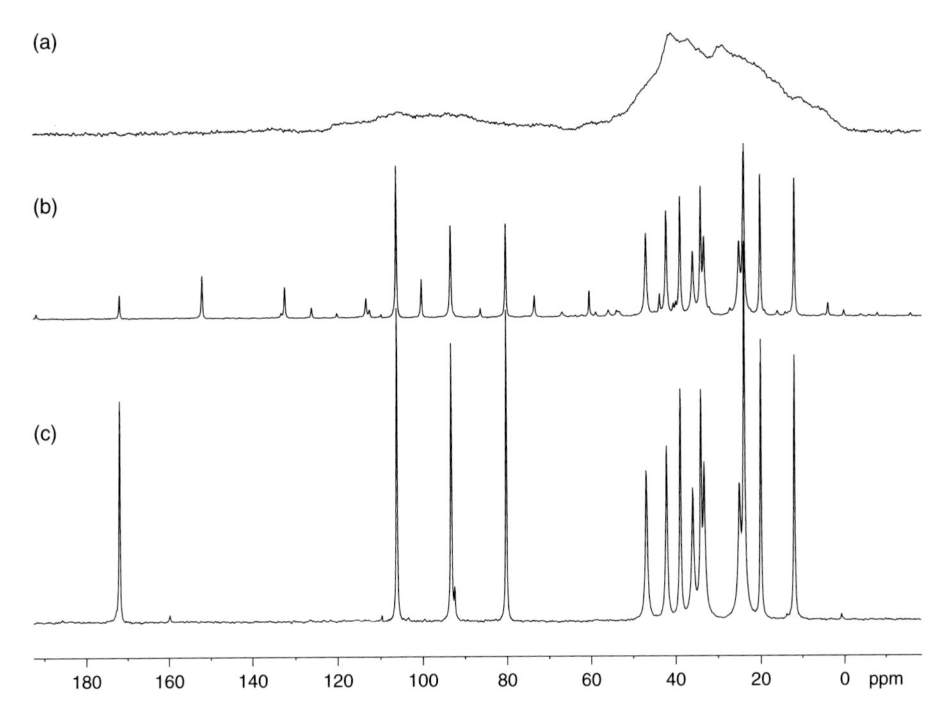

Figure 1 Solid-state ^{13}C NMR spectra of antimalarial drug artemisinin recorded (a) without sample rotation (static), (b) without PD, and (c) using MAS, PD, and CP techniques.

$\eta=(\sigma_{11}-\sigma_{22})/(\sigma_{33}-\sigma_{iso})$ The anisotropy parameter reflects the distortion of geometry from ideal (tetrahedral, octahedral) and the asymmetry parameter characterizes the environment of the nucleus under observation. CSA is an important source of information on molecular structure and interactions. For aliphatic carbons, chemical shift anisotropies are within 15–50 ppm, but reach 120–200 ppm for aromatic and carbonyl sp^2-type carbons.

The rotational speed should be larger than the anisotropy $\Delta\sigma$, given in Hz; if it is not, the signal is flanked by rotational sidebands on both sides. In the case of polycrystalline sample, there is a random distribution of all possible orientations and a broad line results from a superposition of multiple resonances. The intensities of sideband lines reflect the shape of the static spectrum of solids.

The most important analytical feature of NMR is that the signal intensity (integrated signal area) in a spectrum is directly proportional to the number of nuclei generating this signal. For quantifying purposes, one has to add the intensity of center band and the intensities of any spinning sidebands manifolds together. To overcome chemical shift anisotropy (CSA) and have no spinning sidebands for carbonyl carbons, the sample needs to be spun at speeds of over 13 kHz. The intensity of central line (isotropic) increases with increasing spinning speed, reaching 94.5% at 12 kHz, as illustrated in Figure 3.

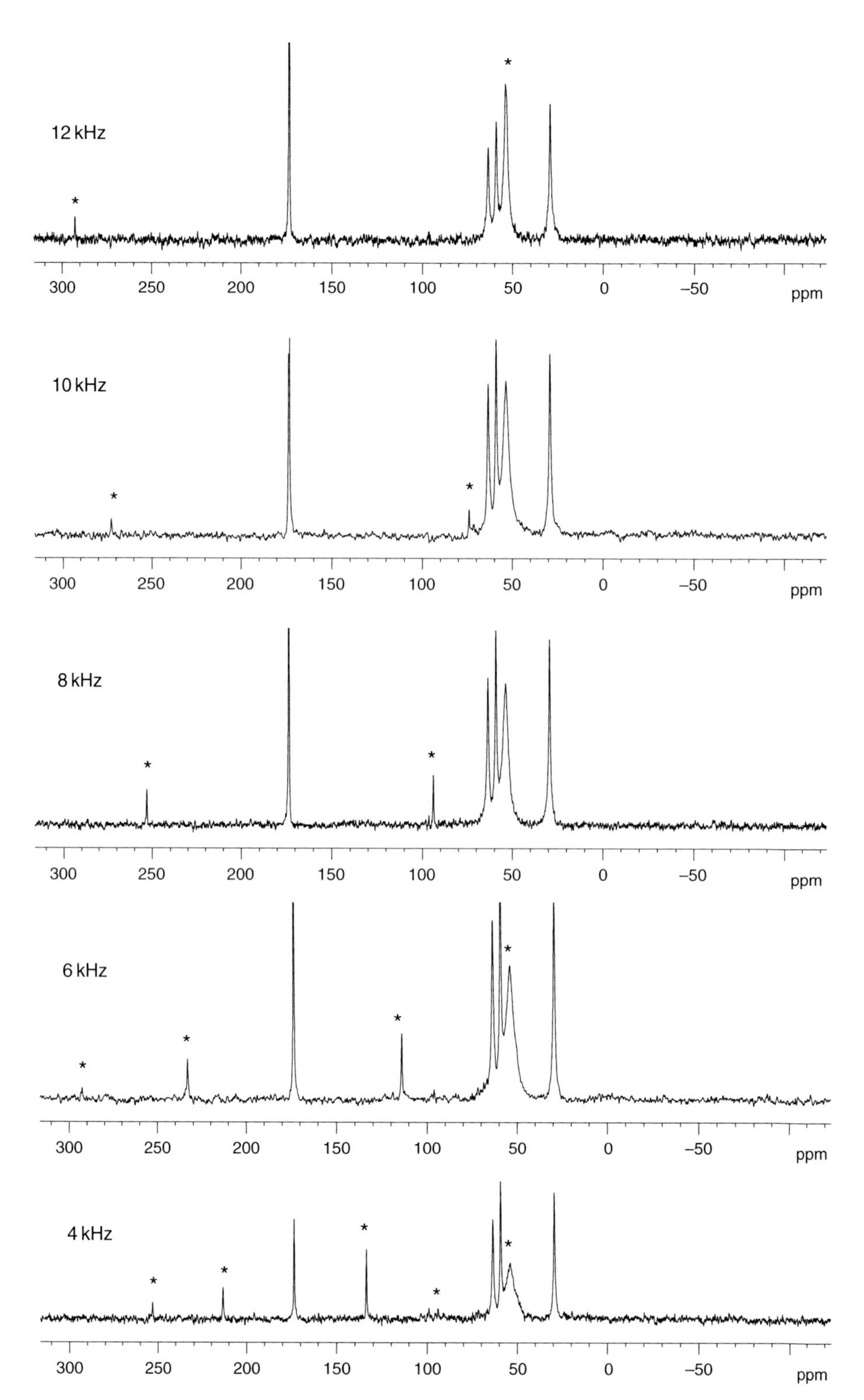

Figure 2 ^{13}C solid-state spectra of suxamethonium chloride recorded with different MAS speeds.

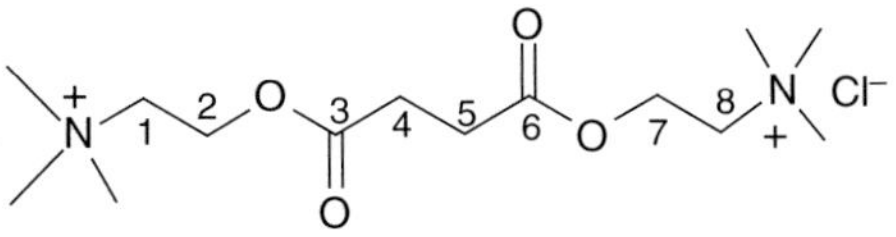

Scheme 1 Suxamethonium chloride.

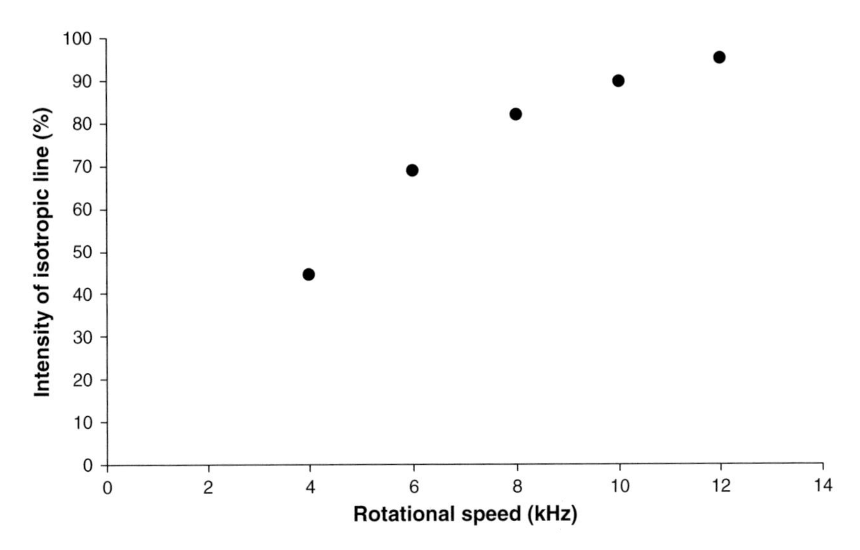

Figure 3 Intensity of central isotropic line versus spinning speed (the sidebands intensity was added back into the central resonance, and taken as 100%).

Solid-state NMR spectra of diluted nuclei, such as ^{13}C or ^{15}N, suffer from low sensitivity and long spin lattice relaxation times. The registration of a spectrum with several thousand scans and long delays take a very long time to acquire. CP technique offers the enhancement of magnetization from the rare spin system ($S = {}^{13}C$) by transfer of magnetization from the abundant spin system ($I = {}^{1}H$). Figure 4 illustrates basic pulse sequence for the CP experiment.

Step I of this sequence involves a $\pi/2$ pulse, which is applied to the ^{1}H spins along the x'-axis of rotating frame. The subsequent step II is known as "spin locking": the pulses are applied to ^{1}H and to ^{13}C spins along the y'-axis, during

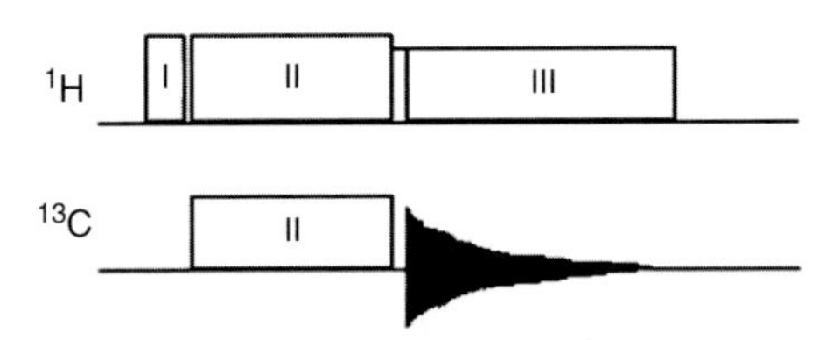

Figure 4 Basic CP pulse sequence: (I) $\pi/2$ pulse is applied to the I (^{1}H) spins along the x'-axis of rotating frame, (II) spin locking pulse is applied to I and to S (^{13}C) spins along the y'-axis, during the contact time, and (III) FID of the S spins is observed under decoupling.

the contact time t_{CP}. The amplitudes of the spin-locking fields B_{1H} and B_{1C} are adjusted to satisfy the Hartmann–Hahn condition:

$$\gamma_H B_{1H} = \gamma_C B_{1C} (\gamma_H \text{ and } \gamma_C \text{ are gyromagnetic ratios of both nuclei}) \quad (2)$$

Finally, during the step III the field applied to the ^{13}C spins is turned off, while the proton field is maintained on, and a free induction decay (FID) is observed under decoupling. The sequence can be repeated with a short recycle time, it is governed by proton relaxation times, not by the long T_1 of ^{13}C. The rate of CP of particular carbons is, in the first approximation, proportional to the number of directly bonded protons[3] and the following relative rates are usually observed: $CH_3(\text{static}) > CH_2 > CH > CH_3(\text{rotating}) > C(\text{nonprotonated})$.

Cross-polarization sequence provides evident advantages, such as three- to fourfold signal enhancement and reduction of the measurement time. It is sensitive to internuclear distances and molecular mobility. The drawback to the CP experiment is that quantitation is more difficult to achieve. The intensities of signals within the same CP spectrum cannot be compared without a careful examination of their CP kinetics.

In the CP experiment, the contact time, i.e., the time during which the magnetization is transferred from ^1H to the ^{13}C nuclei, has a remarkable effect on line intensity. According to the classical model I–S, the intensity $I(t)$ initially increases according to time constant (T_{CP}) and then decreases according to another time constant ($T^H{}_{1\rho}$):

$$I(t) = I_0 \left(\frac{1 - T_{CP}}{T^H_{1\rho}} \right)^{-1} \left[\exp\left(\frac{-t}{T^H_{1\rho}} \right) - \exp\left(\frac{-t}{T_{CP}} \right) \right] \quad (3)$$

The peaks within the same spectrum must be compared on the basis of absolute intensities I_0 obtained from fittings of the kinetic results.

Cross-polarization in the C–H bonds proceeds through transient harmonic oscillations, which are gradually damped with time by the proton spin diffusion. Classical CP theory assumes that spin diffusion is rapid. However, the spin diffusion rate R is not sufficiently fast to prevent oscillations and another model of CP was proposed, known as I–I*–S model. The kinetics within this model is described by the following equation:

$$I(t) = I_0 \exp\left(\frac{-t}{T^I_{1\rho}} \right)^{-1} \left[1 - \lambda \exp\left(\frac{-t}{T_{df}} \right) - (1 - \lambda) \exp\left(\frac{-1.5t}{T_{df}} \right) \exp\left(\frac{-0.5t^2}{T_2^2} \right) \right] \quad (4)$$

The proton diffusion rate $1/T_{df}$ depends on the number of hydrogen atoms in a functional group and on their mobility. T_2 time constant is ~0.1 ms and the fitting gives $\lambda = 1/2$ and $1/3$ (or the values close to 0.5 and 0.3) for the CH and CH_2 groups.

It is not possible to predict which model better describes any particular CP case.[4] In principle, the I–I*–S model applies for strong I*–S interactions when the nuclei are close in space and relatively immobile. CP under MAS in polycrystalline compounds proceeds according to this model for the CH and CH_2 groups. It is frequently assumed that quaternary carbons and rotating methyl groups cross-polarize according to the classical I–S model.

Suxamethonium chloride is a suitable model to illustrate the CP kinetics; kinetic parameters obtained using the I–S and I–I*–S models of CP are summarized in Table 1 (Figures 5 and 6).

Table 1 CP parameters (ms) as obtained using two CP models for carbons of suxamethonium chloride

δ (ppm)	Carbon type	Model I–S		Model I–I^*–S	
		T_{CP}	$T_{1\rho}$	T_2	T_{df}
173	IV (C=O)	1.09 ± 0.07	4.6 ± 0.2	–	–
53	CH_3	0.22 ± 0.01	4.1 ± 0.2	–	–
63	CH_2	–	4.1 ± 0.2	2.6 ± 0.1	22.7 ± 1.6
59	CH_2	–	3.9 ± 0.1	2.4 ± 0.2	19.4 ± 1.2

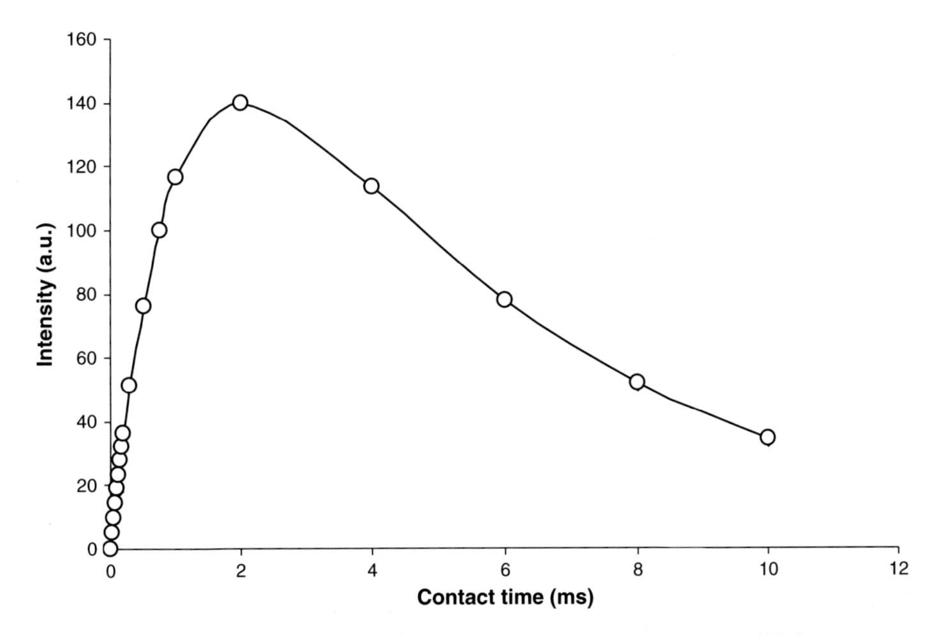

Figure 5 CP kinetics for C=O carbon of suxamethonium chloride. Signal intensity in arbitrary units (a.u.), contact time in ms. Fitting to I–S model using Eq. (3) gives T_{CP} = 1 ms and $T_{1\rho}$ = 4.6 ms.

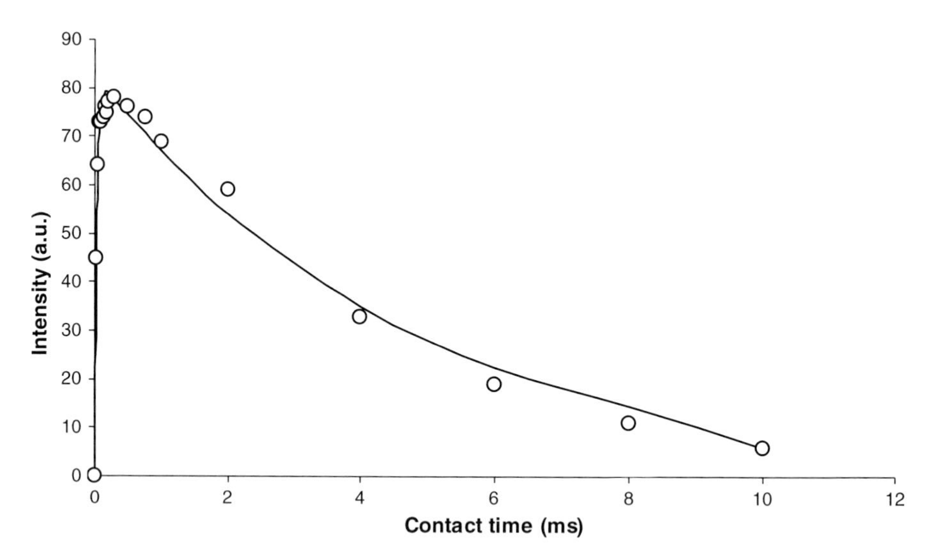

Figure 6 CP kinetics for methylene carbon ($\delta = 59$ ppm) of suxamethonium chloride. Signal intensity in arbitrary units (a.u.), contact time in ms. Fitting to I–I*–S model using Eq. (4) gives $T_2 - 2.5$ ms and $T_{df} - 20$ ms.

2. Quantitative Aspects of Solid-state NMR, Protocol for Quantitation

Although solid-state NMR is now in routine use for the identification and structural studies of organic solids, it is less frequently used in quantitation studies.[5,6,7] To the best of our knowledge, the method has not been validated yet. The validation process requires the testing of linearity, robustness, specificity, selectivity as well as the parameters of accuracy: measurement uncertainty, repeatability, and comparability. A validation of a novel method also requires a comparison of the results from various laboratories. In 2005 a protocol for the experimental setup of the measurement, the processing and evaluation was developed for solution NMR as a result of international efforts and testing (NMR-1 and NMR-2 intercomparisons).[8] Similar work has to be done by the validation of solid-state NMR method.

There is considerable interest in pharmaceutical industry in quantifying situations involving polymorphism in drug substances and for formulated products. A protocol for quantitation[9] should include spectrometer settings, data processing, measuring the peak area, constructing a plot of composition versus peak area, and error analysis.

1. Spectrometer settings:
   - The choice of an appropriate probe head and rotor: experiments on ^{1}H (^{19}F) are performed in small-diameter zirconia rotors (e.g., 2.5 mm outside diameter), for quantitation of ^{13}C nuclei the 4 mm rotors or larger are appropriate.
   - Rotor, end-cap, and stator – these materials contain no carbon and hence should not give a signal in ^{13}C spectra. Nevertheless, some impurities (plastics,

glue) can be a source of a signal. It is advisable to record a spectrum on an empty rotor under standard acquisition conditions to assess the background signal. Background signal can be subtracted from the sample spectrum.

- The magic angle setting is usually stable for weeks but should be verified before quantitative measurements using the ^{79}Br resonance of KBr (as the ^{13}C and ^{79}Br NMR frequencies are close).
- For each sample studied the probe must be correctly tuned, including the sample used to set up the CP. Poor or inconsistent tuning will reduce sensitivity and largely invalidate calibration experiments.
- The Hartmann–Hahn condition and pulse durations (^{1}H 90° pulse, ~5 μs) need to be calibrated to ensure maximum sensitivity. Usually, $^1H \rightarrow {}^{13}C$ cross polarization is set up using a sample of adamantane. Besides the CP setup, this sample can be used for checking field homogeneity and chemical shift referencing.
- The linewidths and signal-to-noise ratio (S/N) from the setup spectrum should be checked to ensure adequate resolution and sensitivity. Linewidths in ^{13}C spectra of adamantane should be <10 Hz; larger linewidths may imply poor shimming.
- The sample packed into a rotor should be powdered and homogeneous. For samples that do not completely fill the rotor, ceramic inserts can be used at the top and bottom of the rotor. Sensitivity is the greatest in the middle third of the rotor, which contributes 50% of the signal.
- The rotor should be spun at a sufficiently high speed to avoid any spinning sidebands overlapping with the peaks of interest. Running a spectrum at lower MAS rate (4–5 kHz) requires correction for sidebands in integral region data.
- The optimal values for the CP contact time (1–5 ms) and recycle delay (5–15 s) are determined to maximize the sensitivity of each sample. Appropriate recycle delay can be determined as follows: (i) 5–10 times the measured T_1 or (ii) running experiments with a set of recycle delays and choosing the shortest delay for which there is no saturation.
- NMR spectrum is then recorded with sufficient acquisition time and number of transients to give both acceptable resolution and S/N in a reasonable total spectrometer time (~3 h). S/N increases as the square root of the number of scans, e.g., to improve the S/N by a factor of 2 requires four times the acquisition time.
- Line intensity and position can change significantly as the temperature varies. At spinning frequency up to 5 kHz, the temperature is typically a few degrees lower. At spinning of 10 kHz, the temperature increases by 5–10°. Higher spinning frequencies result in an exponential increase of the sample temperature. It is advisable to use a temperature control unit and to calibrate the temperature inside the rotor at various speeds.

2. Data processing:
   - The data are zero filled to 2–8 times the acquisition length, and a Lorentzian line broadening is applied (10 Hz) prior to Fourier transformation.
   - All the peaks should appear as purely positive absorption signals on a flat baseline (without any baseline roll) – the spectrum should be "phased" by the operator.

3. Measuring the peak area:
   - Peaks areas can be obtained by integration (manual or automatic). Deconvolution (i.e., fitting to a model lineshape) is necessary if the peak of interest is overlapped.
   - The peaks are fitted to a mixed Lorentzian/Gaussian shape (frequently a 50% Gaussian/50% Lorentzian lineshape provided a good fit).
4. Plot of composition versus peak area:
   - A plot with molar composition (x-axis) of a compound of interest versus the measured ratio of peak areas for selected peaks (y-axis) is made. Such a plot is a prerequisite for applying the protocol to a relevant sample of unknown content. It is expected to be linear, so that only the slope is required accurately.
   - For statistical reasons, each measurement should be repeated 3–4 times.
   - An error analysis should be carried out for the plot to understand the precision of the procedures. The uncertainty arising from sample preparation, spectrometer conditions, and processing the FID should be calculated.
   - Limit of detection (LOD) is usually assumed as the situation when the signal is three times the standard deviation of the noise level.
   - The limit of quantitation (LOQ) is the lowest limit that can be quantified with adequate repeatability.

Any deviation from the precise 54°44′ angle will increase line broadening proportional to the shielding anisotropy, as it was shown by Harris[5] in 1985 for hexamethyl benzene. This sample is useful in tuning, and even more in setting the magic angle. Aliphatic and aromatic carbons should give the signals of equal height, which is usually not feasible to attain. Observable changes appear in the peak–height ratio with deviations of as much as 0.05° from the magic angle.

Quantitation is frequently performed by packing a standard together with an unknown substance in a rotor and a comparison of respective peak integrals. However, such an analysis is complicated by RF inhomogeneity effects. To demonstrate these effects, a 1-mm-thick rubber disk with a diameter equal to that of the rotor was used.[10] Typical experimental parameters employed included a 5 μs 90° pulse width and repetition delays between 2 and 4 s. ^{1}H Bloch decay spectra showed that the spectral sensitivity is a strong function of the position of the sample within the rotor. The largest sensitivity at the central part and its decrease at the top and bottom of the rotor were attributed to the RF field inhomogeneity effects. The problem can be avoided if: (i) the constituents are homogeneously distributed within the rotor and (ii) the spatial distribution of each material is known and superimposed on the sensitivity profile. It is obvious that reduced RF amplitudes and directional effects influence CP experiments.

Cross-polarization MAS spectra sometimes show deviations in intensity distribution of the sideband pattern. The influence of Hartmann–Hahn mismatching and CP dynamics was investigated,[11] and a multiple-contact pulse sequence was proposed to reduce the errors. Sidebands can be eliminated using specially designed pulse sequence, known as TOSS[12] (total suppression of spinning sidebands). Upon utilization of the TOSS sequence, the spinning sidebands are nearly removed and the isotropic line remains. Bugay[13] considered this sequence as advantageous for the assignment of resonances and useful in simplifying the interpretation of the spectra. However, some intensity is also lost in the process, and it is not advisable to use TOSS in qNMR measurements.

Direct comparison of standard and unknown peak integrals that are simultaneously present in the spectrum eliminates most of the problems connected with spectrometer settings and data processing. The use of an internal intensity reference is a reasonable choice but often unsuitable, due to sample preciousness or signal overlap.

Quantitative measurements in solid-state NMR can be performed using a synthetic signal as an intensity reference, i.e., by calibration of the circuit response through a low-power pulse injected during the acquisition (the so-called *ERETIC* method).[14] This method has been in use for some time in liquid-phase and in vivo NMR. In common solid-state MAS probe heads, the ERETIC pulse can be injected using a channel not used for the observation. There is no interference between the ERETIC pulse and the CP sequence or decoupling. Some variation in relative intensity of the ERETIC and signal peaks is observed due to temperature; the reference intensity drops of about 2% over 57°. The temperature dependence of the signal intensity imposes care in the application to MAS experiments; however, the technique offers the precision estimated at about 1%.

Analytical methods based on solid-state NMR are becoming increasingly popular, one can gain quantitative or near-quantitative results for various types of materials: foods,[15] soil, plant, or other naturally derived samples.[16,17] Especially, the characterization of drugs in solid dosage forms by means of MAS NMR spectroscopy is interesting because the identity, quantity, and purity can be determined simultaneously. These activities bring up the issue of how accurately NMR can assess an analyte proportion in a solid sample. Although most drugs are crystalline, other crystalline or amorphous forms often appear during the formulation process. The presence of other forms can affect the physical and chemical stability of the drug as well as its bioavailability. The identification and quantitation of different forms of a drug is an interesting challenge.

3. An Example: Quantitation of Pseudoephedrine in Dosage Form

^{1}H NMR spectroscopy was applied for quantitative determination of the alkaloids from Ephedra species: (−)-ephedrine, (+)-pseudoephedrine, and (±)-norephedrine, either singly or in mixtures with each other. Determination of individual alkaloids was carried out in solution, with acetamide[18] or anthracene[19] as internal standard. The method was specific and accurate. However, there are a number of situations when it is necessary to determine the concentrations of components in solid-state mixtures without dissolving the sample.

Ephedrine alkaloids are used for the treatment of asthma, nasal congestion, and obesity. There are numerous drugs containing pseudoephedrine hydrochloride in combination with paracetamol and other ingredients. The effects of drug, excipients, drug/polymer ratio, and tablet properties were widely studied. A heterogeneous matrix comprising hydroxypropyl methylcellulose (HPMC) and polyvinylpyrrolidone (PVP) in various ratios was granulated using acetaminophen and pseudoephedrine as model drugs.[20,21] It is interesting to note that the presence of pseudoephedrine in the formulations affects the hydration characteristics of the matrix polymer(s) and compression properties of the granules. Formulations containing both acetaminophen and pseudoephedrine required less water to granulate than those containing only acetaminophen. Tablet hardness increased considerably in all formulations containing pseudoephedrine. The results suggest the presence of intermolecular interactions between pseudoephedrine and excipient.

An attempt was made to quantify pseudoephedrine in a solid formulation with microcrystalline cellulose. Samples with different fractions (from 5.09% to 95.97% (g/g)) were prepared by mixing pure (3*S*,2*S*)-(+)-*pseudoephedrine hydrochloride* (Scheme 2) and pure microcellulose (Avicel-105), one of the most commonly used excipients. To obtain homogeneous physical mixtures, the components were mixed well for ~5 min using an electric mixer.

CPMAS solid-state ^{13}C NMR spectra were recorded on a Bruker DSX-400 instrument. Powder samples were spun at 10 kHz in a 4-mm ZrO_2 rotor. Spectrometer settings, acquisition, and data processing parameters are summarized in Table 2.

The appropriate relaxation delay was established in a separate experiment, running the spectra with a set of recycle delays. The CP parameters were optimized from experiments with pure (3*S*,2*S*)-(+)-pseudoephedrine hydrochloride because it is the compound of interest.

Variable contact time experiments were necessary to find "perfect" contact time which ensures quantitation for all carbons of interest (outside the region overlapped by the resonances of cellulose). The peaks chosen for the analysis are these of methyl carbons (C1 and C4) and quaternary carbon C5.

Solid-state ^{13}C NMR chemical shifts and CP kinetic parameters T_{CH} and $T^{H}_{1\rho}$ (ms) for pure pseudoephedrine are summarized in Table 3.

CP experiments on most organic compounds where spin diffusion is fast give a single $T^{H}_{1\rho}$ value. In heterogeneous samples, the relaxation time is not uniform, long $T^{H}_{1\rho}$ and T_{CP} are observed for mobile systems (functional groups). Very

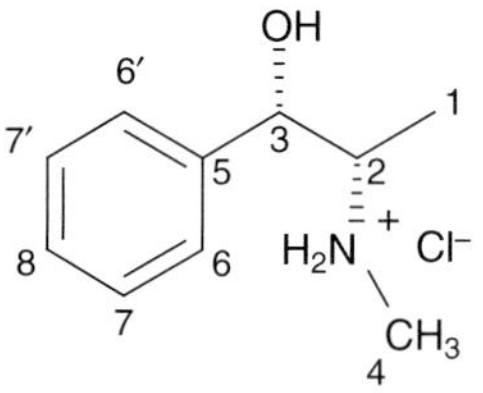

Scheme 2 (3*S*,2*S*)-(+)-pseudoephedrine hydrochloride.

Table 2 Standard parameters for measurements with mixtures of pseudoephedrine with Avicel-105 on a BRUKER DSX-400 instrument

Parameter	^{13}C	^{1}H
Frequency, SF	100.62 MHz	400.13 MHz
Sweep width	44 kHz	–
90° pulse length, P3	–	4.408 μs
90° pulse strength, Pl2	–	10.50 dB
Contact time, P15	2.00 ms	–
Relaxation delay, D1	10 s	
Receiver gain, RG	8192	–
Acquisition time, AQ	0.046 s	–
Decoupling power Pl12	–	7.1 dB
Power level for contact, Pl1	18.6 dB	–
Number of scans, NS	500	
Number of FID points, TD	4096	–
Number of frequency points, SI	8192	
Line broadening, LB	15 Hz	

Table 3 Solid-state ^{13}C NMR chemical shifts and cross-polarization kinetic parameters T_{CH} and $T^{H}_{1\rho}$ (ms) for pure (3*S*,2*S*)-(+)-pseudoephedrine hydrochloride; chemical shifts of pseudoephedrine in its formulation with microcrystalline cellulose are essentially the same

C	δ_{CPMAS} pure	δ_{CPMAS} sample 14.96%	T_{CH}	$T^{H}_{1\rho}$
1	12.5	12.4	0.12 ± 0.02	122 ± 9
2	62.4	62.3	0.03 ± 0.01	69 ± 7
3	76.3	–	0.05 ± 0.02	61 ± 7
4	33.4	33.4	0.08 ± 0.02	24 ± 3
5	139.8	139.7	0.61 ± 0.07	89 ± 8
6	128.1	128.8	0.06 ± 0.02	164 ± 12
6′	128.1	128.8		
7	126.5	126.6	0.06 ± 0.02	159 ± 12
7′	126.5	126.6		
8	131.0	130.9	–	–

long $T^{H}_{1\rho}$ values were obtained for solid β-carotene[22] (all CP curves end with plateaux), indicating that the chain with alternating double and single bonds is fairly mobile. Figure 7 represents typical CP curves for organic crystalline samples. Experimental data were fitted to the classical I–S model. Long $T^{H}_{1\rho}$ of 122–164 ms for aromatic carbons suggest that the ring may rotate (ring flip-flop rotation) in the crystals of pseudoephedrine. Longer $T_{CH} = 0.61$ ms result in slower built up magnetization, as expected for quaternary carbon C5, and also greater signal intensity at longer contact time.

The curves for $T_{CH} = 0.08$ ms (typical of protonated carbons) and that for $T_{CH} = 0.61$ ms (for C5) intersect at a contact time close to 2 ms. Therefore, the optimal contact time is 2 ms because it should result in reliable quantitation

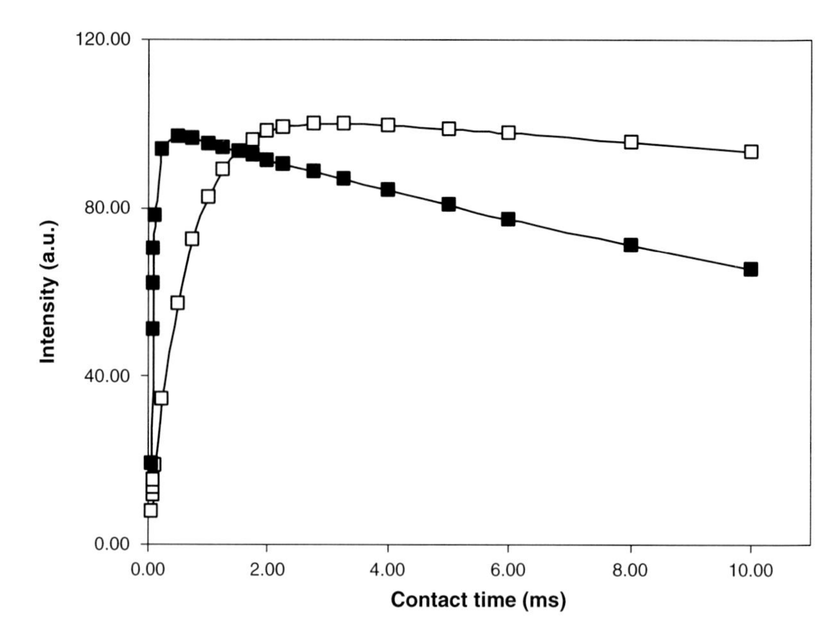

Figure 7 CP kinetics for pseudoephedrine hydrochloride. Fitting to I–S model (Eq. (3)) gives for quaternary carbon C5: $T_{CH} = 0.61$ ms, $T_{1\rho} = 89$ ms, and for methyl carbon C4 (N–CH_3): $T_{CH} = 0.08$ ms, $T_{1\rho} = 24$ ms.

for protonated and non-protonated carbons in this sample, and is the best compromise.

^{13}C CPMAS NMR spectra of pseudoephedrine mixtures (19 samples) with various compositions were recorded. For statistical reasons, each measurement was repeated three times. The spectra for selected samples are illustrated in Figure 8. Spectrum (a) is that of pure pseudoephedrine, and spectrum (e) represents microcrystalline cellulose (Avicel-105), the excipient.

As a next step, a plot was constructed with composition of pseudoephedrine mixtures (x-axis) versus the measured ratio of peak areas (y-axis). The mixtures were prepared gravimetrically; therefore, the concentrations of pseudoephedrine are presented as the weighed fraction X (%) = pseudoephedrine (g) × 100/pseudoephedrine + Avicel (g). Assuming that peak areas in the spectrum of pure pseudoephedrine represent 100%, gravimetric values (%, g/g) plotted versus experimental values (%, mean value of three integrals) yielded a linear dependence with the correlation coefficient $R^2 = 0.9992(3)$. The differences between gravimetric data and experimental values do not exceed 2%, indicating good accuracy of the method.

Parameters of linearity testing for the two methyl group signals (at 12 and 33 ppm) confirm the linearity of the method and its usefulness in quantitative analysis (Figure 9).

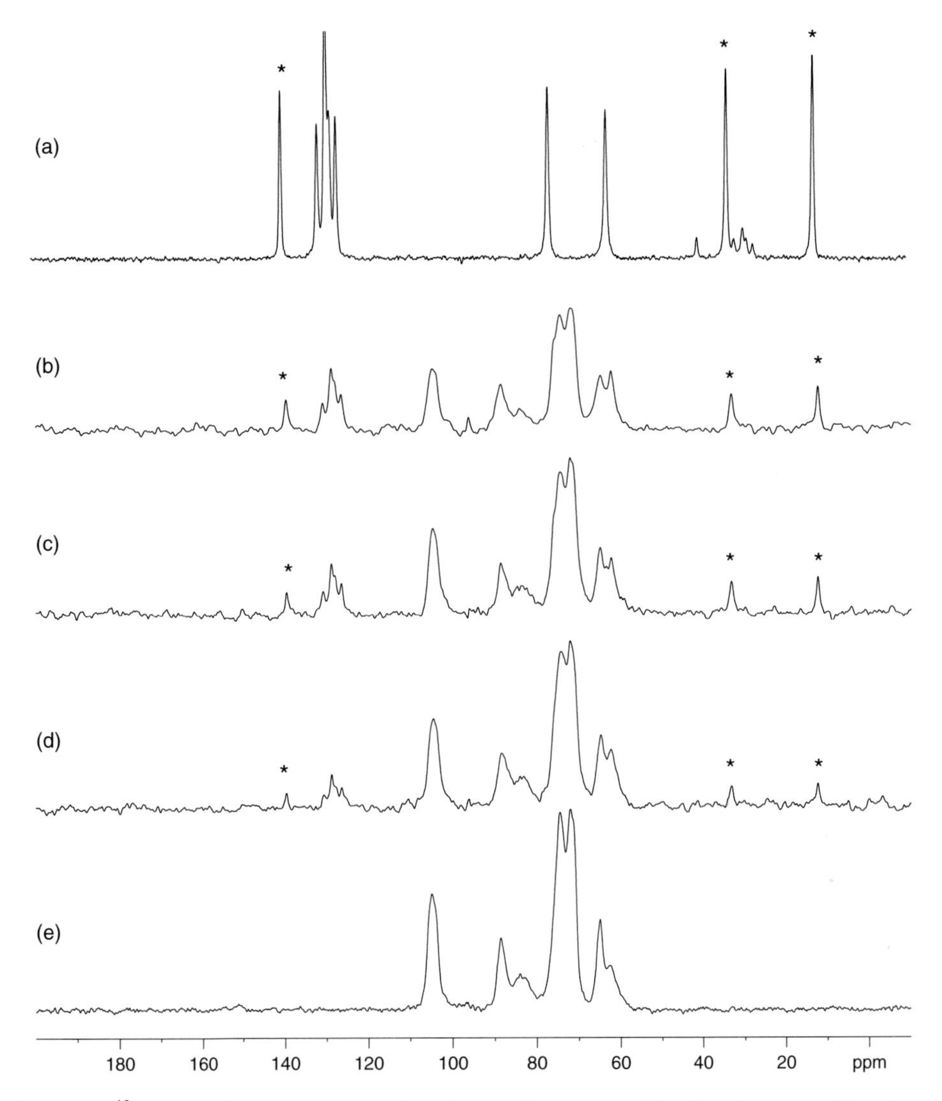

Figure 8 ^{13}C CPMAS NMR spectra of pseudoephedrine hydrochloride mixtures with various compositions. From top to bottom the percentage of pseudoephedrine is (a) 100, (b) 44.78, (c) 25.02, (d) 14.96, and (e) 0 (pure Avicel-105). The peaks used for quantitation measurements are indicated by asterisks.

When performing the next quantitative experiment we would rather reduce the number of samples required for the calibration plot and run the spectra with a larger number of scans (profit from better S/N). However, the total spectrometer time for the analysis should not be longer than 2–3 days.

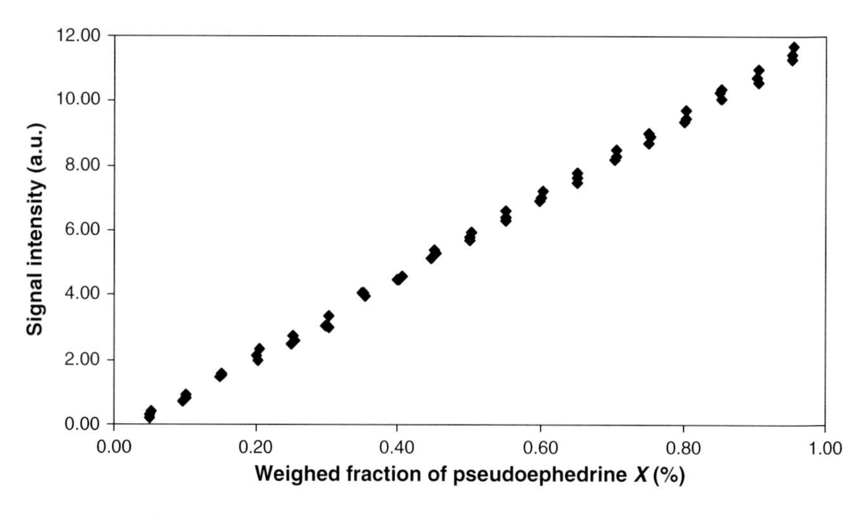

Figure 9 Quantification calibration plot for pseudoephedrine. The ordinate is the peak area (from integrated peak areas) of the specified pseudoephedrine signals. The concentrations are presented as the weighed fraction X (%, g/g).

4. Quantitation of Polymorphs and Formulated Drugs

The main request for the quantitation method is the linearity of NMR response. Special attention has to be paid to the quality of experimental data; in particular, the complete relaxation of all the spectra must be fulfilled. This can sometimes be the main drawback of the method when samples with long T_1 times are studied, for instance this may occur for well-crystallized states. In many cases, quantitative estimation of mixed phases by NMR relies on the comparison of the integrated intensity of two separate lines in the spectrum. This method is often used when the reference spectra of individual constituents are unavailable. The situation can be more favorable in the case of continuous transformations of pure compounds, for instance when following a phase transition over a temperature range, or recrystallization over time.

The amorphization of trehalose under mechanical stress appears as a typical example of such a case. Solid-state NMR and differential scanning calorimetry (DSC) methods were used for quantifying the amorphous content of *trehalose* in solid dosage forms[23] and the accuracy of NMR method was compared with that of the DSC technique. The samples of fully crystalline and fully amorphous trehalose were available. This allowed the authors to implement the simple and precise deconvolution method using the reference NMR spectra. Samples with different amorphous fractions were prepared by physical mixing of purely amorphous and purely crystalline powders. The fraction τ_{NMR} of amorphous trehalose obtained by NMR is in fair agreement with the weighed τ_w values, and the confidence intervals appear remarkably narrow. The results reveal a close correlation between the imposed compositions of the physical mixtures and those determined by NMR and DSC, indicating that both are useful and accurate methods for compositional characterization of powders. This solid-state NMR methodology was then used to determine the evolution of the amorphous fraction in a trehalose powder, during a milling, which leads to a fully amorphous state. The relatively higher uncertainty evaluated for two samples (with $\tau_{NMR} = 39.8 \pm 0.7\%$ and $\tau_{NMR} = 58.8 \pm 1.1\%$) might be artifacted by a small drift of the Hartmann–Hahn conditions, for those spectra were recorded several days after the reference spectra. This indicated that the best results might be obtained when all the spectra (references and samples) are recorded within the same run of experiments, ensuring the best stability of the spectrometer.

The ability of solid-state ^{13}C CPMAS NMR spectroscopy was described[24] to quantify the amounts of three crystalline and amorphous forms of the artificial sweetener *neotame*. It was possible to quantify, in a mixture of two anhydrous polymorphic forms of neotame, the amount of each polymorph within 1–2%. In mixtures of amorphous and crystalline forms of neotame, the amorphous content could be determined within 5%. It was found that the crystalline standards that were used to prepare the mixtures were not pure crystalline forms, but rather a mixture of crystalline and amorphous forms. The amorphous content in the crystalline standards influenced the overall quantitation process of the two crystalline polymorphic forms. The importance of differences in relaxation parameters and CP efficiencies on quantifying mixtures of different forms was discussed.

Quantitative solid-state NMR method was developed[25] to characterize two crystalline forms I and II of *roxifiban*, a cardiovascular antithrombotic prodrug. Two lots of roxifiban, pure form I and II, were used to prepare physical blends used as calibration standards. The samples were prepared at I:II ratios of 100:0, 75:25, 50:50, 25:75, and 0:100. The entire sample was put into a 5-mm rotor thereby eliminating artifacts associated with sampling only a portion of the blend. Form I had a lower symmetry structure and the signals of methyl and methylene carbons of the n-butyl group are split. The n-butyl group of form II resides in a single defined location. The differences in the NMR spectra of the polymorphs were utilized in the analyses of physical blends of the pure crystalline forms to establish a calibration curve. Pure form I exhibited peaks at 66, 63, 21, and 19 ppm, whereas pure form II had peaks at 66 and 19 ppm only. The solid-state NMR spectra peak areas at 66 and 19 ppm remained relatively constant, whereas the peak areas at 63 and 21 increased proportionally as the concentration of form I increased. The data analysis showed that the CPs and relaxation times for form I (including the split peaks) and form II were equivalent, based on the fact that the calibration curve was linear and the slope was equal to 1. To estimate the detection limit (DL), a physical blend of form I and form II was prepared at a I:II = 90:10 ratio. The results of the analysis (in triplicate) were $87.1 \pm 3.1\%$ form I, which would indicate a DL of form II in form I of 9% with the DL defined as 3 times the intrasample standard deviation of the solid-state NMR assay. This study indicates that solid-state NMR was a valuable technique to quantify the polymorphic purity of samples, where other techniques such as DSC could not be used for this purpose.

The polymorphic content of a drug in its formulation can be determined as shown by Apperley et al.[26] ^{13}C CPMAS spectrum of anhydrous *formoterol fumarate* and its dihydrate (total drug concentration in lactose was 2%) showed overlapped

C–O resonances of the two forms. In the deconvoluted simulated spectrum, the areas of anhydrate and dihydrate can be determined from which a molar concentration ratio of 43:57 (%) was obtained.

Quantitation experiments can be performed on whole tablets inserted into an MAS rotor, as demonstrated by Harris et al.[9] in 2005. *Bambuterol hydrochloride* (BHC) and *terbutaline* (TBS) are used in asthma therapy; three polymorphs of these drugs have been identified so far. The samples including only one form of BHC and TBS formulated together in α-lactose monohydrate, as a filler, and magnesium stearate, as a lubricant, have been studied. The aim of the study was to obtain a robust protocol for quantifying the amount of BHC. ^{13}C CPMAS spectrum (recorded at 4.7 T, accumulation time 3 h) of this simulated drug formulation exhibited a cluster of peaks at 20–40 ppm (arising from stearate) and at 140–160 ppm (aromatic carbons region). Selected single peaks of bambuterol and stearate were used for quantitation. The LOD was 1% and the LOQ was 0.5% (w/w).

The idea to use one of the common excipients, the stearate, to provide an internal reference standard, can be recommended (if mass % of this additive is known).

^{13}C CPMAS NMR spectra[27] of the *amphetamines* and lactose amphetamine mixtures were recorded. The study was an extension of a forensic investigation concerned with quantitatively measuring amphetamine derivatives in tablets, and the experiment mimicked the illicit production of "Ecstasy" tablets. The observed solid-state chemical shifts of *R*,*S*-3,4-methylenedioxy-*N*-methylamphetamine × HCl changed when lactose monohydrate was added as a dry powder and mixed at room temperature.

^{13}C CPMAS NMR method has been proposed[28] as a fast method for authentication and quantitation of *sildenafil citrate* (Scheme 3) in commercial products, tablet dosage forms with Viagra. ^{13}C CPMAS spectra of the tablets of sildenafil citrate were recorded under identical experimental conditions (Figure 10).

Large peaks in the 50–110 ppm range are due to excipients, and the CH resonances of carbohydrates may be suppressed using the dipolar dephasing pulse sequence. Sildenafil citrate is easily detected in the pharmaceutical dosage forms since only two of its carbon resonances (OCH_2 and quaternary carbon of the citrate anion) fall into carbohydrate-type region.

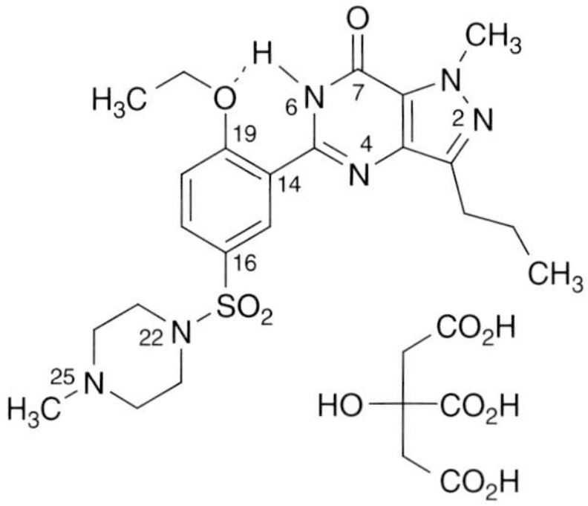

Scheme 3 Sildenafil citrate.

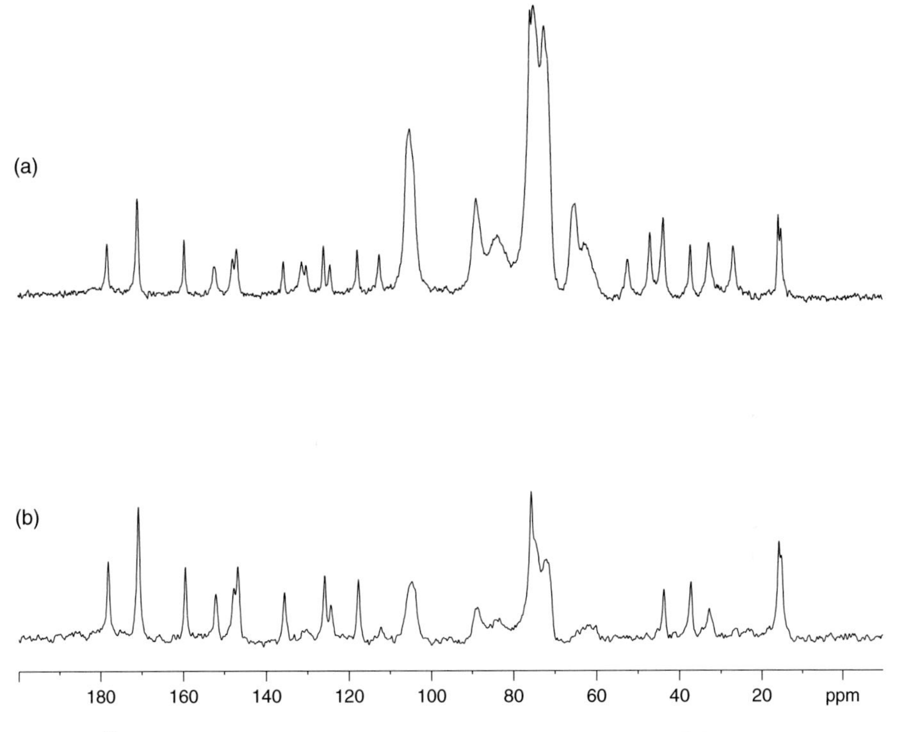

Figure 10 ^{13}C CPMAS NMR spectra of sildenafil citrate (Viagra) of the 50 mg commercial tablet: (a) standard spectrum and (b) spectrum recorded with dipolar dephase pulse sequence (50 μs delay).

While measuring the spectra for a series of commercial tablets it was found that the signals in the spectra of 50 and 25 mg tablets were of the same intensity as those from a 100 mg tablet. It was evident that commercial tablets of Viagra containing 25, 50, and 100 mg of sildenafil citrate in the dosage form differ in the tablet's mass but not in the concentration of the active substance. Therefore, a calibration plot was established using the mixtures of sildenafil citrate with hydroxymethyl cellulose. The signals of C═O carbons of citrate (δ 178.5 and 170.7 ppm) and the signals of C19, C16, and C14 (δ 159.5, 125.8, and 117.6 ppm, respectively) of sildenafil are distinct in each spectrum and can be used for quantitative analysis. Their intensities are linear functions of the amount of the drug in these solid mixtures. Thus, solid-state ^{13}C NMR can be applied to perform quantitative analysis of this drug in commercial tablets. The quantitation of solid-state NMR method for Viagra (establishing the limits of detection and quantitation as well as the accuracy and reproducibility) requires proper statistical evaluation. Actually, the method may be a valuable tool for fast identification of illegally produced tablets (sometimes sold in sex shops), which may contain only blue painted placebo or analogues of sildenafil (e.g., homosildenafil added to a functional food marketed for penile erectile dysfunction in Korea).

In conclusion, NMR spectroscopy is a simple and reliable means of quantifying a substance, no matter whether it exists as a pure drug or in dosage forms. Microcrystalline solids, amorphous or heterogeneous materials can be examined and its components can be quantified, down to ~1% of the total sample. The powder substance can be used without further preparation, which is the main advantage of MAS NMR. Ordered, proton-rich organic solids yield high-resolution ^{13}C spectra in a very short time (even 0.5 h); sometimes packing the rotor needs more time than registering one standard MAS NMR spectrum.

REFERENCES

1. J. Klinowski, ed., New Techniques in Solid-state NMR, Springer, Berlin, 2005.
2. M. Duer, ed., Solid-state NMR Spectroscopy, Principles and Applications, Blackwell Science, Oxford, 2002.
3. L.B. Alemany, D.M. Grant, D.M. Pugmire, T.D. Alger, K.W. Zilm, J. Am. Chem. Soc. 105 (1983) 2133–2142.
4. W. Kołodziejski, J. Klinowski, Chem. Rev. 102 (2002) 613–628.
5. R.H. Harris, Analyst 110 (1985) 649–655.
6. R. Suryanarayanan, T.S. Wiedmann, Pharm. Res. 7 (1990) 184–187.
7. J.Z. Hu, X. Wu, N. Yang, L. Li, C. Ye, K. Qin, Solid State NMR 6 (1996) 187–196.
8. F. Malz, H. Jancke, J. Pharm. Biomed. Anal. 38 (2005) 813–823.
9. R.K. Harris, P. Hodgkinson, T. Larsson, A. Muruganantham, J. Pharm. Biomed. Anal. 38 (2005) 858–864.
10. G.C. Campbell, L.G. Galaya, A.J. Beeler, A.D. English, J. Magn. Reson. A 112 (1995) 225–228.
11. G. Jeschke, G. Grossmann, J. Magn. Reson. A 103 (1993) 323–328.
12. W.T. Dickson, J. Chem. Phys. 77 (1982) 1800–1809.
13. D.E. Bugay, Magnetic Resonance Spectrometry, in: Physical Characterization of Pharmaceutical Solids, ed. H. G. Brittain, Marcel Dekker Inc., New York, 1995, 94–125.
14. F. Ziarelli, S. Caldarelli, Solid State NMR 29 (2006) 214–218.
15. P. S. Belton, B. P. Hills, G. A. Webb, eds, Advances in Magnetic Resonance in Food Science, RSC, Cambridge, 1999.
16. A.L. Bates, P.G. Hatcher, Org. Geochem. 18 (1992) 407–416.
17. C.E. Snape, D.E. Axelson, R.E. Botton, J.J. Delpuech, P. Tekely, B.C. Gerstein, M. Prusik, G.E. Maciel, M.A. Wilson, Fuel 68 (1989) 547–560.
18. G.M. Hanna, J. AOAC Int. 78 (1995) 946–954.
19. H.K. Kim, Y.H. Choi, W.T. Chang, R. Verpoorte, Chem. Pharm. Bull. (Tokyo) 51 (2003) 1382–1385.
20. N.K. Ebube, A.H. Hikal, C.M. Wyandt, D.C. Beer, L.G. Miller, A.B. Jones, Drug Dev. Ind. Pharm. 22 (1996) 561–568.
21. N.K. Ebube, A.H. Hikal, C.M. Wyandt, D.C. Beer, L.G. Miller, A.B. Jones, Int. J. Pharm. 156 (1997) 49–57.
22. W. Kołodziejski, T. Kasprzycka-Gutman, Solid State NMR 11 (1998) 177–180.
23. R. Lefort, A. De Gusseme, J.F. Willart, F. Danede, M. Descamps, Int. J. Pharm. 280 (2004) 209–219.
24. T.J. Offerdahl, J.S. Salsbury, Z. Dong, D.J. Grant, S.A. Schroeder, I. Prakash, E.M. Gorman, D.H. Barich, E.J. Munson, J. Pharm. Sci. 94 (2005) 2591–2605.
25. R.D. Vickery, G.A. Nemeth, M.B. Maurin, J. Pharm. Biomed. Anal. 30 (2002) 125–129.
26. D.C. Apperley, R.K. Harris, T. Larsson, T. Malstrom, Pharm. Sci. 92 (2003) 2496–2503.
27. G.S.H. Lee, R.C. Taylor, M. Dawson, G.S.K. Kannangara, M.A. Wilson, Solid State NMR 16 (2000) 225–237.
28. I. Wawer, M. Pisklak, Z. Chilmonczyk, J. Pharm. Biomed. Anal. 38 (2005) 865–870.

CHAPTER 4

Microcoil Nuclear Magnetic Resonance Spectroscopy

A.G. Webb

Contents

Abstract

Nuclear magnetic resonance (NMR) spectroscopy is one of the most widely used techniques for chemical and biochemical analysis. In comparison with many analytical techniques, NMR has a low intrinsic sensitivity, and many studies are precluded by the limited available quantity of synthesized or isolated sample. Recently, there has been considerable progress in developing technology that can be used to study very small amounts of material. This chapter specifically describes the development and applications of small NMR coils.

Keywords: Saddle/Helmholtz coils, solenoidal coils, RF microcoils, superconducting microprobes, cITP–NMR, multiple coil probehead

1. Introduction

Nuclear magnetic resonance (NMR) spectroscopy is one of the most widely used and versatile analytical techniques, applicable to gaseous, liquid, and solid samples. It can be used to study, for example, chemical structure, molecular dynamics, and binding kinetics. Many potentially interesting scientific studies, however, cannot be performed using NMR due to the inherent low sensitivity of the technique. In particular, when the mass of a particular sample is limited, the data acquisition times required to obtain useful spectra can become unrealistically long. Such situations exist in the fields of structural analyses of the products of combinatorial and multistage chemical syntheses, extracts from rare plants with pharmacological properties, and certain "metabonomic" samples. The low sensitivity also precludes many studies of dynamic processes on short timescales. Technological efforts to improve the sensitivity of NMR have included increases in the strength of static magnetic fields,[1,2] the development of cryogenically cooled detectors,[3,4] and the use of hyperpolarization techniques.[5,6] This chapter concentrates on the approach which uses very small, highly sensitive radiofrequency (RF) detectors. In this approach, mass-limited samples are dissolved in very small volumes (typically nanoliters to microliters) of solvent to match the volume of the RF coil. Although simple in concept, there are many challenges, including the optimization of static (B_0) and radiofrequency (B_1) magnetic field homogeneity over the sample and efficient loading of small sample volumes into the detector.

In addition to their high sensitivity, small coils have several other advantages. For example, signals from solvent impurities are much less problematic with smaller sample volumes and the amount of deuterated solvents can be reduced significantly for NMR-coupled separations. In the case of electrically conductive samples, there is also a reduction of the loading effects of the sample compared to larger coils. Small coils produce very high B_1 fields (in the MHz range) and so can excite very large bandwidths, which is particularly useful for solid samples. Finally, small coils allow the design of probeheads containing more than one coil, enabling a new range of NMR experiments to be performed.

2. Intrinsic NMR Sensitivity

The limit of detection (LOD) of a particular method is defined as the smallest measure that can be detected "with reasonable certainty."[7] Mathematically,

$$x_{\mathrm{L}} = x_{\mathrm{b}} + k\sigma_{\mathrm{b}} \tag{1}$$

where x_{L} is the smallest measure, x_{b} is the mean of "blank" measurements, σ_{b} is the standard deviation of the blank measurements, and k is a numerical factor which determines the confidence level. IUPAC suggests a value of $k = 3$, corresponding to a confidence level of ~90%. In comparison with many other analytical techniques, NMR suffers from poor LODs. For example, laser-induced fluorescence (LIF) has

LODs of 10^{-13} mol, Fourier transform ion cyclotron resonance (FTICR) 10^{-20} mol, Fourier transform infrared spectroscopy and Raman spectroscopy 10^{-12}–10^{-15} mol, and mass spectrometry 10^{-19} mol, but LODs for NMR are several orders of magnitude poorer.

The NMR signal depends upon the net magnetization, M_0, of the sample. For a two-level system with spin quantum number, $I = 1/2$:

$$M_0 = \frac{\gamma^2 h^2 B_0 N_s}{16\pi^2 kT} \tag{2}$$

where γ is the gyromagnetic ratio of the nucleus, h is the Planck's constant, B_0 is the static magnetic field, N_s is the total number of nuclei in the sample, k is Boltzmann's constant, and T is the temperature. For protons at a static magnetic field of 11.7 T there is only a factor of 5×10^{-6} difference in the two energy levels, and this very small value is the reason for the intrinsic low sensitivity of NMR. The problem becomes even more pronounced if a nucleus such as ^{13}C, which has small gyromagnetic ratio, roughly one-quarter that of protons, and a natural abundance of ~1.1%, is to be detected.

The signal-to-noise (S/N) of an NMR measurement can be expressed as[8]

$$\mathrm{S/N} \propto \frac{k_0 \dfrac{B_1}{i} V_s N \gamma \dfrac{h^2}{4\pi^2} I(I+1) \dfrac{\omega_0^2}{kT3\sqrt{2}}}{\sqrt{4kT\Delta f \left(R_{\text{coil}} + R_{\text{sample}}\right)}} \tag{3}$$

where V_s is the sample volume, k_0 a constant that accounts for spatial inhomogeneities in the B_1 field produced by the RF probe, N the spin density, ω_0 the Larmor frequency, Δf the measurement bandwidth, R_{coil} and R_{sample} the coil and sample resistances, respectively, and the factor of $\sqrt{2}$ is introduced as the noise measure is root mean square (rms). The factor B_1/i, the magnetic field per unit current, is defined to be the coil sensitivity. Because the NMR signal for successive measurements is coherent, and the noise is incoherent, the S/N increases by a factor proportional to the square root of the number of coadded measurements.

Equation (3) suggests a number of approaches to maximize the S/N of an NMR experiment. The first is to use as high a static magnetic field as possible. Currently, commercial superconducting magnets for high-resolution NMR studies at 21.1–21.5 T (proton Larmor frequency >900 MHz) represent the highest field available. A second approach is to reduce the noise voltage from the NMR coil via the use of high-temperature superconducting (HTS) materials: this has resulted in the development of "cryoprobes" from all of the major NMR manufacturers.[3,9–14] Typical S/N increases of a factor of 4–5 are possible for samples with low conductivity, with these factors becoming smaller as the sample conductivity increases. A third possibility is to increase the intrinsic coil sensitivity. As shown by Hoult and Richards,[8] the value of B_1/i for both saddle and solenoid coil geometries is inversely proportional to the diameter of the coil. Analysis by Peck et al.[15] showed that, for very small solenoids, this relationship holds for coil diameters as low as ~100 μm, below which the sensitivity increases proportional to the square root of the coil diameter. It is clear, then, that the maximum sensitivity can be realized by making the coil as small as possible to accommodate the particular sample. For high-resolution NMR spectroscopy, this corresponds to using the minimum volume of solvent based on the solubility constant of the solute.

3. Development of Small Coils for High-Resolution NMR

As for RF coils of conventional diameters (3–10 mm), the aim in designing small RF coils is to produce a strong, homogeneous B_1 field within the sample, and to minimize line-broadening effects due to magnetic susceptibility mismatches. This section summarizes the main approaches taken to achieve these goals.

3.1. Saddle/Helmholtz coils

For a vertically oriented sample placed within an RF coil of matched geometry, electromagnetic analysis can be used to derive the condition necessary to produce a perfectly uniform B_1 field in a direction perpendicular to that of the B_0 field. The condition corresponds to an infinitely long cylinder carrying a surface current, J_s, given by

$$J_s = \hat{z} J_0 \sin\phi \tag{4}$$

where ϕ is the azimuthal angle subtended at the center of the cylinder. The field, $B(\rho, \phi)$, inside the cylinder is given by

$$\begin{aligned} B(\rho,\phi) = & -\hat{\rho}\sum_{m=1}^{\infty} m\rho^{m-1}(A_m \cos m\phi + B_m \sin m\phi) \\ & + \hat{\phi}\sum_{m=1}^{\infty} m\rho^{m-1}(A_m \sin m\phi - B_m \cos m\phi) \end{aligned} \tag{5}$$

By applying boundary conditions at the cylinder surface (considering the B_1 field created outside the cylinder), all the B_m coefficients must be zero, and the only nonzero A_m coefficient is given by $A_1 = -\mu_0 J_0/2$. Therefore,

$$B(\rho,\phi) = \frac{\mu_0 J_0}{2}\left(\hat{\rho}\cos\phi - \hat{\phi}\sin\phi\right) = \hat{x}\,\frac{\mu_0 J_0}{2} \tag{6}$$

This shows that a perfectly uniform B_1 field, directed along the x-direction, is produced by a sinusoidal current along the surface of a cylinder of infinite length. One practical realization in high-resolution NMR is the "saddle" or "Helmholtz" coil (Figure 1), which can be thought of as a "six-point" approximation to the sinusoidal current distribution, given the optimum placement of the conductor elements.[16]

The earliest work demonstrating the potential of high-resolution NMR spectroscopy using a small diameter (1.7 mm) saddle coil was performed by Shoolery,[17] who

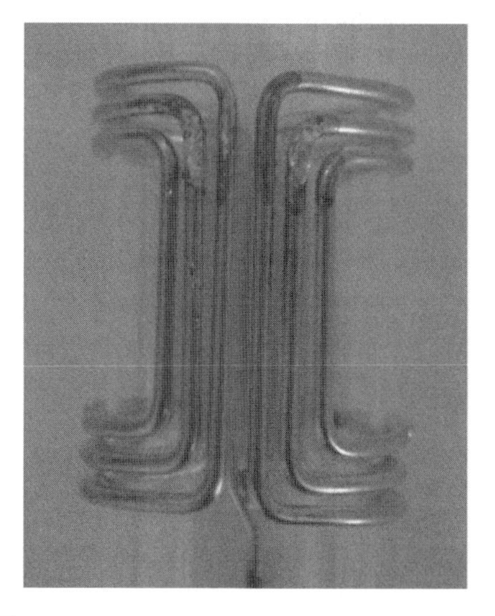

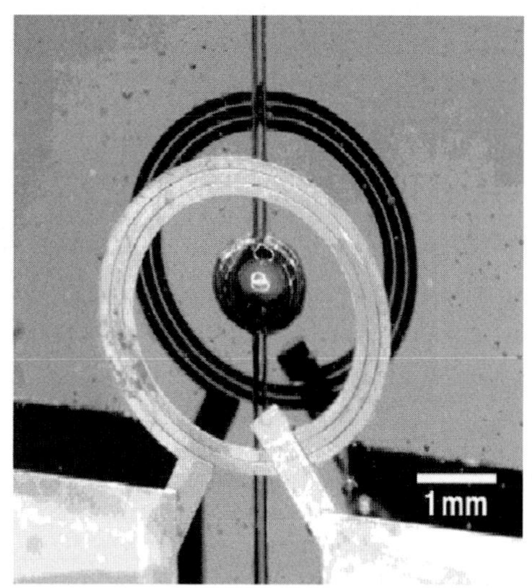

Figure 1 (left) Conventional saddle coil used for high-resolution NMR wound with susceptibility-matched wire. (right) Microfabricated Helmholtz coil with an integrated spherical sample chamber. Reproduced with permission from Walton et al.[20] Copyright 2003 American Chemical Society.

published a number of applications for both ^{1}H and ^{13}C at relatively low magnetic fields. The development of 3 and 2.5 mm micro-NMR probes with sample volumes of 150 and 100 μl, respectively, at 11.7 T was reported by Bruker in 1992.[18] In 1998, Nalorac developed a 1.7 mm submicroprobe with a 30 μl volume: a comparison with previous 3-mm diameter probes showed the expected gains in S/N.[19] Bruker (Switzerland) developed a 2.5 μl active volume TXI probe in 2002, which uses disposable 1-mm diameter capillaries. In addition to commercial developments, small saddle coils have also been produced lithographically.[20–23] Figure 1 shows photographs of conventional wire-wound and microfabricated saddle coils.

3.2. Solenoidal coils

An alternative approach for achieving a relatively homogeneous B_1 field is to use a horizontally oriented sample with a solenoidal coil. For a coil of radius a, length l, and n turns, the B_1 field in the x-direction is given by

$$B_{1,x} = \frac{\mu_0 n I}{2l}\left[\frac{x + l/2}{\sqrt{a^2 + (x + l/2)^2}} - \frac{x - l/2}{\sqrt{a^2 + (x - l/2)^2}}\right] \quad (7)$$

At the center of the solenoid ($x = 0$), assuming $l >> a$, this simplifies to

$$B_{1,x} = \frac{\mu_0 n I}{l_{\mathrm{D}}} \quad (8)$$

where l_D is the length of the solenoid diagonal. If the coil length-to-diameter ratio is large, then $l_D \sim l$. The on-axis magnetic flux density ($B_{1,\text{end}}$) at either end of the coil is given by

$$B_{1,\text{end}} = \frac{\mu n i}{2\sqrt{a^2 + l^2}} \tag{9}$$

Again, using the simplification that $l >> a$:

$$B_{1,\text{end}} = \frac{\mu n i}{2l} \tag{10}$$

Therefore, the magnetic field at the ends of a solenoid is one-half that at the center. The B_1 field produced by a solenoidal coil is shown schematically in Figure 2.

If Eq. (7) is re-expressed, then the sensitivity (B_1/i) can be directly related to the diameter (d) of the coil:

$$\frac{B_1}{i} = \frac{\mu n}{2a\sqrt{1+[l/2a]^2}} \tag{11}$$

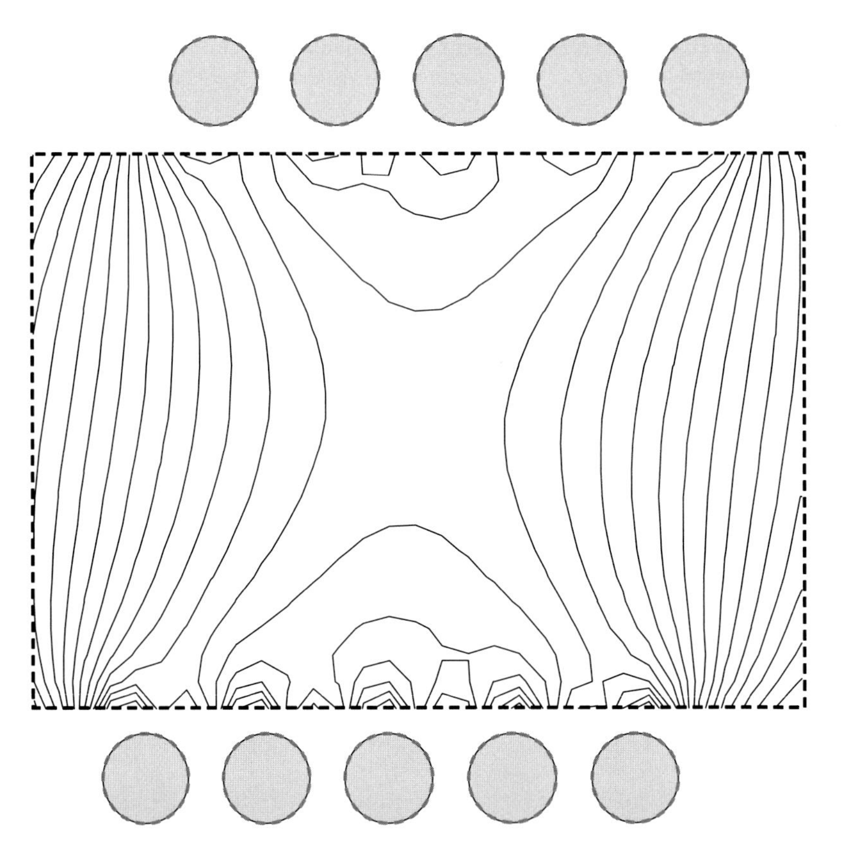

Figure 2 Contour plot of the magnitude of the B_1 field produced by a five-turn solenoidal coil. Each line represents a 5% contour from the maximum field at the center of the coil.

Therefore, assuming that the length-to-diameter ratio is kept constant, the coil sensitivity increases as the inverse of the coil diameter. Within reasonable error, these calculations remain valid whatever the size of the coil, and the exact form of the current distribution within the windings of the coil, due to the inherent symmetry afforded by the cylindrical geometry of the solenoid. Exact calculations of noise in solenoidal coils at high frequencies and small diameters are complicated.[15] The net result of such analysis is that, for coils of diameter greater than about 100 μm, the S/N per unit volume of sample is inversely proportional to the coil diameter. For coils of diameter below 100 μm, the inverse dependence is reduced to the square root of the coil diameter.[15]

The sensitivity of a saddle coil was calculated by Hoult and Richards[8] by considering the spatial configuration of the magnetic vector potentials:

$$\frac{B_1}{i} = \frac{n\mu\sqrt{3}}{\pi}\left[\frac{4al}{(4a^2 + l^2)^{3/2}} + \frac{2l}{2a\sqrt{4a^2 + l^2}}\right] \tag{12}$$

The authors showed that the sensitivity of the solenoid coil is approximately three times that of the saddle coil, and this was confirmed experimentally by measuring the 90° pulse lengths for each coil.[8]

In 1993, a magic angle spinning "nanoprobe" with a 40 μl sample volume was introduced by Varian.[24,25] The RF coil was a 4 mm diameter solenoid coil oriented at the magic angle with respect to B_0. Several studies were performed in the area of combinatorial chemistry of single beads. Developing a number of academic ideas using small solenoid coils surrounded by a perfluorinated fluid for susceptibility matching, Protasis/MRM produced a 1.5 μl active volume solenoid-based probe, which has been used extensively for hyphenation to capillary separations[26] and for flow-through NMR.[27]

The normal method of producing submillimeter solenoidal coils is simple hand winding. However, these can also be produced by lithographic techniques[28] as well as sophisticated micromachining.[29,30] The wire coil is often formed around a "bubble cell" formed, for example, by hydrofluoric acid etching to increase the filling factor of the coil.[31,32] Figure 3 shows photographs of both hand-wound and micromachined solenoidal coils.

Small solenoids can be wound using wire of either round or rectangular cross section. The relative sensitivity of each geometry has been analyzed by Li et al.[33] Using the geometries shown in Figure 4(a) and (b), and assuming that wire dimensions, including the diameter of the round wire and the thickness and width of the rectangular wire, which are at least 20 times the skin depth, currents in the coil can essentially be treated as surface currents.[34,35] For perfect conductors of finite dimension, the magnetic fields normal to the surface vanish, giving the boundary condition:[34]

$$\vec{B} \cdot \vec{n} = 0 \tag{13}$$

In general, the current distribution on the cross-sectional area of the wire varies along the wire from the center to the two ends of the coil due to finite coil length and associated wave phenomena[36,37] and there is no general analytical solution to

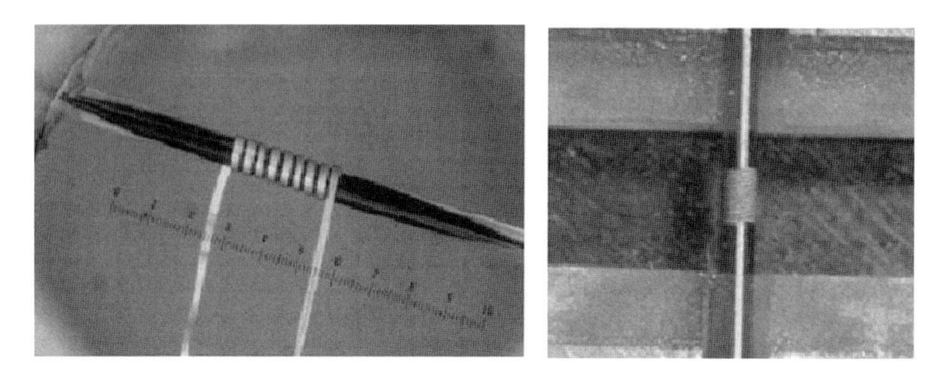

Figure 3 (left) Solenoidal coil wound with copper wire of a rectangular cross section around an elongated "bubble cell" produced by hydrogen fluoride etching of a fused silica capillary. (right) A solenoidal coil produced by laser lathe technology. Figure reproduced from Demas et al.[30] Copyright (2007) with permission from Elsevier.

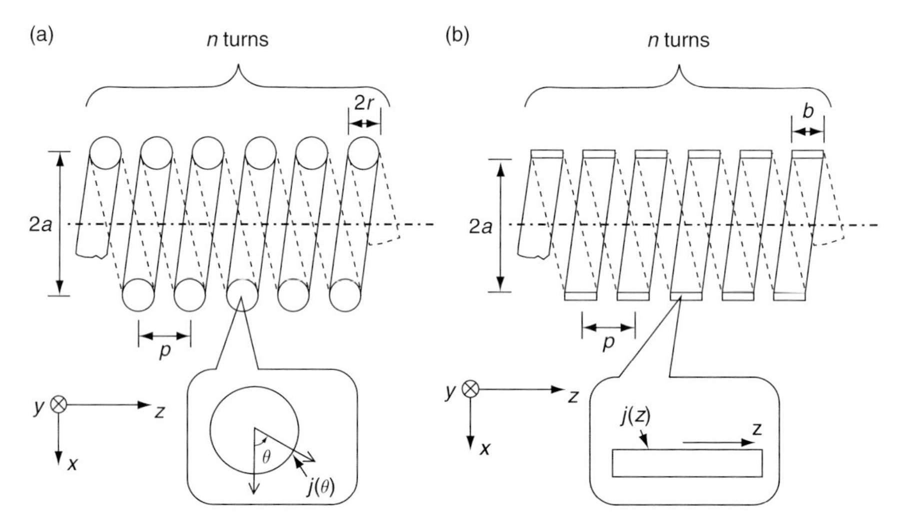

Figure 4 Schematic of a solenoid coils wound with round (a) and rectangular (b) wire. The origin of the Cartesian coordinate system is at the center of the long axis of the coil. Parameters: a, coil radius; p, pitch; r, radius of round wire; b, width of rectangular wire; n, number of turns. Reproduced with permission from Li et al.[33] Copyright 2003 Wiley.

determine the exact B_1 distribution. To develop an analytical solution, two approximations were made. The first was that the current distribution in each turn is the same: this clearly becomes more accurate for a larger number of turns and/or a greater length-to-diameter ratio. The second assumption is that wave phenomena are negligible because the coil is well balanced and the conductor length for these small solenoids is typically much less than one wavelength. The power loss in each coil given by

$$P_{\text{loss,round}} = \frac{\rho L}{r\delta} \int_0^{2\pi} j^2(\theta)\, d\theta, \quad P_{\text{loss,rectangular}} = \frac{\rho L}{\delta} \int_{-b/2}^{b/2} j^2(z)\, dz \tag{14}$$

The relative S/N can be expressed in terms of a parameter q, given by

$$\frac{\mathrm{S}}{\mathrm{N}} \propto \sqrt{q} = \frac{B_1(0)}{\sqrt{P_{\text{loss}}}} \tag{15}$$

The q values of a series of solenoid coils with different winding parameters are plotted vs length-to-diameter ratio in Figure 5. It can be seen that there is an optimum length-to-diameter ratio of ~0.7 for a round-wire coil and ~1.0 for a

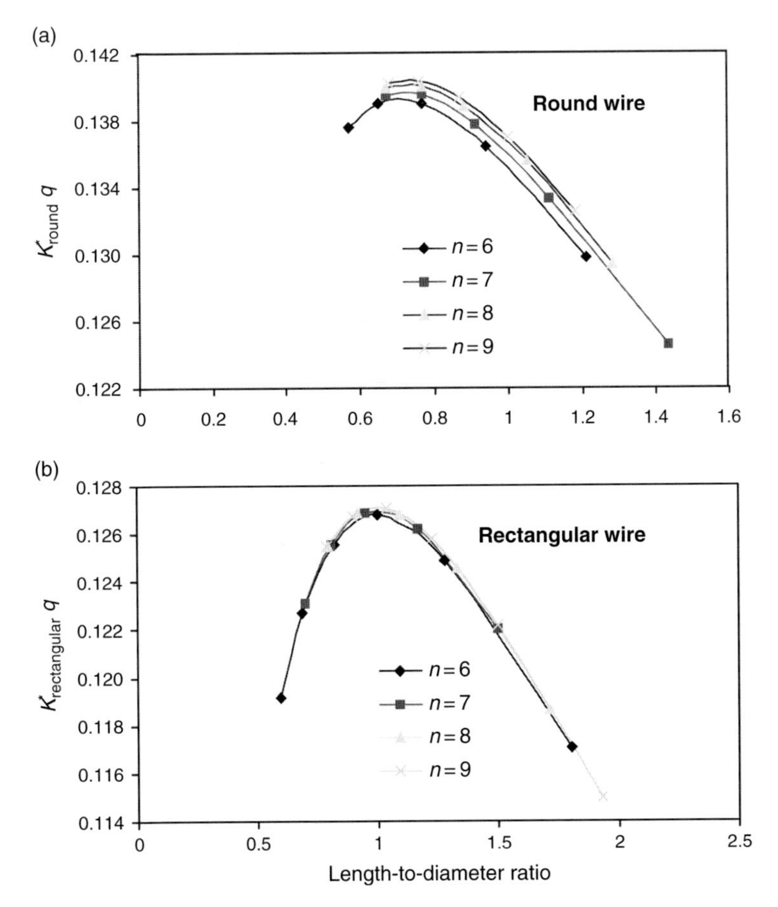

Figure 5 Plots of calculated K^*q values vs length-to-diameter ratios and turn numbers for small solenoid coils built from (a) round wires, and (b) rectangular wires. Reproduced with permission from Li et al.[33] Copyright 2003 Wiley.

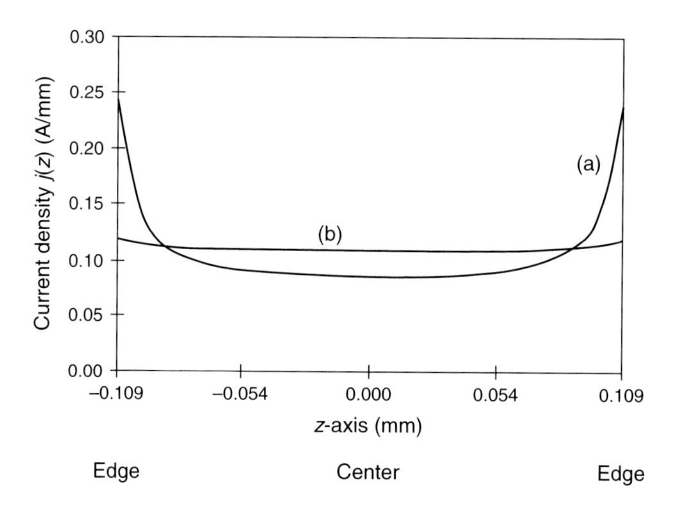

Figure 6 (a) Current distribution pattern of a spaced-turn solenoid coil of rectangular wires with winding parameters: $2a = 1$ mm, $b = 0.23$ mm, $p = 0.36$ mm, and $n = 6$. Severe currents crowding occurs at the two edges of the wire. (b) Current distribution pattern of a close-turn solenoid coil of rectangular wires with winding parameters: $2a = 1$ mm, $b = 0.23$ mm, $p = 0.23$ mm, and $n = 6$. Currents are relatively uniform across the wire. Reproduced with permission from Li et al.[33] Copyright 2003 Wiley.

rectangular-wire coil. Figure 5 also shows that the q values have a weak dependence on the number of turns. The results in Figure 5(a) match well with two previous studies on Q optimization of solenoid coils.[38,39]

Figure 6 shows the current distribution patterns in spaced and close-turn rectangular-wire solenoids. For the spaced turns (Figure 6(a)), the current density difference between the center and edges of the conductor surface is as large as 50%. These edge effects can be suppressed by winding a close turn, i.e., with near-zero inter-turn distance, coil. A current distribution plot in such a coil with the same length-to-diameter ratio is also shown in Figure 6(b). The flat curve implies a relatively uniform current distribution across the wire. These results agree well with a previous study on current distribution in saddle coils built from rectangular wire.[34] The results also suggest that a zero inter-turn distance in rectangular-wire coil should be used for the optimum S/N performance. With such a configuration, the solenoid coils built from rectangular wire have a more uniform current distribution than those constructed using round wires if the length-to-diameter ratio is large. For this reason, one expects a smaller power loss and hence a higher S/N in rectangular coils than that in round-wire coils for a high length-to-diameter ratio. Experimental results presented in this paper confirmed the theoretical predictions.

3.3. Planar RF coils

Although hand-wound solenoidal coils give high NMR sensitivity, reproducibility in coil construction is challenging, and there is a lower limit in diameter that can be achieved. There are natural reasons, therefore, for considering microfabrication

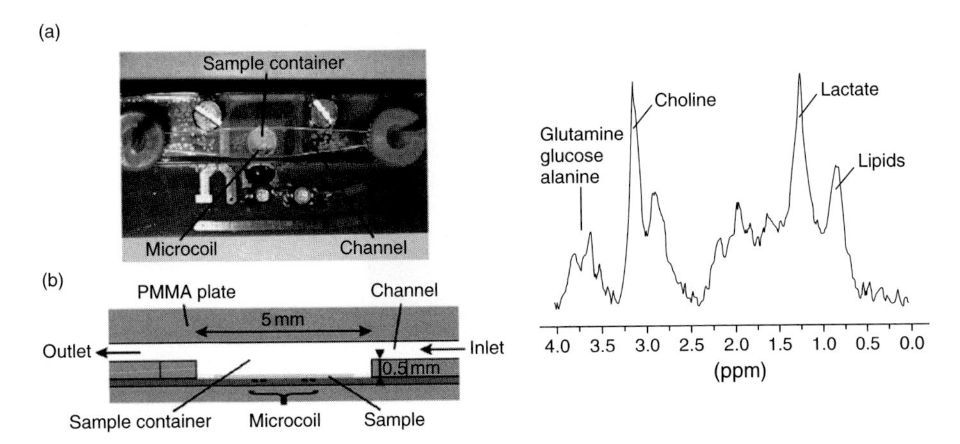

Figure 7 (left) (a) A photograph and (b) a schematic of a planar microprobe for sample perfusion within the NMR-magnet during analysis. (right) Spectrum of CHO cells that are uniformly immobilized on the microprobe surface. Reproduced from Ehrmann et al.[42] with permission of The Royal Society of Chemistry.

techniques, provided that it is borne in mind that the conductor dimensions should be larger than the skin depth to minimize coil loss. As microfabrication techniques are usually, though not exclusively, planar a number of reports have used various microfabrication techniques to construct planar detectors for small-volume NMR spectroscopy.[40–44] It should be noted that saddle[20] and solenoidal[28,45] geometries have also been produced. In all of these cases the quality of the spectra, both in terms of linewidth and sensitivity, has not yet been as high as that obtained using the more labor-intensive hand-wound coils. Nevertheless, microfabrication is a promising area particularly in terms of integration of NMR detection with microfluidics. Figure 7 shows recent results from an integrated planar RF coil and sample perfusion chamber.[42]

3.4. Novel microcoil designs

Saddle, solenoidal, and planar RF geometries have been used essentially because the inception of magnetic resonance. Recently, designs, used extensively in the microwave and satellite community, have been tested as NMR probes. Maguire et al.[46] reported recently on a microslot waveguide, which consists of a microstrip with a small hole in the center, as shown in Figure 8. A slot whose size is smaller than the height of the dielectric produces an increased inductance ΔL given by

$$\Delta L = \frac{\mu_0 \pi}{2} h \left[1 - \frac{Z_0}{Z_0'} \sqrt{\frac{\varepsilon_0}{\varepsilon_0'}} \right]^2 \tag{16}$$

where h is the thickness of the dielectric, Z_0 and ε_0 are the characteristic impedance and dielectric constant of the microstrip, and Z_0' and ε_0' are the impedance and

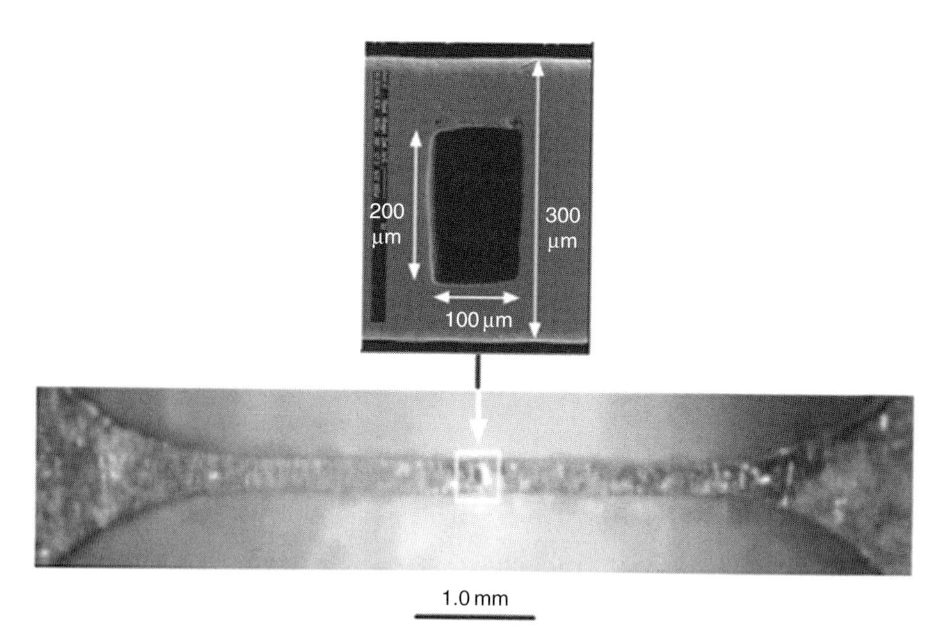

Figure 8 (bottom) A microslot resonator manufactured on a planar substrate, with the detector area denoted by the box outline. (top) Expansion of the actual microslot, fabricated by using a 248-nm excimer laser, just after laser exposure. Reproduced with permission from Maguire et al.[46] Copyright 2007, National Academy of Sciences.

dielectric constants of the microstrip at the location of the microslot. The change in inductance upon introducing a slot can also be expressed in terms of the microslot flux, ϕ, over an enclosed surface S:

$$\Delta L = \frac{\phi}{i} = \frac{\int_S B\,\mathrm{d}A}{i} \tag{17}$$

From Eq. (17) the effect of the microslot is to increase the overall flux density B. Local to the slot, the B_1 field is highly homogeneous, with the electric field being largely confined to the dielectric substrate, as in a microstrip. The authors used a conventional macroscopic stripline to impedance match with an adiabatic taper to the microfabricated microslot, with a sample placed slightly above the slot. Using various samples, the authors determined that the microslot was ~36% less sensitive than a solenoidal microcoil of the same dimensions, but gave both sensitivity and spectral linewidths far better than previous microfabricated planar microcoils.

A related design was recently proposed by van Bentum et al.,[47] who manufactured a resonator out of a stripline structure, shown in Figure 9. This is a very easy structure to produce lithographically, and has several desirable properties including high power capability, relatively good B_1 homogeneity, and easy scalability to very small sizes. Figure 10 shows a high-resolution spectrum acquired with this new stripline design.

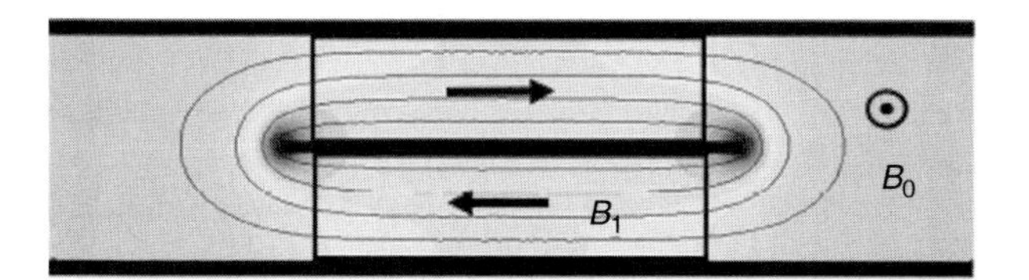

Figure 9 Schematic cross-section of a stripline design and associated B_1 field lines. Because of the boundary conditions imposed by the metallic planes above and beneath the strip, the magnetic field lines are forced parallel to the surface. Suitable sample chambers are indicated by the two black rectangles where the B_1 field is homogeneous within about 10%. Also, the current distribution in the strip is homogeneous thus minimizing electrical losses. The static field B_0 is oriented perpendicular to the cross-section shown in the figure (along the stripline axis). Figure reproduced from Van Bentum et al.[47] Copyright (2007) with permission from Elsevier.

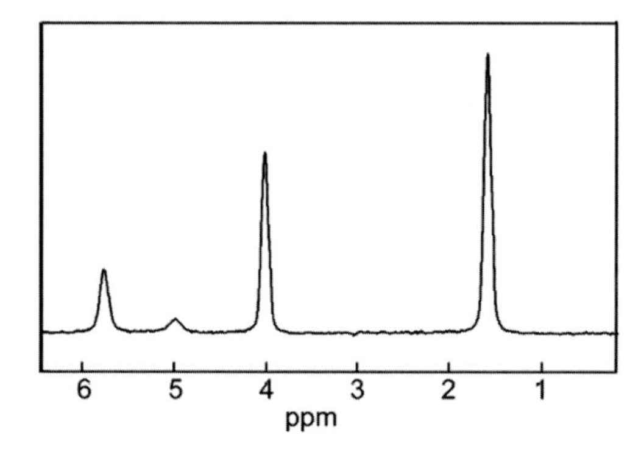

Figure 10 Single-scan proton spectrum of a 12 nl ethanol sample obtained at 600 MHz (14.1 T) in the prototype stripline probe. The spectrum was obtained by subtracting a 180° spectrum divided by two from a single scan spectrum excited under normal 90° pulse conditions, in this way signals from material above contact pads that experience a rather low RF field strength are eliminated from the overall signal. Figure reproduced from Van Bentum et al.[47] Copyright (2007) with permission from Elsevier.

3.5. High-temperature superconducting microprobes

The introduction of cryogenically cooled RF coils represents one of the most important technological developments in NMR in recent years. A recent review article[4] describes many of the relevant design features, and also applications of cryogenic coils. Simple analysis can be used to understand the origin of the significant increases in S/N. The signal voltage, V_s, is given by

$$V_s \propto \frac{B_1}{I\sqrt{R_{\text{coil}} + R_{\text{sample}}}} \tag{18}$$

where

$$R_{\text{coil}} = \frac{\omega_0 L_{\text{coil}}}{Q_{\text{unloaded}}}, \quad R_{\text{sample}} = \omega_0 L_{\text{coil}} \left(\frac{1}{Q_{\text{loaded}}} - \frac{1}{Q_{\text{unloaded}}} \right) \tag{19}$$

and the coil sensitivity, B_1/I, is inversely proportional to the coil diameter. The noise voltage, V_{noise}, is given by

$$V_{\text{noise}} \propto \sqrt{4k\Delta f \left(T_{\text{N,probe}} + T_{\text{N,preamplifier}} \right)} \tag{20}$$

where $T_{\text{N,probe}}$ and $T_{\text{N,preamplifier}}$ are the noise temperatures of the devices, defined as

$$T_{\text{N,probe}} = \frac{T_{\text{coil}} R_{\text{coil}} + T_{\text{sample}} R_{\text{sample}}}{R_{\text{coil}} + R_{\text{sample}}}, \quad T_{\text{N,preamplifier}} = 290 \left(10^{(\text{NF}/10) - 1} \right) \tag{21}$$

where NF is the noise figure of the preamplifier. A narrow-band GaAsFET- or HEMT-based preamplifier can readily be constructed with a noise figure of ~0.3 dB, which corresponds to an equivalent noise temperature of ~20 K. Cooling the preamplifier to liquid nitrogen temperatures and below can reduce the preamplifier noise figure to ~0.1 dB, an equivalent noise temperature of ~10 K. Equations (18)–(20) show that the effectiveness of the cryogenic approach depends upon the relative values of the sample and coil resistances. An increase in the absolute signal (Eq. (18)) and a decrease in the noise level (Eq. (20)) combine to give an increased S/N. Sample loading reduces the gain, up to the point at which the sample noise dominates that of the coil. S/N gains for nonlossy samples are around a factor of 4, while for protein studies in aqueous solutions with a high concentration of salt, typically a factor of 2 increase results.

In addition to the S/N advantages afforded by the use of high-temperature superconductors, reducing the coil diameter increases the sensitivity in exactly the same way as for standard coils, and therefore the highest possible sensitivity coil is a small, superconducting structure. Brey et al.[48] have designed and constructed such a coil, a 1-mm triple resonance probe with z-axis gradient for operation at 600 MHz. The probe has an active volume of 6.3 μl and a required sample volume of ~7.5 μl. It consists of four nested Helmholtz pairs of resonant HTS coils for ^{1}H, ^{2}H, ^{13}C, and ^{15}N, as shown in Figure 11. All the coils are cooled to ~20 K, and the ^{1}H, ^{2}H, and ^{13}C preamplifiers were cooled to 77 K using a commercial cryoplatform. The 1 mm sample was vacuum isolated from the HTS coils and warmed by an airstream. The ^{1}H resonators were based on two distributed interdigital capacitors, with a very small spatial periodicity of 125 μm, which produces a very low electric field within the sample, thus allowing it to be placed very close to the resonators and achieve a high filling factor. In addition, the current-carrying fingers were slit to reduce any shielding currents that would otherwise decrease the B_0 homogeneity within the sample. The ^{2}H, ^{15}N, and ^{13}C coils were formed as spiral resonators.

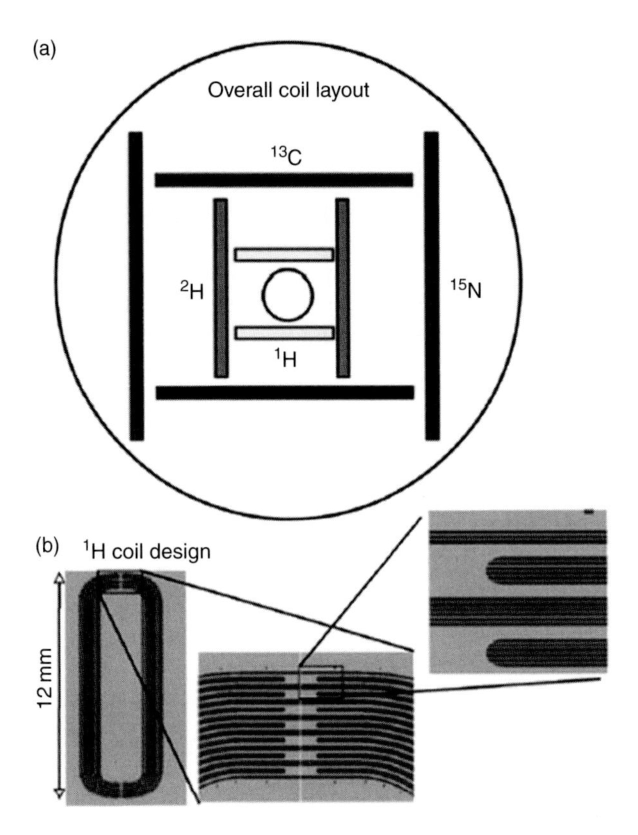

Figure 11 Design of a 1 mm HTS triple-resonance probe. Coils were constructed by depositing a thin coating of $Y_1Ba_2Cu_3O_{7-\delta}$ (YBCO) on a sapphire supporting surface. (a) Overall coil layout. (b) Details of the interdigitated design of the ^1H coil to create the correct resonance at 600 MHz while minimizing parasitic currents. Figure reproduced from Brey et al.[48] Copyright (2006) with permission from Elsevier.

4. Sensitivity Comparisons

NMR sensitivity can be defined in terms of mass sensitivity, S_m, or concentration sensitivity, S_c:[27]

$$S_c = \frac{S/N}{C}\,\mu M^{-1}, \quad S_m = \frac{S/N}{m}\,ng^{-1} \tag{22}$$

The S/N per unit mass increases as the size of the coil decreases, and therefore in situations where the mass of the sample is limited, and high solute concentrations are possible, the S/N achieved with a small coil is higher than using a larger coil. If only low solute concentrations are possible, due to either solubility constraints or concentration-dependent biological activity, and the total mass is not limited, then larger coils give a higher S/N. In the intermediate regime, where mass and concentration are both limited, then the optimum coil size is dictated by the relative values of these two variables.

A number of studies have been performed comparing the sensitivity of small coils with standard larger coils and each other. In one such study by Schlotterbeck et al.[49] the mass sensitivity of the Bruker 1-mm TXI microliter probe with a sample in a 1-mm diameter capillary tube was shown to be five times greater than a 5 mm conventional TXI probe at 600 MHz. The small TXI probe also showed a factor of 1.7 enhancement over a 5 mm TXI cryoprobe with the sample in a 5 mm tube, and a factor of 1.3 over a 5 mm TXI cryoprobe with the sample contained in a 1 mm capillary. A second detailed sensitivity analysis of different probes at 600 MHz was published by Olson et al. [27] Although some data had to be inferred from the literature, the general results show that the mass sensitivity of the 1.5 μl CapNMR probe was approximately 10 times greater than that of a standard 5 mm probe. The concentration sensitivity of the CapNMR probe was found to be approximately 15 times poorer than the 5 mm probe. Bruker has developed a 1.7 mm microcryoprobe, which has 6 times the mass sensitivity of a conventional 1.7 mm probe, and 10–14-fold increase over a conventional 5 mm probe. For the 1 mm HTS probe described previously[48] the mass sensitivity was ~20 times greater than a conventional 5 mm triple resonance probe, four times greater than a 5 mm cryogenic probe, and slightly greater than a commercial 1-mm diameter solenoidal coil.

5. Nanoliter Volume Applications of RF Microcoils – Hyphenated cITP–NMR

Although the original demonstration of high-resolution microcoil NMR used an active volume of 5 nl,[50] in practice, this volume is too small for realistic concentrations of most analytes. One exception is the coupling of microcoil NMR and capillary isotachophoresis (cITP).[51–56] cITP can be used either as a separation method or also as a way of concentrating dilute solutions. cITP experiments typically use a fused silica capillary which has been specially modified to reduce or eliminate electro-osmotic flow (EOF). Application of a high voltage across the two ends of the capillary causes the sample components to concentrate into separate zones depending on their individual electrophoretic mobilities. cITP uses a discontinuous buffer system, which is comprised of a leading electrolyte (LE) of higher and a trailing electrolyte (TE) of lower electrophoretic mobility, with the sample introduced between the LE and the TE. Using the appropriate concentration of LE, it is possible to concentrate charged analytes by up to two to three orders of magnitude into nanoliter volumes. In coupled cITP–NMR experiments, the concentrated analytes move through the separation capillary as distinct-focused cITP bands and enter the NMR coil as bands with very sharp edges. The physical setup is shown in Figure 12: nanomole quantities of analyte can be detected using this technique.[51–56]

For the cITP–NMR spectra shown in Figure 13, a multiple coil probe with two separate 1-mm-long microcoils arranged 1 cm apart was developed to facilitate peak trapping and sample band positioning. A 9.4 μl injection of 200 μM (1.9 nmol)

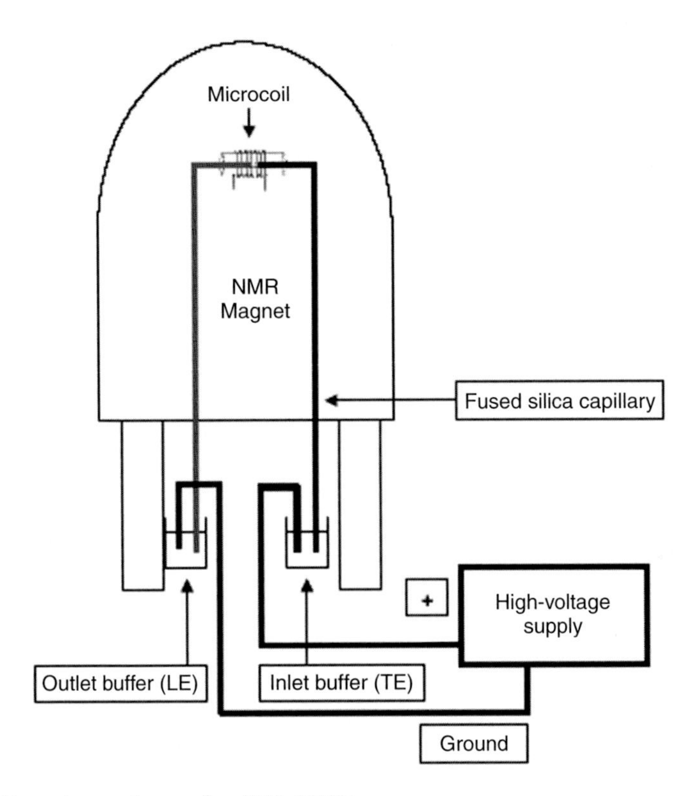

Figure 12 Experimental setup for cITP–NMR.

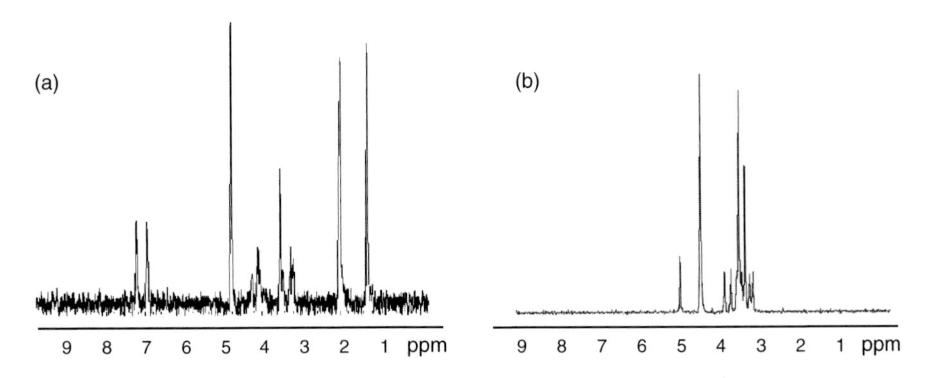

Figure 13 (a) On-flow cITP–NMR spectrum depicting the atenolol sample band at peak maximum during analysis of the trace impurity sample (200 μM atenolol and 200 mM sucrose in 50% TE–D_2O solution). No sucrose peaks can be observed. The S/N for the atenolol methyl peak at 1.3 ppm is 34. (b) Stopped-flow cITP–NMR spectrum of sucrose at peak maximum from the trace impurity sample. Reproduced with permission from Kautz et al.[51] Copyright 2002, American Chemical Society.

atenolol in a 1000-fold excess of sucrose (200 mM) was analyzed by cITP–NMR. As shown in Figure 13, cITP successfully separated the atenolol from the sucrose while also concentrating the atenolol ~200-fold from 200 μM to ~40 mM. The active volume of the receive coil was 30 nl, insufficient to detect the initial concentration of 200 μM, but spectra of the concentrated atenolol were acquired in a relatively short acquisition time. For this particular probe and sample, the experimentally obtained stacking efficiency was close to the theoretical value of 67%.

cITP–NMR can also be used for more fundamental studies of the concentration and separation process. As an example, Almeida and Larive[52] studied the interaction of related pharmaceuticals *S*-alprenolol, *S*-atenolol, *R*-propranolol, *R*-salbutamol, and *S*-terbutaline with β-cyclodextrin (β-CD) during cITP concentration. Most of the analytes studied were concentrated by cITP sample stacking by a factor of ~300. For analytes that formed a strong inclusion complex, β-CD coconcentrated during cITP sample stacking. However, once the focusing process was complete, a discrete diffusional boundary formed between the cITP-focused analyte band and the LE and TE, which restricted diffusion into and out of the analyte band. Figure 14 shows the variation in analyte concentration as a function of time in the cITP–NMR experiments. As observed in Figure 14(a), the concentration of β-CD in the LE and TE solutions adjacent to the focused propranolol band was significantly less than 10 mM. This suggests that as the propranolol concentration increased during the cITP stacking process, β-CD was pulled into the focused analyte band from both the LE and the TE. This is consistent with the idea that the β-CD co-concentrated with the propranolol at the start of the experiment before a discrete diffusion boundary formed around the analyte band. It is also interesting to

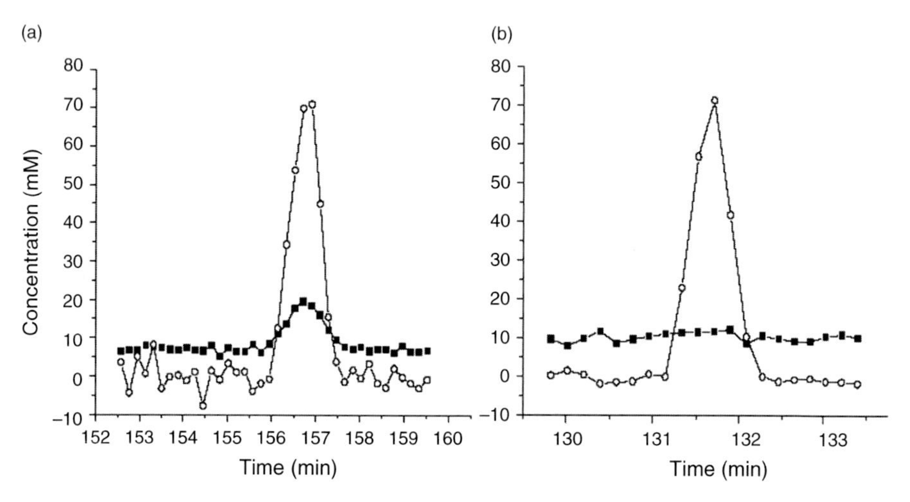

Figure 14 Variation in analyte concentration as a function of time in the cITP–NMR experiments; (a) propranolol (o), β-CD (□) and (b) salbutamol (o), β-CD (■). The initial concentrations of propranolol and salbutamol were 250 μM. The initial β-CD concentration in all solutions (sample, LE and TE) was 10 mM. Reproduced with permission from Almeida and Larive.[52] Copyright 2005 Wiley.

note that the time required for the propranolol band to focus and pass through the detection coil in this experiment, 156 min, is significantly longer than in an identical cITP–NMR experiment for propranolol alone, 88 min. This increase in experimental time reflects the greater size, and hence, smaller electrophoretic mobility of the propranolol–β-CD complex relative to that of propranolol. The results obtained in a similar cITP–NMR experiment with salbutamol are shown in Figure 14(b). Although salbutamol concentrated as well as propranolol, increasing from 250 μM to 72 mM because of cITP sample stacking, a significant increase in the β-CD concentration was not observed in the salbutamol cITP–NMR spectra, reflecting its lower binding affinity for β-CD.

6. Microliter Volume Applications of RF Microcoils

With the exception of cITP–NMR described above, most of the practical applications of small-coil NMR use sample volumes in the low microliter range. This allows acquisition of spectra within a few minutes, for 1D, or a few hours, for 2D or heteronuclear scans, using analyte concentrations in the low millimolar range. Examples of studies using the different forms of small coil described in Section 3 are given below.

6.1. Helmholtz/saddle coils

A Bruker 1 mm TXI probe with an active volume of 2.5 μl has been used for metabolic profiling of rodent biological fluids.[57] Because only ~2 μl of fluid is needed for this probe, this volume could be removed from the animals without the need for euthanasia, unlike the case for the larger volumes needed for larger probes. Figure 15 shows results, together with spectral assignments, obtained at 600 MHz.

6.2. Solenoidal coils

In[58] a solenoidal NMR microprobe with an observe volume of 800 nl was used to acquire high-resolution ^{1}H NMR spectra from the cleaved product of individual 160-μm-diameter Tentagel beads, which are often used in solid-phase synthesis for combinatorial chemistry. This work focused on one compound discovered in a series of serine protease inhibitors with leukotriene B_4 receptor binding affinity. The cleaved product was dissolved in dimethylsulfoxide and sandwiched between two perfluorinated organic liquid plugs. NMR spectra of the product cleaved from single beads were acquired at 600 MHz in ~1 h. Carr–Purcell–Meiboom–Gill (CPMG) experiments were performed to remove broad baseline components in the spectrum. Representative spectra are shown in Figure 16. Based on calibration experiments, the total amount of material cleaved from a single bead was estimated to be 540 ± 170 pmol, with approximately 180 pmol within the observe volume of the probe.

The second example illustrates the utility of small flow-through probes for high-throughput analysis of small amounts of sample. Eldridge et al.[59] used a Protasis/MRM

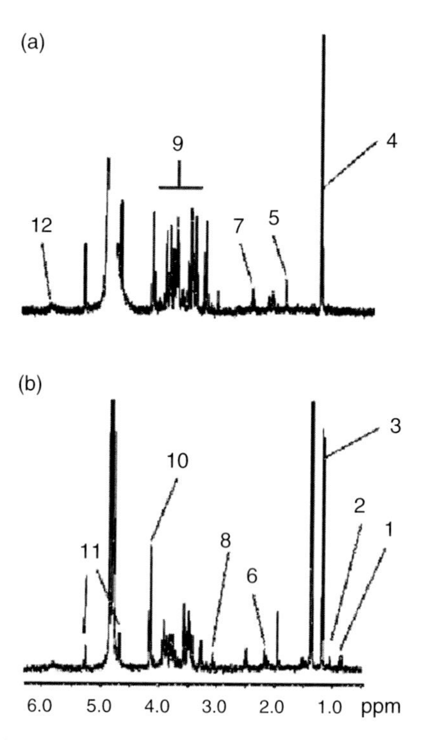

Figure 15 The 600 MHz ^{1}H NMR spectra of rat (a) and mouse (b) cerebrospinal fluid (CSF) using a 1 mm TXI probe with 128 scans. Both spectra were acquired using a water presaturation sequence based on the start of the NOESY pulse sequence. CSF of 2 ml were diluted with 3 ml of D_2O. Key: 1, leucine + valine; 2, valine; 3, propandiol (vehicle for anesthetic); 4, lactate; 5, acetate; 6, glutamine; 7, glutamine; 8, creatine; 9, glucose; 10, lactate; 11, H1 glucose; 12, amino groups. Reproduced from Griffin et al.[57] with permission of The Royal Society of Chemistry.

5 μl proton indirect carbon gradient flow probe with an active volume of 1.5 μl in the analysis of a large natural product library. The particular illustrative compound was an extract of *Taxus brevifolia*, the pacific yew tree containing paclitaxel (Taxol) and its derivatives. In all, 5–10 μg of material was sufficient for 2D COSY (correlation spectroscopy) experiments, with ~50 μg needed for experiments such as gradient heteronuclear multiple quantum coherence (HMQC) and heteronuclear multiple bond correlation (HMBC). Figure 17 shows a ^{1}H spectrum of paclitaxel which, in combination with the COSY data, allowed the identification of major peaks from a series of related compounds.

Small coils have been also been developed for protein experiments: this is likely to be particularly important in the case of proteins produced in eukaryotic cells rather than bacteria, where isotopic enrichment is considerably more difficult and expensive. The electrical circuit used in a 2.5 mm diameter, solenoid TXI probe[60] is shown in Figure 18. The L1–C1 trap presents a very low impedance at low frequency and high impedance at high frequency. The L2–C6–C7 tank circuit appears as a high impedance at the proton frequency, but as a very low impedance path at the ^{15}N frequency. The proton channel has the shortest electrical path to the

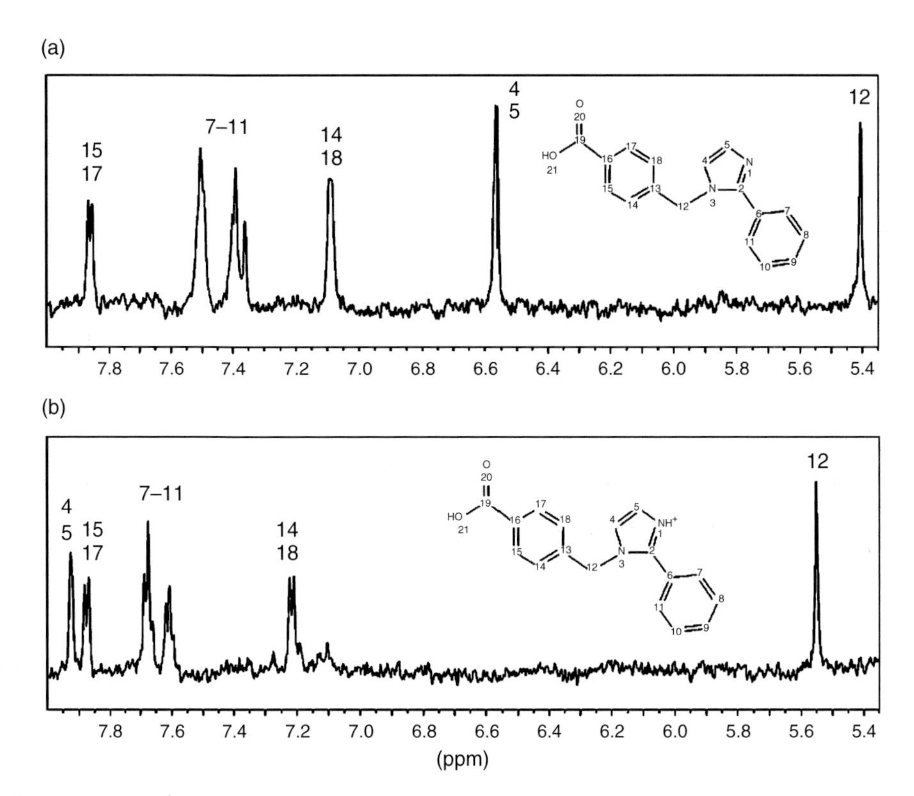

Figure 16 The ^{1}H CPMG spectra obtained from the cleaved product of a single bead from a combinatorial chemistry synthesis. (a) dissolved in neat DMSO-d_6 and (b) dissolved in acidified DMSO-d_6. Figure reproduced from Lacey et al.[58] Copyright (2001) with permission from Elsevier.

sample coil to minimize signal loss. LC trap circuits at both ^{15}N and ^{13}C frequencies are used between the ^{15}N and ^{13}C channels, and the lock channel is attached to the ^{13}C channel with a trap circuit at the ^{13}C frequency. The 90° pulse widths for all channels were compared with those from a commercial 5 mm TXI probe: the values were 4.0 vs 12 μs (^{1}H, 50 W amplifier, 6 dB attenuation), 3.8 vs 43 μs (^{15}N, 300 W amplifier, full power), and 1.8 vs 14 μs (^{13}C, 300 W amplifier, full power). The much shorter ^{13}C pulses are particularly important for experiments at high B_0 fields as they allow a much higher excitation bandwidth. Figure 18 shows specific 2D ^{1}H, ^{15}N planes extracted at different ^{13}C=O frequencies from a 3D HNCO spectrum collected on a 1 mM, double-labeled ^{15}N/^{13}C IA-3 sample, an intrinsically unstructured 68 amino acid protein inhibitor of yeast proteinase A, using the TXI solenoidal probe.

A second protein study[61] used a commercial microcoil probe from Protasis/MRM. The volume of the TXI HCN z-gradient microcoil NMR probe is 5 μl with an active volume of 1.5 μl. Most of the experiments were performed using proteins from the *Thermotoga maritima* proteome, in particular, the conserved hypothetical protein, TM0979. HNCA/HNCOCA spectra were acquired to test the ability to perform the sequential backbone assignment of TM0979. In the

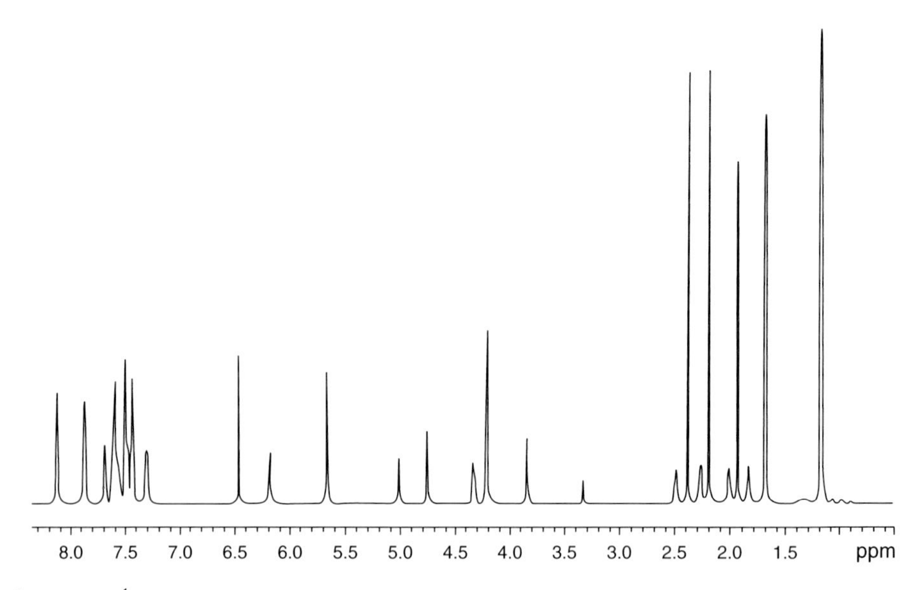

Figure 17 ^{1}H NMR spectrum of 50 μg of paclitaxel in 3 μl CD_3OD acquired using a 5 μl microcoil flow probe at 600 MHz. Reproduced with permission from Eldridge et al.[59] Copyright 2002 American Chemical Society.

HNCA spectrum, all inter- and intraresidue peaks were detectable, nearly all C_α and C_β peaks were observed in the CBCACONH spectrum, and all CO correlation peaks could be identified from the HNCO spectrum, allowing complete backbone assignment of TM0979. As outlined previously, the very short ^{13}C pulse widths are particularly important in a number of pulse sequences: in this particular study it was possible to record a single HCCH–TOCSY (total correlation spectroscopy) spectrum[62,63] across the full aliphatic and aromatic side-chain carbon range. This experiment allows complete side-chain assignment of all amino acids in a protein within a single spectrum. The correlation between the aliphatic and aromatic carbons is hindered due to the large carbon chemical shift ranges (aliphatic carbons, 0–75 ppm; aromatic carbons, 115–140 ppm) that correspond to a bandwidth of roughly 20 kHz at 14.1 T. Standard 5 mm NMR probes or cryoprobes are not rated for the high-power levels required to produce Hartman–Hahn mixing over this broad chemical shift range. An HCCH–TOCSY spectrum with a z-filter FLOPSY-16 mixing sequence using a 20 kHz spin lock field was used to acquire the spectra shown in Figure 19. This type of spectrum can be used for the assignment of connectivities between the aliphatic C_α and C_β atoms and the rest of the aromatic side chain, and also within the aromatic ring itself, in one measurement thus accelerating greatly aromatic side-chain assignment.

6.3. Superconducting microcoils

Using the unique 1 mm HTS probe[48] described previously, Dossey et al.[64,65] were able to perform detailed molecular studies of a mixture of natural products from

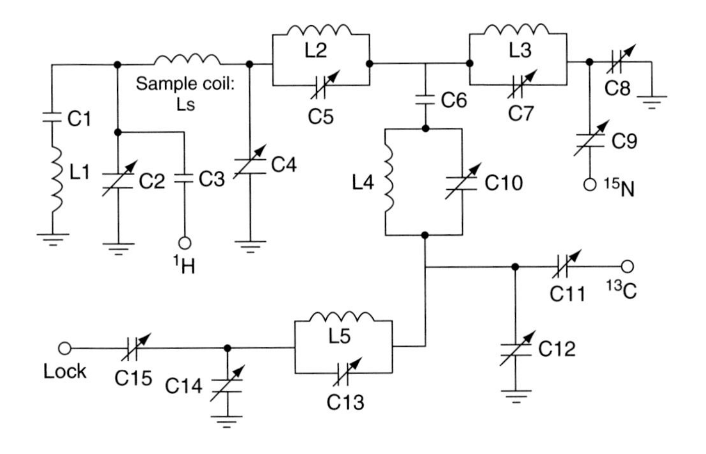

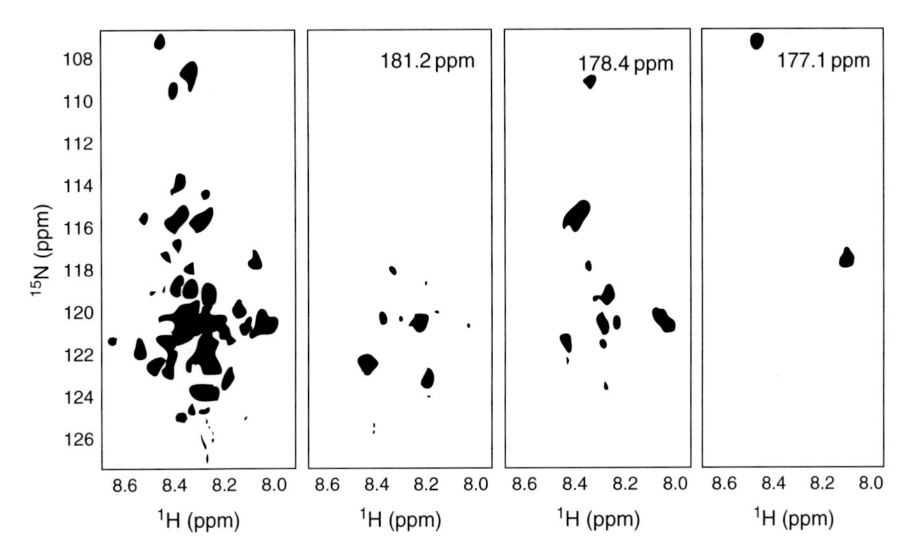

Figure 18 (top) Circuit diagram for the TXI solenoidal probe used to acquire the 3D HNCO data (bottom) of double-labeled 1 mM $^{15}N/^{13}C$ IA-3 using the triple-resonance solenoidal probe. Experimental parameters: sw 6614 Hz, sw1 (^{15}N) 1302 Hz, sw2 (^{13}C) 3001 Hz, 32 signal averages, 64 real data points in the $^{13}C{=}O$ dimension, 60 real data points in the 15N dimension, 4096 complex acquisition data points, total data acquisition time 59 h. Solvent suppression used presaturation. The 2D projection of all the $^{13}C{=}O$ frequencies is shown in the leftmost panel, and the plots of selected single $^{13}C{=}O$ slices are shown in the other three panels. Figure reproduced from Li et al.[60] Copyright (2003) with permission from Elsevier.

single milkings from an individual insect for the first time. They examined the chemical composition of defensive secretions from walking stick insects, *Anisomorpha buprestoides* (order Phasmatodea). The authors found that the species secretes similar quantities of glucose and mixtures of monoterpene dialdehydes which are

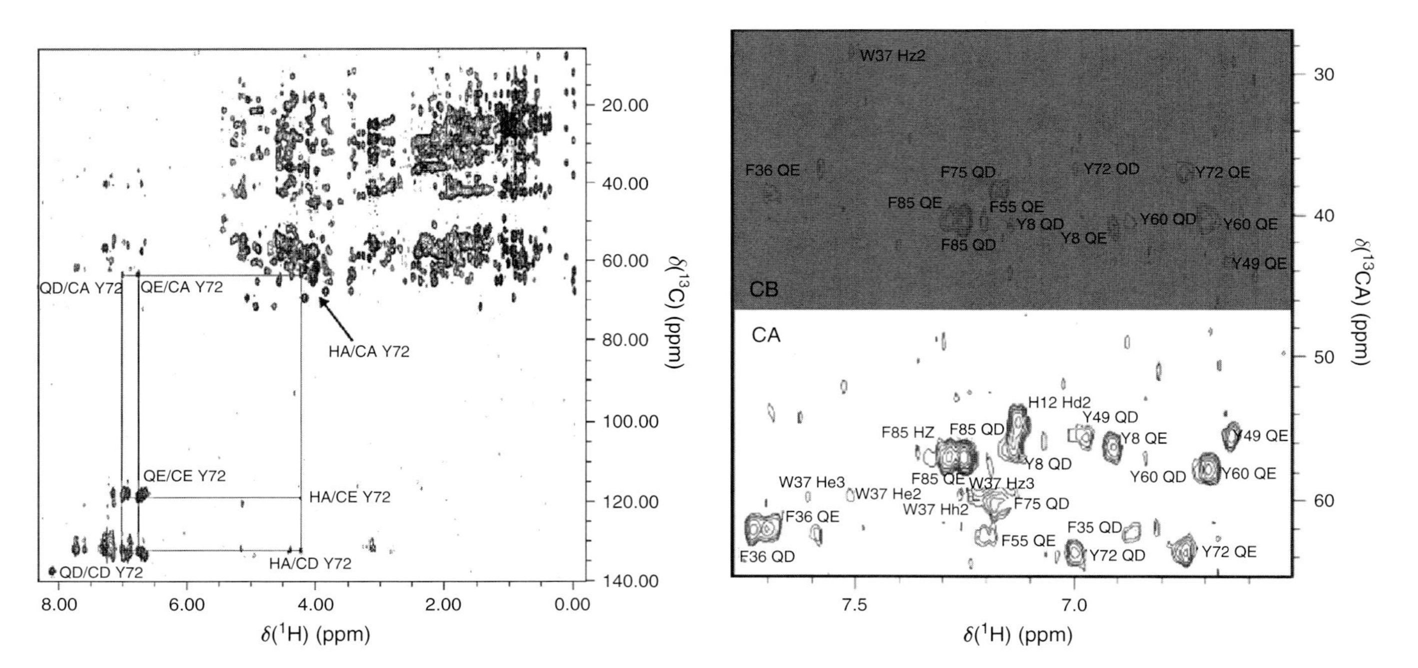

Figure 19 (left) 2D [^{1}H, ^{13}C] aliphatic–aromatic HCCH–TOCSY FLOPSY-16 spectrum recorded with a mixing time of $\tau_m = 11.77$ ms. The assignment of tyrosine 72 based on the C_α chemical shift is indicated in the spectrum (10 mM ^{13}C/^{15}N TM0979, 313 K, measurement time 30 h). (right) Enlargement of the C_α/C_β carbon chemical shift–proton aromatic chemical shift region. All cross-peaks are annotated, and the complete aromatic chemical shift assignment can be achieved. The C_β cross-peak area is highlighted with a gray box. Reproduced with permission from Peti et al.[61] Copyright 2004 American Chemical Society.

stereoisomers of dolichodial. Different animals were found to produce different mixtures of stereoisomers. They also found that a second walking stick insect, *Peruphasma schultei*, also secretes glucose and a single, unique stereoisomer termed "peruphasmal". Figures 20 and 21 shows results from single *A. buprestoides* milkings.

An interesting approach[66] to expand the use of this high-sensitivity probe was to combine it with covariance NMR, a technique used to separate out NMR spectra from a mixture. Figures 22 and 23 show results from a single milking from a defensive spray from *A. buprestoides*, with excellent separation of the individual components of the mixture.

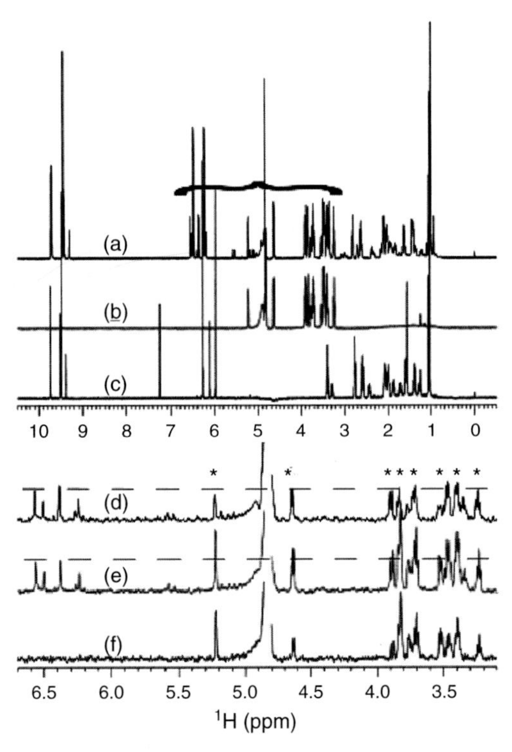

Figure 20 1D ^{1}H NMR spectra of single *A. buprestoides* milkings. All spectra were collected at 600 MHz using a 1 mm HTS probe, and sample temperatures were 27°C. Each spectrum was collected with eight scans. (a) About 1 μl was collected from a single insect on a glass pipet tip and added to 10 μl of D_2O containing 0.11 mM TSP. The sample was loaded into a 1 mm capillary NMR tube without purification, and the spectrum was obtained within 10 min of the sample collection. Sample (a) was extracted with 15 μl of chloroform-d_3, and the aqueous (b) and organic (c) fractions were collected and recorded. (d) Expansion of a second sample that includes the aqueous component and the vinyl organic region of the spectrum. (e) 0.9 μl of pure 50 mM D-glucose was added to sample (d). The region of the expansions in (d–f) is indicated by a bracket in spectrum (a). The horizontal dashed lines in spectra d and e indicate the constant vinyl peak intensities, and the asterisks indicate peaks that increased in intensity. (f) NMR spectrum of pure glucose. Reproduced with permission from Dossey et al.[65] Copyright 2006 American Chemical Society.

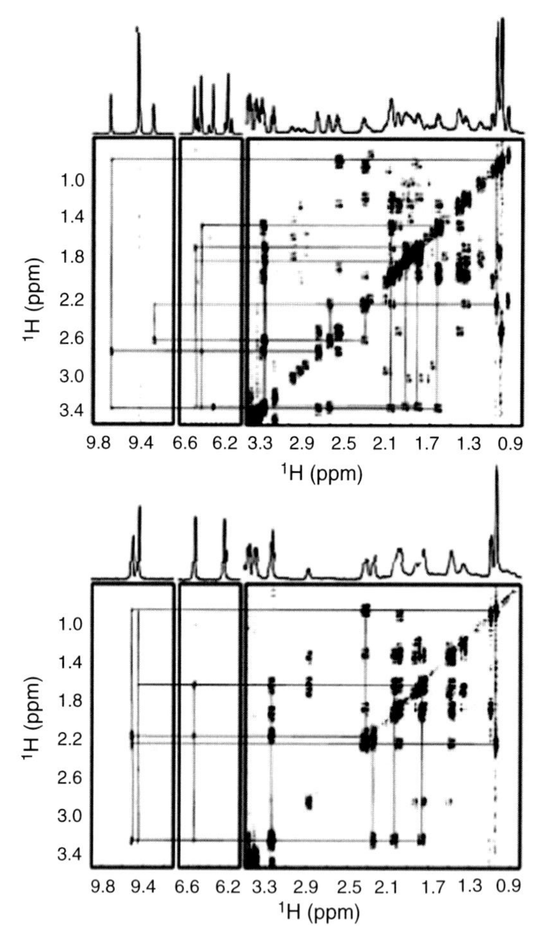

Figure 21 2D expansions of COSY (right panels) and ROESY (left and center panels) from a single milking of *A. buprestoides* (top) and a pooled sample from three *P. schultei* milkings (bottom). 1D ^{1}H spectra from the same samples are shown along the top. All data were collected at 600 MHz using the 1 mm HTS probe. The COSY experiments were collected in ~2.5 h with 8 scans and 512 complex indirect data points. The ROESY experiments were collected in ~9 h with 32 scans, 512 complex points, and a 400 ms mixing time. The defensive secretion was found to contain glucose, water, and a new monoterpene (4*S*)-(3-oxoprop-1-en-2-yl)cyclohex-1-enecarbaldehyde (parectadial). Reproduced with permission from Dossey et al.[64] Copyright 2006 American Chemical Society.

7. Hyphenation of Microseparation Techniques with Microliter NMR Detection

The small detection volumes associated with NMR microcoils make it a natural choice for coupling with many chemical microseparation techniques such as capillary liquid chromatography (cLC), as well as capillary electrophoresis (CE)

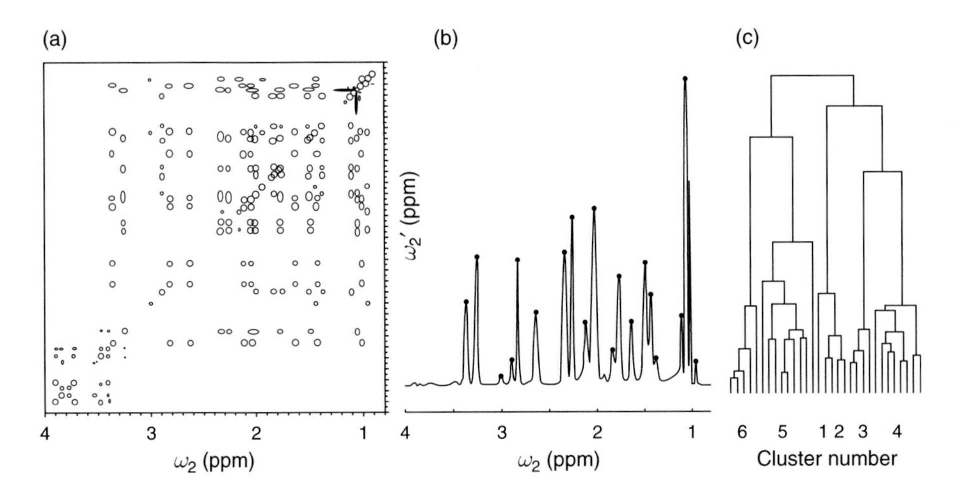

Figure 22 (a) Aliphatic section of covariance proton TOCSY spectrum of defensive secretion of a single walking stick insect. (b) Importance index profile. (c) Dendrogram representation of cluster analysis according to similarity of the traces picked in (b). Reproduced with permission from Zhang et al.[66] Copyright 2007 American Chemical Society.

and capillary electrochromatography (CEC). In general, microseparation techniques enable faster analysis, higher concentration elution peaks, and less chromatographic dilution than their larger scale counterparts. The use of small coils is ideally suited to small total sample amounts, present as relatively high concentrations in small volumes. The majority of hyphenated NMR-detected studies carried out thus far have used cLC as the separation technique. The mode of hyphenation has been described in detail,[67] and consists of a long transfer capillary from the cLC separation column (which is usually magnetic), up through the magnet bore to the NMR probe. Typical transfer capillary inner diameters are narrow, between 50 and 100 μm, to minimize peak dispersion, and broadening after the separation: flow rates are usually ~5 μl/min. To increase the filling factor of the probes, and therefore the S/N of the NMR experiment, a "bubble cell" or flowcell is often used.[31]

cLC–NMR can be performed in either continuous-flow or stopped-flow mode. In continuous-flow NMR, the sample remains inside the detection coil only for a limited time period, termed "the residence time" (τ), which is determined by the volume of the detection cell and the flow rate. The residence time alters the effective relaxation times according to the following relationships:

$$\frac{1}{T_{1,\text{effective}}} = \frac{1}{T_1} + \frac{1}{\tau}, \quad \frac{1}{T_{2,\text{effective}}} = \frac{1}{T_2} + \frac{1}{\tau} \tag{23}$$

As the flow rate increases (τ decreases), the effective T_1 decreases, allowing shorter recycle delays to be used. However, the linewidth of the observed NMR resonance, limited by $1/T_{2,\text{effective}}$, is broadened significantly at high flow rates. The NMR line not only broadens as the flow rate increases, but its intrinsic shape also

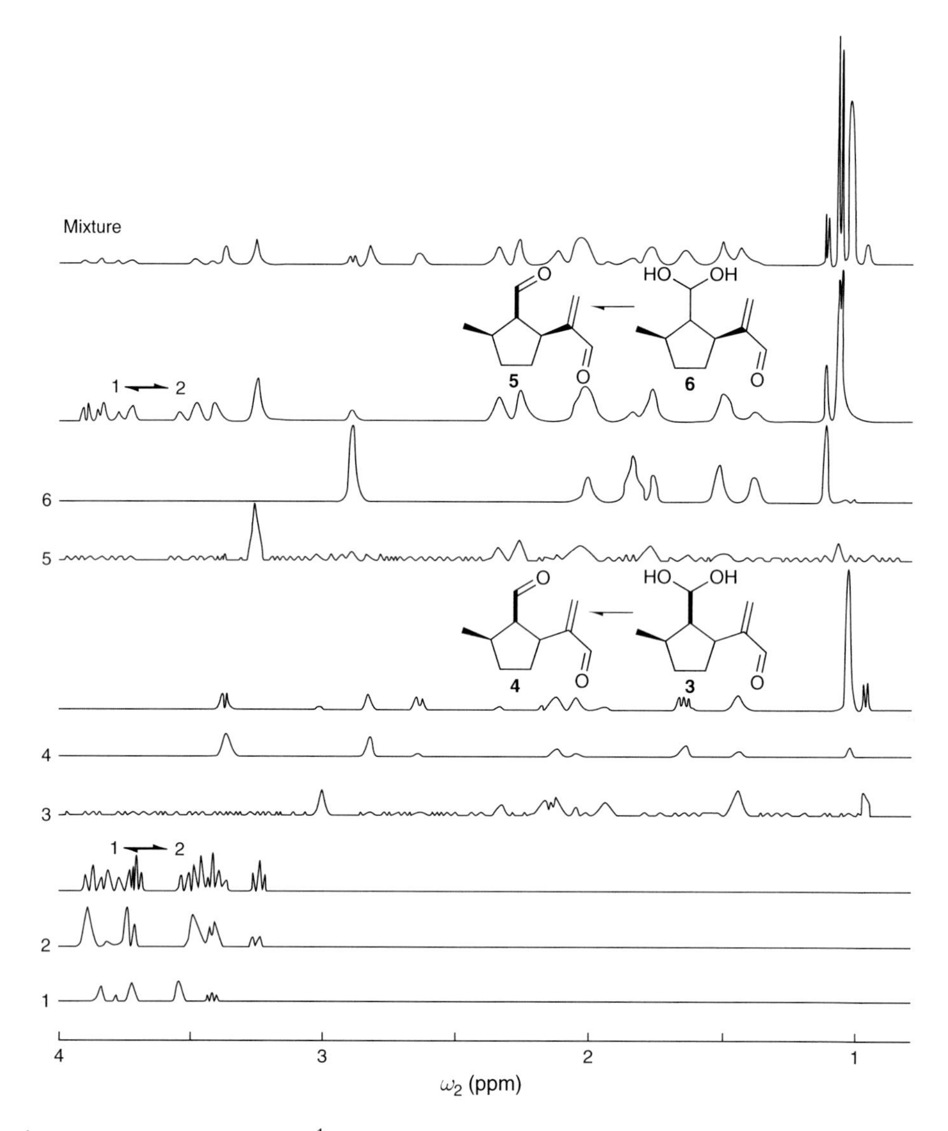

Figure 23 (a) Top spectrum: ^{1}H NMR spectrum of the mixture from a milking of a single insect. Spectra 1–6 are the covariance TOCSY traces representing the clusters of Figure 22(c) with the numbers corresponding to the cluster numbers at the bottom of Figure 22(c). The three spectra in between are reference 1D spectra of purified components. Each reference spectrum contains two species, α-glucose (trace 1) and β-glucose (trace 2); dialdehyde and diol forms of the anisomorphal (traces 4 and 3, respectively), and the peruphasmal (traces 5 and 6, respectively) monoterpenes. The top reference trace also contains some glucose signals (on the left side). Chemical structures of the anisomorphal and peruphasmal and their corresponding geminal diols are shown as insets. Reproduced with permission from Zhang et al.[66] Copyright 2007 American Chemical Society.

changes. Whereas for stopped flow the line shape is ideally a pure Lorentzian, as the flow rate increases the line shape is best described by a Voigt function, defined as the convolution of Gaussian and Lorentzian functions. Quantitative NMR measurements under flow conditions must take into account these modifications in line shape. One other limitation of continuous-flow NMR, illustrated in detail elsewhere,[67] is a distortion in spectral lineshape associated with the changes in magnetic susceptibility when solvent-gradient separations are used.

Given the problems outlined above, stopped-flow detection is often used. Usually the peaks are detected before entering the probe via a conventional UV detector, and flow is stopped after a suitable delay to allow a particular peak to enter the active volume of the NMR probe. The major disadvantage of this mode of operation is the loss in chromatographic resolution due to diffusion, and also the loss in NMR S/N due to the same diffusive process. Figures 24 and 25 show two examples, one in which a 1-mm-long plug is trapped within a 1-mm-long coil, and the other with a 3 mm plug within a 3-mm-long coil.[68] The results show that considerable signal loss occurs within a time frame of a few minutes. In this situation, the use of a tapered flowcell helps considerably, as shown by the simulations presented in Figure 26.[68]

An example of the use of cLC–NMR is the characterization of bixin by Rehbein and colleagues.[69] Bixin is the major carotenoid of the seeds of the plant annatto or urucum. It is widely used for coloring foods, especially dairy products. Carotenoids are extremely sensitive to air and UV exposure, and so for unambiguous assignment of natural carotenoid stereoisomers, closed-loop cLC–NMR techniques are much more desirable than offline isolation and identification, as the use of offline techniques often results in carotenoid isomerization and degradation. In this study, the capillary separation column was packed on site, using the slurry packing procedure with Bischoff Pronto-Sil C_{30} (3 μm, 200 Å, 15 cm × 250 μm) and equipped with end fittings consisting of 2SR1 filter screens in zero-dead volume unions ZU1C (ViciAG, Schenkon, Switzerland).

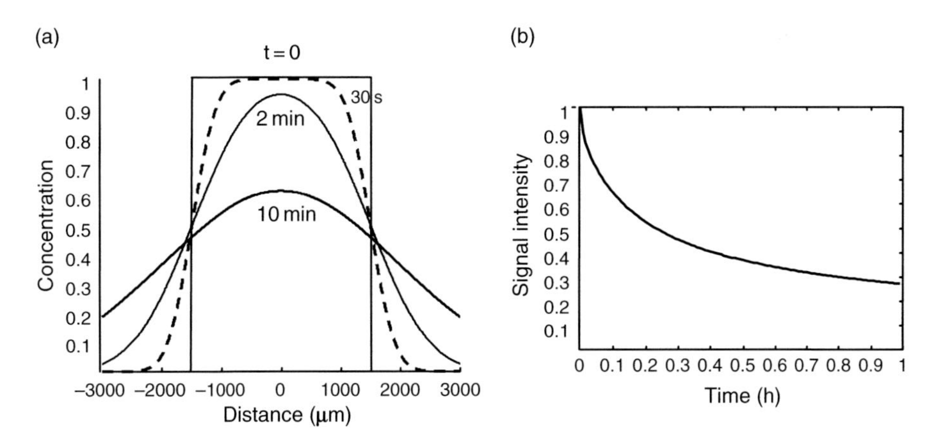

Figure 24 (a) Relative concentration profiles of a 1 mm long plug of sample as a function of time. (b) Relative signal intensities from the material detected by a 1 mm long solenoidal coil (assuming a uniform B_1 distribution) as a function of time. The diffusion coefficient of water at room temperature was used for the calculations. Reproduced with permission from Webb.[68] Copyright 2005 Wiley.

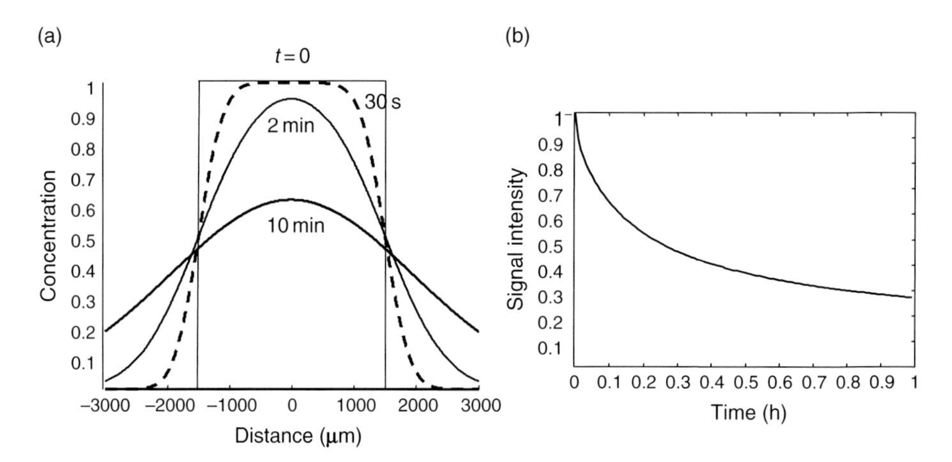

Figure 25 As in Figure 24 for a 3 mm plug of sample and a 3-mm-long solenoidal coil. Reproduced with permission from Webb.[68] Copyright 2005 Wiley.

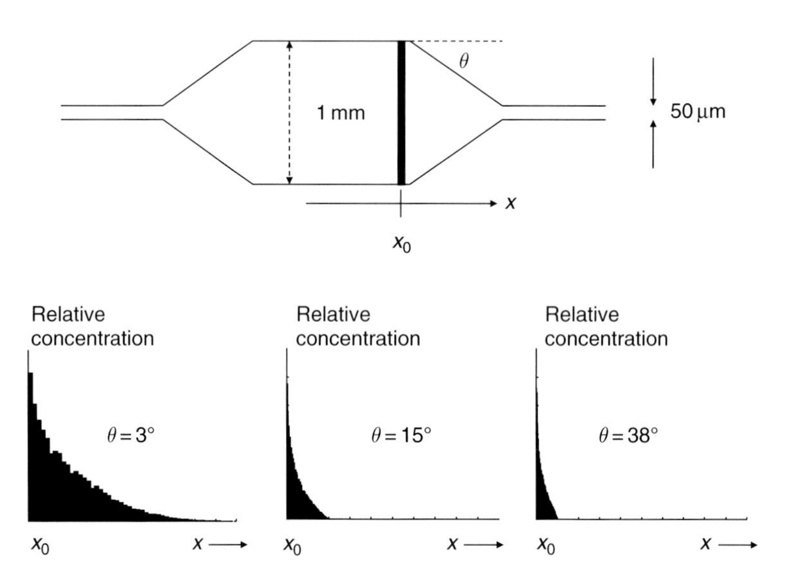

Figure 26 Concentration profiles for different taper angles of a flowcell. The figures show the concentration of material initally at the edge of the central part of the flowcell as it diffuses into the tapered region. Reproduced with permission from Webb.[68] Copyright 2005 Wiley.

The chromatographic separation utilized an isocratic elution method with mixtures of acetone-d_6 and D_2O (acetone-d_6/D_2O = 92:8, v/v), injecting 500 nl of the sample into the system. UV detection was carried out on the capillary at 450 nm. All transfer capillaries consisted of fused silica with a dimension of (50 μm i.d./360 μm o.d.). Unlike most of the known natural carotenoids, where the major isomer is the all-E form, the major isomer of bixin is the 9′-Z bixin. Using the cLC–NMR setup, Figure 27 shows stopped-flow 1D spectra from 9′-Z bixin and all-E bixin.

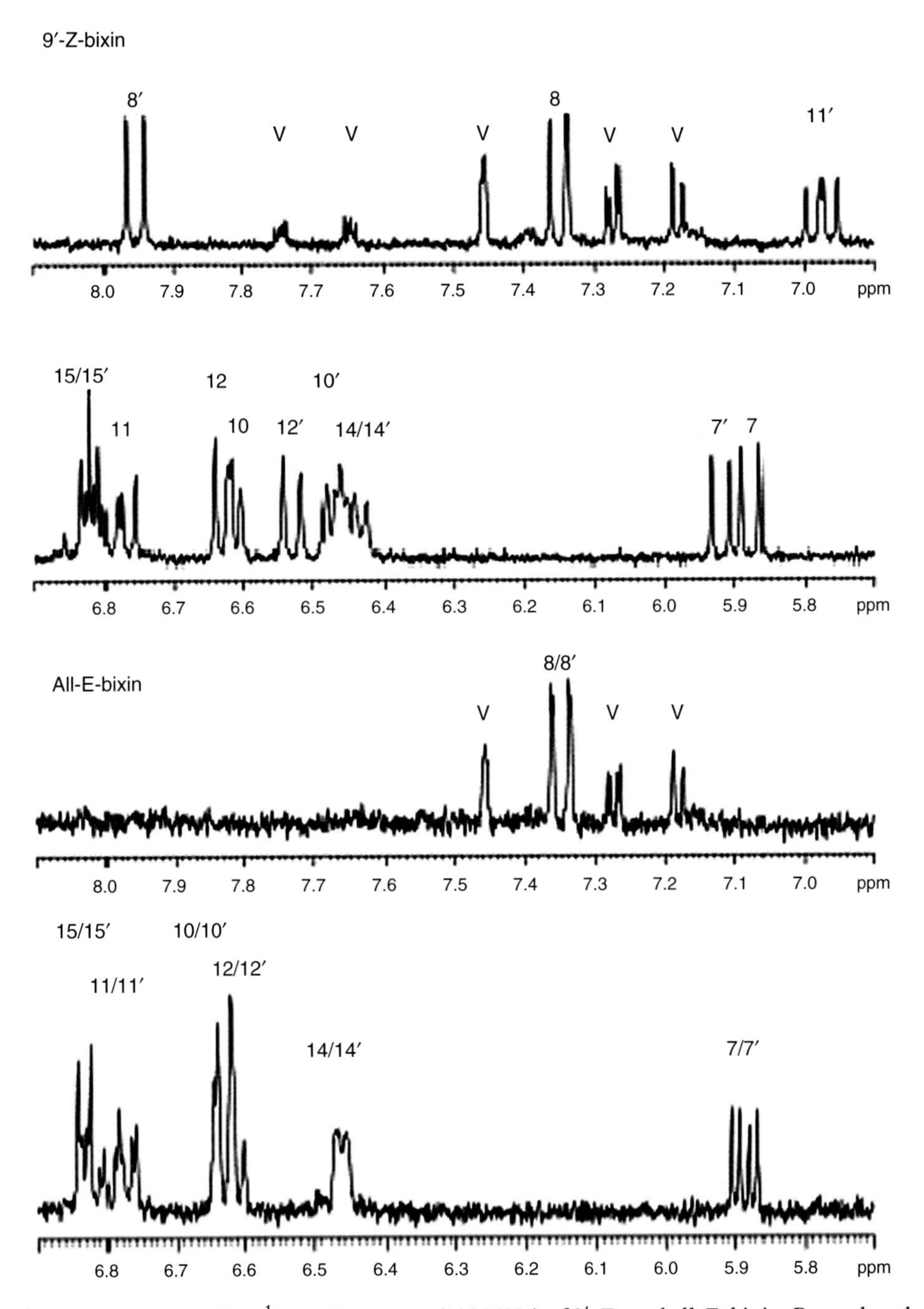

Figure 27 Stopped-flow ^{1}H NMR spectra (600 MHz) of 9′-Z- and all-E-bixin. Reproduced with permission from Rehbein et al.[69] Copyright 2007 Wiley.

The issue of column overloading is a major challenge in coupling microseparations with NMR detection. It is well-known that polymer-based chromatographic materials have a higher loadability than silica-based sorbents. Unfortunately, these polymer materials cannot be used under high pressure, which is necessary to obtain high flow rates, and hence long times are needed to perform a separation. However, by immobilizing a polymer on a mechanically stable porous silica core, this problem can be circumvented and higher flows become feasible on these materials. Grynbaum et al.[70] have developed a highly shape-selective chromatographic sorbent by covalently immobilizing a poly(ethylene-co-acrylic) acid copolymer ($(—CH_2CH_2—)x[CH_2CH(CO_2H)—]y$ ($x = 119$, $y = 2.4$) with a mass fraction of acrylic acid of 5% as stationary phase on silica via a spacer molecule (3-glycidoxypropyl-trimethoxysilane). Figure 28 shows that chromatographic resolution is not degraded for a high concentration, compared to a much lower concentration, of tocopherol homologues.

Using this setup, Grynbaum et al.[70] were able to separate tocopherol homologues in a far shorter retention time than in previous publications[71] using standard separation columns. Figure 29 shows spectra extracted from the 2D chromatograph, corresponding to 16 coadded scans.

Another approach to overcome the loading limitations of small-scale columns was shown by Lambert et al.,[72] in which they combined the advantages of the high loadability of normal-bore HPLC columns with capillary NMR detection by an offline hyphenated HPLC-solid-phase extraction (SPE)-capNMR sequence. SPE allows analytes to be concentrated into much smaller solvent volumes, and also enables a switch from protonated solvents used in conventional HPLC separation to deuterated solvents for NMR detection. This overall hyphenation was used for

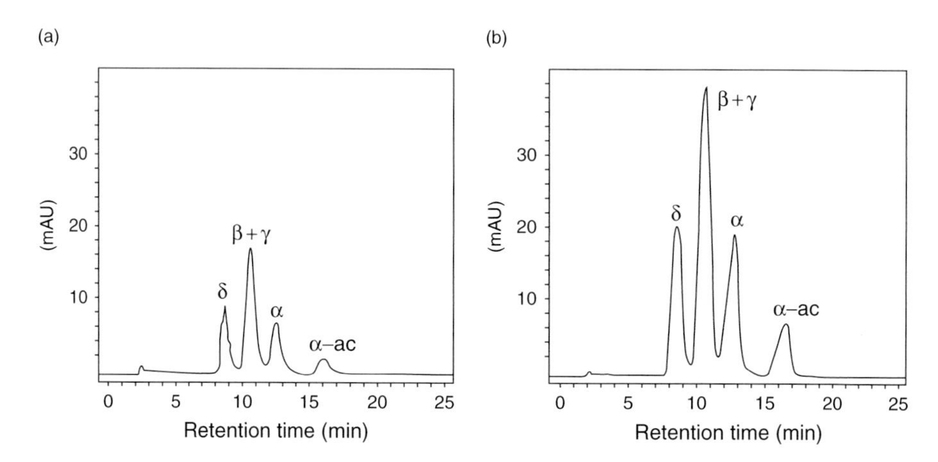

Figure 28 Chromatograms of the capillary HPLC separation of tocopherol homologues eluted from the polymer-based chromatographic sorbent (15 cm × 250 μm), UV detection at 285 nm, mobile phase: methanol/water 90/10 (v/v), flow rate: 5 μl/min, (a) $[c_1] = 1.6$ mg/ml 200 nl injected, (b) $[c_2] = 5.66$ mg/ml 500 nl injected. δ, δ-tocopherol; γ, γ-tocopherol; β, β-tocopherol; α, α-tocopherol; α-ac, α-tocopherol acetate. Figure reproduced from Grynbaum et al.[70] Copyright (2007) with permission from Elsevier.

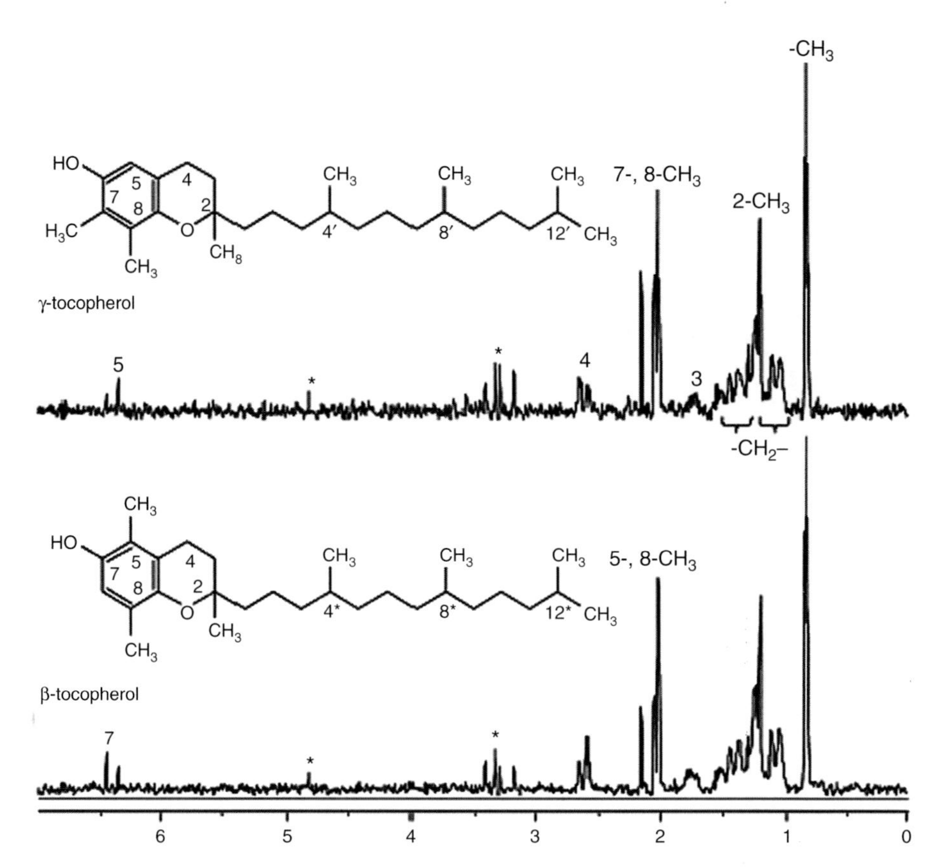

Figure 29 Extracted NMR spectra of the tocopherol isomers at the corresponding signal maxima. Figure reproduced from Grynbaum et al.[70] Copyright (2007) with permission from Elsevier.

rapid identification of sesquiterpene lactones and esterified phenylpropanoids in a toluene fraction of an ethanolic extract of *Thapsia garganica* fruits. A hand clamp was connected to the 1.5 μl active volume NMR probe through a 50 μm i.d. fused silica capillary and used to elute the dried poly(divinylbenzene) resin SPE cartridges. Figure 30 shows typical proton spectra acquired from one of the separated fractions.

Grynbaum et al.[73] have shown the first hyphenation of gas chromatography (GC) to microcoil NMR at 400 MHz., using a solenoidal coil with an active volume of 2 μl. The experiments were performed by connecting a GC column to a transfer capillary, which was in turn connected to an inlet capillary around which the coil was constructed. A splitless injection of 10 μl of diethyl ether, dichloromethane, and tetrahydrofuran (3.3 μl of each) gave the pseudo-2D plot shown in Figure 31. The flow rate in the GC column was 0.72 ml/min. The residence time τ in the detection cell was determined from the measured signal linewidths from the diethyl ether spectra recorded in stopped-flow and continuous-flow mode to be 0.625 s. The flow rate was a compromise between high

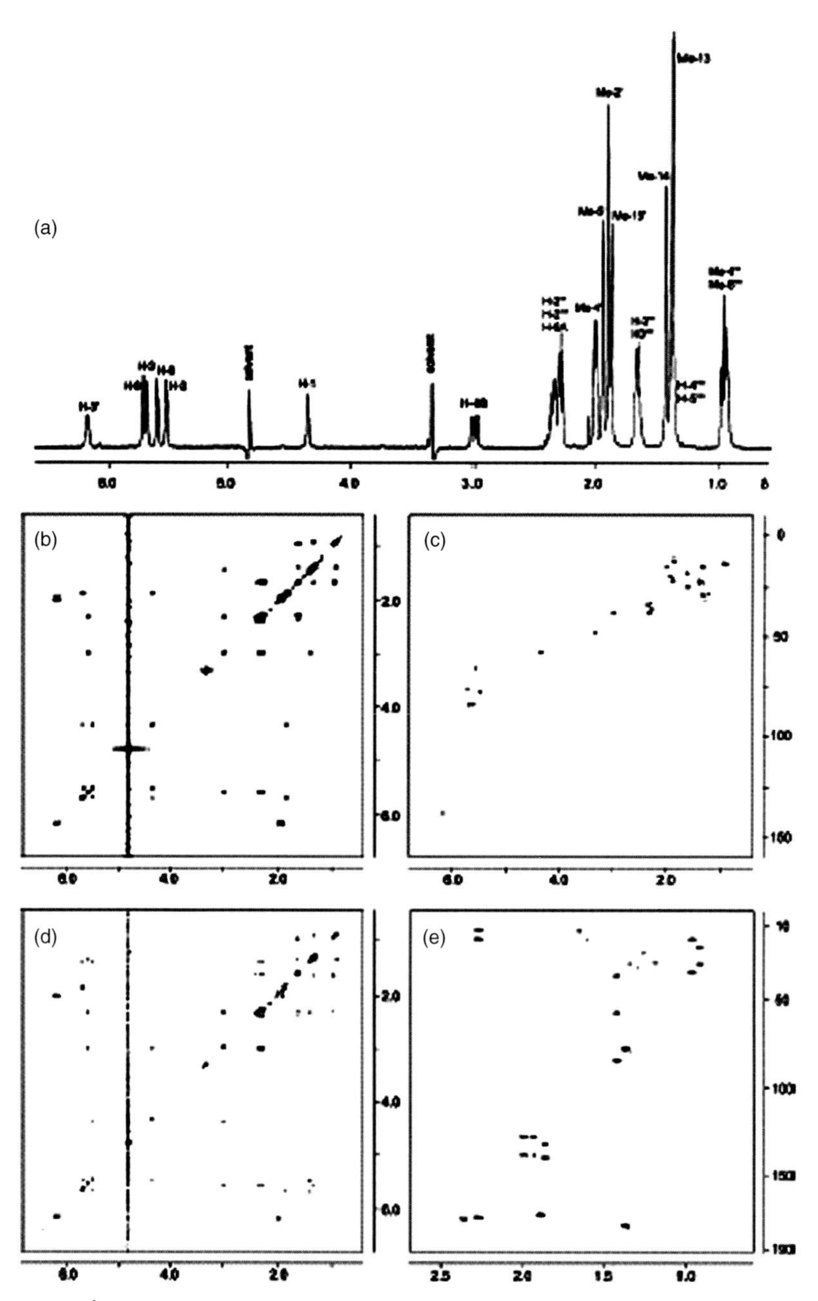

Figure 30 (a) ^{1}H NMR spectrum (256 transients) acquired from a single eluting peak containing thapsigargicin using the capNMR probe after SPE sample concentration. Shown is a difference spectrum obtained by subtraction of a spectrum obtained by eluting an empty cartridge from the actual spectrum to eliminate solvent resonances (both spectra were acquired without solvent peak suppression). (b) COSY spectrum acquired with peak 9 (containing 2); 8 transients, 512 increments, total acquisition time 1.5 h. (c) HSQC spectrum acquired with peak 11 containing thapsigargin (1); 4 transients, 512 increments, total acquisition time 43 min. (d) NOESY spectrum acquired with peak 11 (1); 16 transients, 1024 increments, total acquisition. Reproduced with permission from Lambert et al.[72] Copyright 2007 American Chemical Society.

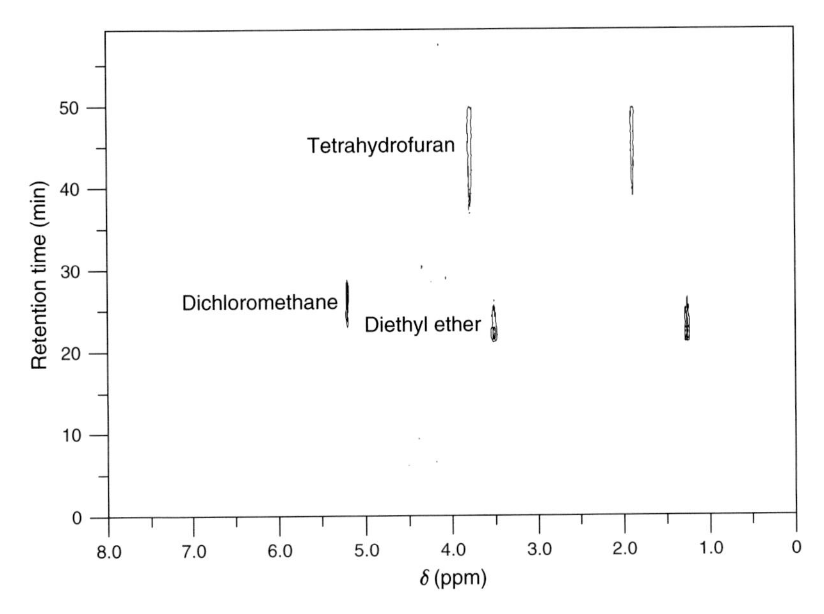

Figure 31 Contour plot of the GC–NMR separation of diethyl ether, dichloromethane, and tetrahydrofuran. Data acquisition parameters: 32 transients with 4k time-domain points and a spectral width of 5618 Hz were accumulated with a relaxation delay of 500 ms. During the separation, 128 rows with an acquisition time of 28 s per row were recorded. Reproduced with permission from Grynbaum et al.[73] Copyright 2007 American Chemical Society.

chromatographic resolution and sufficient S/N for NMR detection. The relatively slow flow rate is below the van-Deemter optimum, but higher flow rates would lead to too short residence times in the detection cell and very broad NMR linewidths. Averaging 32 transients (over 28 s) leads to partial loss of chromatographic resolution. The total amounts of analytes were 32 μmol of diethyl ether, 52 μmol of dichloromethane, and 40 μmol of tetrahydrofuran. The contour plot in Figure 31 also showed that the analytes do not condense in the flow cell (in which case the NMR spectra would remain at constant amplitude over time).

Figure 32 shows the extracted spectra of diethyl ether, dichloromethane, and tetrahydrofuran. Even in the gaseous continuous-flow mode, the multiplets of diethyl ether are resolved and the coupling constants can be derived from the extracted spectrum.

8. Multiple Coil Probeheads

The small size of a microcoil, compared to the extent of homogeneous B_0 field within the bore of a standard NMR magnet, enables multiple coils to be incorporated into a single probehead. The technical challenges in producing a

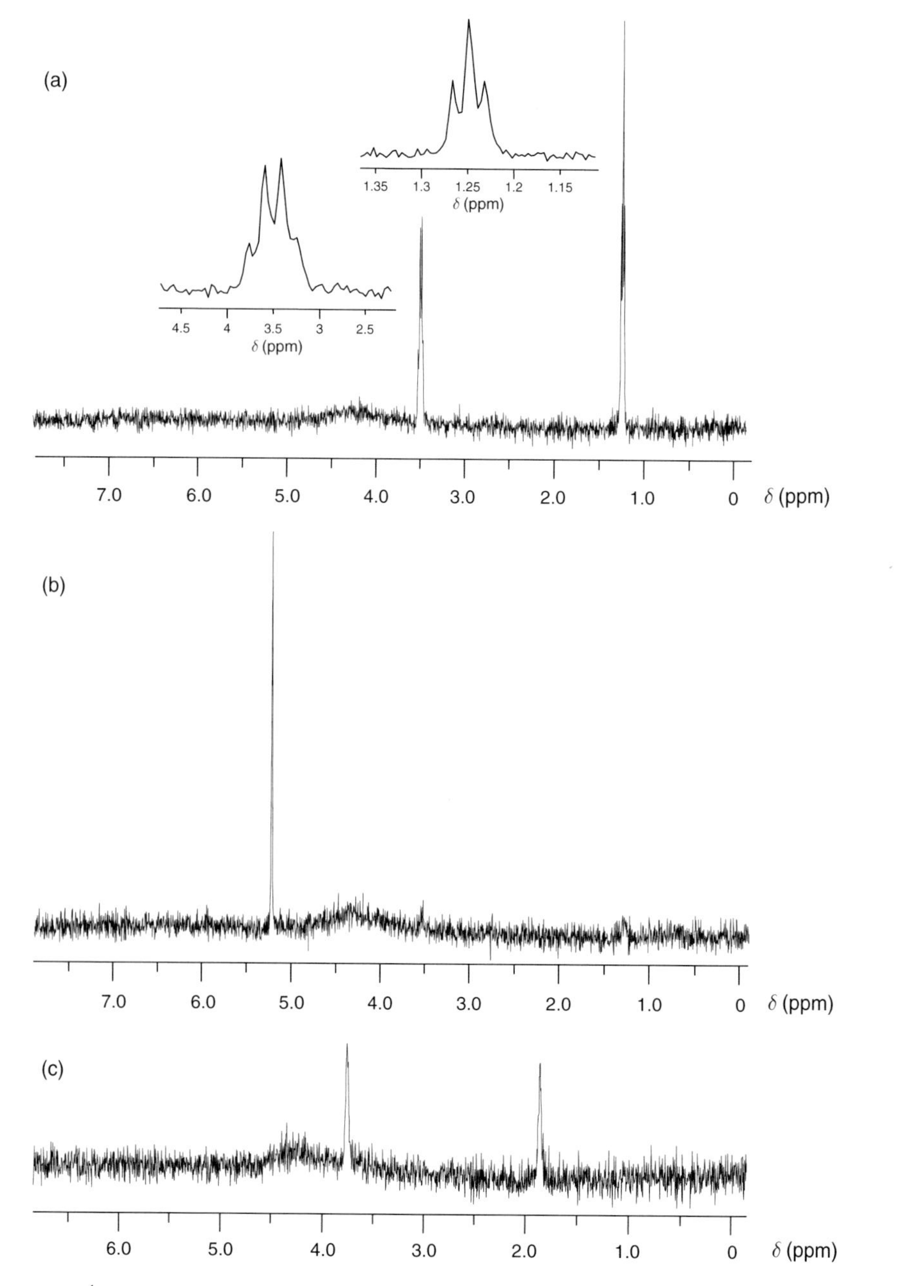

Figure 32 ^{1}H NMR spectra extracted from Figure 31 of (a) diethyl ether, (b) dichloromethane, and (c) tetrahydrofuran. Reproduced with permission from Grynbaum et al.[73] Copyright 2007 American Chemical Society.

practical multicoil probe include maintaining high local B_0 homogeneity for each sample despite the presence of the other coils, achieving maximum sensitivity for each coil, and separating the signals from each sample through either hardware or software. Applications of multicoil technology include increasing the throughput of NMR by acquiring data from more than one sample simultaneously,[74–84] as well as enabling other specialized types of experiment such as removing line-broadening effects in NMR-detected electrophoretic separations,[83] monitoring rapid chemical reactions[84] and performing solvent suppression.[85,86]

Two basic approaches have been used in designing probeheads with multiple coils. In the first, the coils are connected in parallel,[77,87,88] effectively forming a single resonant circuit, and the signals from each sample are separated through the use of spatially selective pulse sequences. The alternative approach is to construct a number of separate RF circuits, one for each coil,[74,75,78,81,83,84] and then either to use multiple receiver channels or time-domain multiplex the signals into a single receiver, to separate the signals from each coil.

The most promising application of increasing NMR throughput is in the area of screening large compound libraries with target protein molecules[89,90]. In terms of increasing NMR throughput, the maximum number of coils that have been incorporated into a single probehead is currently eight,[81] with each coil having an observation volume of ~35 nl. In this probehead, Teflon flow tubes were attached to both the ends of the capillary for sample loading. The coils were mounted one above the other with a vertical spacing of approximately 3 mm and alternate coils were rotated 90° with respect to each other to minimize the coupling. Using a vertical separation smaller than 3 mm resulted in substantial distortions of the local static magnetic field. The coils were surrounded by an 18 mm i.d. container filled with FC-43. The hardware additions to the standard spectrometer consist of a four-way power splitter that is placed between the transmitter and the coils, and four RF switches, shown in Figure 33. The position of the switch is controlled by one of the five TTL outputs from the spectrometer. The four outputs of the switches are connected to the four receiver channels, each of which consists of an independent preamplifier, transmit/receive switch, and analog-to-digital converter. The timing diagram describing the pulse sequence and data acquisition scheme is also shown in Figure 33. Because the spectrometer has only four receiver channels, the eight coils were divided into two groups, each containing four coils. The RF pulse sequence was transmitted to, and data acquired from, one coil group at a time. With a "compromise" value of the shim currents the linewidths for all eight coils were between 3 and 6 Hz. The 2D COSY, TOCSY, and gradient COSY experiments were run on eight different samples at 600 MHz. Figure 33 shows the results from the COSY experiments.

A key component in increasing NMR throughput is integration with multiple separation columns or fluidic devices, which can load multiple samples in a reproducible fashion. In a recent paper, Macnaughtan et al.[79] showed that a four-coil "Multiplex probe" consisting of four solenoidal coils connected in parallel could be interfaced with a robotics liquid sampler and 96-well plate. Simultaneous injection of four compounds allowed an analysis rate of 34 s per

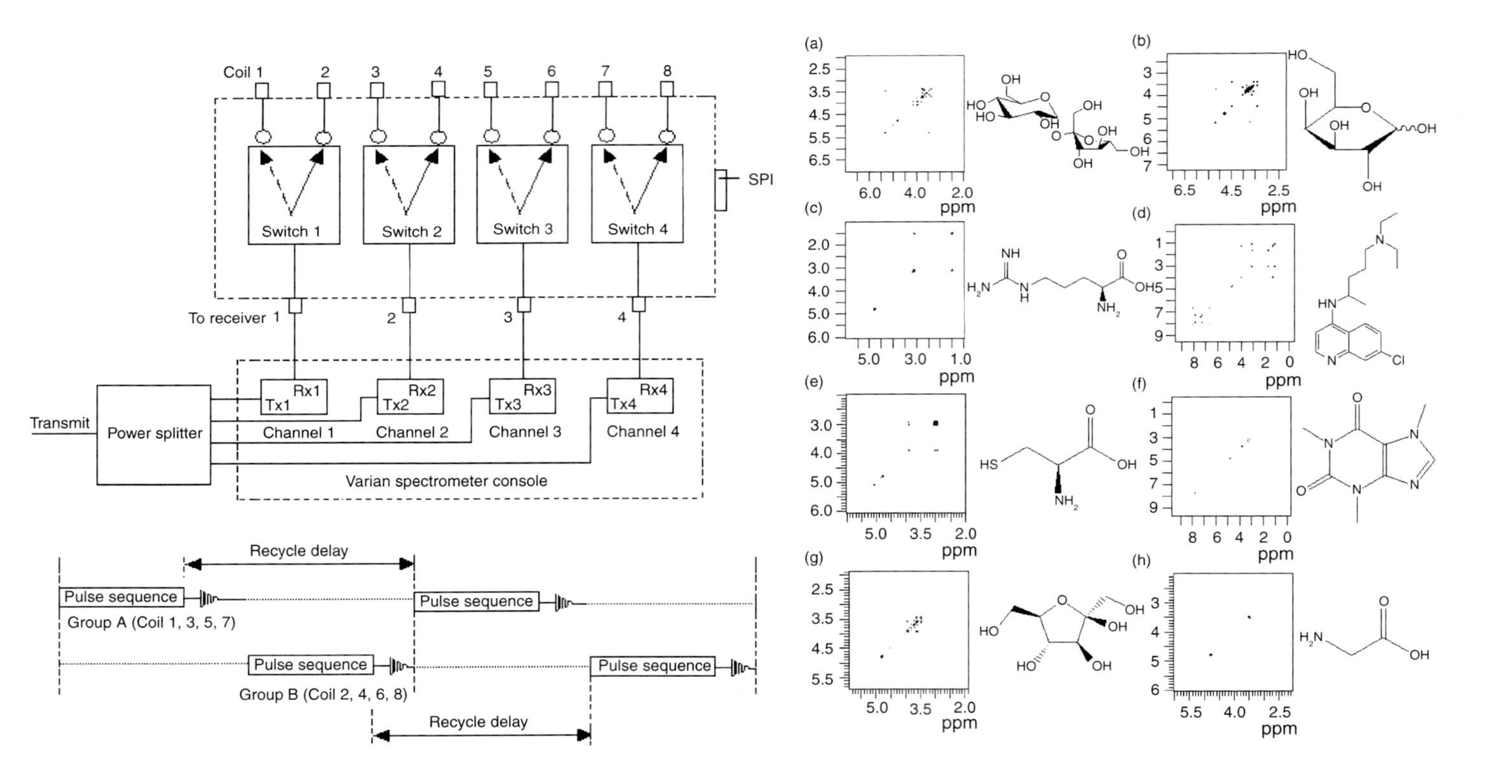

Figure 33 (Top left) Schematic showing a transmit and receive scheme for multicoil operation. The four switching networks were controlled by TTL signals from the Varian Unity console. The eight coils were divided into two groups, A (coils 1, 3, 5, 7) and B (coils 2, 4, 6, 8). (Bottom left) Timing diagram showing the pulse sequence and data acquisition scheme used. (right) COSY spectra acquired with the eight-coil probe and the chemical structures of the compounds used. Each sample (10 mM solution in D_2O) was loaded into the coil via the attached Teflon tubes. (a) sucrose, (b) galactose, (c) arginine (d) chloroquine, (e) cysteine, (f) caffeine, (g) fructose, and (h) glycine. Data acquisition parameters: data matrix 2048×256, 8 scans, sw = 6000 Hz, $sw_1 = 6000$ Hz. Data were zero filled in t_1 to 2048 points, processed with shifted sine-bell window functions applied in both dimensions, symmetrized and displayed in magnitude mode. Figure reproduced from Wang et al.[81] Copyright (2004) with permission from Elsevier.

sample for single scan 1D proton spectroscopy. In this setup, plugs of D_2O separated by air bubbles were introduced around the samples, which were transferred to the probe via 320 μm i.d. fused silica capillaries: 35 μl of D_2O was injected, followed by 25 μl of sample, a further 35 μl of D_2O, and 15 μl of air; 125 μl of H_2O was then used to push the samples into the center of each of the coils of the Multiplex probe. Spectra were collected at 300 MHz with a selective excitation sequence following water presaturation. The linewidths in each spectrum were between 1 and 2 Hz, and the metabolite concentrations used were 100 mM. Figure 34 shows the physical setup of the system, and two sets of four spectra acquired using the probe.

In terms of extending multiple coil probeheads for applications such as protein/ligand binding or protein structure studies, multifrequency coils must be designed with high isolation between all the resonant frequencies of all of the coils. One such probe, capable of acquiring two simultaneous ^{1}H–^{15}N HSQC spectra, has been designed for operation at 500 MHz, with two 3.5 mm long coils wrapped on a 2.6 mm o.d. polyimide sheath: a glass tube of 2.5 mm o.d., 2.2 mm i.d. containing the protein sample can be slid into each polyimide sheath. Each of the two coils was double tuned to ^{1}H and ^{15}N frequencies, and an external lock coil was incorporated to compensate for field drift: this lock coil consisted of a three-turn solenoid with 1 mm o.d. and contained a capillary of D_2O, and was placed approximately 5 mm from the two detection solenoidal coils. The two sample coils were situated 6 mm apart, separated by a thin copper shield for increased electrical isolation (Figure 35). Electrical measurements showed that there was negligible crosstalk (<-40 dB) between sample coils at all frequencies. FC-43 was used for magnetic susceptibility matching to improve sample shimming. Two low-loss single-pole-five-throw (SP5T) switches are used to multiplex the signal into a single receiver.

To demonstrate the operation of this probe, two proteins with widely different chemical shifts were used. This was most easily realized by using one protein which is unfolded (characterized by poorly dispersed chemical shifts and often having several high-intensity narrow peaks from overlapping resonances) and one which is folded (usually characterized by a large chemical shift dispersion). The first coil was loaded with 1.25 mM ^{15}N-labeled ubiquitin (VLI, Malvern, PA) in 90% H_2O/10% D_2O and 50 mM phosphate buffer at a pH of 5.5, and the second coil was loaded with 1 mM ^{15}N-labeled IA-3 in 90% H_20/10% D_2O, 50 mM phosphate buffer, also at a pH of 5.5. Shimming was performed on the water-free induction decay for each coil, using a low tip angle to avoid radiation damping, and gave linewidths of the water peaks for each sample of ~20 Hz (a value that was very similar to that of either sample in a 5 mm commercial probehead). Figure 35 shows two ^{1}H–^{15}N HSQC spectra of the two proteins collected at the same time. The Varian spectrometer had only a single receiver channel, so data collection was staggered: pulse transmission and data acquisition for one sample were performed during the relaxation delay of the other sample. This relaxation delay was used for one-second presaturation of the water signal using a shaped RF pulse. Immediately following data acquisition from the first coil,

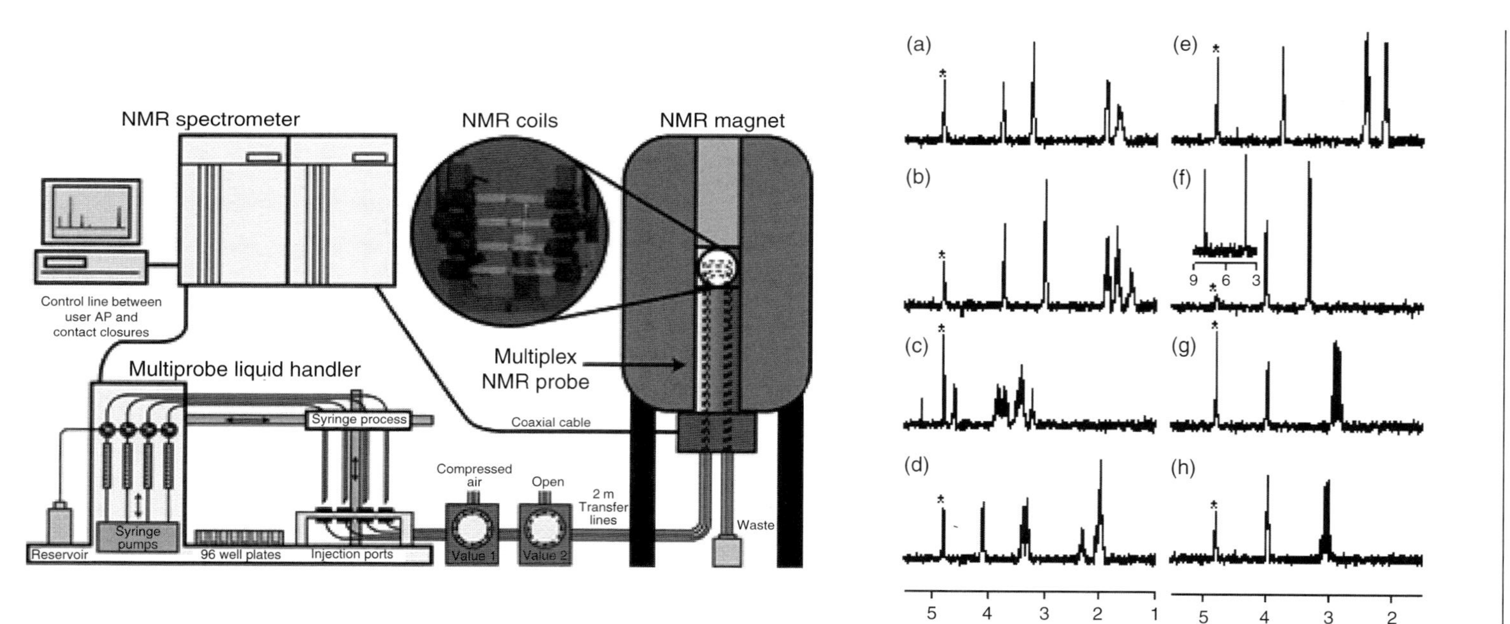

Figure 34 (left) An automation system includes a multiprobe liquid handler, two 12-port valves, the Multiplex NMR probe, and the NMR spectrometer. The liquid handler and spectrometer are electrically connected and communicate through the contact ports on the liquid handler and the user analogue port on the spectrometer. The 12-port valves are controlled by the liquid handler and are used to control and remove the samples during the automation routine. The liquid handler can inject four samples at a time into the Multiplex probe through four transfer lines. The Multiplex flow probe, which is placed inside an NMR magnet, has four sample capillaries each with an NMR excitation/detection coil as shown in the photograph. (right) Spectra of eight samples automatically injected into the Multiplex NMR probe were acquired with the automated flow injection routine: (a) L-arginine (b) L-lysine, (c) D-(+)-glucose, (d) L-proline, (e) L-glutamine, (f) L-histidine, (g) L-asparagine, and (h) L-cysteine, all at 100 mM. The asterisks indicate the residual HDO peaks after suppression with a presaturation pulse. Reproduced with permission from Macnaughtan et al.[79] Copyright 2003 American Chemical Society.

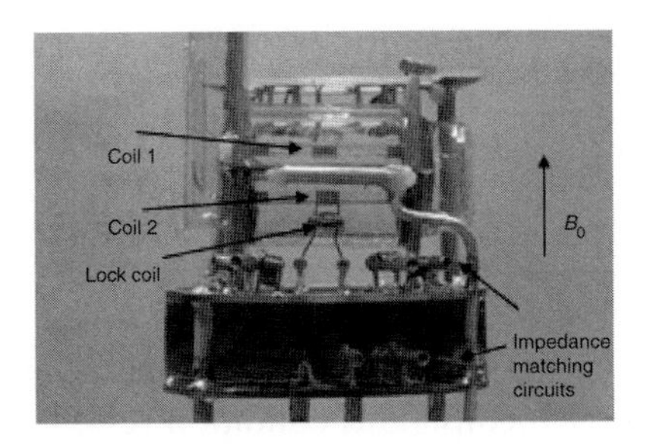

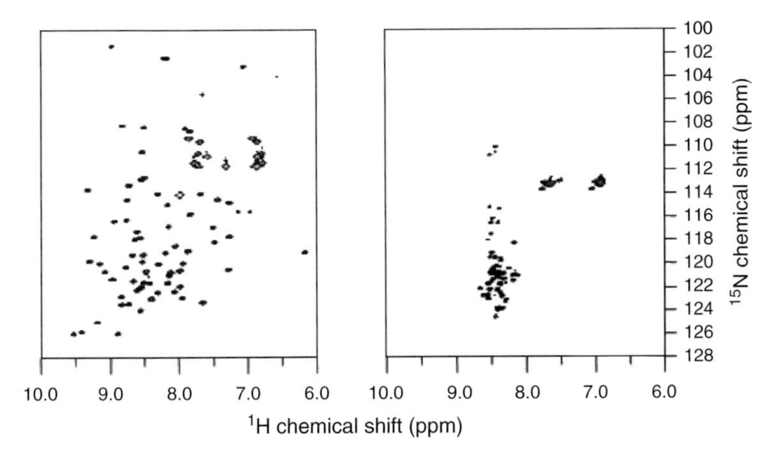

Figure 35 (top) A photograph of a two-coil probe assembly used for protein studies, together with the lock coil and impedance matching circuits. The samples are slid horizontally into the thin clear tubes around which the coils are formed. (bottom left) A ^1H–^{15}N HSQC spectrum of 1.25 mM ^{15}N-labeled ubiquitin in 90% H_2O/10% D_2O, 50 mM phosphate buffer, pH 5.5. Data acquisition parameters: sw = 4000 Hz, sw_1 = 1600 Hz, 1024 complex data points, 192t_1 increments acquired in States mode, 1 s water presaturation, 32 signal averages. Total data acquisition time 3.5 h. (bottom right) A ^1H–^{15}N HSQC spectrum of 1 mM ^{15}N-labeled IA-3 in 90% H_2O/10% D_2O, 50 mM phosphate buffer, pH 5.5. Identical data acquisition parameters were used. Data were acquired in interleaved fashion with pulse transmission and data reception routed through an RF switch controlled from the console. Reproduced with permission from Webb.[68] Copyright 2005 Wiley.

the two transmit/receive switches were shifted automatically to the second coil, and vice versa.

9. Solid-State Applications of Small Coils

Sakellariou et al.[91] have shown that microcoils can also be used for high-resolution magic angle spinning (MAS) spectroscopic experiments. The approach used was to wind a microcoil of 750 μm diameter around a capillary holding the sample, resonate the coil close to the Larmor frequency using a fixed value chip capacitor, and then place the assembly inside a much larger (7 mm diameter) solenoid which made up the conventional, commercial coil. The two coils are inductively coupled, with the sizes and Q values of the two coils resulting in effective overcoupling, which corresponds to the condition that

$$\frac{V_2}{V_1} > \frac{1}{Q_1 Q_2} \tag{24}$$

where V_1 is the volume of the larger coil, V_2 is that of the microcoil, with the respective Q values of the two coils. This condition results in maximum sensitivity (compared to undercoupling). The physical setup is shown in Figure 36.

In addition to being able to use solid samples, MAS also overcomes most of the susceptibility-induced line broadening that occurs due to the coil being so close to the sample. This is because broadening due to isotropic susceptibility effects transforms under rotation similarly to a rank 2 tensor, and its time-dependent component scales with the Legendre polynomial $P_2(\cos\theta)$, thus averaging to zero under MAS conditions.[91] Figure 37 shows results from a biological tissue at a low spinning frequency. The sensitivity using the inductively coupled microcoil was over 20 times that of using the larger coil alone.

The strong B_1 field produced by small coils has been exploited to cover the very large bandwidths required for static solid-state applications.[92] In this study, solenoidal coils of between 300 and 400 mm i.d. were mechanically integrated into a

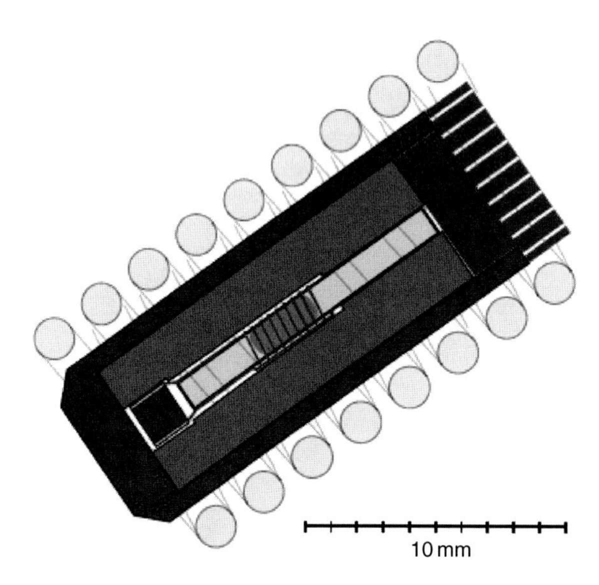

Figure 36 Schematic diagram of the magic angle coil spinning (MACS) insert. A tuned microcoil can be tightly wound around the sample, which is placed inside a glass capillary. A cylindrical ceramic insert is used to keep the capillary and the tuning capacitor centred while spinning. Inductive coupling between the tuned circuit and the probe electronics generates a high RF field and enhances the detection sensitivity. Reproduced with permission from Sakellariou et al.[91] Copyright 2007, Nature Publishing Group.

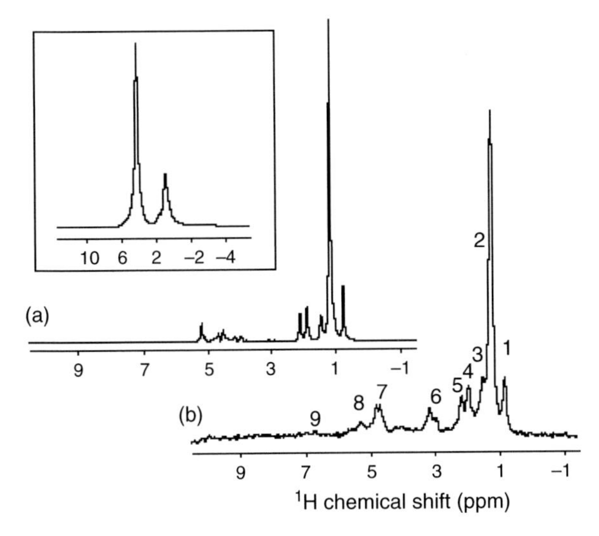

Figure 37 Proton NMR spectra of bovine muscle tissue. (a) The high-resolution spectrum from ~365 mg of bovine muscle tissue using a full 7 mm rotor spinning at 3000 Hz was acquired in 33 s (eight scans). Presaturation of the water resonance (4.7 ppm.) was achieved using a 2 s irradiation before the $\pi/2$ hard pulse. (b) High-resolution spectrum from ~0.3 mg of bovine muscle tissue using the MACS insert under the same experimental conditions. The susceptibility broadening is distributed in spinning sidebands outside the range of the proton chemical shifts and the line width is of the order of 0.05 ppm. for all resonances (no deuterium lock was used). This width is mainly attributed to temperature gradients (temperature difference less than 3–4°C). In the absence of sample spinning, the susceptibility broadening hides most of the isotropic chemical shift information, as seen from the spectrum (inset) of the static sample used in (a). Reproduced with permission from Sakellariou et al.[91] Copyright 2007, Nature Publishing Group.

custom-made capacitor to minimize lead lengths and also to provide mechanical stability. ^{27}Al spectra obtained at 14.1 T are shown in Figure 38.

Multinuclear studies of small crystals have also recently been presented[93] using designs in which the goniometer is moved outside the RF coil, enabling much higher filling factors and S/N to be achieved than previously possible. Results are shown in Figure 39.

10. Conclusion

After initial work in the 1970s, the past decade has seen a rapid expansion in small coil development for high-resolution NMR spectroscopy. Higher B_0 fields, cryogenic probe technology, and "microcoils" have enabled successful realization of a number of NMR applications which were previously not possible due to low S/N. With comparable mass sensitivities, microprobes and cryoprobes play complementary roles in NMR studies: the facility of microprobe use is advantageous in applications where relatively high concentrations are possible, whereas cryogenic

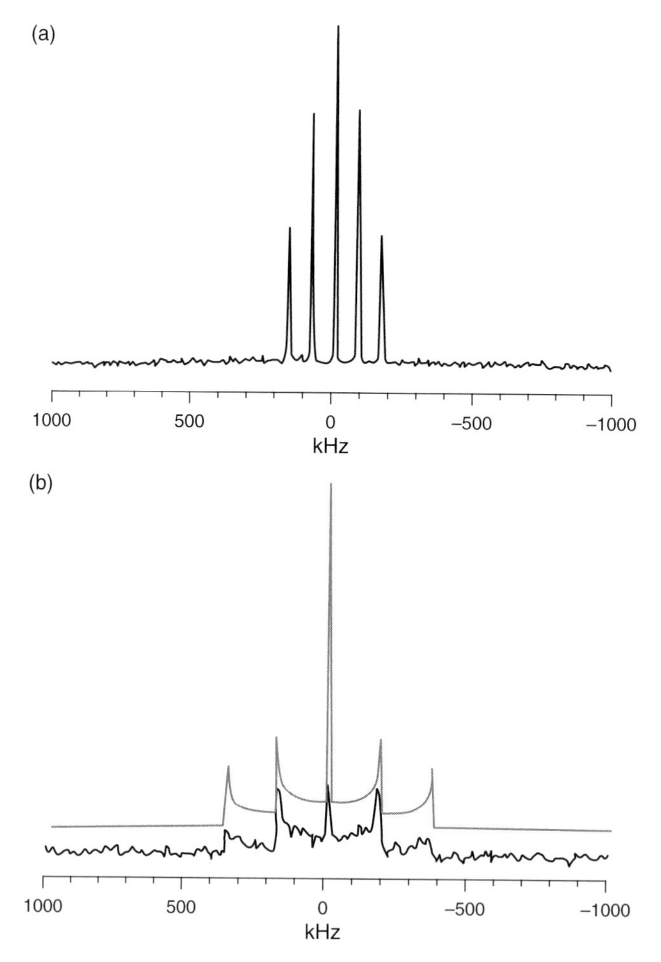

Figure 38 (a) ^{27}Al spectrum of a sapphire single crystal obtained with a broadband microcoil probehead at 14.1T. This spectrum is obtained by single pulse experiment averaging 1024 scans. A 60 s pulse delay was used. The experimental ratio of single-quantum peak integrals 5.0:7.8:9.0:7.6:4.7 is very close to the theoretical 5:8:9:8:5. (b) ^{27}Al spectrum of powdered sapphire obtained in the same probehead (black line). A Hahn echo pulse sequence is averaging over 4096 accumulations using a 60 s pulse delay. The powder lineshape of the satellite transitions agree with theoretically calculated spectrum (gray line) from literature data. Figure reproduced from Yamauchi et al.[92] Copyright (2004) with permission from Elsevier.

probes come into their own where lower concentrations are necessary. The relatively new area of multiple coil probeheads shows great promise, not only in enabling high throughput NMR to become a practical reality, but also in enabling completely new types of experiment to be performed. Novel geometries such as stripline and microslot probes continue to drive the innovative use of magnetic resonance spectroscopy for mass-limited samples.

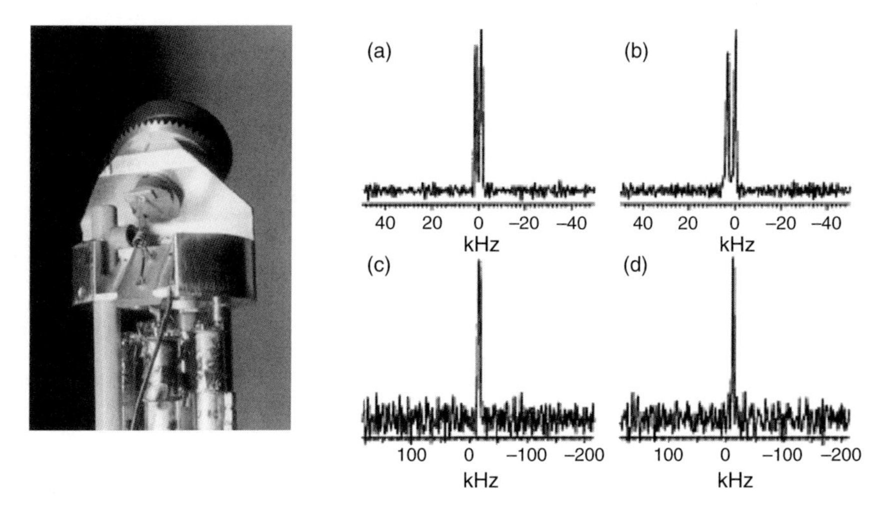

Figure 39 (left) Single-crystal variable-temperature NMR probe equipped with an optimized three-turn 2 mm inner diameter RF coil for studies of submillimeter crystals. (right) (a) and (b) 242.9 MHz ^{31}P CP NMR spectra (proton decoupled) of two different small crystals of $(NH_4)_2HPO_4$. (a) Crystal size $0.25 \times 0.5 \times 0.7$ mm = 0.088 mm^3, RF coil of 3.8 mm i.d., and 512 scans in 9 min. (b) Crystal size $0.2 \times 0.4 \times 0.4$ mm = 0.032 mm^3, RF coil of 2.0 mm i.d., and 128 scans in 2 min. (c) and (d) 196.3 MHz ^{87}Rb NMR spectra (proton decoupled) of two different small crystals of $RbZn_2(HPO_4)PO_4$. (c) Crystal size $0.05 \times 0.38 \times 2.5$ mm = 0.048 mm^3, RF coil of 3.8 mm i.d., and 8192 scans in 55 min. (d) Crystal size $0.05 \times 0.20 \times 1.5$ mm = 0.015 mm^3, RF coil of 2.0 mm i.d., and 4096 scans in 27 min. Figure reproduced from Vosegaard et al.[93] Copyright (2000) with permission from Elsevier.

REFERENCES

1. R.F. Service, Science 279 (1998) 1127–1128.
2. R. Fu, W.W. Brey, K. Shetty, P. Gor'kov, S. Saha, J.R. Long, S.C. Grant, E.Y. Chekmenev, J. Hu, Z. Gan, M. Sharma, F. Zhang, T.M. Logan, R. Bruschweller, A. Edison, A. Blue, I.R. Dixon, W.D. Markiewicz, T.A. Cross, J. Magn. Reson. 177 (2005) 1–8.
3. P. Styles, N.F. Soffe, C.A. Scott, D.A. Cragg, F. Row, D.J. White, P.C.J. White, J. Magn. Reson. 60 (1984) 397–404.
4. H. Kovacs, D. Moskau, M. Spraul, Prog. NMR Spectrosc. 46 (2005) 131–155.
5. C.R. Bowers and D.P. Weitekamp, Phys. Rev. Lett. 57 (1986) 2645–2648.
6. J.H. Ardenkjaer-Larsen, B. Fridlund, A. Gram, G. Hansson, L. Hansson, M.H. Lerche, R. Servin, M. Thaning, K. Golman, PNAS 100 (2003) 10158–10163.
7. V. Thomsen, D. Schatzlein, and D. Mercuro, Spectroscopy, 18 (2003) 112–114.
8. D.I. Hoult, R.E. Richards, J. Magn. Reson. 24 (1976) 71–85.
9. W.H. Wong, R.S. Withers, R. Nast, V.Y. Kotsubo, M.E. Johansson, H.D.W. Hill, L.F. Fuks, K.A. Kelin, B. Cole, W.W. Brey, A. Barfknecht, W.A. Anderson, Adv. Cryog. Eng. 42 (1996) 953–959.
10. T.M. Logan, N. Murali, G.S. Wang, C. Jolivet, Magn. Reson. Chem. 37 (1999) 762–765.
11. D.J. Russell, C.E. Hadden, C.E. Martin, A.A. Gibson, A.P. Zens, J.L. Carolan, J. Nat. Prod. 63 (2000) 1047–1049.
12. R.C. Crouch, W. Llanos, K.G. Mehr, C.E. Hadden, D.J. Russell, G.E. Martin, Magn. Reson. Chem. 39 (2001) 555–558.

13. H.C. Keun, O. Beckonert, J.L. Griffin, C. Richter, D. Moskau, J.C. Lindon, J.K. Nicholson, Anal. Chem. 74 (2002) 4588–4593.
14. J.L. Griffin, H. Keun, C. Richter, D. Moskau, C. Rae, J.K. Nicholson, Neurochem. Int. 42 (2003) 93–99.
15. T.L. Peck, R.L. Magin, P.C. Lauterbur, J. Magn. Reson. B, 108 (1995) 114–124.
16. D.M. Ginsberg, M.J. Melchner, Rev. Sci. Instrum. 41 (1970) 122–123.
17. J.N. Shoolery, Top. Carbon-13 NMR Spectrosc. 3 (1979) 28–38.
18. R.C. Crouch, G.E. Martin, Magn. Reson. Chem. 30 (1992) S66–S70.
19. R.C. Crouch, G.E. Martin, Magn. Reson. Chem. 37 (1999) 721–729.
20. J.H. Walton, J.S. de Ropp, M.V. Shutov, A.G. Goloshevsky, M.J. McCarthy, R.L. Smith, S.D. Collins, Anal. Chem. 75 (2003) 5030–5036.
21. A.G. Goloshevsky, J.H. Walton, M.V. Shutov, J.S. de Ropp, S.D. Collins, M.J. McCarthy, Rev. Sci. Instrum. 76 (2005) 024101–024105.
22. A.G. Goloshevsky, J.H. Walton, M.V. Shutov, J.S. de Ropp, S.D. Collins, M.J. McCarthy, Meas. Sci. Technol. 16 (2005) 505–512.
23. A.G. Goloshevsky, J.H. Walton, M.V. Shutov, J.S. de Ropp, S.D. Collins, M.J. McCarthy, Meas. Sci. Technol. 16 (2005) 513–518.
24. T.M. Barbara, C.E. Bronnimann, J. Magn. Reson. 140 (1999) 285–288.
25. W.L. Fitch, G. Detre, C.P. Holmes, J.N. Shoolery, P.A. Keifer, J. Org. Chem. 59 (1994) 7955–7956.
26. K. Putzbach, M. Krucker, M.D. Grynbaum, P. Hentschel, A.G. Webb, K. Albert, J. Pharm. Biomed. Anal. 38 (2005) 910–917.
27. D.L. Olson, J.A. Norcross, M. O'Neil-Johnson, P.F. Molitor, D.J. Detlefsen, A.G. Wilson, T.L. Peck, Anal. Chem. 76 (2004) 2966–2974.
28. J.A. Rogers, R.J. Jackman, G.M. Whitesides, D.L. Olson, J.V. Sweedler, Appl. Phys. Lett. 70 (1997) 2464–2466.
29. V. Malba, R. Maxwell, L.B. Evans, A.E. Bernhardt, M. Cosman, K. Yan, Biomed. Microdevices 5 (2003) 21–27.
30. V. Demas, J.L. Herberg, V. Malba, A. Bernhardt, L. Evans, C. Harvey, S. C. Chinn, R. S. Maxwell, and J. Reimer, J. Magn. Reson., 189 (2007) 121–129.
31. K. Albert, E. Bayer, Anal. Meth. Instrum. 2 (1995) 302–311.
32. X.F. Zhang, A.G. Webb, Anal. Chem. 77 (2005) 1338–1344.
33. Y. Li, A.G. Webb, S. Saha, W.W. Brey, C. Zachariah, A.S. Edison, Magn. Reson. Chem. 44 (2006) 255–262.
34. J.W. Carlson, Magn. Reson. Med. 3 (1986) 778–790.
35. J. Guo, D. Kaifez, A.W. Glisson, Electron. Lett. 33 (1997) 966–967.
36. P.F. Ryff, IEEE Trans. Industrial App. IA-8 (1972) 485–490.
37. F. Engelke, Conc. Magn. Reson. 15 (2002) 129–155.
38. S. Kan, P. Gonord, Magn. Reson. Med. 23 (1992) 372–375.
39. R.W. Wiseman, T.S. Moerland, M.J. Kushmerick, NMR Biomed. 6 (1993) 153–156.
40. C. Massin, F. Vincent, A. Homsy, K. Ehrmann, G. Boero, P.A. Besse, A. Daridon, E. Verpoorte, N.F. de Rooij, R.S. Popovic, J. Magn. Reson. 164 (2003) 242–255.
41. J.E. Stocker, T.L. Peck, A.G. Webb, M. Feng, R.L. Magin, IEEE. Trans. Biomed. Eng. 44 (1997) 1122–1127.
42. K. Ehrmann, K. Pataky, M. Stettler, F.M. Wurm, J. Brugger, P.A. Besse, R. Popovic, Lab Chip 7 (2007) 381–383.
43. B. Sorli, J.F. Chateaux, M. Pitaval, H. Chahboune, B. Favre, A. Briguet, P. Morin, Meas. Sci. Technol. 15 (2004) 877–880.
44. J.D. Trumbull, I.K. Glasgow, D.J. Beebe, R.L. Magin, IEEE Trans. Biomed. Eng. 47 (2000) 3–7.
45. K. Ehrmann, N. Saillen, F. Vincent, M. Stettler, M. Jordan, F.M. Wurm, P.A. Besse, R. Popovic, Lab Chip 7 (2007) 373–380.
46. Y. Maguire, I.L. Chuang, S.G. Zhang, N. Gershenfeld, PNAS 104 (2007) 9198–9203.
47. P.J.M. Van Bentum, J.W.G. Janssen, A.P.M. Kentgens, J. Bart, and J.G.E. Gardeniers, J. Magn. Reson., 189 (2007) 104–113.

48. W.W. Brey, A.S. Edison, R.E. Nast, J.R. Rocca, S. Saha, R.S. Withers, J. Magn. Reson. 179 (2006) 290–293.
49. G. Schlotterbeck, A. Ross, R. Hochstrasser, H. Senn, T. Kuhn, D. Marek, O. Schett, Anal. Chem. 74 (2002) 4464–4471.
50. D.L. Olson, T.L. Peck, A.G. Webb, R.L. Magin, J.V. Sweedler, Science 270 (1995) 1967–1970.
51. R.A. Kautz, M.E. Lacey, A.M. Wolters, F. Foret, A.G. Webb, B.L. Karger, J.V. Sweedler, J. Am. Chem. Soc. 123 (2001) 3159–3160.
52. V.K. Almeida, C.K. Larive, Magn. Reson. Chem. 43 (2005) 755–761.
53. L.A. Cardoza, V.K. Almeida, A. Carr, C.K. Larive, D.W. Graham, Trends Anal. Chem. 22 (2003) 766–775.
54. A.M. Wolters, D.A. Jayawickrama, C.K. Larive, J.V. Sweedler, Anal. Chem. 74 (2002) 4191–4197.
55. A.M. Wolters, D.A. Jayawickrama, C.K. Larive, J.V. Sweedler, Anal. Chem. 74 (2002) 2306–2313.
56. A.K. Korir, V.K. Almeida, D.S. Malkin, C.K. Larive, Anal. Chem. 77 (2005) 5998–6003.
57. J.L. Griffin, A.W. Nicholls, H.C. Keun, R.J. Mortishire-Smith, J.K. Nicholson, T. Kuehn, Analyst 127 (2002) 582–584.
58. M.E. Lacey, J.V. Sweedler, C.K. Larive, A.J. Pipe, R.D. Farrant, J. Magn. Reson. 153 (2001) 215–222.
59. G.R. Eldridge, H.C. Vervoort, C.M. Lee, P.A. Cremin, C.T. Williams, S.M. Hart, M.G. Goering, M. O'Neil-Johnson, L. Zeng, Anal. Chem. 74 (2002) 3963–3971.
60. Y. Li, T.M. Logan, A.S. Edison, A. Webb, J. Magn. Reson. 164 (2003) 128–135.
61. W. Peti, J. Norcross, G. Eldridge, M. O'Neil-Johnson, J. Am. Chem. Soc. 126 (2004) 5873–5878.
62. L.E. Kay, G.Y. Xu, A.U. Singer, D.R. Muhandiram, J.D. Formankay, J. Magn. Reson. B 101 (1993) 333–337.
63. W. Peti, C. Griesinger, W. Bermel, J. Biomol. NMR 18 (2000) 199–205.
64. A.T. Dossey, S.S. Walse, J.R. Rocca, A.S. Edison, ACS Chem. Biol. 1 (2006) 511–514.
65. A.T. Dossey, S.S. Walse, O.V. Conle, A.S. Edison, J. Nat. Prod. 70 (2007) 1335–1338.
66. F. Zhang , A. T. Dossey, C. Zachariah, A. S. Edison, and R. Bruschweiler, Anal. Chem., 79 (2007) 7748–7752.
67. D.A. Jayawickrama, J.V. Sweedler, J. Chromatogr. A 1000 (2003) 819–840.
68. A.G. Webb, J. Pharm. Biomed. Anal. 38 (2005) 892–903.
69. J. Rehbein , B. Dietrich, M. D. Grynbaum, P. Hentschel, K. Holtin, M. Kuehnle, P. Schuler, M. Bayer, and K. Albert, J. Sep. Sci., 30 (2007) 2382–2390.
70. M.D. Grynbaum, C. Meyer, K. Putzbach, J. Rehbein, K. Albert, J. Chrom. A 1156 (2007) 80–86.
71. M. Krucker, A. Lienau, K. Putzbach, M.D. Grynbaum, P. Schuger, K. Albert, Anal. Chem. 76 (2004) 2623–2628.
72. M. Lambert, J.L. Wolfender, D. Staerk, S.B. Christensen, K. Hostettmann, J.W. Jaroszewski, Anal. Chem. 79 (2007) 727–735.
73. M.D. Grynbaum, D. Kreidler, J. Rehbein, A. Purea, P. Schuler, W. Schaal, H. Czesla, A. Webb, V. Schurig, K. Albert, Anal. Chem. 79 (2007) 2708–2713.
74. G. Fisher, C. Petucci, E. MacNamara, D. Raftery, J. Magn. Reson. 138 (1999) 160–163.
75. Y. Li, A.M. Wolters, P.V. Malawey, J.V. Sweedler, A.G. Webb, Anal. Chem. 71 (1999) 4815–4820.
76. E. MacNamara, T. Hou, G. Fisher, S. Williams, D. Raftery, Anal. Chim. Acta 397 (1999) 9–16.
77. T. Hou, J. Smith, E. MacNamara, M. Macnaughtan, D. Raftery, Anal. Chem. 73 (2001) 2541–2546.
78. X. Zhang, J.V. Sweedler, A.G. Webb, J. Magn. Reson. 153 (2001) 254–258.
79. M.A. Macnaughtan, T. Hou, J. Xu, D. Raftery, Anal. Chem. 75 (2003) 5116–5123.
80. D. Raftery, Anal. Bioanal. Chem. 378 (2004) 1403–1404.
81. H. Wang, L. Ciobanu, A.S. Edison, A.G. Webb, J. Magn. Reson. 170 (2004) 206–212.
82. H. Wang, L. Ciobanu, A. Webb, J. Magn. Reson. 173 (2005) 134–139.
83. A.M. Wolters, D.A. Jayawickrama, A.G. Webb, J.W. Sweedler, Anal. Chem. 74 (2002) 5550–5555.

84. L. Ciobanu, D.A. Jayawickrama, X.Z. Zhang, A.G. Webb, J.V. Sweedler, Ang. Chemie. Int. Ed. 42 (2003) 4669–4672.
85. M.A. Macnaughtan, T. Hou, E. MacNamara, R.E. Santini, D. Raftery, J. Magn. Reson. 156 (2002) 97–103.
86. M.A. Macnaughtan, A.P. Smith, P.B. Goldsbrough, R.E. Santini, D. Raftery, Anal. Bioanal. Chem. 378 (2004) 1520–1527.
87. E. MacNamara, T. Hou, G. Fisher, S. Williams, D. Raftery, Anal. Chim. Acta 397 (1999) 9–16.
88. M.A. Macnaughtan, T. Hou, E. MacNamara, R.E. Santini, D. Raftery, J. Magn. Reson. 156 (2002) 97–103.
89. B.J. Stockman, C. Dalvit, Prog. NMR Spectrosc. 41 (2002) 187–231.
90. C. Dalvit, D.T.A. Hadden, R.W. Sarver, A.M. Ho, B.J. Stockman, Combust. Chem. High Throughput Screen. 6 (2003) 445–453.
91. D. Sakellariou, G. Le Goff, J.-F. Jacquinot, Nature 447 (2007) 694–697.
92. K. Yamauchi, J.W.G. Janssen, A.P.M. Kentgens, J. Magn. Reson. 167 (2004) 87–96.
93. T. Vosegaard, P. Daugaard, E. Hald, H.J. Jakobsen, J. Magn. Reson. 142 (2000) 379–381.

CHAPTER 5

qNMR Spectroscopy in Drug Analysis – A General View

U. Holzgrabe

Contents

Abstract

^{1}H and ^{13}C NMR (nuclear magnetic resonance) spectroscopy can be used in a quantitative manner to identify a drug or excipient, to evaluate their quality, i.e., impurity profiling, and to quantify the content. However, international pharmacopoeias make only rarely use of the technique.

Keywords: drugs, international pharmacopoeia, validation, peptides, vaccines, excipients

1. Introduction

The regulatory dossier of a medicinal product, which is necessary for the application of a license, consists amongst others of (1) control of the starting material (active substance, excipient and immediate packing material), (2) control tests on intermediate products, (3) the control tests on the finished product, and (4) stability tests on the active substance and on the finished products. The scheme of the control, of the active substance and the excipient follows the philosophy of the monographs of international pharmacopoeias, such as European Pharmacopoeia (PhEur),[1] United States Pharmacopoeia (USP),[2] and Japanese Pharmacopoeia (JP)[3] and is governed by the guideline of the

International Conference on Harmonization (ICH).[4] The drug or excipient has to be identified, which is often done by means of IR spectroscopy and thin-layer chromatography (TLC) as well as wet chemistry methods for the identification of characteristic moieties of the drug molecule or a counter ion. The quality of the drugs is often evaluated by chromatographic methods, especially high-performance chromatography (HPLC), and some specific tests. Finally, the content of a drug is mostly assessed by a titration, rarely by HPLC or UV spectroscopy.

^{1}H and ^{13}C NMR (nuclear magnetic resonance) spectroscopy is routinely used for the elucidation of structures of newly synthesized compounds, natural products, and semisynthesized compounds. Utilizing Correlation Spectroscopy (COSY), Heteronuclear Multiple Quantum correlation (HMBC), Heteronuclear Single-Quantum correlation (HSQC), Total Correlation Spectroscopy (TOCSY), Nuclear Overhauser Effect Spectroscopy (NOESY), and Rotating-frame Overhauses Effect Spectroscopy (ROESY) experiments the constitution, configuration and conformation of small molecules, polymers, peptides, sugars, or nucleotides, can be elucidated.[5] ^{19}F, ^{15}N, and ^{31}P NMR were additionally employed in structure determination.

Because the NMR spectroscopy is a primary method of measurement,[6,7] it can be also used to evaluate the quality of drugs and excipients besides the possibility to identify a drug by the comparison of the ^{1}H or ^{13}C NMR spectra of the substance to be examined and a reference substance.[8,9]

The following items can be controlled by NMR spectroscopy using various quantification designs:

- identification of a drug or excipient
- the level of impurities (i.e., related substances)
- the content of residual solvents
- the isomeric composition:
  - the ratio of diastereomers
  - the ratio of enantiomers and the enantiomeric excess (ee)
- assay of single drugs or drug composition

When quantifying the impurities of a drug, the spectrum measured always comes with the entire information. Thus is can be also used for the structure elucidation of new unknown impurities. Additionally, it is often easy to observe the course of decomposition of a drug by the measurement of time-dependent spectra.

Even though, countless examples of application of quantitative NMR spectroscopy are reported (see Chapter 1, Part I), International pharmacopoeias make only limited use of qualitative NMR spectroscopy for identification purposes and of quantitative NMR spectroscopy for evaluation of the impurities in drugs and of the composition of polymers (see Chapter 2, Part II).

2. NMR Spectroscopy in International Pharmacopoeias

All pharmacopoeias describe the physical and methodological principles of the NMR spectroscopy and their quantitative application in a general chapter.

However, the number of applications in each pharmacopoeia is low because the pharmaceutical companies do not use NMR spectroscopy in routine analysis. Currently, the following examples can be found. For identification: buserelin, goserelin, tobramycin (PhEur 6.0), hydrocortisone sodium phosphate (BP 1998), amylnitrite isomers (USP 30), heparins-low molecular mass, *Haemophilus* Type b conjugate vaccine, meningococcal Group C conjugate vaccine, pneumococcal polysaccharide conjugate vaccine (adsorbed) (PhEur 6.0), salmon oil farmed (PhEur 6.0). For tests: poloxamer: ratio of oxypropylene/oxyethylene(PhEur 6.0); hydroxypropylbetadex: molar substitution (PhEur 6.0); lauromacrogol 400 known as polidocanol 9 or macrogol 9 lauryl ether: average chain length of fatty alcohol and average number of moles of ethylene oxide (PhEur 6.0), orphenadrine citrate: *meta/para* isomer (USP2007). For assay: amylnitrite isomers (USP 30).

Furthermore, solid-state NMR spectroscopy is mentioned for the study of polymorphism besides other techniques such as microcalorimetry, IR, and Raman spectrometry. Further applications are under discussion, especially in the field of vaccines (see Chapter 3, Part II). The European Directorate for Quality of Medicines and HealthCare (EDQM) is currently performing an interlaboratory trial to evaluate the NMR as a tool for the batch consistency control for polysaccharide vaccines. Additionally, EDQM had send out a questionnaire to find out whether pharmaceutical industries are interested to replace the amino acid analysis (AAA), which is used for the identification of small peptides such as oxytocin, with ^{1}H or ^{13}C NMR spectroscopy. These activities underline the increasing interest in NMR spectroscopy in drug analysis. Therefore, the PhEur will include a completely revised general chapter on NMR spectroscopy soon, which describes methods for qualitative and quantitative analysis including solid-state NMR.

However, whereas some industrial laboratories perform NMR analysis routinely others refused to use the method due to the expensive instrumentation (but did not calculate the saving due to short analysis time and less consumption of solvents) and to supposedly problems with the validation and interpretation.

2.1. Identification of drugs

The pharmacopoeias are using the ^{1}H and ^{13}C NMR spectroscopy especially identification of drugs of complex structures, e.g., peptides, steroids, or aminoglycosides, and vaccines in a similar way as the IR spectroscopy.[8,9] The spectra of the substance to be examined are compared with either a reference spectrum delivered by the EDQM or the spectrum of a reference substance, measured by the user. A typical example is given in Figure 1 displaying the ^{1}H NMR spectrum of bussereline acetate. In the case of the vaccines, the NMR spectroscopy is mostly an additional method beside immunological methods. For further details, see Jones (see Chapter 2, Part III).[10]

2.2. Tests

The most important section in the paragraph "Tests" in a monograph is dealing with related substances, i.e., impurities originated from the synthesis pathways, decomposition reactions, etc. Because the intensity I_A of a signal is directly

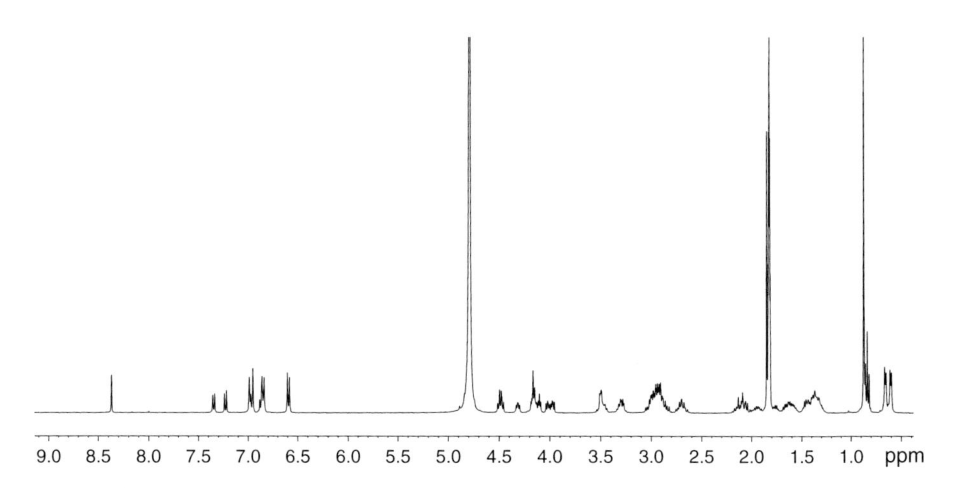

Figure 1 ^{1}H NMR spectrum of busereline acetate in D_2O/acetic acid-d_4 (400 MHz).

proportional to the number of nuclei *N* evoking the signal, the intensities of NMR signals (=areas under specific signals) can be taken for quantification of related substances in the presence of the drug substance (see Chapter 1, Part II). There is no example in the international pharmacopoeias of impurity profiling by means of any type of NMR spectroscopy. This may be due to the fact that (1) NMR spectroscopy is believed not to be sensitive enough and (2) sufficient separation of signals between the drug substance and the impurities is missing. However, both problems can be overcome.

As the sensitivity depends on the magnetic field strength, the sensitivity can be enhanced by using high field instruments, e.g., with a 600 MHz instrument a limit of detection (LOD) in a nanomolar range of concentration can be achieved. Besides the high field strength, the low sensitivity can be overcome by (1) the increase of the number of scans (accumulations) leading to a higher signal-to-noise (S/N) ratio, (2) precise (gradient) shimming techniques which increases the quality of the spectra, (3) using inverse and cryoprobes which increase the S/N ratio (by a factor of about 10), e.g., for ^{1}H measurements, and (4) maximizing of concentration, and optimizing solvent volume (microcoil technology, e.g., nano-NMR probes or microcryoprobes).

To achieve a proper separation of the signals to be considered for quantification purposes, the kind of solvent or the pH value may be changed, or auxiliary reagents ("shift reagents") added (cf. Chapter 1, Part I). In some cases, the temperature influences the signal separation and may contribute to a better separation. However, in contrast to chromatographic methods, there is only little possibility to influence this separation because the chemical shift is directly related to the molecular structure. Thus, the missing separation of signals of the main component and the impurities is one of the main limitations of the method.

The USP 30 makes use of the relative method of quantification to determine the content of *m*- and *p*-methylphenyl isomer in the *o*-methylphenyl-substituted orphenadrine citrate. The signals of the benzylic methine hydrogen atoms are well separated and, hence, from the areas of the signal of the *m*/*p*-methyl-substituted compound resonating at about 5.23 ppm and of the orphenadrine signal appearing at 5.47, the content of the impurities can be determined.

Interestingly, the European Pharmacopoeia makes use of the NMR spectroscopy to characterize the composition of oligo- and polymer excipients, i.e., the determination of the molar substitution of hydroxypropylcyclodextrin and the determination of the oxyethylene/oxypropylene ratio in the different types of poloxamer, respectively. In these cases, signals of similar size are compared. Thus, there is no sensitivity problem and the signals compared are properly separated.[8,9] Further applications can be found in Chapter 2, Part II.

2.3. Assay

qNMR is only applied once in an assay. For quantification of amyl nitrite, the absolute method using benzyl benzoate as an internal standard is used in the USP 30. The quantity of amyl nitrite is calculated from the signal area of the α-methylene group of the drug (at 4.8 ppm) and the signal area of the methylene hydrogens of benzyl benzoate at 5.3 ppm.

3. VALIDATION

For validation, the accuracy, precision, and robustness have to be studied.

The precision of the integrals determines the accuracy of quantification depending (a) on the noise level of the spectrum, (b) on the line shape, (c) quality of shimming, (d) choice of the window function, and (e) phase, baseline, and drift corrections.[11] Considering these parameters the qNMR methods can be validated with regard to linearity, robustness, parameters of accuracy (repeatability, comparability and measurement uncertainty), specificity, and selectivity.

The linearity is often found to be characterized by a correlation coefficient higher than 0.999.[7] Because of the fact that NMR signals are Lorentzian lines, the LOD and LOQ cannot be determined by means of the S/N ratio. Therefore, a calibration curve has to be studied using samples containing the analyte in the range of the LOD. The LOD based on the standard deviation of the response and the slope may be expressed as $\mathrm{LOD} = 3.3\,\sigma/S$, where σ is the standard deviation of the response and S the slope of the calibration curve. The LOQ may be expressed as $\mathrm{LOQ} = 10\,\sigma/S$.[12]

As key prerequisites, the specificity and selectivity must be checked prior to qNMR investigations. Specificity means the ability to assess unambiguously the analyte of interest in the presence of other components; thus, all NMR lines have to be assigned to the structure of the analyte. The selectivity of a method is given by the ability to determine analytes to be examined in a complex mixture without interference from other components in the mixture. This can be checked by homonuclear correlation experiments.

Further details of the validation are given in Malz and Jancke[7] and Chapter 1, Part II.

4. Conclusions

The decisive advantage of NMR spectroscopy over chromatographic and electrophoretic methods is because of the fact that almost no preparation time exists. Whereas in HPLC much time has to be spent for equilibration of the columns or derivatization of the analyte in the case of UV, fluorescence, or electrochemical detection, the NMR spectrometer is always ready to measure and needs experimental time depending on analyte concentration and experimental mode only. The substance has to be dissolved in a proper deuterated solvent. The amount of solvent needed is about 0.7 ml (using common 5-mm probe heads). So, facing the larger solvent volume in HPLC the higher costs for deuterated solvents are compensated. Sample amounts of 20 mg or less are approximately needed to achieve a good S/N ratio and an adequate experimental time.

Moreover, qNMR analysis is often more accurate and precise than standard HPLC methods.[13] Normally, no isolation of the impurity is necessary, no expensive chemical reference substances are necessary, and additional structural information of impurities, isomers, etc. are available. Another advantage of NMR is the fact that for the purity assay of a drug not all impurities (used for the 100% method in HPLC) need to be identified.

The following chapters of this book will clearly demonstrate the suitability of NMR spectroscopy for drug analysis.

Acknowledgment

Thank are due to the Federal Ministry of Economics and Labour (BMWA), the German Federation of Industrial Cooperative Research Associations "Otto von Guericke" (AiF), and the German Research Association of Medicine Manufacturers (FAH) for support of the AiF project.

References

1. European Pharmacopoeia, 6th edition, European Department for the Quality of Medicines, Strasbourg, France, 2008. www.pheur.eu.
2. United States Pharmacopoeia (USP) 30 (2007), No <761>, the United States Pharmacopoeia Convention, Rockville, MD, USA.
3. Japanese Pharmacopoeia (JP) XIV, 2004.
4. ICH Secretariat, Geneva, Switzerland. www.ICH.org.
5. T.D.W. Claridge, High-Resolution NMR Techniques in Organic Chemistry, Pergamon, Amsterdam, 1999.
6. T.J. Quinn, Metrologia 34 (1997) 61–65.
7. F. Malz, H. Jancke, J. Pharm. Biomed. Anal. 38 (2005) 813–823.
8. U. Holzgrabe, I. Wawer, B.W.K. Diehl, NMR Spectroscopy in Drug Development and Analysis, Wiley-VCH, Weinheim, 1999.

9. U. Holzgrabe, R. Deubner, C. Schollmayer, J. Pharm. Biomed. Anal. 38 (2005) 806–812.
10. C. Jones, J. Pharm. Biomed. Anal. 38 (2005) 840–850, and Chapter 2, Part III in this book.
11. C. Szantay, A. Demeter, in: Identification and Determination of Impurities in Drugs, Ed. S. Görög, Elsevier, Amsterdam, 2000, chapter 2.4, pp. 109–145.
12. Guide lines Q2 (R1), Validation of Analytical Procedures, Methodology, International Conference on Harmonization 2005.
13. R.J. Wells, J.M. Hook, T. Tareq Al-Deen, D.B. Hibbert, J. Agric. Food Chem. 50 (2002) 3366–3374.

PART II

GENERAL APPLICATIONS

CHAPTER 1

Investigation of Multi-Component Drugs by NMR Spectroscopy

T. Beyer *and* U. Holzgrabe

Contents

Abstract

Quantitative nuclear magnetic resonance (NMR) spectroscopy can be used for the evaluation of drugs consisting of multiple components that share certain structural similarities. To achieve optimal signal separation of all components for a valid integration, different solvents and mixtures of solvents with different ratios can be applied. This approach is demonstrated for codergocrine mesylate. The results obtained by means of ^{1}H and ^{13}C NMR spectroscopy were found to be in good agreement with the results of a high-performance liquid chromatography (HPLC) method that is described in the European Pharmacopoeia.

Keywords: Solvents, auxiliary reagents, codergocrine mesylate, ^{1}H NMR, ^{13}C NMR, HPLC

1. Introduction

Besides the well-known application of nuclear magnetic resonance (NMR) spectroscopy in structural elucidation, especially for newly synthesized, natural, and semisynthesized products, nowadays this technique is of growing importance in the field of quantitative drug analysis. Taking gentamicin as an example, the group of Holzgrabe has shown that NMR spectroscopy can be used for the identification of the main components C1, C1a, C2, and C2a+b and its impurities and that this technique is also suitable for quantitative purposes.[1] A comparison of the results obtained from NMR spectroscopy with the MEKC (micellar electrokinetic chromatography) results confirms that NMR spectroscopy is a very good orthogonal and complementary method to MEKC.

The prerequisite for quantitative NMR spectroscopy are clearly separated signals of certain protons which can be reliably interpreted. To achieve the optimal signal separation, a couple of parameters can be adjusted. In this chapter we will examine the influence of a variation of such parameters in general and take a closer look to the special case of the multi-component drug codergocrine mesylate.

To influence the signal separation, the kind of solvent or the pH value of the solution can be changed, different sample concentrations can be applied or auxiliary reagents like cyclodextrins or shift reagents can be added. In some cases the temperature also influences the signal separation.

1.1. Solvent

The chemical shifts of protons depend on the solvent; the interactions between solute and solvent are responsible for these solvent effects. These effects are due to hydrogen bonding, the anisotropy of the solvent molecules, van der Waals interactions, and steric effects.[2,3]

Abraham et al.[4] presented a study comparing ^{1}H chemical shifts of 124 components containing different functional groups using DMSO-d_6 and $CDCl_3$ as solvent. The components were divided into four major groups: (1) nonpolar groups like alkanes and aromatics, (2) polar aprotic groups such as aliphatic or aromatic tertiary amines and esters, (3) protic groups like aliphatic and aromatic primary and secondary amines and alcohols, and (4) polyhydroxy compounds such as diols and inositols. It was demonstrated that the $\Delta\delta$ value ($\Delta\delta = \delta(\text{DMSO}) - \delta(\text{CDCl}_3)$) of nonpolar components is small, while for protons of polar aprotic components larger $\Delta\delta$ values were found ($\sim\pm 0.1$ ppm). For protic components, especially components with labile hydrogens such as NH protons of amines and OH protons of alcohols, very large positive $\Delta\delta$ values (1–4 ppm) were found using DMSO-d_6 or $CDCl_3$.

The number of deuterated solvents is poor for NMR spectroscopy compared with chromatographic methods like high-performance liquid chromatography (HPLC). Moreover, the solubility of the analyte further determines the selection of suitable solvents, because the most important requirement for quantitative NMR spectroscopy is complete solubility. Thus, mixtures of two solvents have to be often used. Additionally, the chemical shift of the solvent signal and the HOD signal (residual water of the solvent) can put a limit to the application of a certain solvent because these signals may overlap with signals of interest.

At the beginning of the measurements, pure solvents can be used in which the analyte is completely soluble. To reliably assign the ^{1}H and ^{13}C signals, DEPT-135, COSY-90, HMQC, and HMBC experiments are carried out. These first measurements should specify which protons will be suitable for quantification by means of NMR spectroscopy. Changing from a polar to an apolar solvent, from a nonaromatic to an aromatic one, or from a protic solvent to an aprotic one can be very useful for finding the optimal signal separation. Often the required signal separation cannot be achieved using only pure solvents. A poor solubility of the analyte can further limit the choice of solvents. So it is often necessary to use mixtures of solvents.

Our experiments have shown that when adding a second solvent to achieve complete dissolution of the analyte, it is not only responsible for complete dissolution, it can also influence the signal separation. For example, when using only solvent A for the first sample and only solvent B for the second sample, in some cases a complete signal separation of the components analyzed cannot be achieved. However, using an appropriate mixture of solvents A and B, an optimal signal separation is possible. This approach will be demonstrated for codergocrine mesylate.

1.2. pH value

It is well known that the chemical shift of atoms such as hydrogen, fluorine, and carbon is related to the amount of charge present at the atom.[5] Therefore, NMR spectroscopy can be used to determine the pK_a values and additionally to allocate the atom that is protonated or deprotonated.[6] In turn, solutions with different pH values can be applied to achieve an optimal signal separation.

Taking the quinolone antibiotic norfloxacin as an example, the influence of the pH value on the chemical shift of certain protons and carbon atoms can be observed.[7,8]

Three solvents, 0.01 M DCl, 5 M DCl, and 0.01 M NaOD, were used to examine the influence of the deprotonation of the carboxylic group in position 3 and the protonation of the external nitrogen of the piperazine ring in position 7 on the chemical shifts of certain protons (Figure 1). ^{1}H NMR measurements in acidic solutions mainly affect the chemical shifts of H2 and the piperazine protons H15/19 and H16/18. Decreasing the pH value (change from 0.01 M DCl to 5 M DCl), the resonances of H2 and the piperazine protons are strongly downfield shifted. The chemical shifts of the quinolone hydrogens H5 and H8 are only slightly influenced. 0.01 M NaOD caused a deprotonation of the carboxylic group and the resonance of H2 was upfield shifted, analogous to the signals of the protons of the piperazine ring. The resonances of H5 and H8 were downfield shifted in comparison with the measurements using DCl.

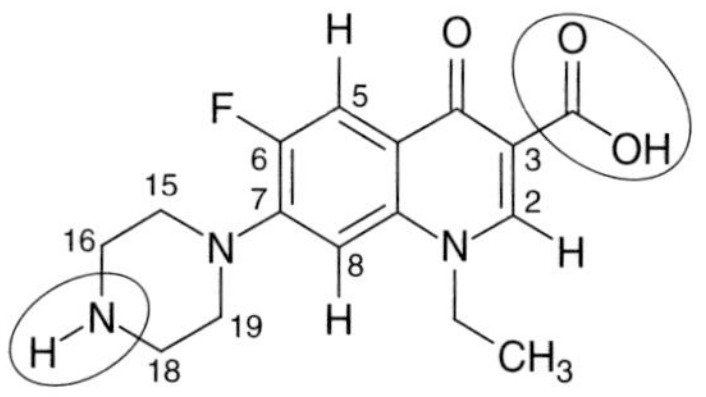

Figure 1 Structural formula of norfloxacin.

In analogy to ^1H experiments, the pH value also influences the chemical shift of certain carbon atoms, e.g., C3, COOH group, and piperazine carbon atoms and also the chemical shift of the fluorine atom in position 6.

1.3. Temperature

In some cases, the HOD signal of the applied solvent overlaps with signals in the region of interest. Varying the temperature can contribute to a better signal separation. Owing to the temperature dependence of the HOD signal, the HOD resonance can be shifted relative to the signals of the studied molecule. The shift corresponds to approximately 5 Hz/K at a 500 MHz NMR spectrometer.[9]

The temperature also plays an important role for the investigation of molecules with hindered rotations by means of NMR spectroscopy. Taking a closer look at the classical example of *N,N*-dimethylformamide (Figure 2: $R_1 = R_2 = CH_3$), the C—N bond between the carbonyl group and the nitrogen atom shows significant double-bond character.

At or below room temperature two CH_3 signals can be recognized in the ^1H spectrum, because of their different chemical environments relative to the carbonyl group (*cis/trans* isomerism). At room temperature, the rotation around the C—N bond is hindered because of the high rotational barrier (about 90 kJ/mol); thus, two isomers can be separated. When increasing the temperature, the two CH_3 signals broaden and finally merge to a single CH_3 signal, because, exceeding the coalescence temperature, the free rotation around the C—N bond predominates and the ability to distinguish the two CH_3 groups is lost (see also Chapter 1, Part I).[10,11,12]

1.4. Auxiliary reagents

In some cases signal separation can be achieved using the so-called "shift reagents". These reagents, containing a strong anisotropic paramagnetic centre in the molecule, can affect the chemical shift of certain protons in the order of several ppm, due to the ability to reversibly form complexes with organic molecules containing ion pairs.

Of interest is especially the chiral discrimination of enantiomers by means of NMR spectroscopy. Since the ^1H spectra of enantiomers are identical, there is no evidence whether the component exists as pure enantiomer or as racemate. Using chiral auxiliary reagents such as chiral lanthanide shift reagents, the determination of optical purity of enantiomeric mixtures is possible. The prerequisite for chiral discrimination is the formation of diastereomeric complexes between the enantiopure shift reagent and the component analyzed, leading to different chemical resonances for diastereomeric complexes of R and S species.[3,13]

$R^2R^1N{-}CH{=}O \rightleftharpoons R^1R^2N{-}CH{=}O$

Figure 2 Isomerism of the C—N bond.

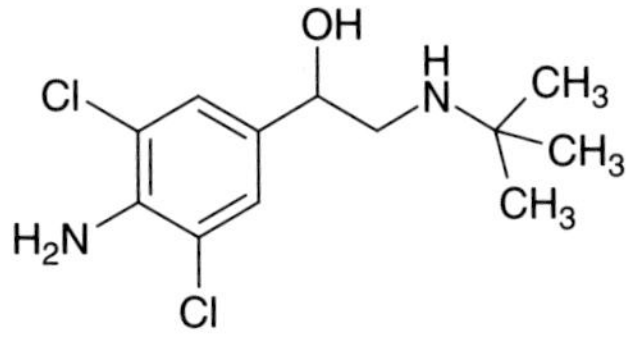

Figure 3 Structural formula of clenbuterol.

The group of Rothchild presented a study using a chiral lanthanide shift reagent, tris[3-(heptafluoropropylhydroxymethylene)-(+)-camphorato]europium (III), known as $Eu(HFC)_3$ or $Eu(HFBC)_3$, for enantiomeric determination of clenbuterol[14] (Figure 3).

For racemic clenbuterol, different chemical shifts for signals of NH_2, tertiary-butyl, methine, and aryl protons can be observed. The best signal separation was achieved for the aryl protons; here the singlet resonances are duplicated and a complete baseline separation was possible, facilitating the determination of the enantiomeric ratio of clenbuterol.

Antipodal analysis by means of NMR spectroscopy can be also carried out using chiral solvating agents such as cyclodextrins.[13] Owing to the existence of inclusion complexes formed between annular cyclodextrins and the molecules analyzed, a duplication of certain resonances can be achieved. When the included molecules are chiral, diastereomeric pairs can be formed, and due to the different physical properties of diastereomers, different ^1H spectra can be observed.

The group of Casy has investigated the phenomenon of inclusion complexes formed with different cyclodextrin molecules like α-, β-, and γ-cyclodextrins for a variety of chiral drugs, e.g., antihistaminics and analgetics by means of NMR spectroscopy.[15] The use of cyclodextrins for chiral discrimination has several distinct advantages over methods using lanthanide shift reagents. Cyclodextrins are soluble in water (and D_2O) and can be applied directly to chiral molecules, which are soluble in water. In contrast to lanthanide shift reagents, no broadening effects on the signals in the ^1H spectrum occur using cyclodextrins, and the chemical shifts of cyclodextrins (about 3.5–5.1 ppm) do not overlap in most cases with protons of interest, e.g., aromatic or alkenic protons.

1.5. Limitations

In contrast to chromatographic methods such as HPLC, NMR spectroscopy offers only a few parameters which can be optimized to achieve optimal signal separation.

When using HPLC, there are a variety of parameters which influence the separation.[16] A multitude of columns with different packing materials (normal phase ↔ reversed phase) and modifications of the stationary phase (e.g., endcapped), particle size, and shape, dimensions like length and internal diameter of the column can be chosen for the separation. Also the composition of the mobile phase offers a large number of possibilities to achieve optimal conditions for separation. In contrast to NMR spectroscopy, a variety of solvents, organic, and aqueous solutions especially aqueous buffer solutions with fixed pH value can be used, and different flow rates and elution programs (isocratic elution ↔ gradient elution) can be applied. The value of the column temperature and the addition of auxiliary reagents like cyclodextrins for enantiomeric determination or ion-pairing agents for the separation of polar components can also be relevant for optimal separation.

Using quantitative NMR spectroscopy the primary parameter which influences the signal separation is the kind of the solvent. In some cases pure solvents can provide baseline separation but sometimes also mixtures of two different solvents are necessary. For some applications also pH value, temperature, or auxiliary reagents can be of importance.

In the literature a lot of advantages of quantitative NMR spectroscopy over conventional techniques like HPLC are summarized.[17] In NMR spectroscopy the chemical shift is directly related to the structure of the molecule. Thus, for compounds, which show only slight differences in their chemical structure, it can be difficult to separate signals in spite of suitable parameter settings.

2. Codergocrine Mesylate

Codergocrine mesylate is a mixture of the methanesulfonate salts of three dihydroergopeptide alkaloids, namely dihydroergocornine (Cor), dihydroergocristine (Cr), and dihydroergocryptine, in a ratio of 1:1:1. Dihydroergocryptine is composed of two regio isomers, called α-dihydroergocryptine (α) and β-dihydroergocryptine (β). The structures of all these components are nearly identical, they only differ in the side chain in position 5′ (Figure 4).

2.1. ^{1}H NMR spectroscopy

In this section we will take a closer look to the steps to be taken to achieve the optimal signal separation of the multi-component drug codergocrine mesylate.

For optimal signal separation of all four components of codergocrine mesylate, a variety of solvents (Table 1) and mixtures thereof with different polarity, proticity, and aromaticity were applied to show their different influences concerning the chemical shifts (Table 1).

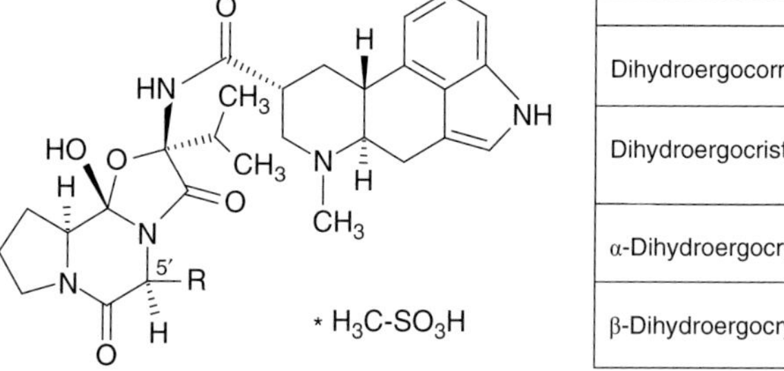

	—R
Dihydroergocornine	CH_3, CH_3
Dihydroergocristine	
α-Dihydroergocryptine	CH_3, H_3C
β-Dihydroergocryptine	H, CH_3, CH_3

Figure 4 Structural formula of codergocrine mesylate.

Table 1 Solvents used for signal separation of codergocrine mesylate, solubility drug

Solubility	Apolar solvents		Polar solvents	
	Aromatic	Non-aromatic	Protic	Aprotic
Good			Methanol-d_4	DMSO-d_6
Poor	Benzene-d_6	$CDCl_3$ dichloromethane-d_2	D_2O	

Codergocrine mesylate is completely soluble in methanol, acetone, and DMSO only. In solvents such as water, acetonitrile, chloroform, dichloromethane, and benzene the analyte shows a poor solubility. Therefore, small amounts of DMSO-d_6 or methanol-d_4 had to be added to achieve complete solubility of the analyte.

To reliably assign the ^1H and ^{13}C signals in the NMR spectra, DEPT, COSY, HMQC, and HMBC experiments were carried out. For all experiments, the standard pulse programs provided by Bruker were used.

The first measurements in DMSO-d_6 made clear that a determination of the four components of codergocrine mesylate on the basis of their different side chains in position 5′ is impossible. Strong signal overlapping of the protons of the CH, CH_2 and CH_3 groups (Figure 5: CH_X) and even the aromatic protons of

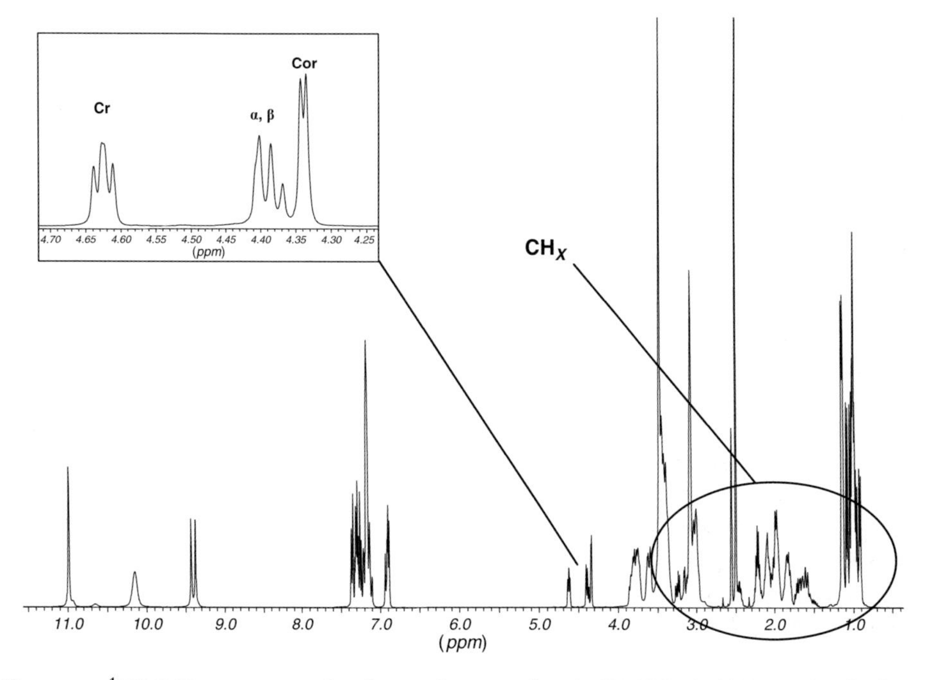

Figure 5 ^1H NMR spectrum of codergocrine mesylate in DMSO-d_6 (400 MHz). The inset shows an expanded view of the H5′ region which contains the signals of dihydroergocristine (Cr), α- and β-dihydroergocryptine (α, β), and dihydroergocornine (Cor).

dihydroergocristine showed that these protons are not suitable for quantification (Figure 5).

However, looking at the ^1H spectrum of codergocrine mesylate in DMSO-d_6 (Figure 5: inset), the signals of the tertiary proton in position 5′ could be suitable for quantification by means of NMR spectroscopy. The signals of the protons in position 5′ of dihydroergocristine and dihydroergocornine are clearly separated, only the signals of the two protons of α- and β-dihydroergocryptine overlap. The H5′ resonances of dihydroergocristine and dihydroergocornine are clearly recognized as triplet and doublet, respectively.

Using methanol-d_4 or D_2O the ^1H spectra are similar to the ^1H spectrum recorded in DMSO-d_6 (Figure 6(a)). The H5′ signals of α- and β-dihydroergocryptine overlap, only the signals of dihydroergocristine and dihydroergocornine are clearly separated. Nevertheless, DMSO-d_6 is preferred because there are not any solvent signals in the region of the H5′ resonances which can interfere with the signals of H5′. Applying methanol-d_4, no complete baseline separation between the HOD signal of methanol-d_4 and that of dihydroergocristine could be achieved. Using D_2O, the solvent signal overlaps with the H5′-signal of dihydroergocristine.

Using $CDCl_3$ a separation of the proton in position 5′ of α- and β-dihydroergocryptine is visible and the H5′ signals of dihydroergocristine and dihydroergocornine are clearly separated, too (Figure 6(b)). Experiments in $CDCl_3$–DMSO-d_6 as well as experiments in $CDCl_3$–methanol-d_4 did not show any differences in signal separation; therefore, DMSO-d_6 or methanol-d_4 can be used to dissolve the analyte in combination with an apolar solvent. Varying the amount of the polar component, one finds that signal separation is best for low amounts of DMSO-d_6 or methanol-d_4, respectively (Figure 6(b)). Nevertheless, a baseline separation of α- and β-dihydroergocryptine was not observed.

Replacing $CDCl_3$ by dichloromethane-d_2, the obtained ^1H spectra were similar to the ^1H spectrum recorded in $CDCl_3$. There was no improvement in signal separation by using dichloromethane-d_2.

Applying benzene-d_6 in combination with small amounts of DMSO-d_6, a complete baseline separation between the H5′ resonances of α- and β-dihydroergocryptine was possible. Due to benzene-d_6, the H5′ signal of α-dihydroergocryptine is downfield shifted relative to the other resonances of the proton in position 5′. When using a mixture with only small amounts of DMSO-d_6, the two H5′ signals of α-dihydroergocryptine and dihydroergocristine overlap (Figure 6(c): left expansion). Increasing the amount of DMSO-d_6, the signal of α-dihydroergocryptine is upfield shifted relative to the other resonances (Figure 6(c): right spectrum). Optimal signal separation could be achieved using a mixture of benzene-d_6 and DMSO-d_6 in a ratio of 10:1 v/v (Figure 6(d)). We observed similar qualitative behavior when replacing DMSO-d_6 with methanol-d_4. However, at a ratio of 1:1 v/v of benzene-d_6 and methanol-d_4 a separation was possible, whereas in DMSO-d_6 it was not (data not shown).

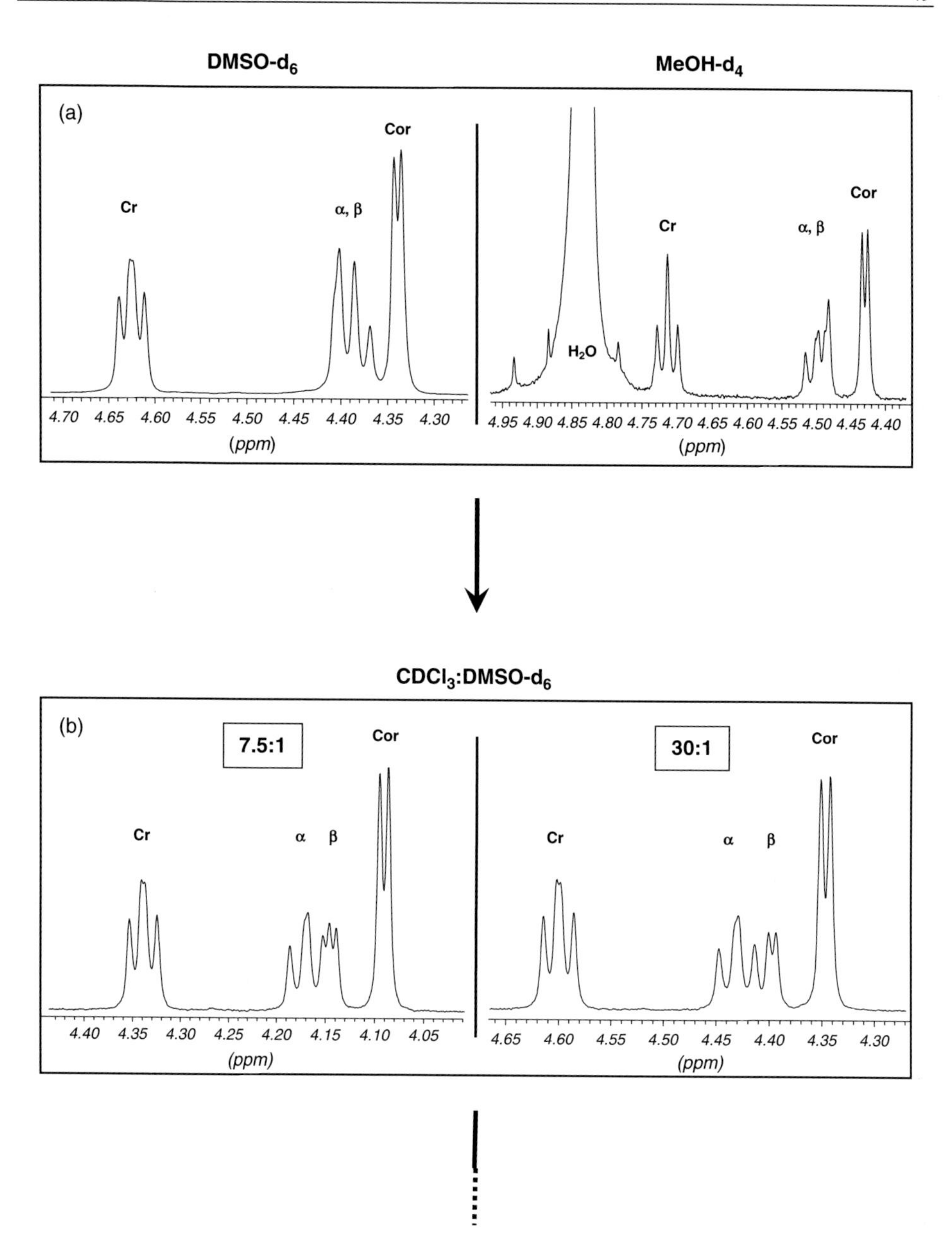

Figure 6 Partial ^1H NMR spectra of the H5′ region of codergocrine mesylate (400 MHz), ordered by improved signal separations, for the following solvents: (a) DMSO-d_6 (left), methanol-d_4 (right); (b) $CDCl_3$:DMSO-d_6; (c) benzene-d_6:DMSO-d_6; and (d) optimal signal separation is achieved with benzene-d_6:DMSO-d_6 in a ratio of 10:1 v/v. In the case of mixtures of different solvents, the corresponding ratio is given in boxes.

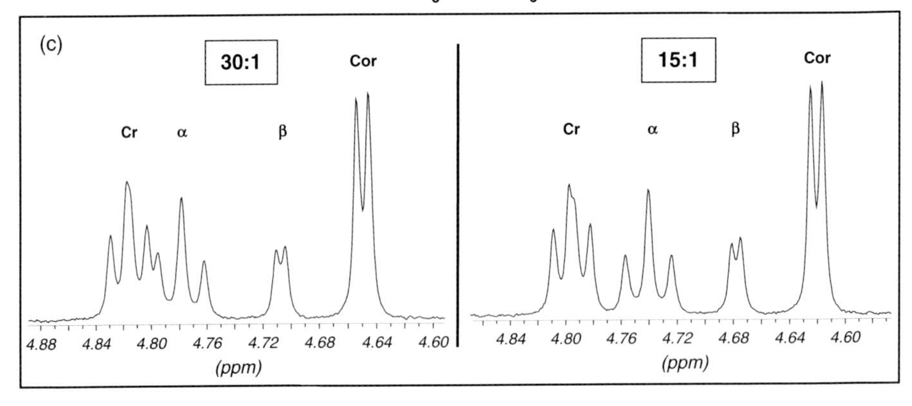

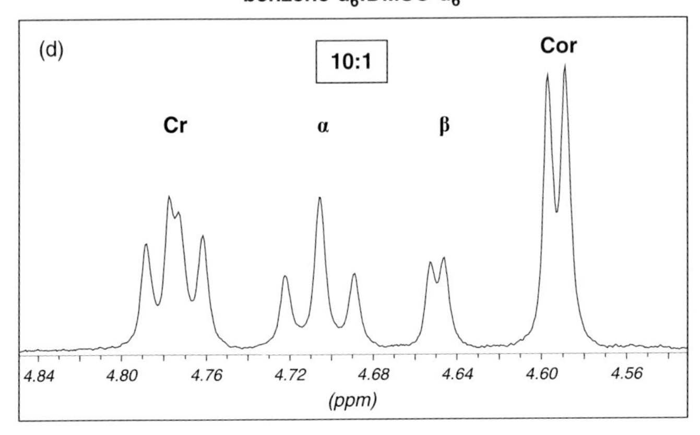

Figure 6 (*Continued*)

2.2. ^{13}C NMR spectroscopy

Taking a closer look to the ^{13}C NMR spectra of codergocrine mesylate, complete signal separation of the carbon atom in position 5′ of the four components could also be achieved. In comparison to the ^{1}H NMR

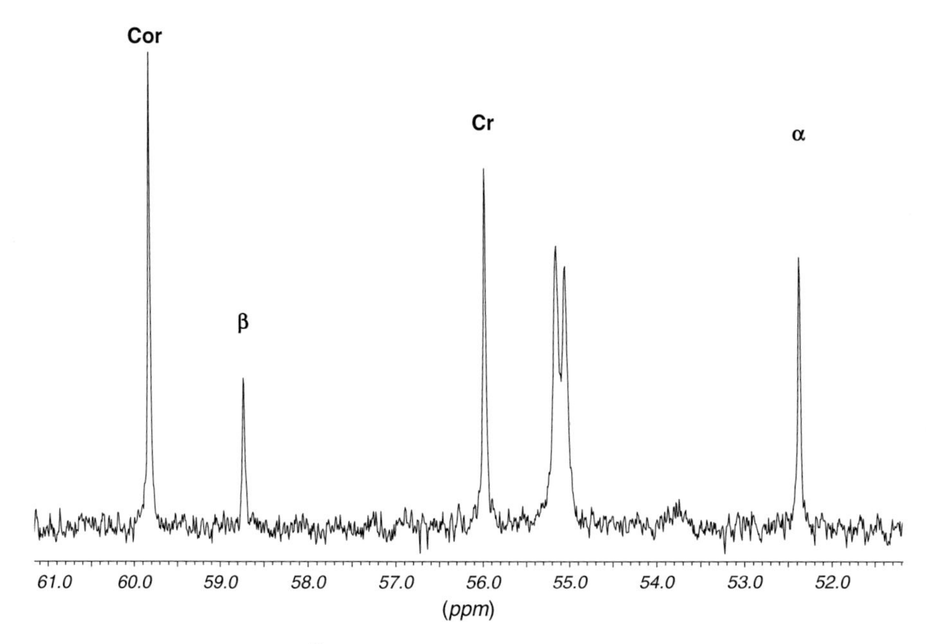

Figure 7 C5′ region of the ^{13}C NMR spectrum of codergocrine mesylate in DMSO-d_6 (400 MHz).

spectroscopy, clear signal separation was observed using pure DMSO-d_6 (Figure 7) or a mixture of $CDCl_3$–DMSO-d_6 (7.5:1, v/v). Therefore, the ^{13}C spectra could be used to determine the proportions of the four components in codergocrine mesylate.

In the introductory literature it is widely described that it is impossible to integrate ^{13}C signals and insert this technique in the field of quantitative analysis because of the nuclear Overhauser effect (NOE). When recording carbon spectra, broadband proton decoupling experiments are used and therefore the NOE appears. Owing to this phenomenon, the intensity of carbon signals can increase up to 200%.[3] Quantification by means of NMR spectroscopy takes advantage of the fact that the intensity of a signal (integrated signal area) is proportional to the number of contributing nuclei. This prerequisite is not fulfilled due to the phenomenon of NOE. Therefore, the applicability of ^{13}C NMR spectroscopy in the field of quantitative analysis is limited.

Looking at codergocrine mesylate, the carbon resonances of the carbon atoms in position 5′ were used to estimate the proportions of the four components. As all these carbon atoms are in the same position, we can therefore assume similar influence of the NOE on the carbon resonances in position 5′ and the intensity of the C5′ signals of the four components to increase in the same manner. The results obtained for codergocrine mesylate using ^{13}C NMR spectroscopy are in accordance with the results obtained using ^{1}H NMR spectroscopy.

2.3. High-performance liquid chromatography

HPLC is used as an orthogonal method to determine the accuracy of the NMR method. Two similar HPLC methods, which are prescribed in the European Pharmacopoeia (PhEur 6.0)[18] and in the United States Pharmacopoeia (USP 30)[19], were used for comparison. Both methods can be regarded as validated, so consequently, they are suitable for the intended use, namely, the quantification of the four components of codergocrine mesylate.

The aforementioned methods are very similar. The separation of codergocrine mesylate is carried out using reversed-phase columns packed with octadecylsilane chemically bonded phase particles (RP 18) in both cases, only the dimensions of the columns like length and internal diameter vary. The mobile phase consists of water, acetonitrile, and triethylamine (pH > 12), their quantitative composition slightly varies. In the European Pharmacopoeia (PhEur) method, higher amounts of acetonitrile are used; therefore, the retention times can be reduced. The HPLC system operates at a flow rate of 1.5 ml/min and the detection wavelength is fixed at 280 nm. For sample preparation, codergocrine mesylate is dissolved in a mixture of ethanol and (+)-tartaric acid (1:2)[18] or in a mixture of water and acetonitrile (1:1)[19]. Both methods achieve complete baseline separation. The order of elution is dihydroergocornine, α-dihydroergocryptine, dihydroergocristine, and β-dihydroergocryptine (Figure 8).

To determine the content of the four components the peaks were integrated and the percentage contents were calculated in two ways:

1. In the PhEur method, the content of codergocrine mesylate is determined using the normalization method, e.g., the content of one component is calculated as the peak area of this component divided by the sum of the peak areas of all components.[20]

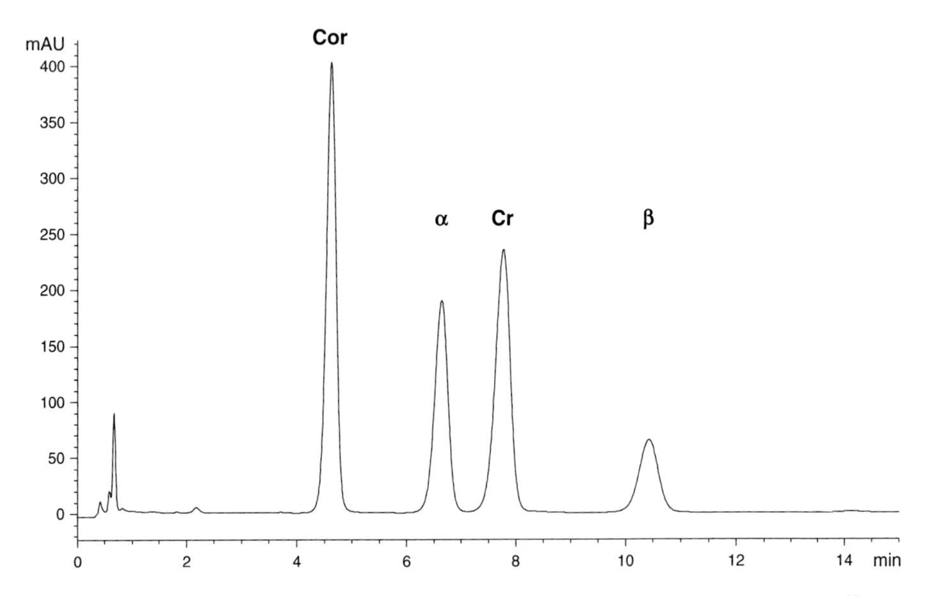

Figure 8 HPLC chromatogram of codergocrine mesylate obtained by the EP method.[18]

2. In the USP method, the peak areas are weighted with the corresponding molecular weights.[19] Consequently, the results are stated as mass percentage.

All HPLC experiments were utilized using both the methods, the corresponding results were in perfect agreement.

2.4. HPLC versus NMR spectroscopy

The results obtained from NMR spectroscopy in comparison to the HPLC results confirm that NMR spectroscopy is a very good orthogonal and complementary method to conventional HPLC (Table 2).

A comparison of NMR spectroscopy with HPLC shows a variety of advantages of NMR spectroscopy over the HPLC method, both generally and in the special case of the determination of the four components of codergocrine mesylate.

The primary advantage of NMR spectroscopy is its efficiency due to the lack of any preparation times. The analyte has to be weighed and dissolved in the solvent only, which is in our case a mixture of benzene-d_6 and DMSO-d_6 (10:1, v/v), and afterwards the analyte can be measured immediately. The experimental time depends on the concentration of the analyte. Using sample concentrations of codergocrine mesylate of approximately 15 mg/ml and 128 numbers of scans, the experiment is finished within 10 min. Using common NMR probe heads (5 mm) small amounts of deuterated solvent are needed (about 700-750 μl).

Using HPLC for the determination of an analyte much time has to be spent for the equilibration of the column. Complete baseline separation of codergocrine mesylate is only achieved using a mobile phase at pH > 12. To increase the stability of the column, an intermediate stage at pH 9 is inserted in the equilibration process. The stability of reversed-phase systems is given only over a pH value of 2–9. Using a mobile phase at pH > 12, the stationary phase is not stable over a long duration. Therefore, the column has to be washed every day after the measurements to neutral milieu with a mixture of water and acetonitrile to prolong the lifetime of the column. Some authors take the view that when using organic bases such as triethylamine instead of inorganic salts like NaOH, the column is at least as stable as under conditions of pH 8 although the pH value is about 12, and so the analytical column can be used for hundreds of routine analyses.[21,22,23] Other authors have the opinion that the lifetime of the column is limited to a few days only.[24] Our experiments have shown that the lifetime of the column can be prolonged when

Table 2 Results obtained by the four applied methods – for each component of codergocrine mesylate, the relative amount is given in mass percentage: dihydroergocornine (Cor), dihydroergocristine (Cr), α-dihydroergocryptine (α), and β-dihydroergocryptine (β).

	Cor (%)	Cr (%)	α (%)	β (%)
HPLC PhEur 6.0[18]	34.11 ± 0.09	32.76 ± 0.10	22.15 ± 0.04	10.98 ± 0.04
HPLC USP 30[19]	34.00 ± 0.11	32.85 ± 0.08	22.33 ± 0.11	10.83 ± 0.13
^{1}H NMR	33.57 ± 0.08	32.67 ± 0.04	22.80 ± 0.06	10.96 ± 0.06
^{13}C NMR	33.79 ± 0.27	32.97 ± 0.49	22.20 ± 0.20	11.04 ± 0.16

the column is washed to neutral milieu after measurements every day. However, we could not reach the mentioned lifetime.

When using the HPLC technique, often much time has to be spent for sample preparation, e.g., derivatization of the analyte. In the case of codergocrine mesylate, the preparation of the sample is very simple because the analyte only has to be dissolved in a mixture of ethanol and (+)-tartaric acid[18] or in a mixture of water and acetonitrile.[19] A further disadvantage is the large amount of solvent necessary for the HPCL separation.

The discussed example proves NMR spectroscopy to be a valid independent alternative to HPLC for quantification, as the corresponding results are in agreement. Additionally, the NMR method turns out to be more efficient than the conventional HPLC techniques.

ACKNOWLEDGMENT

Thank are due to the Federal Ministry of Economics and Labour (BMWA), the German Federation of Industrial Cooperative Research Associations "Otto von Guericke" (AiF), and the German Research Association of Medicine Manufacturers (FAH) for support of the AiF project 14842 N.

REFERENCES

1. R. Deubner, U. Holzgrabe, J. Pharm. Biomed. Anal 35 (2004) 459–467.
2. A.D. Buckingham, T. Schaefer, W.G. Schneider, J. Chem. Phys 32 (1960) 1227–1233.
3. H. Friebolin, Ein- und zweidimensionale NMR-Spektroskopie, 4. Auflage, Wiley-VCH, Weinheim, 2006.
4. R.J. Abraham, J.J. Byrne, L. Griffiths, M. Perez, Magn. Reson. Chem. 44 (2006) 491–509.
5. D. Farcasiu, A. Ghenciu, J. Progr. NMR 29 (1996) 129–168.
6. C. Hägele, U. Holzgrabe, in: Holzgrable, I. Wawer, B. Diehl (Eds.), pH-dependent NMR Measurements in NMR Spectroscopy in Drug Development and Analysis. Wiley-VCH: Weinheim, 1999, Chapt. 4, 61–70.
7. S.K. Branch, U. Holzgrabe, Magn. Reson. Chem. 32 (1994) 192–197.
8. D.A. Buckingham, C.R. Clark, A. Nangia, Aust. J. Chem. 43 (1990) 301–309.
9. T.D.W. Claridge, High-Resolution NMR Techniques in Organic Chemistry, Pergamon, Amsterdam, 1999.
10. H.A. Staab, D. Lauer, Chem. Ber. 101 (1968) 864–878.
11. H. Kessler, Angew. Chem. internat. Edit. 9 (1970) 219–235.
12. H. Günther, NMR-Spektroskopie, 3, Auflage, Georg Thieme Verlag, Stuttgart, 1992.
13. A.F. Casy, TrAC, 12 (1993) 185–189.
14. R. Martin, R. Rothchild, Spectrosc. Lett., 22 (1989) 511–532.
15. A.F. Casy, A.D. Mercer, Magn. Reson. Chem. 26 (1988) 765–774.
16. S. Kromidas, HPLC richtig optimiert, Wiley-VCH, Weinheim, 2006.
17. U. Holzgrabe, R. Deubner, C. Schollmayer, B. Waibel, J. Pharm. Biomed. Anal. 38 (2005) 806–812.

18. Monograph "Codergocrine mesylate", in European Pharmacopoeia. 6th edition, suppl. 6.0 European Directorate for Quality of Medicines, Strasbourg, France, 2008.
19. Monograph "Dihydroergotoxine mesylate" in United States Pharmacopoeia (USP) 30, States Pharmacopoeia Convention, Rockville, MD, USA, 2007.
20. Chapter 2.2.46, European Pharmacopeia, 5th edition. European Directorate for Quality of Medicines, Strasbourg, France, 2005.
21. V. Hartmann, N. Rödiger, W. Ableidinger, H. Bethke, J. Pharm. Sci. 67 (1978) 98–103.
22. V. Hartmann, M. Rödiger, Chromatographia 9 (1976) 266–272.
23. E. Papp, J. Chromatogr., 502 (1990) 241–242.
24. T. Dankházi, A. Oszlánczy, Á. Kósa, G. Farsang, ACH-Models in Chemistry 133 (1996) 351–355.

CHAPTER 2

NMR Applications for Polymer Characterisation

B. Diehl

Contents

Abstract

Spectroscopy of polymers is most practicable using solid-state nuclear magnetic resonance (NMR) and NMR in solutions. This chapter is limited to the one-dimensional NMR applications of polymers in solution. For a better overview, a subdivision of the different categories of chemical substructures is discussed.

1. Introduction

Polymers in drugs are widely used as formulation aids and as active ingredients. Because of their polymer structures, conventional chromatographic methods often fail for different reasons and the lack of suitable standards. Nuclear magnetic resonance (NMR) spectroscopy provides an alternative for qualitative and quantitative characterisation, as these analytical drawbacks do not exist. Although NMR spectroscopy is often regarded to be an expensive technique, it must be noted that the high investment costs for NMR spectrometers are compensated by the possibility of fast and powerful analysis.

In the literature, numerous examples of NMR in polymer analysis have been published.[1,2] In this analytical field, the classification into liquid- and solid-state

NMR is gaining momentum and this method of differentiation can be considered as a continuation of the pioneering days of NMR spectroscopy. The research groups of Bloch und Purcell had been working on the first successful NMR experiments at the end of 1945. At the Massachusetts Institute of Technology, Purcell[3] used an electromagnetic cavity filled with sodium paraffin – a solid – and on the other side of the United States at the Columbia University in California, Bloch[4] was working with water – a liquid. The result of the contest was an even score. In 1952 both got the Nobel Prize for physics.

Rather the solid-state NMR became a tool for physicists. NMR spectroscopy in liquids has been developed to become a tool mainly for chemists and nowadays plays a major rule in structure elucidation. Because of the evolution of magnets and data processing, meanwhile, the solid-state NMR is often used in material research. However, this chapter deals with the more common liquid NMR techniques. By means of applied examples of routine analysis a summary is given about the ability and the scope of NMR spectroscopy in polymer analysis. Thus, different chemical types of polymers shall be distinguished to get a better overview about the wide field of applications. The analysis of the following polymer classes will be presented: polydimethyl siloxane (PDMS); polysaccharides (modified starches, glucan, heparin, chondroitin sulphate, etc.); polyether-, polyethylene glycol (PEG)- and polypropylene glycol (PPG)-based emulsifiers; polyester, poly- and oligopeptides (goserelin, etc.); polyvinyl compounds (polyvinylpyrolidone (PVP), polyvinyl chloride (PVC), etc.)

2. Polydimethyl Siloxane

The ^{1}H NMR chemical shifts of different silicone molecules are located between 0 and 0.5 ppm. Within this frequency area, interferences with signals of other functional groups are uncommon. Besides this selectivity, the spectral dispersion allows the distinction between several cyclic and linear siloxane types. Quantitative analysis is enabled by the use of external and internal standards and correlates well with the documented Fourier transform (FT)/IR (infrared). The detection limit of the NMR quantification at the ppm level and the analysis in complex matrices enhances NMR to be the most powerful method. A second attribute of NMR analysis is the possibility to detect different chain lengths and end groups of polymers, e.g. trimethylsilyl (TMS) or hydroxy groups as well as other chemical modifications. Heteronuclear NMR spectroscopy of ^{13}C or ^{29}Si molecules completes the application table.

Classically, quantification of polydimethylsiloxanes (PDMS) is done by IR spectroscopy. A cross validation[5] shows the compatibility of the quantitative NMR (qNMR) to the common IR method. Furthermore, the sensitivity of qNMR is two orders of magnitude higher. The typical chemical shift of O—Si$(CH_3)_2$ groups about 0 ppm causes the enhanced selectivity and in addition the singlet signal structure of six protons represents one monomer unit of 75 d. This enables the quantification of PDMS in complex matrices even in trace amounts.

Usually for qNMR different internal standards can be selected. Diphenylmethylsilyl ethanol (DPSE) is very suitable because of its high molecular weight

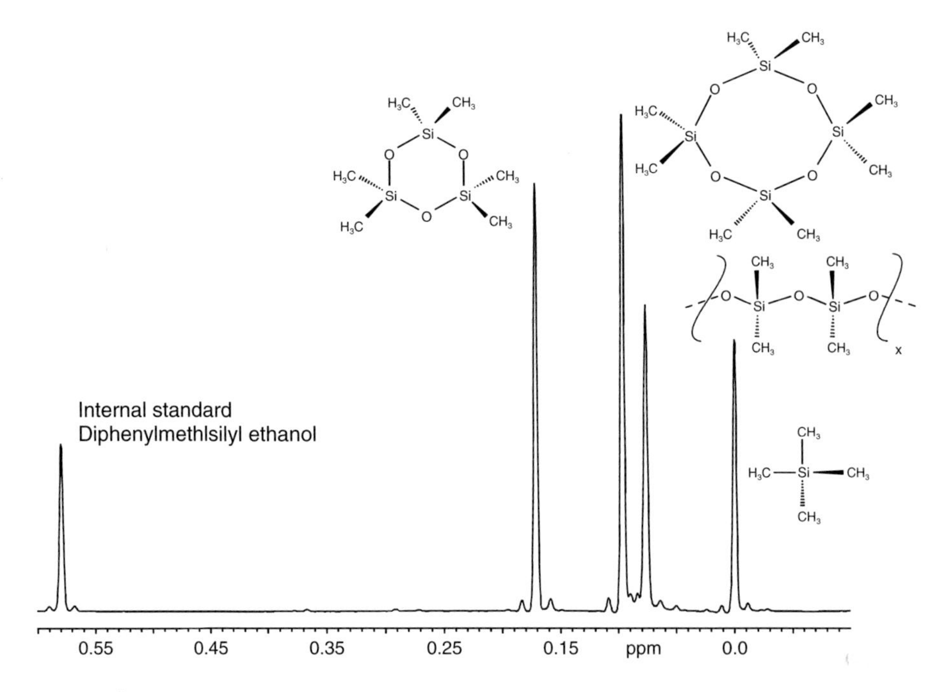

Figure 1 ^{1}H NMR spectrum of a mixture of different silicone types.

(convenient for an exact initial weight) and the chemical shift of its Si—CH_3 signal at $\delta = 0.55$ ppm, which does not interfere with ^{1}H NMR signals neither with silanes nor with other common molecules. Mostly PDMS are of analytical interest; however, different other silane signals are separated for simultaneous quantification (Figure 1).

Some other features of PDMS analysis can be demonstrated as follows. Typically, PDMS is terminated by TMS- or OH-chain ends (Figure 2). Both the TMS and the last chain link O—$Si(CH_3)_2$—OH are separated from the summarised signals of PDMS chain and the ^{29}Si satellites. Using the signal area ratio of the responding chain end groups and the summarised integral of all silane signals, a mean chain length can be calculated. Thus, the ^{29}Si- or ^{13}C NMR satellites with their defined natural contribution of 4.7% and 1.108%, respectively, are useful.

The total PDMS integral area can be normalised. This normalised sum integral divided by the integral of a single terminated $SiO(CH_3)_2$ unit multiplied by two due to the symmetry of the PDMS 2 results in the mean chain length. By multiplying the calculated result by 75 g/mol (one monomer dimethylsilane) the mean molecular weight is available. The chain end group is thereby neglected. Signal separation of the chain end groups is better for OH compared to TMS-terminated PDMS in $CDCl_3$ (cf. Figure 2 with Figure 3). For TMS-terminated PDMS benzene-d_6 is an useful alternative solvent (Figure 4).

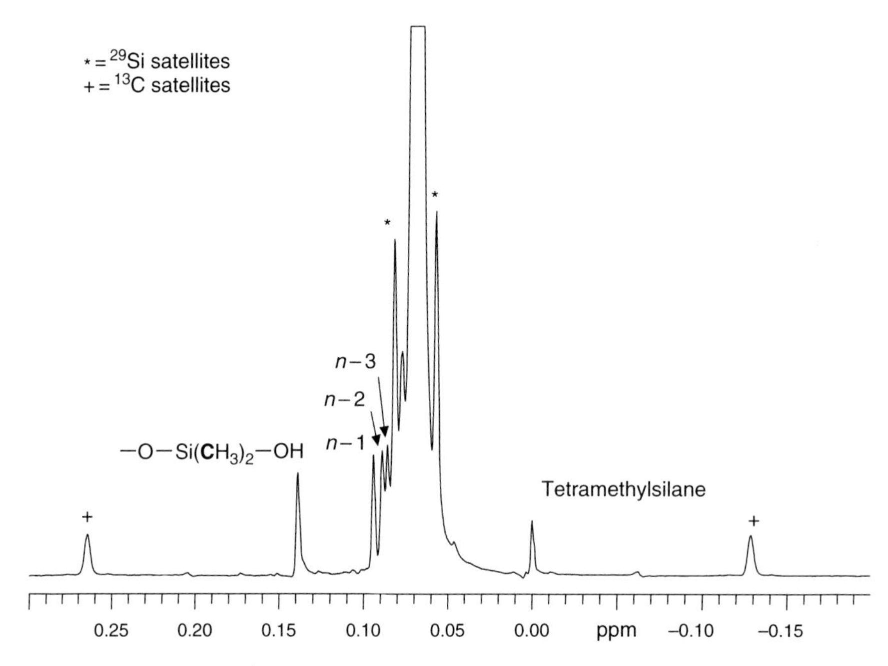

Figure 2 Highly resolved ^{1}H NMR spectrum of PDMS containing OH terminals.

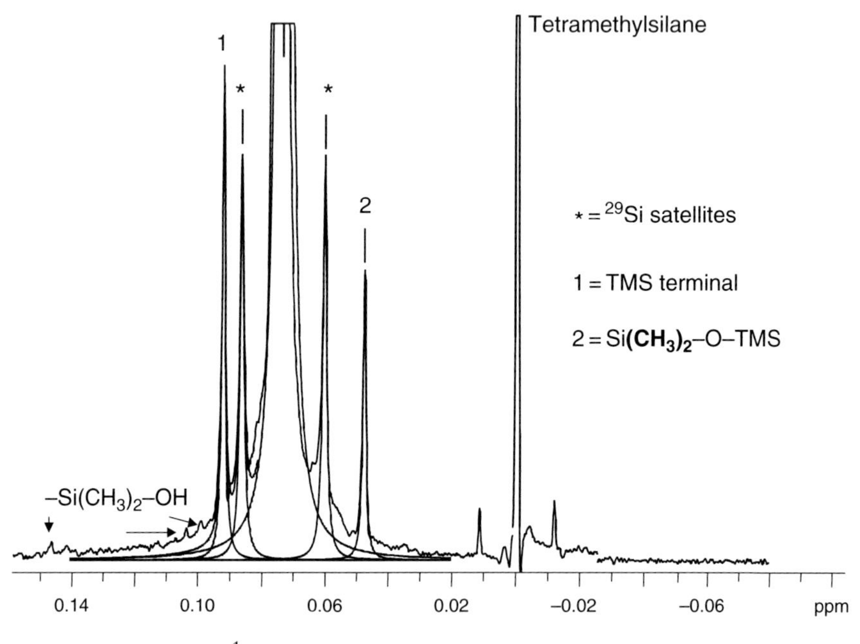

Figure 3 Highly resolved ^{1}H NMR spectrum of PDMS, TMS terminated in $CDCl_3$.

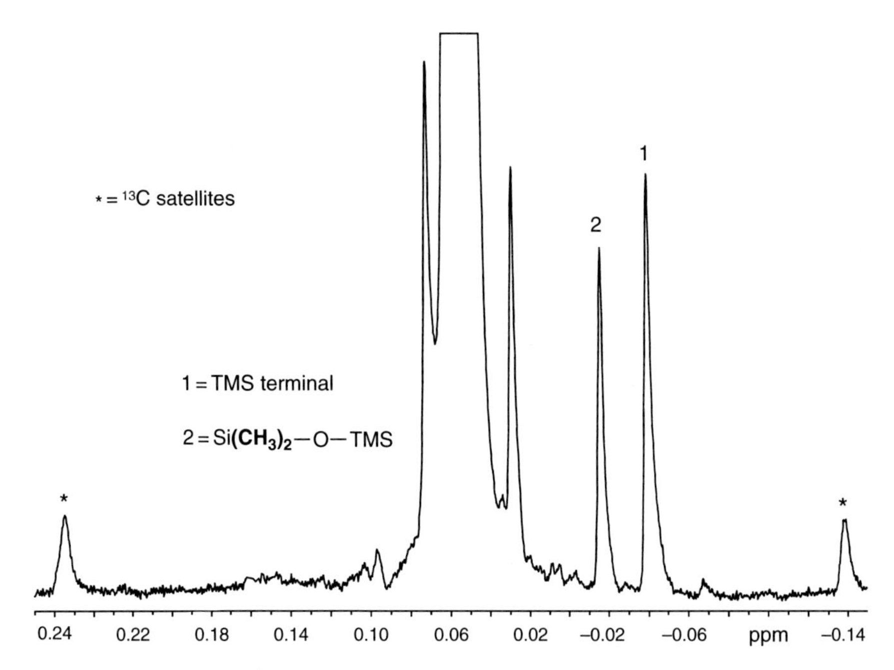

Figure 4 Highly resolved ^{1}H NMR spectrum of PDMS, TMS terminated in benzene-d_6.

3. Polysaccharides

Modified starches, glucan, heparin, chondroitin sulphate, etc. show significant signals in ^{1}H NMR spectra. This allows the detection of artificial groups, e.g. in modified starches (hydroxyethyl, etc.), the amount of branching as well as the identification of natural species in mucopolysaccharides such as heparin and chondroitin sulphate, where even the differentiation of animal origins (bovine, porcine, shark, chicken, etc.) is possible. In many plant extracts, the polysaccharide pattern can be used as principal ingredient for qualitative and quantitative origin test, e.g. aloverose – an acetylated polymannose for *Aloe vera* – or different glucans.[6,7]

To obtain a better signal resolution and a more detailed spectral information, it is recommended to run NMR spectra at high temperature, e.g. at 353 K. A second positive effect of the higher measuring temperature is the shift of the water signal to $\delta = 4.0$ ppm at room temperature; the corresponding signal appears at $\delta = 4.8$ ppm and interferes with the important anomeric protons of the carbohydrates.

Aloverose, a partly acetylated polymannose, stands exemplarily for plant polysaccharides and is characteristic and specific for *Aloe vera*. A ^{1}H NMR spectrum of the isolated polysaccharide is given in Figure 5. For qNMR spectroscopy, the acetyl proton signals can be used as a fingerprint for identification and their integral area for quantification. The signal defines the total amount of aloverose independent of the chain length and the molecular weight. A qNMR of *Aloe vera* including the quantification of aloverose is given in Part II, Chapter 3, Section 5.1.

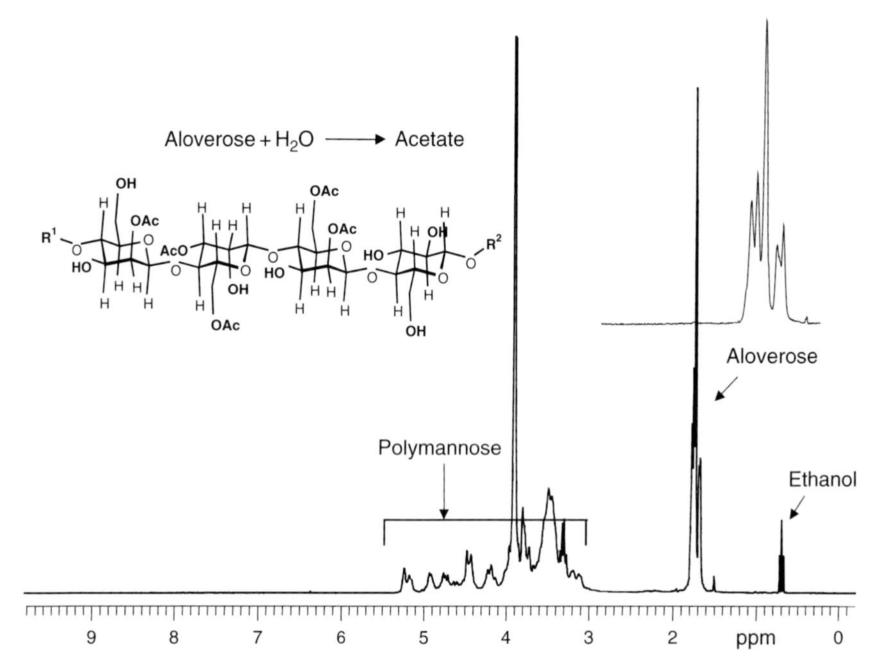

Figure 5 ^{1}H NMR spectrum of isolated Aloverose at 353 K.

The high-temperature ^{1}H NMR spectra of starch and modified starches gives the amount of the branching by integrating the signal at $\delta = 4.95$ ppm (Figure 6) and the reductive terminals of the chain at $\delta = 5.2$ and 4.6 ppm, respectively. The signal distribution of the proton signals between $\delta = 5.2$ and 5.7 (Figure 7) is an assignment of the modification by chemical substitution.

Chondroitin sulphate and heparin are examples for mucopolysaccharides from animal sources used for inner and outer medical applications. The chemical composition of these polymers depends strongly on the derived source (animal or plant species). NMR spectroscopy enables the identification of these sources;[8] thus it is possible to distinguish between chondroitin sulphate from different sources, e.g. porcine, bovine or shark (Figures 8–10). The signals of the specific molecules are used for the characterisation of both the source and the amount. For further details on the two-dimensional techniques, see Part III, Chapter 4. The importance of a proper analysis of heparins recently became apparent due to adulterations of heparin with oversulfated chondroitin sulphate (OSCS) which caused approximately 100 cases of death worldwide.*

The fingerprint-like information a high-temperature ^{1}H NMR spectrum of a polysaccharide provides can be completed, if the D_2O solution of the analysed sample is hydrolysed (~1–3 h at 80° C) after adding trifluoroacetic acid. A quantitative data

* T. Beyer, B. Diehl, G. Randel, E. Humpfer, H. Schäfer, M. Spraul, C. Schollmayer and U. Holzgrabe, Quality assessment of unfrictionated heparin using ^{1}H nuclear magnetic resonance spectroscopy J. Pharm. Biom. Anal. (2008), in press.

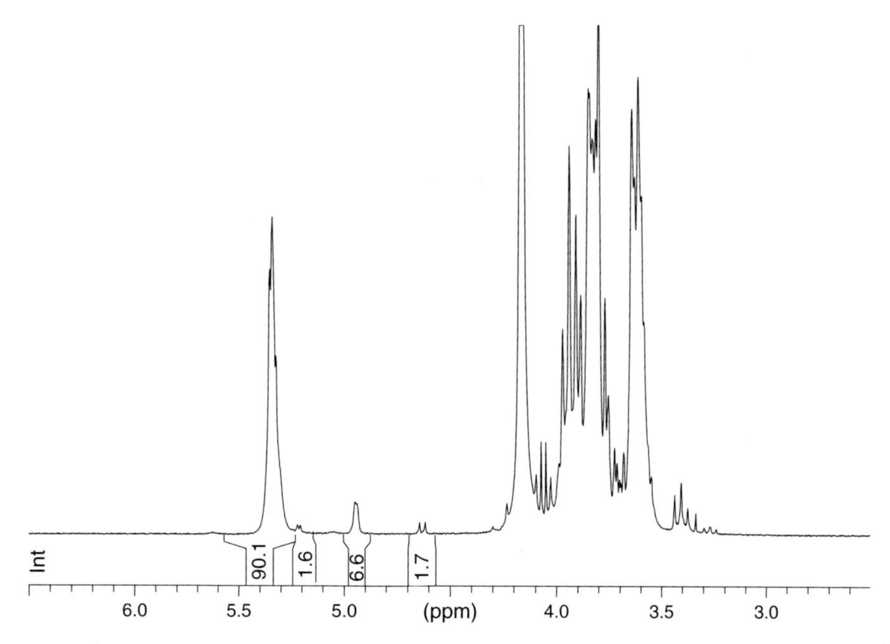

Figure 6 ^{1}H NMR spectrum of starch at 353 K in D2O.

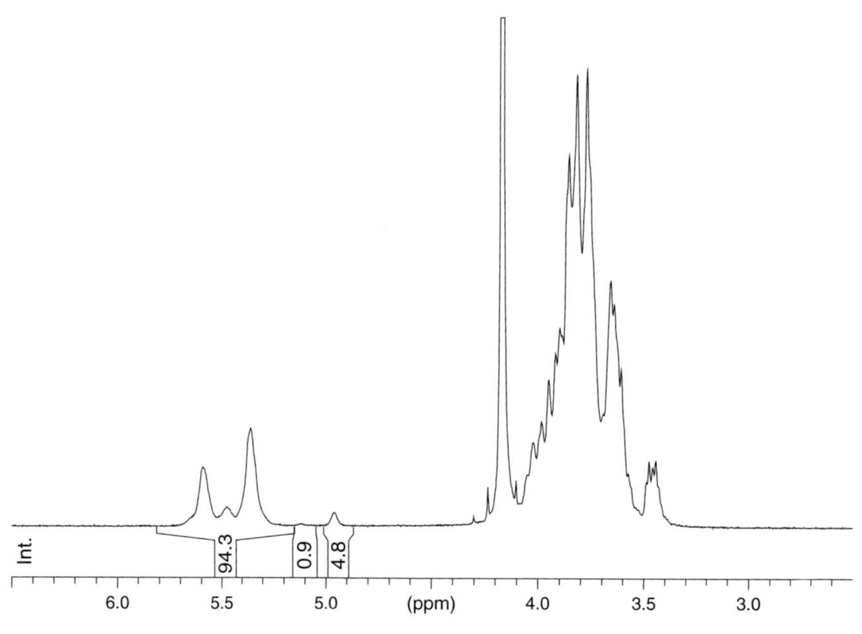

Figure 7 ^{1}H NMR spectrum of hydroxylethyl starch at 353 K in D_2O.

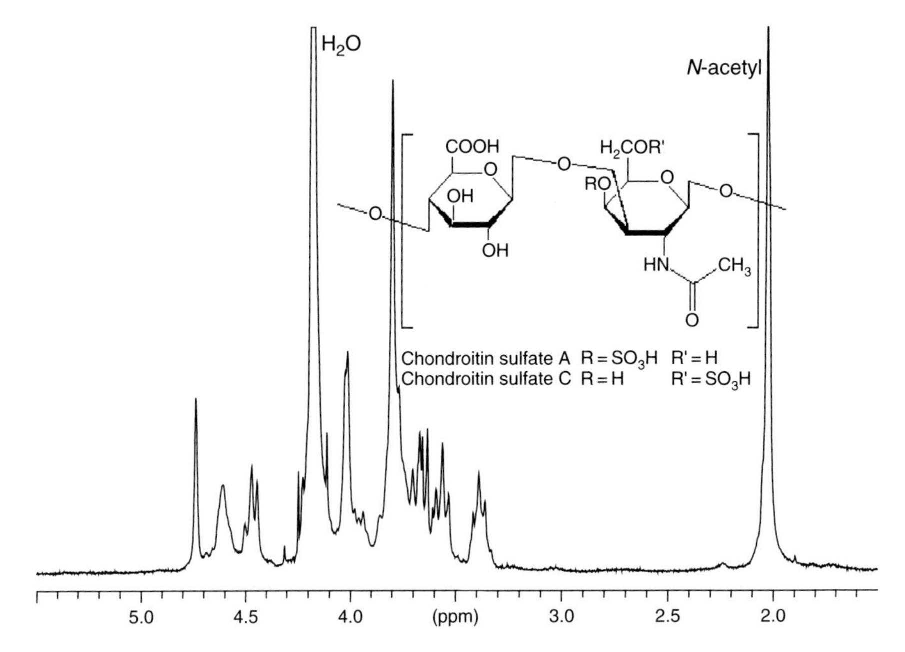

Figure 8 ^{1}H NMR spectrum of heparin of bovine origin at 353 K in D_2O.

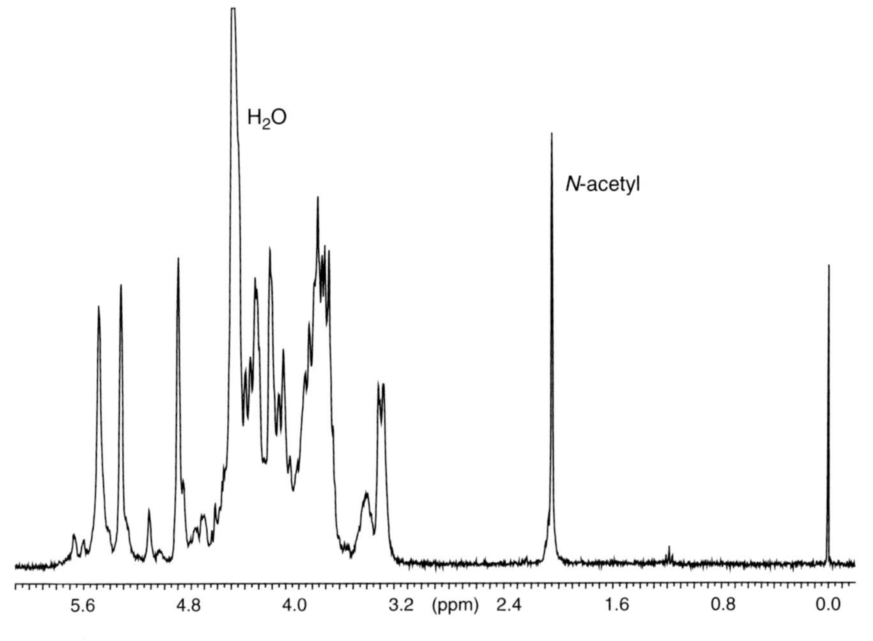

Figure 9 ^{1}H NMR spectrum of chondroitin sulphate of bovine origin at 353 K in D_2O.

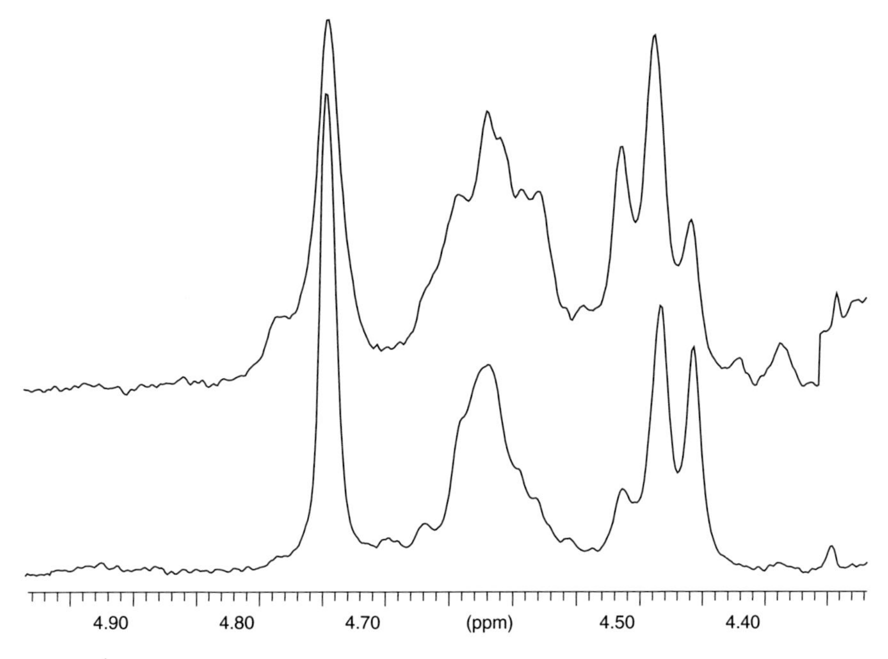

Figure 10 ^{1}H NMR spectra of chondroitin sulphate from different sources at 353 K in D_2O, shark (top) and bovine (bottom).

evaluation is possible to analyse the percentage composition of the monomeric sugar units. Enzymatic degradation and adjacent analysis by ^{1}H NMR spectroscopy is another tool to determine the source of chondroitin sulphates.[9,10]

4. Polyether Formulation Aids

Polyethylene glycol (PEG) and polypropylene glycol (PPG) based emulsifiers and detergents are analysed with regard to their mean chain length, the modification of their functionality (alkyl and aryl, sulphated and phosphated, etc.) and the ratio, respectively, during the sequence of copolymerisation of PEG/PPG. Because of the higher signal dispersion, ^{13}C NMR is preferred. ^{1}H NMR is useful for quantitative analysis of polyethers in matrices. Qualitative NMR methods are already part of the European Pharmacopoeia.[11]

A quantitative characterisation of emulsifiers by NMR based on the ratio of oxyethylene- and oxypropylene-containing copolymers is documented in the official method USP 30.[12] Furthermore, qNMR enables the determination of polyoxyethylene chain length in PEG or modified PEG ab initio, the identification of different emulsifier types in complex matrices and their absolute quantification. Figure 11 shows the ^{13}C NMR spectrum of PEG 400. The integral areas of the terminal methylene groups marked with a and b compared to the sum of the chain methylene groups (c and d) are the basis for chain length determination.

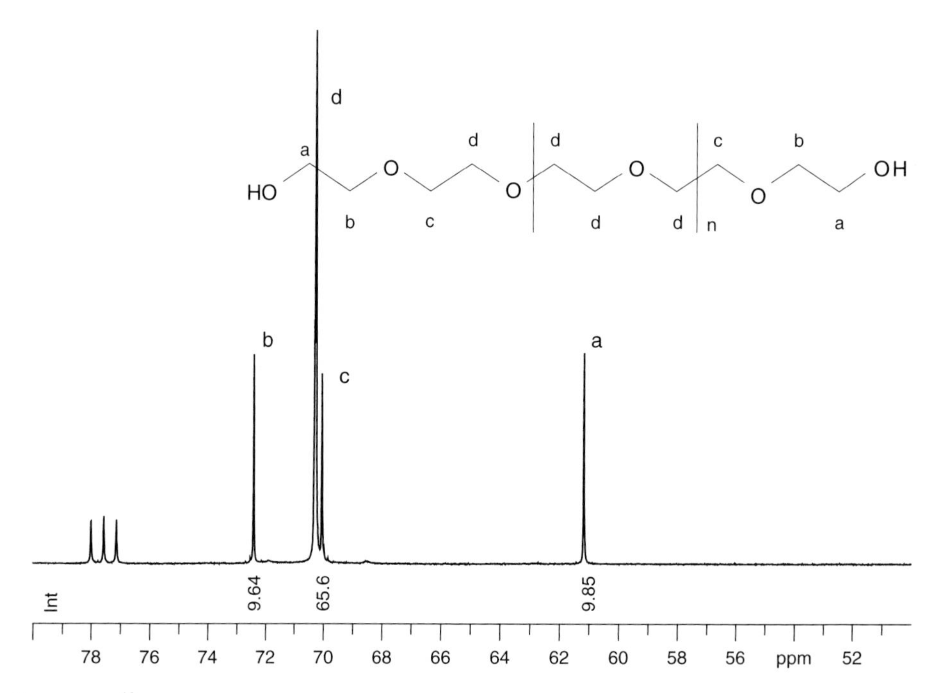

Figure 11 ^{13}C NMR spectrum of PEG 400.

Besides its classical domain of qualitative analysis, qNMR is a suitable method to quantify an emulsifier in a drug formulation. Emulsifiers often consist of a basis structure that represents a mixture of similar molecules (isomers or homologues, e.g. *para*-isononylphenol, fatty acid mixtures). By reaction of these basic materials (alcohols, amines, fatty acids, triglycerides, carbohydrates, phenols, etc.) with ethylene oxide and/or propylene oxide, polymeric structures are formed with a wide range of molecular weight distribution. Nonylphenol PEG (NPE) represents an emulsifier type that is predestined for ^{1}H NMR analysis. The functional groups of these mixtures accumulate to separate the signal areas in the NMR spectra whose characteristics are shown in Figure 12.

By normalisation of the integral areas to 10 for a single proton, the mean ethylene glycol (EO) distribution as well as the alkyl chain lengths are directly detectable. The mean number n of EO units (—CH_2—CH_2—O—) in the respective sample is

$$n = \frac{19.95 + 350}{40} = 9.25 \tag{1}$$

The integral value of the alkyl chain between $\delta = 2.0$ and 0.5 confirms the presence of 19 protons in one part of the alkyl group. Quantification of NPE from complex formulations is possible by standard addition or by using the specific integrals at $\delta = 6.8$ ppm (b) or at $\delta = 4.1$ ppm (c), respectively.

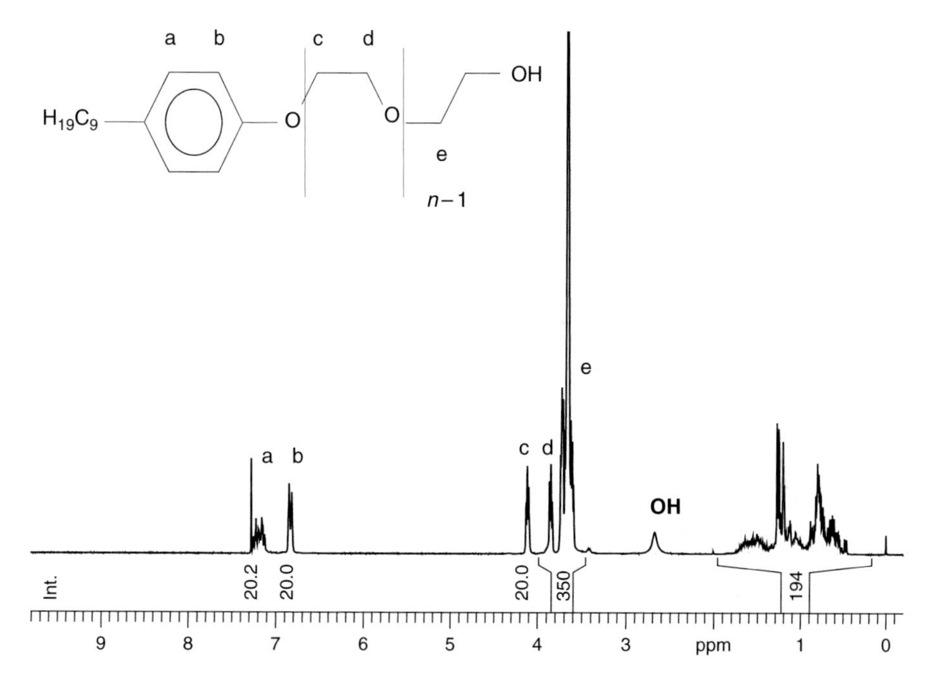

Figure 12 ^{1}H NMR spectrum of NPE IN $CDCl_3$.

By knowledge of the mean molecular weight, the quantification of NPE in complex formulations is possible by using the specific integrals at $\delta = 6.8$ ppm (b) or $\delta = 4.1$ ppm (c) with an internal standard or alternatively by standard addition. Figure 13 shows the NMR spectra observed by the standard addition of NPE in a formulation. Increasing integral areas are observed for different NPE signals. However, for calculation the aromatic protons at $\delta = 6.8$ ppm are most suitable.

Besides NPE, there are more complex emulsifiers based on alkylated phenols (Figure 14). Differences in their molecular structure can be identified by NMR spectroscopy. The terminal parts of the respective emulsifier molecules, e.g. the aromatic protons, enable the calculation of the PEG and PPG chain lengths and so the mean molecular weight.

The higher spectroscopic dispersion of ^{13}C NMR enables the distinction between ethylene glycol- and propylene glycol-terminated systems as shown in Figure 15.

Finally, the following example documents the power of ^{31}P NMR in quantitative analysis. The quantification of mono- and diesters of phosphorous acid with modified polyethylene glycol (e.g. Fosfodet®) and non-reacted phosphorous acid and the pyrophosphate analogues (at $\delta = -9$ to -11 ppm), respectively, are shown in Figure 16. The molar ratio of all difference species can easily be calculated from integral areas. A suitable internal standard can be used for an absolute quantification.

By changing the pH value, the spectroscopic dispersion of the phosphorous monoester allows the quantification of single species with different chain lengths (Figure 17).

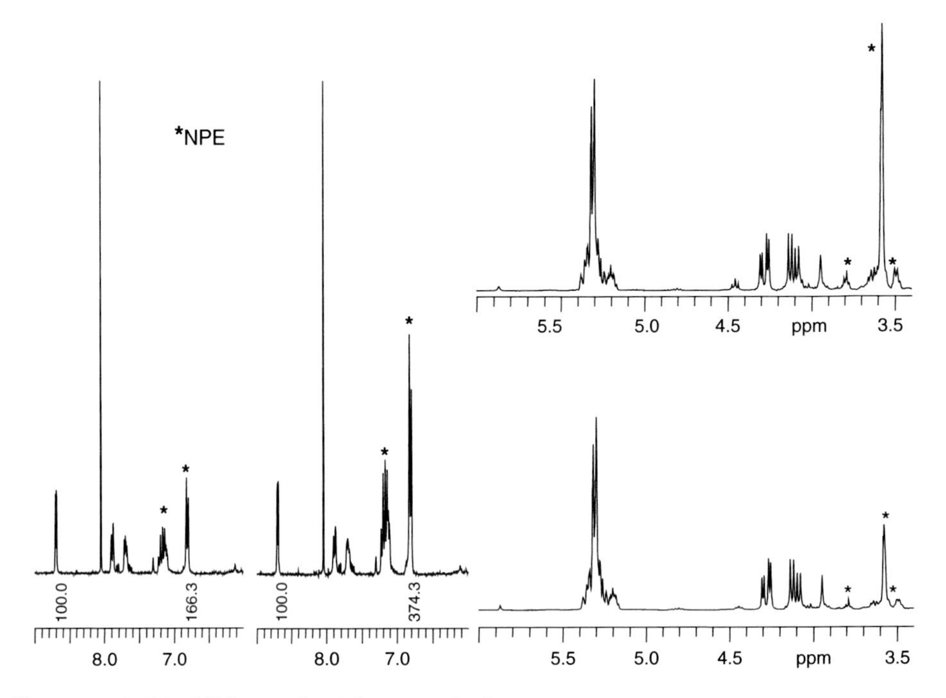

Figure 13 NPE addition on the right respectively upper trace.

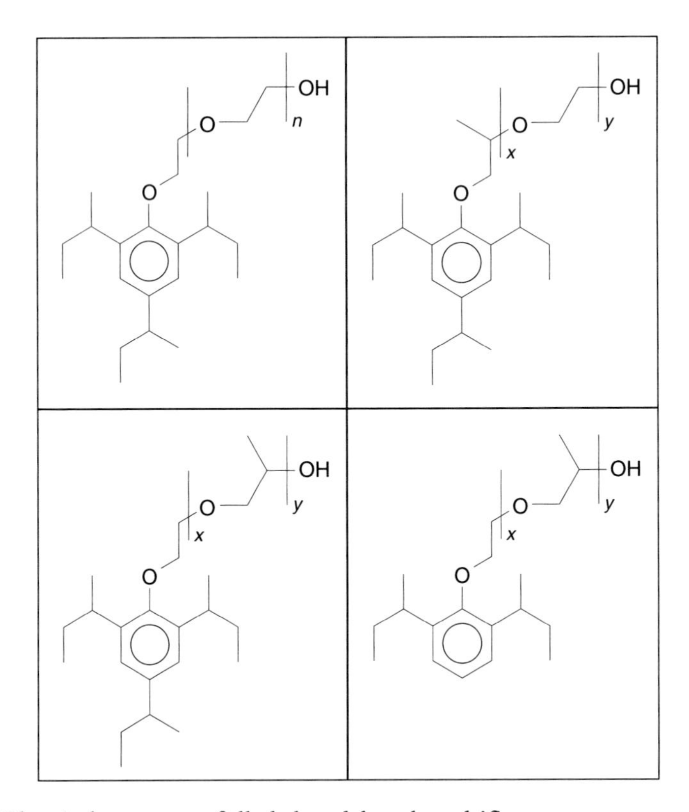

Figure 14 Chemical structures of alkyl phenol-based emulsifiers.

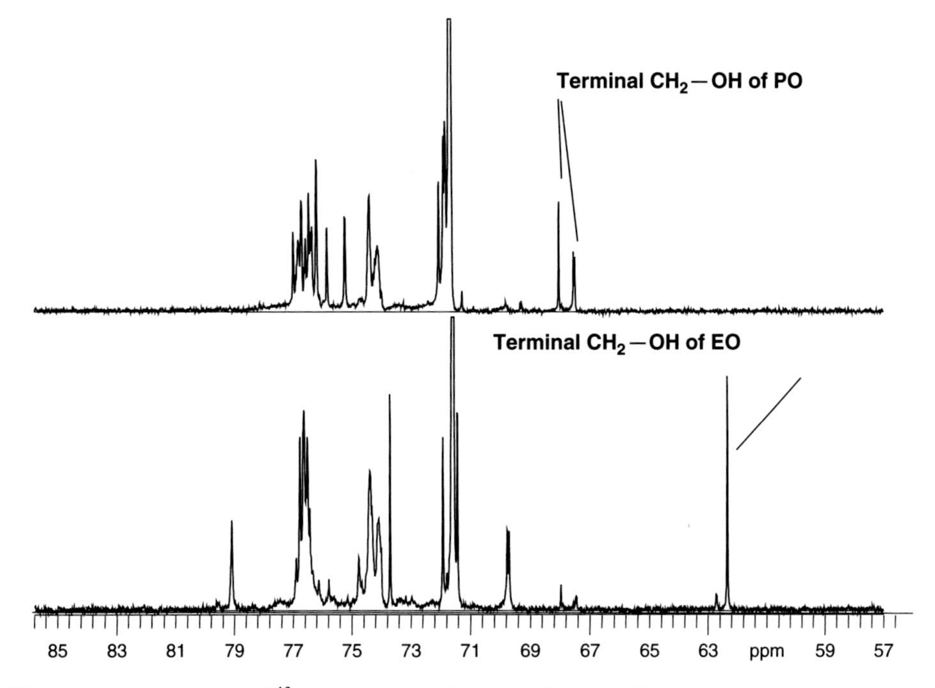

Figure 15 Comparison of ^{13}C spectra of different EO/PO copolymers.

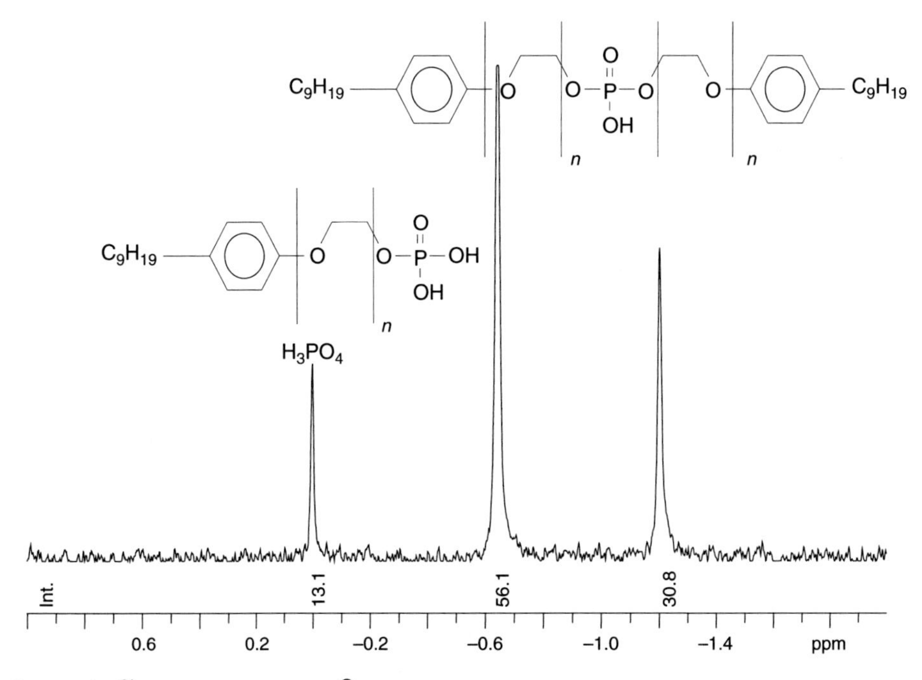

Figure 16 ^{31}P NMR of a Fosfodet® sample, pH ~1 in D_2O.

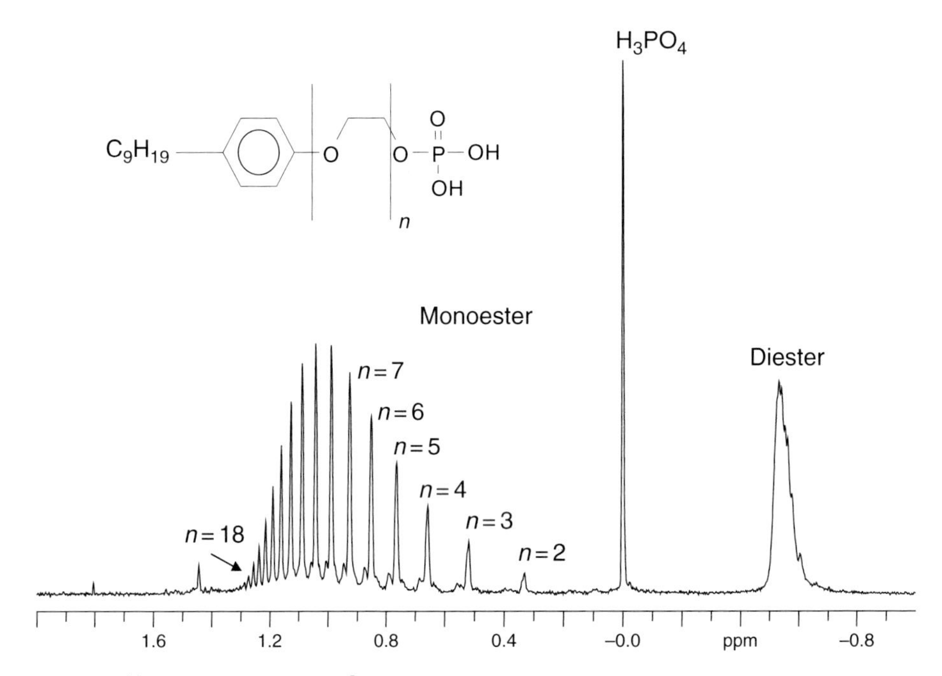

Figure 17 ^{31}P NMR of a Fosfodet® sample, pH ~8.5 in D_2O.

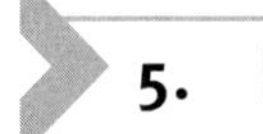

5. FORMALDEHYDE

Formaldehyde (FA) seems to be a monomeric molecule with a simple chemical structure. A quantification of free FA in several cleaning products and disinfectants, however, is an analytical challenge.[13] Owing to its hazardous effects, many different types of FA donor systems, binding methylol groups on glycols, amines or urea are commercially available, which should release only small amounts of free FA when dissolved in water.

The composition of corresponding adducts and free FA strongly depends on the concentration because of the equilibrium reactions. Every analytical step of extraction or concentration necessary for chromatographic analysis will disturb this equilibrium system. Headspace GC is the common quantification method but it is an indirect method that only detects the gaseous $H_2C{=}O$.

^{13}C NMR spectroscopy of neat liquids avoids an equilibrium disturbance and enables the quantification of different FA adducts and the hydrate even in complex matrices. The necessary lock is provided externally by the use of a glass capillary containing a deuterated solvent.

Figure 18 shows a ^{13}C NMR spectrum of commercially available FA solution (37%) containing ~15% mol methanol ($\delta = 50$ ppm) and its adducts ($\delta = 55$–60 ppm). At $\delta = 130$ ppm, the triplet of benzene-d_6 can be seen, which represents the signals of the external lock. More detailed information about the composition of FA adducts is shown in Figure 19. The FA hydrate signal appears at $\delta = 84$ ppm ($n = 0$).

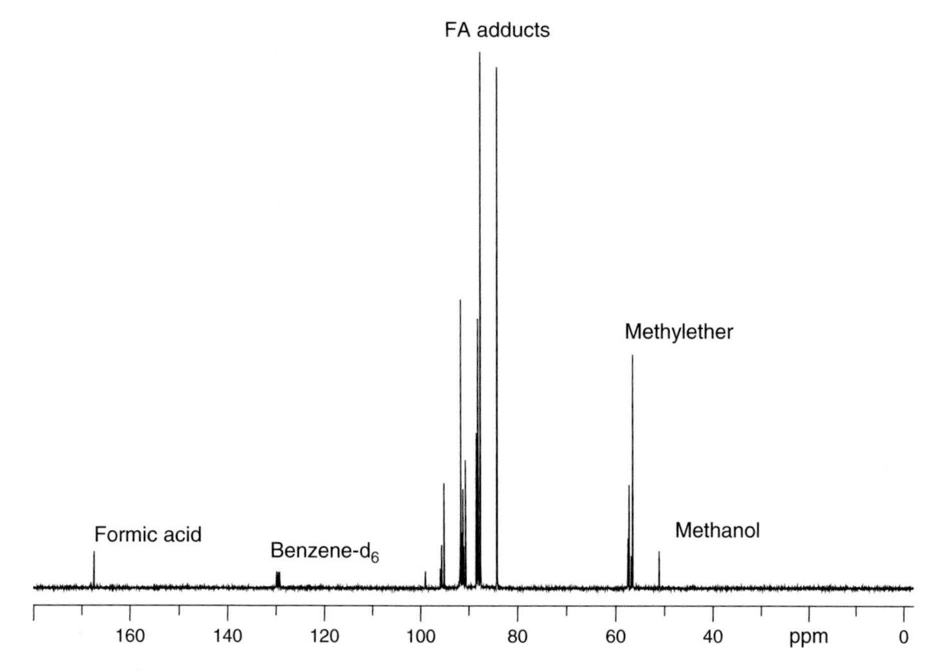

Figure 18 ^{13}C NMR spectrum of FA solution (37%).

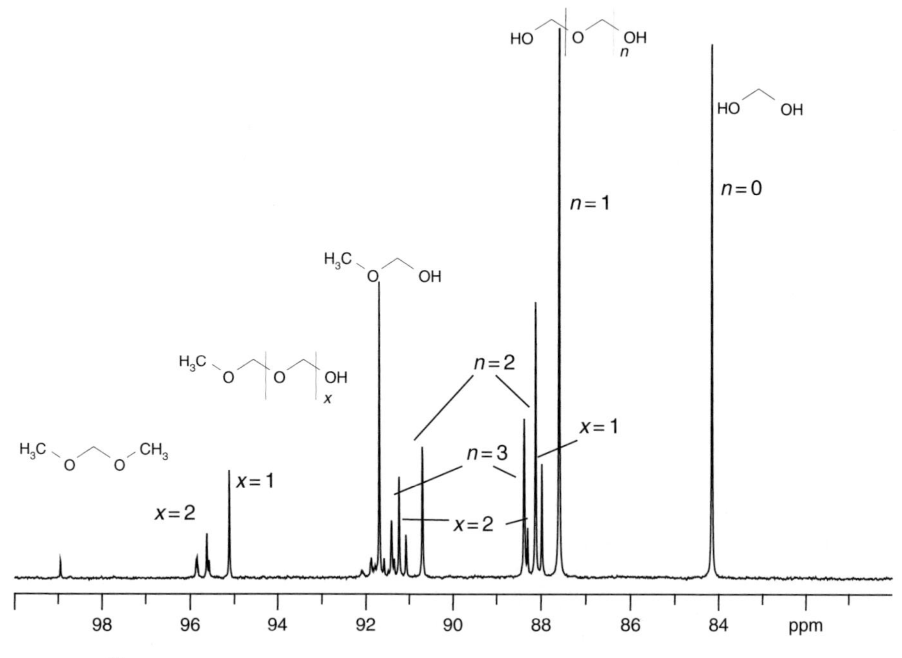

Figure 19 ^{13}C NMR spectrum of FA solution (37%), detail.

Figure 20 shows the same region of a 2% FA solution. The relative amount of FA hydrate now dominates the composition. Using a non-reactive substance as internal standard, the absolute amount of FA hydrate can be quantified. Finally, the absolute amount of free FA hydrate depending on the concentration is analysed using the internal standard method (Figure 21).

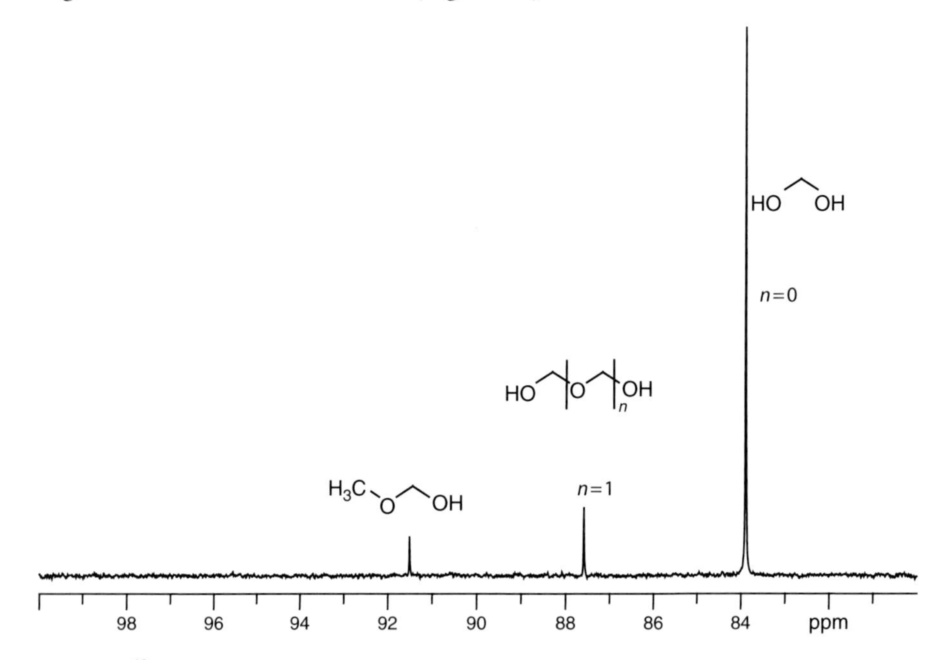

Figure 20 ^{13}C NMR spectrum of FA solution (2%), detail.

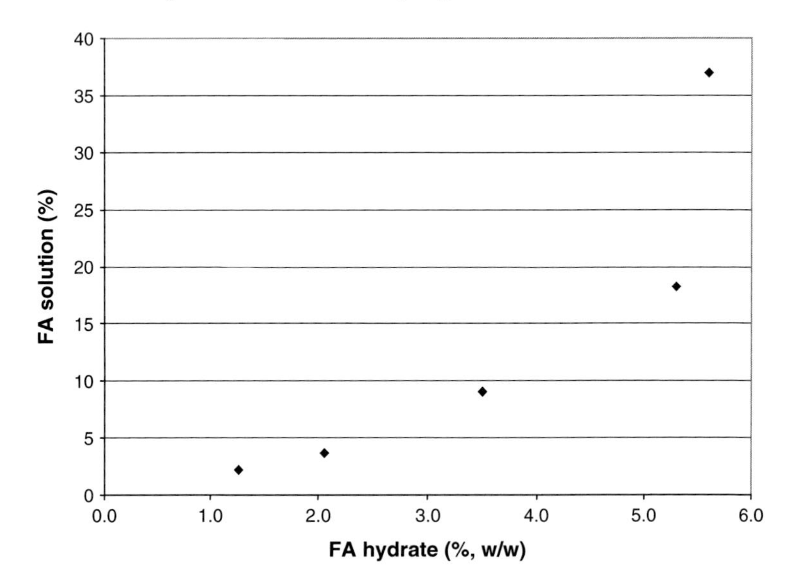

Figure 21 Concentration of FA hydrate depending on the concentration in water.

6. Polyester

The NMR analysis of polyesters is quite similar to polyether analysis. An example is the analysis of polylactide (Resomer R 202 H, Figure 22) and polylactide/polyglycolide copolymers (Resomer RG 503 H, Figure 23) respectively, where not only the amount but also the stereochemistry of the copolymer composition can be characterised.

The integral areas of the lactide methine proton at $\delta = 5.2$ ppm (1 H) is compared with the methylene signal area of the glycolide copolymer part at $\delta = 4.8$ ppm (4 H). Taking the numbers of protons of the corresponding signals into account, a ratio of 1:0.48 can be calculated from the spectrum shown in Figure 23.

For pure isotactic, syndiotactic and atactic polymerisation, see results in the formation of defined ^{13}C-signal groups[2] that can be read like a fingerprint in the NMR spectra. For example, see the carbonyl region of a ^{13}C NMR spectrum of atactic resomer R 202 H (Figure 25). The corresponding isotactic shows a single signal.[14] The triad structures and substructures of the stereochemistry can be clearly observed (Figure 25).

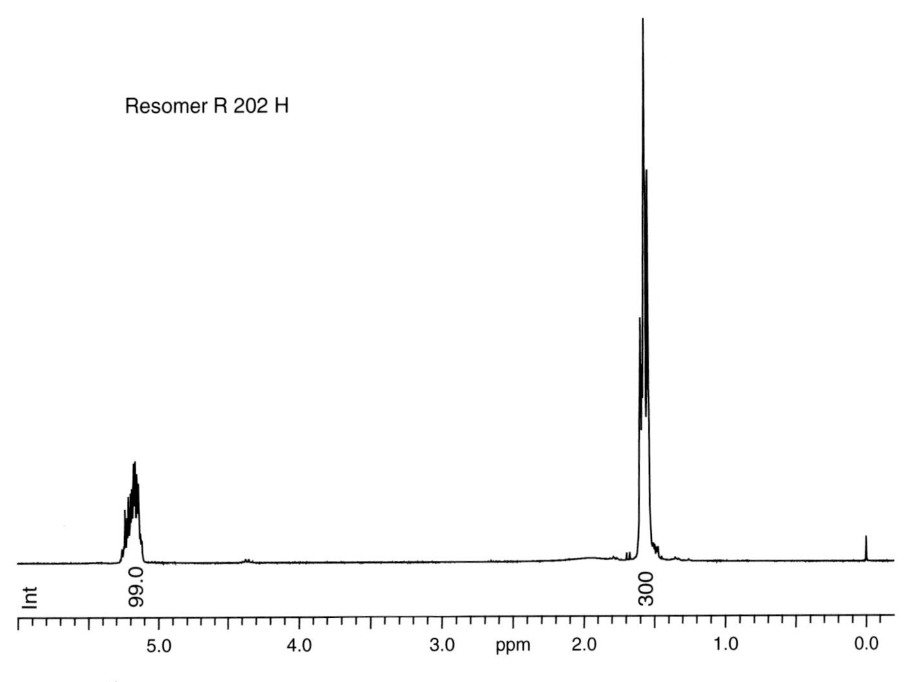

Figure 22 ^{1}H NMR spectrum of resomer R 202 H in $CDCl_3$.

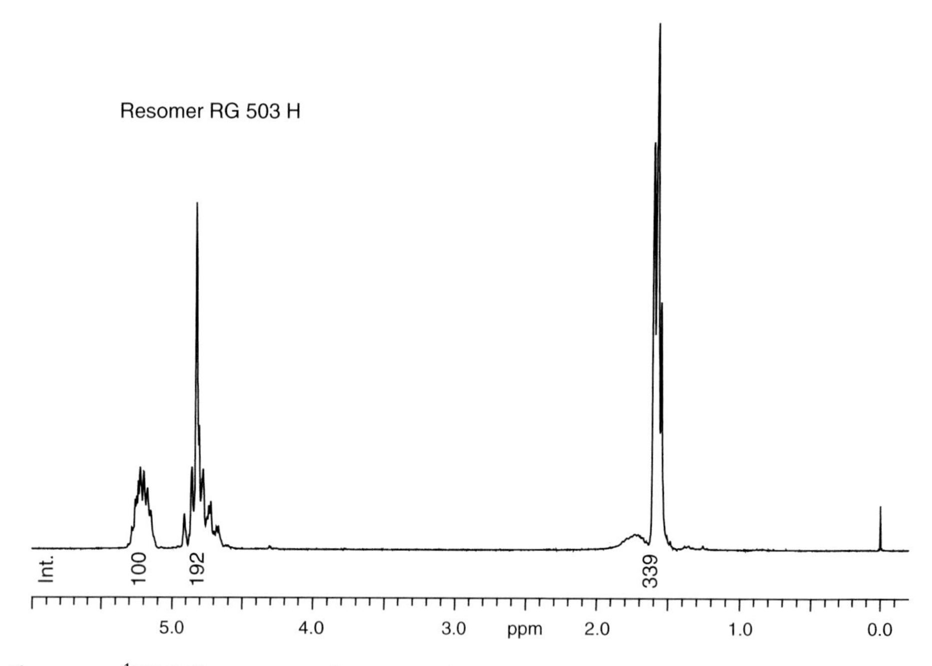

Figure 23 ^{1}H NMR spectrum of resomer RG 503 H in CDC_{13}.

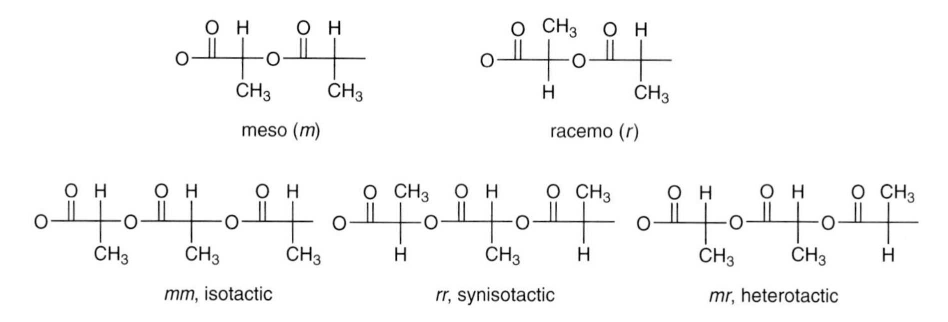

Figure 24 Rotated Fischer projections of polylactide-forming triads.

7. Poly- and Oligopeptides

Currently, the analysis of peptides[15] comprises the most sophisticated NMR techniques. A citation from a book promotion "NMR is better suited than any other experimental technique for the characterisation of supramolecular systems in solution"[16] meets this conclusion. In Figure 26, the ^{1}H NMR spectrum of albumin is shown as an example for an unsorted type, which is not of much interest in pharmaceutical analysis.

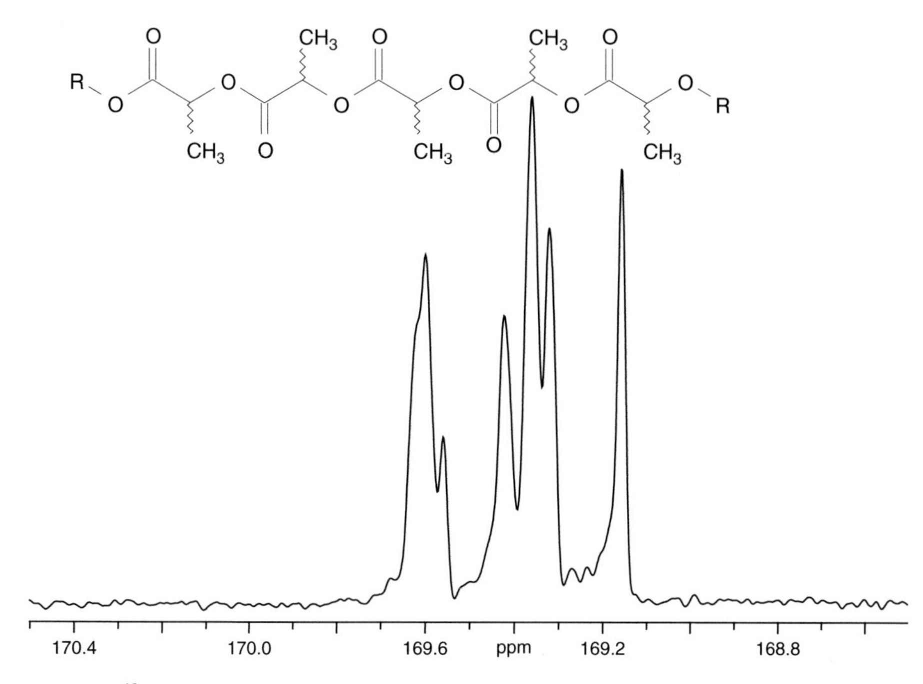

Figure 25 ^{13}C NMR spectrum of an atactic polylactide in $CDCl_3$, carbonyl region.

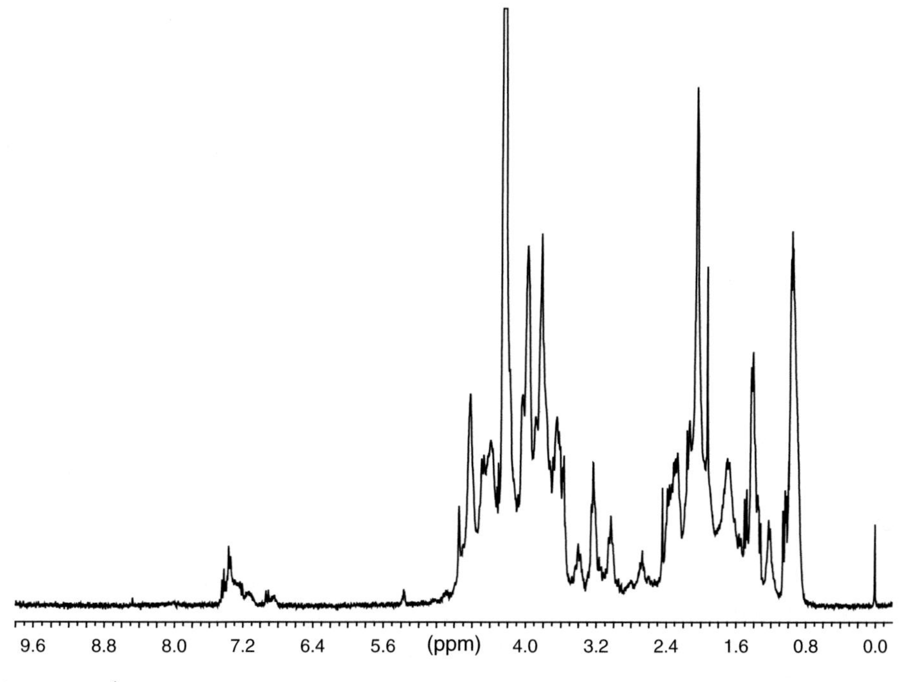

Figure 26 ^{1}H NMR spectrum of albumin at 353 K in D_2O.

In turn, the oligopeptides are of very high interest and described in a separate chapter. An example of structure elucidation will be given below. By a combination of one (Figures 27 and 28)- and two-dimensional NMR spectroscopy, the sequence and the signal assignment of an oligopeptide (Figure 29[17]) were performed. For example, a detail of the H,H COSY (correlation spectroscopy) spectrum is shown in Figure 30.

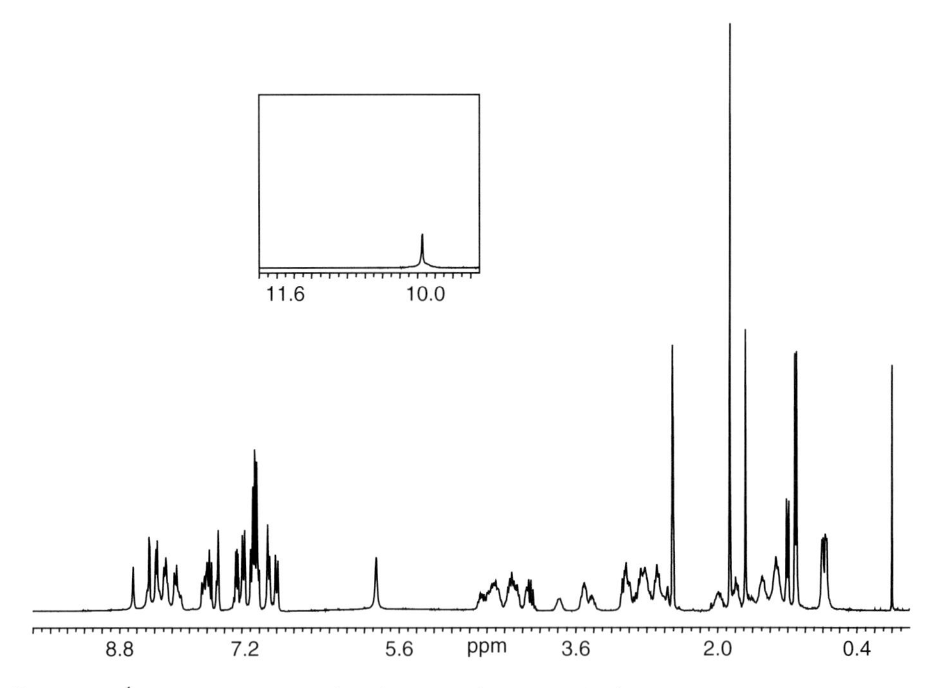

Figure 27 ^{1}H NMR spectrum of an oligopeptide in DMSO-d_6.

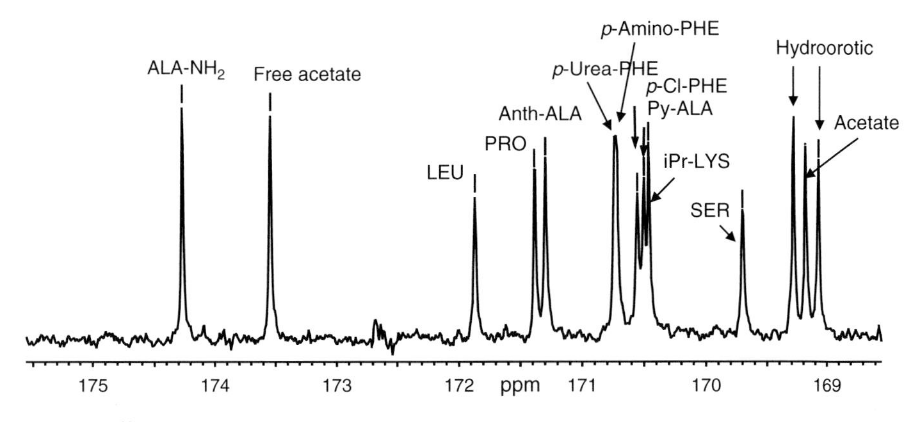

Figure 28 ^{13}C NMR spectrum of an oligopeptide in DMSO-d_6, carbonyl region.

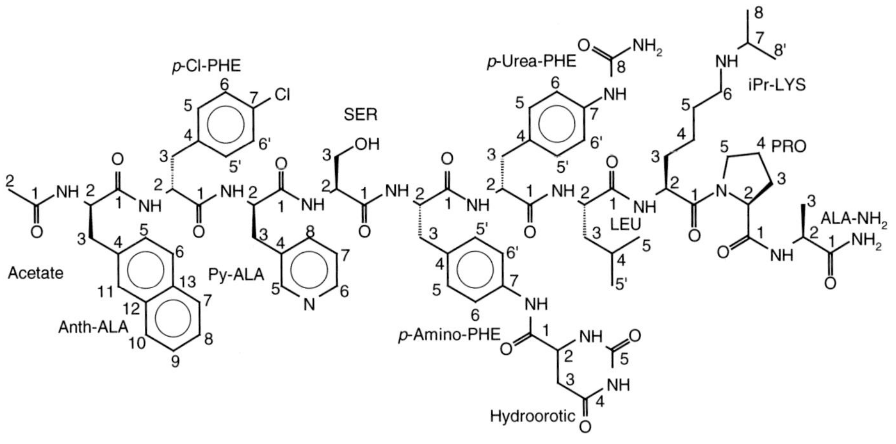

Figure 29 Chemical structure of an oligopeptide.

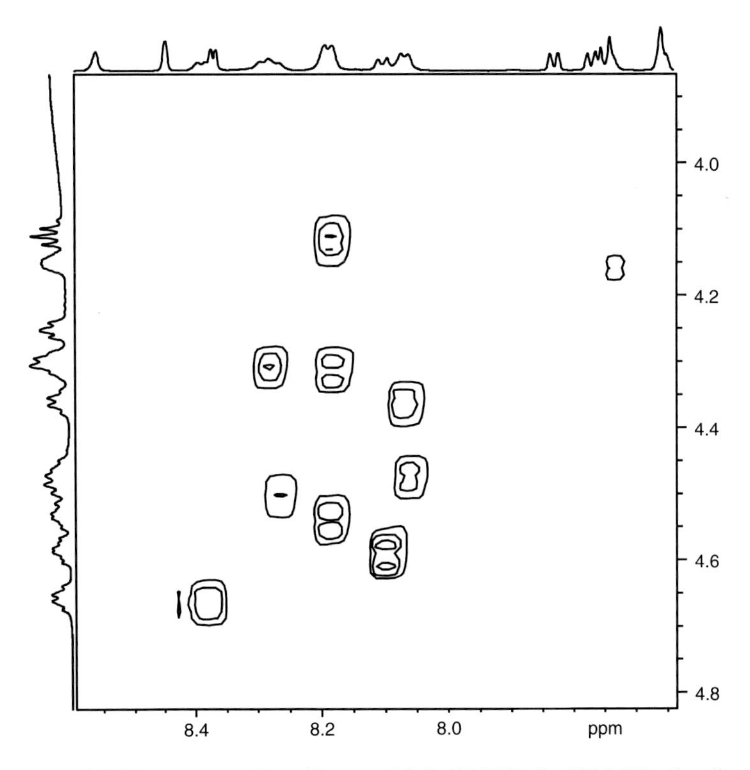

Figure 30 H,H COSY spectrum of an oligopeptide in DMSO-d_6, 600 MHz, detail.

8. Polyvinyl Compounds

Polyvinyl compounds are often used as excipients in tablets. Various copolymers, e.g. of vinylacetate (VA) and vinylpyrolidone (VP), can be characterised by ^{1}H NMR. The quantification in a drug formulation can be performed using the standard addition method.

Quantitative analysis of PVC demonstrates another feature of qNMR. Non-volatile and non-chromatographible polymers are analysed in one step together with low-molecular-weight compounds, e.g. plasticiser. Figure 31 shows the ^{1}H NMR spectrum of the PVC sheathing of a common electric cable dissolved in THF-d_8. Adding an internal standard, both the compounds – the PVC and the plasticiser – can be quantified simultaneously. The triplet character of the PVC CH signals at $\delta = 4.6$–4.1 ppm is caused by the rotation–isomeric structures. This CH signal group interferes with that of the ester CH_2-group of the plasticiser. Because of the molar integral values, this fact easily can be corrected by a corresponding integral subtraction.

Polyvinylpyrolidone represents another polymer. It is used in drug formulations. Different types of the PVP material can be identified by their characteristic signal fingerprints. After identification, a corresponding reference material can be used for standard addition. Figure 32 shows a ^{1}H NMR spectrum of a drug containing PVP as a formulation aid. The broad typical polymeric signals between $\delta = 1$ and 2.5 ppm are used for identification and quantification, other PVP signals interfere with signals

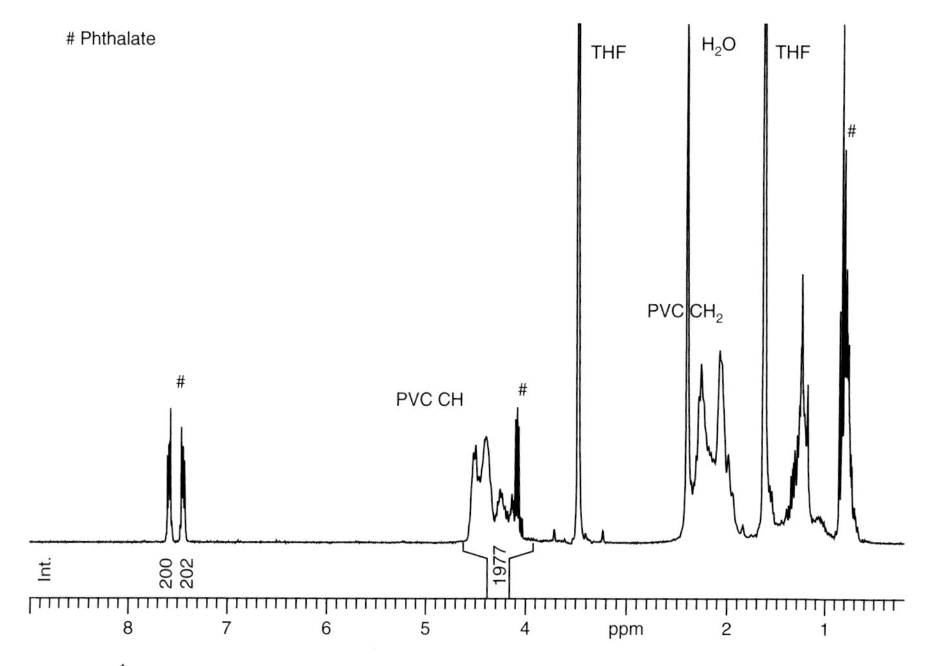

Figure 31 ^{1}H NMR spectrum of PVC containing phthalate plasticiser in THF-d_8.

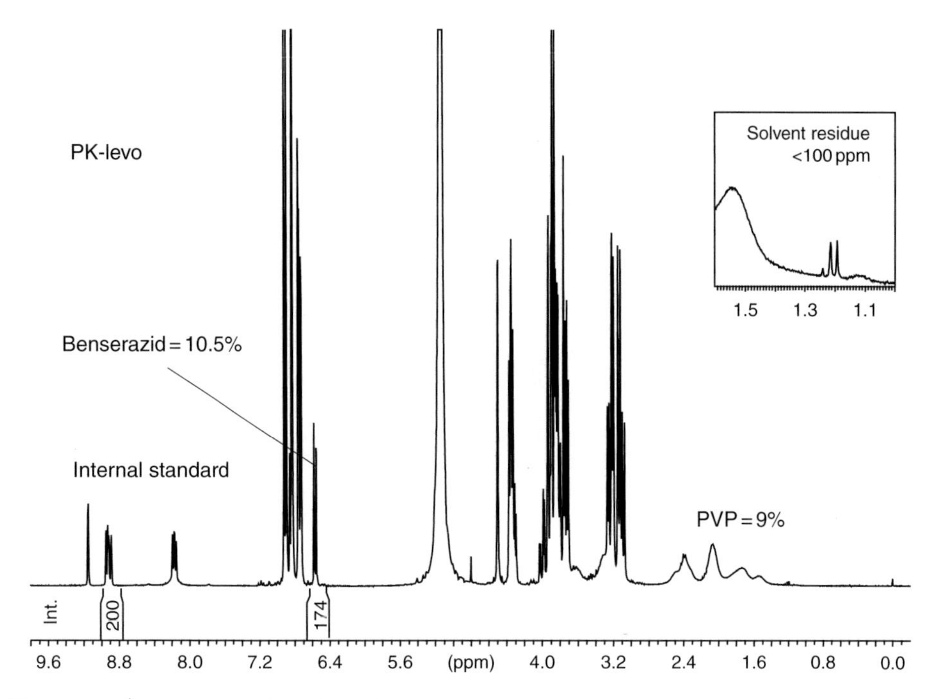

Figure 32 ^{1}H NMR of a drug formulation containing PVP.

caused by the active ingredients (see inset in Figure 32). In addition, the solvent residue consisting of a mixture of isopropanol and ethanol is quantified simultaneously. Other polymers may be analysed in the same or similar way depending on their solubility in a solvent useful for NMR spectroscopy, e.g. polystyrene in acetone-d_6.

9. Experimental Data

All NMR spectra given in this chapter were performed using a Bruker 300 MHz Avance 1 spectrometer equipped with a BBI probe. Data evaluation was performed using WIN-NMR.

REFERENCES

1. K. Schmidt-Rohr, H.W. Spiess, Multidimensional Solid State NMR and Polymers, Academic Press, London, 1999.
2. K. Hatada and T. Kitayama, NMR Spectroscopy of Polymers, Springer, Berlin, 2004.
3. E.M. Purcell, H.C. Torrey, and R.V. Pound, Phys. Rev., 69 (1946), 37–38.
4. F. Bloch, W.W Hansen, and M. Packard, Phys. Rev., 70 (1946), 460–474.
5. Data produced by Spectral Service, not published, available by request.
6. C. Jones, B. Mulloy, and A.H. Thomas, (Eds.) Methods, Mol. Biol., 17 (1993), 1–169.

7. P.K. Agrawal, Phytochem., 31 (1992), 3307–3330.
8. V. Ruiz-Calero, J. Saurina, T.M. Galceran, S. Hernández-Cassou, and L. Puignou Anal. Bioanal. Chem., 373, 4–5 (2002), 1618–2642.
9. T.N. Huckerby, R.M. Lauder, G.M. Brown, I.A. Nieduszynski, K. Anderson, J. Boocock, P.L. Sandall, and S.D. Weeks, Eur. J. Biochem., 268 (2001), 1181–1189.
10. S. Shinobu, O. Eri, T. Toshihiko, and G. Yukihiro, Chem. Pharm. Bull., 55 (2007), 299–303.
11. European Pharmacopoeia, 6th edition, European Department for the Quality of Medicines, Strasbourg, France, 2008. www.pheur.eu.
12. United States Pharmacopoeia (USP) 30 (2007), No <761>, the United States Pharmacopoeia Convention, Rockville, MD, USA.
13. M. Maiwald, H.H. Fischer, M. Ott, R. Peschla, C. Kuhnert, C.G. Kreiter, G. Maurer, and H. Hasse, Ind. Eng. Chem. Res., 42 (2003), 259–266.
14. U. Holzgrabe, I. Wawer, and B.W.K. Diehl, NMR Spectroscopy in Drug Developement and Analysis, Wiley-VCH, Weinheim, 1999.
15. H. Kessler and W. Bermel, "Conformation Analysis of Peptides by Two-Dimensional NMR Spectroscopy," in Applications of NMR Spectroscopy to Problems in Stereochemistry and Conformational Analysis, Y. Takeuchi and A.P. Marchand, Eds., VCH Publishers, Deerfield Beach, FL, Ch. 6, 1986.
16. M. Pons, (Ed.) NMR in Supramolecular Chemistry, Proceedings of the NATO Advanced Research Workshop on Applications of NMR to the Study of Structure and Dynamics of Supramolecular Complexes, Sitges, Spain, 5–9 May 1998, Series: NATO Science Series C, Vol. 526 (1999).
17. Spectral Service, internal study.

CHAPTER 3

NMR Spectroscopy of Natural Substances

B. Diehl

Contents

Abstract

The chapter provides an overview of several options in natural substances analysis by means of examples from routine nuclear magnetic resonance (NMR) spectroscopy. The objects are different chemical classes consisting of reference materials up to complex mixtures of plant extracts and lipids. One- and two-dimensional spectra were applied to demonstrate the characterisation and quantification of natural substances.

Keywords: finger print analysis, primary reference standard, multi component analysis, lipids, lecthins, fats, oils

1. Introduction

Analysis of natural substances is a traditional domain of nuclear magnetic resonance (NMR) spectroscopy and the quantitative applications are in progress.[1] Most of all compounds of interest for pharmaceutical approaches are derived from structures containing one or more chiral centres. This is obvious for the chemical structures of amino acids, carbohydrates or terpenoids, e.g. even the huge class of lipids based on the glycerol backbone undergo a chiral principle although glycerol itself is not chiral (see Part I, Chapter 1, Section 9). The high pharmaceutical potential of natural compounds often is confronted with an insufficient availability of corresponding primary reference standards. The synthesis of analytical standards to quantify these natural substances is very complicated and expensive or even not possible. To produce an analytical standard, the active compounds must be extracted from their plant or animal sources and purified as much as possible. Following these procedures, there it is a need for an analytical method that enables an ab initio determination of the purified active compound to create a primary standard, e.g. chromatography. ^{1}H NMR spectroscopy was proved to fulfil this request. A round robin test for several natural substances[2] was accomplished successfully.

Besides the definition of standards for chromatographic purposes, NMR spectroscopy is a suitable qualitative and quantitative method to analyse complex mixtures based on natural extracts without any standard of the corresponding natural compounds – only by using one well-defined artificial standard. The molar response in NMR spectra gives the condition for this procedure. Both the application models – the creation of primary reference standards and the analysis with unrelated artificial reference material – will be discussed in this chapter. Owing to the enormous number of papers concerning NMR spectroscopy of natural substances, a complete overview goes beyond the scope of this chapter.

This chapter provides three different approaches in natural compound analysis. The first is a fingerprint-like analysis for characterisation, e.g. plant extracts. The second part deals with the qualitative and quantitative analysis of single natural substances, and the third with a quantitative or semi-quantitative analysis of multi-component systems.

2. Characterisation of Natural Substances as a Finger Print Analysis

The characterisation of pharmacologically relevant plant extracts with chromatographic methods is often complicated because of the heterogeneous composition. Carbohydrates, lipids, amino acids, terpenoids and other polymeric compounds cannot be detected using only one chromatographic method for different reasons. On the one hand, the volatility necessary for gas chromatography is not given; on the other hand, the target molecules have no chromophore as required for high-performance liquid chromatography (HPLC) analysis with UV/VIS detection. Chromatography-based methods only provide an information of a minor or a major part of the whole composition. In a ^{1}H NMR spectrum of a plant extract almost all soluble molecules can be determined, because NMR detection is based on a molecular response. A proton of a mono- or oligosaccharide can be detected in the same analysis like a sugar molecule, which is part of a glycolipid or a flavanoid. This fact may be demonstrated by a ^{1}H NMR spectrum of a plant–oil mixture of lemon peel and olive leaves. The observed NMR spectrum is characteristic and can further be used for the identification of this pharmaceutical raw material. In serial analysis, a computer-based automatic spectra analysis is possible by different pattern recognition techniques. The knowledge of all signals is not necessary for this easy procedure.

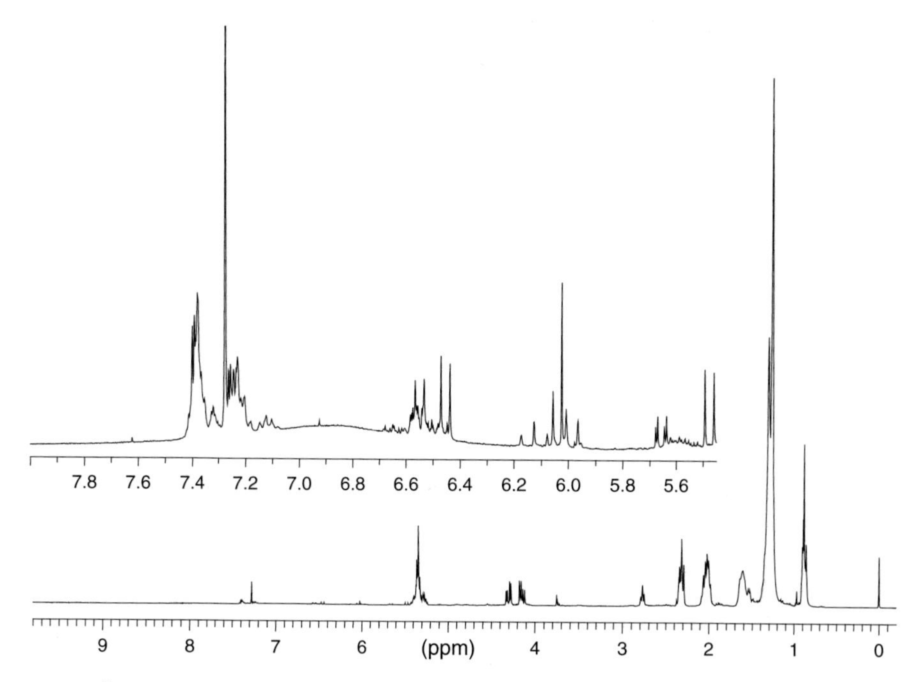

Figure 1 ^{1}H NMR spectrum of *Calophyllum inophyllum* and a detail of the aryl region.

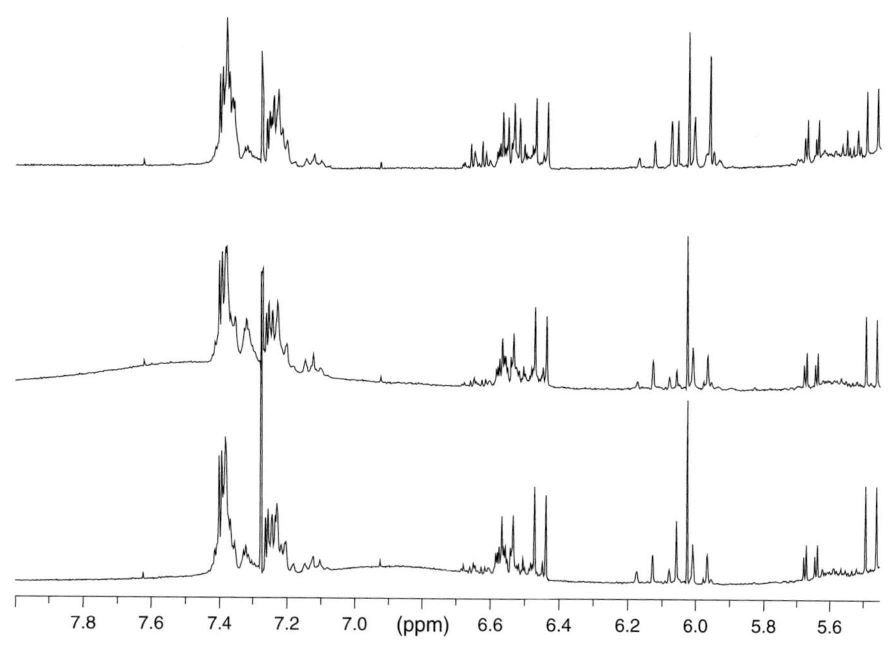

Figure 2 ^{1}H NMR spectra of different oil batches, downfield details in comparison.

In Figure 1 a ^{1}H NMR spectrum of a vegetable oil derived from *Calophyllum inophyllum* is given. The typical signals of triglycerides dominate the spectrum. Furthermore, a typical pattern of calophyllolide and inophyllum in addition to other complex polyphenols can be observed especially in the aromatic region of the spectra. Detailed structure analysis and NMR data are published.[3,4]However, the unassigned ^{1}H NMR data provide a fast tool for testing the origin and the quality of the oil (Figure 2). The simultaneous detection of both the fatty acid composition and the secondary plant compounds using the NMR method provides more information compared to chromatographical methods which of course are used for validation purposes.

3. Characterisation of Natural Substances According to Single Target Molecules

The quantification of the thermo instable molecule allicin in garlic products is only one routinely performed application[5] of NMR spectroscopy within a number of plant extracts. Analysing mixed extracts of lemon peel and olive leaves demonstrates the NMR tool in a more detailed way (Figure 3). The main target molecules of the plant extracts of citric peel is naringin and of the olive part oleuropein (Figure 4), both members of the family of flavanoids. Identification of these leading compounds is possible by ^{1}H NMR and ^{13}C NMR spectroscopy. NMR data of many different types of flavanoids are available from the literature.[6]

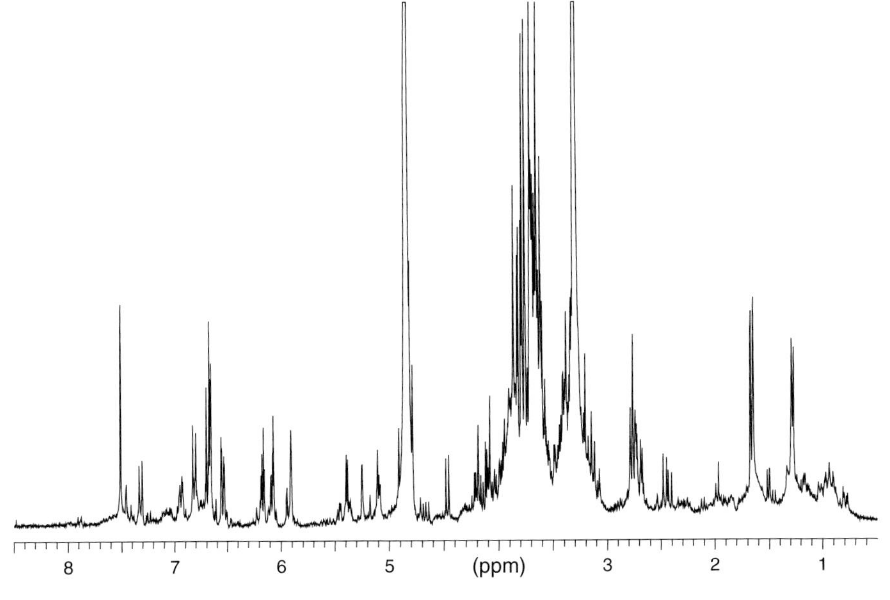

Figure 3 ^{1}H NMR spectrum of a plant extract mixture of lemon peel and olive leaves.

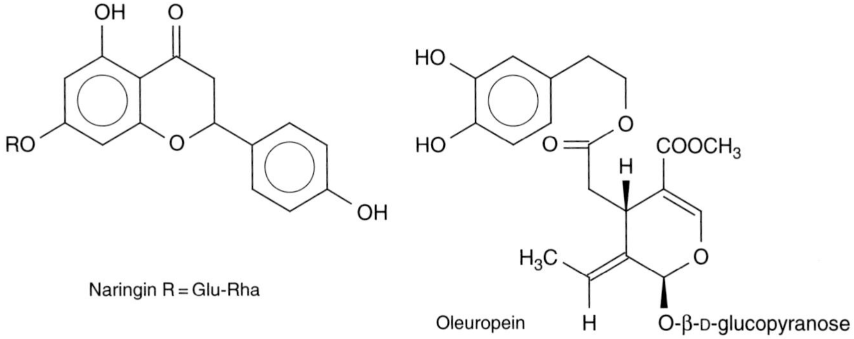

Naringin R = Glu-Rha

Oleuropein

O-β-D-glucopyranose

Figure 4 Chemical structures of the leading compounds of lemon peel and olive leaves extracts.

In the expansion of the ^{1}H NMR spectrum (Figure 5) the specific signals can be assigned by comparison with the spectra of reference materials. In case that minimum one characteristic signal of a target molecule does not interfere with signals of other molecules in the mixture, a direct quantification can be done by standard addition or by the use of a suitable internal standard.

These conditions are given within some restrictions in the above-mentioned example. A high field strength of the NMR instrument is a helpful component

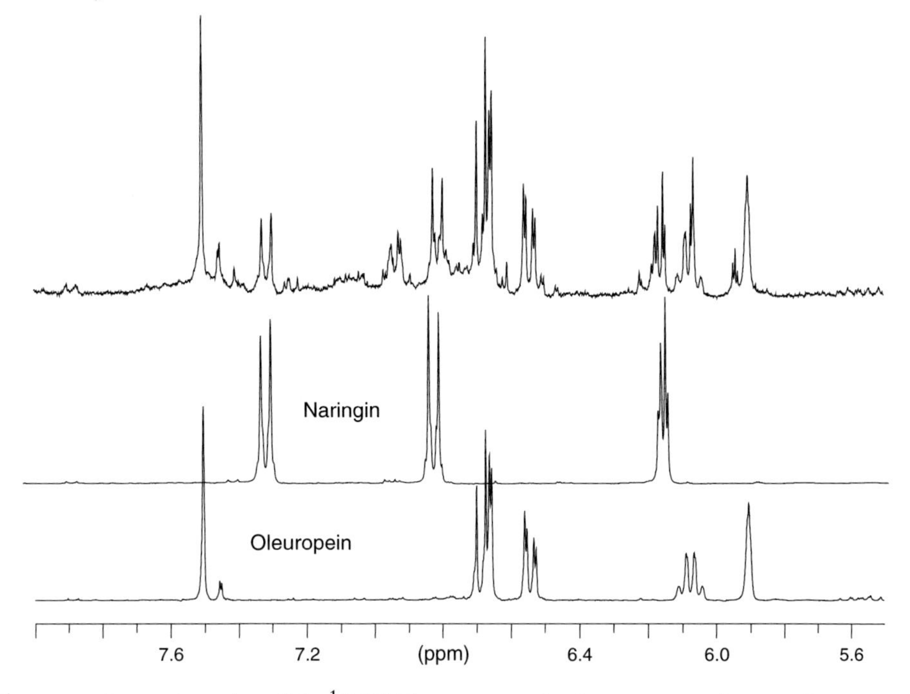

Figure 5 Aromatic region of the ^{1}H NMR spectrum of a plant extract mixture of lemon peel and olive leaves (top) compared with the corresponding spectra of naringin and oleuropein.

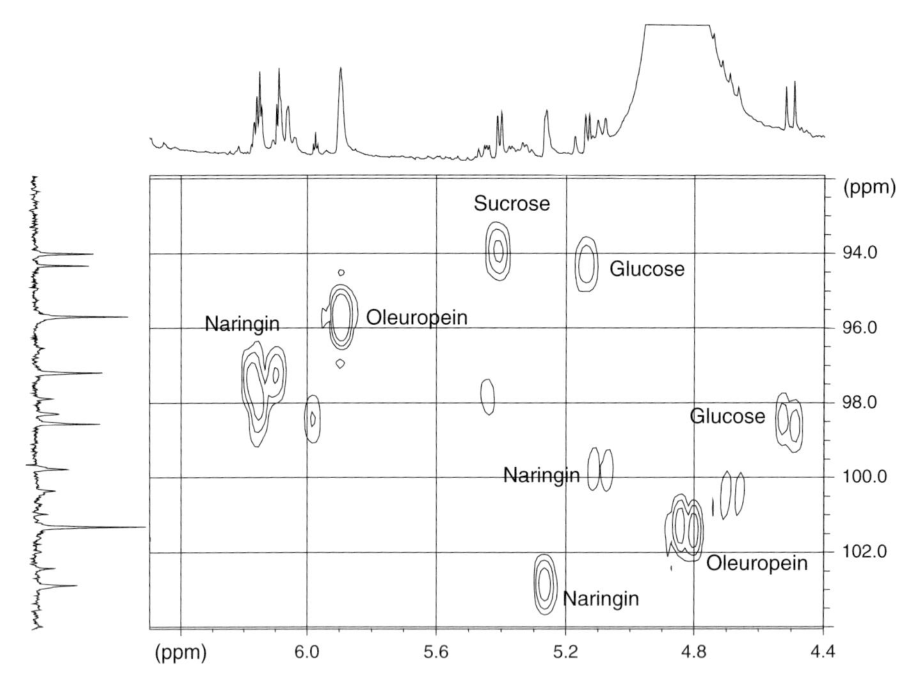

Figure 6 HMQC spectrum of a plant extract mixture of lemon peel and olive leaves

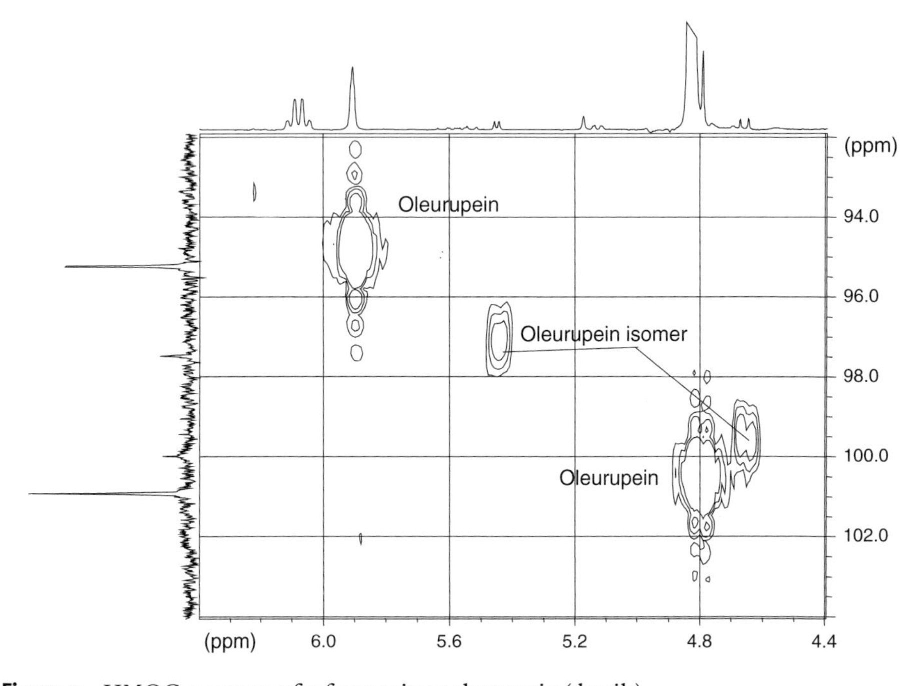

Figure 7 HMQC spectrum of reference item oleuropein (details).

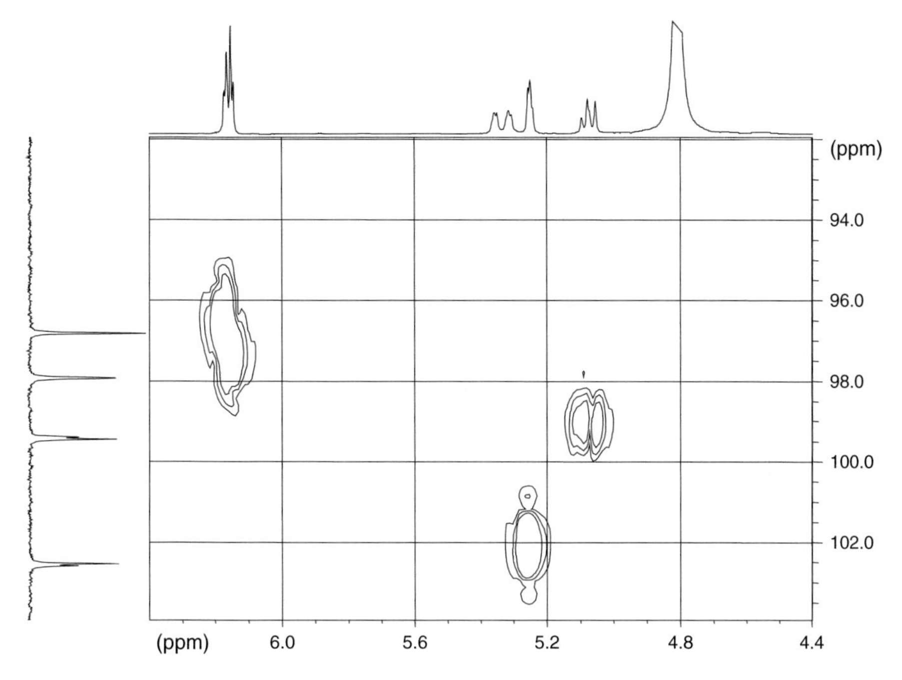

Figure 8 HMQC spectrum of reference item naringin (details).

in this sensible type of analysis. In the near future, modern cryoprobes with a magnitude better signal-to-noise ratio will enable this complex analysis by ^{13}C NMR spectroscopy, which shows a much higher signal dispersion.

In addition, two-dimensional analysis is very helpful in structure elucidation or even only as a fingerprint analysis as demonstrated on the discussed plant extract mixture in Figures 6–8.

4. Characterisation and Definition of Primary Reference Standards

One step further in NMR analysis of natural substances is the characterisation and the definition of primary and secondary reference standards.[7] Corresponding standards of natural compounds are often very expensive or the purities respectively the contents are low. Again the molar response of ^{1}H NMR signals can be used to qualify natural substances as reference standards. Commercially available standards of naringin and oleuropein were characterised by ^{1}H NMR spectroscopy (Figures 9 and 10) and the contents of both were evaluated by ^{1}H NMR using the internal standard method.[8] The detail of the aromatic region may demonstrate the procedure by the integration of the characteristic signals of the test item naringin and the internal standard 4-*N*,*N*-dimethylaminobenzoic acid ethylester (DBEE) (Figure 11). Oleuropein shows an isomer structure of approximately 15% that was

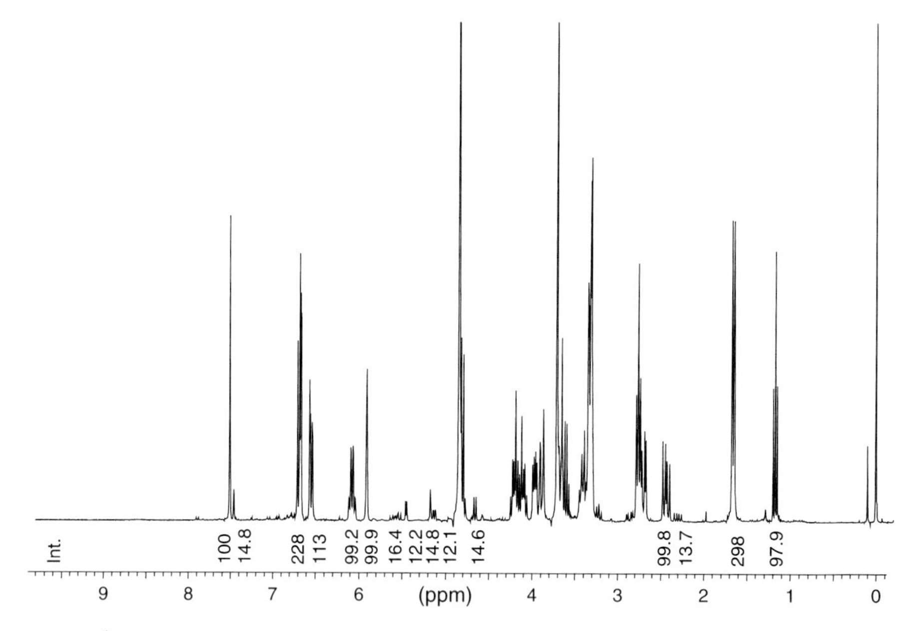

Figure 9 ^{1}H NMR spectrum of reference item oleuropein.

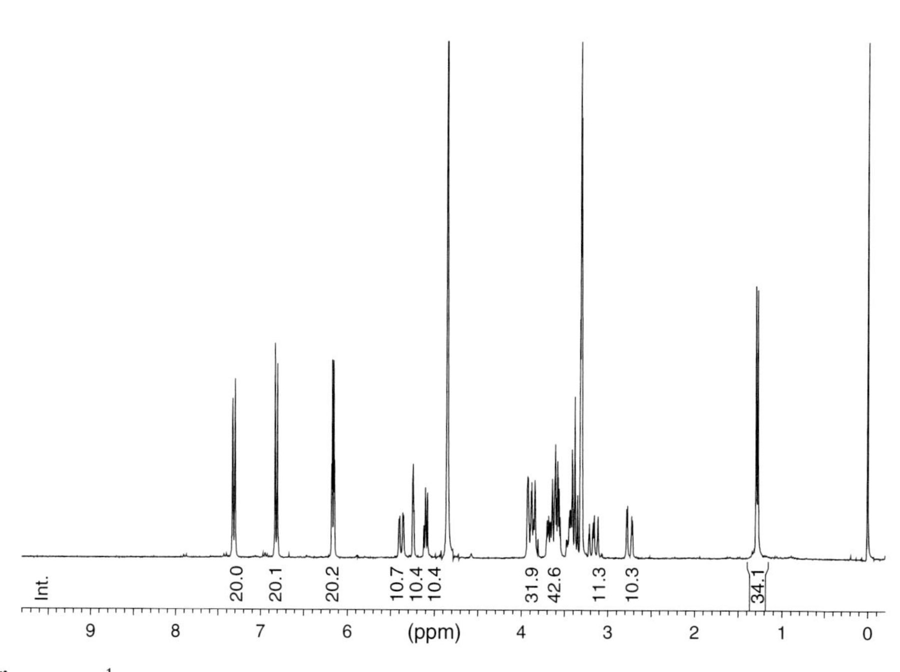

Figure 10 ^{1}H NMR spectrum of reference item of naringin.

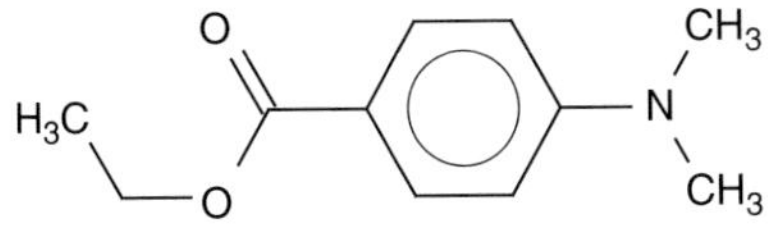

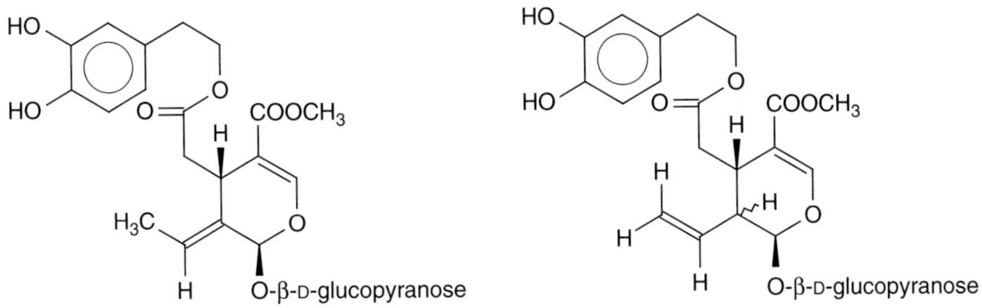

Figure 11 Chemical structure of DBEE used as internal standard (top) and the oleuropein isomers.

confirmed by HPLC/MS (mass spectrometry). The quantification was done in the same analysis.

Natamycin is an antimycotic fungicide. It is commercially available in a formulation with glucose. Amounts of glucose and natamycin, both very different in their chemical structures, can be quantified after dissolving in a mixture of DMSO-d_6 and methanol-d_4 (MeOD). The latter is used for deuterium exchange to avoid integration disturbances by sugar OH signals. A comparison of natamycin and glucose integral areas enables the evaluation of relative amounts of both compounds, which can be sufficient as a quality test (Figure 13).

For exact quantification a suitable internal standard, e.g. nicotinic amide (NA), must be used in a single or multiple analysis. In the present study the downfield signals (11H) of the fungicide are used for quantification as a sum of several proton signals. Additionally, other integral areas may be used, e.g. signals at $\delta = 2.8$ ppm representing 1 proton and $\delta = 0.9$–1.3 ppm representing 10 protons. Simultaneous signal integration of monomeric carbohydrates is complicated due to the equilibrium of different anomeric molecules. For glucose, the sum of one proton of each type α-D-glucose ($\delta = 5.0$ ppm) and β-D-glucose ($\delta = 4.35$) must be used for quantification (Figure 13).

In case that the signal of β-D-glucose interferes with the water signal, alternatively the double doublet at $\delta = 3.0$ ppm can be used for quantification. Furthermore, the water signal can be shifted by changing the solvent ratio between DMSO-d_6 and MeOD, as DMSO-d_6 is shifting the water signal upfield.

To enhance the confidence in this quick analysis, a defined amount of a glucose standard can be added (standard addition method) for further analysis under the same conditions. The increase of the signals of one or both anomeric glucose types enables an absolute quantification of glucose independent of the use of an internal standard. A combination of the internal standard and the glucose standard addition

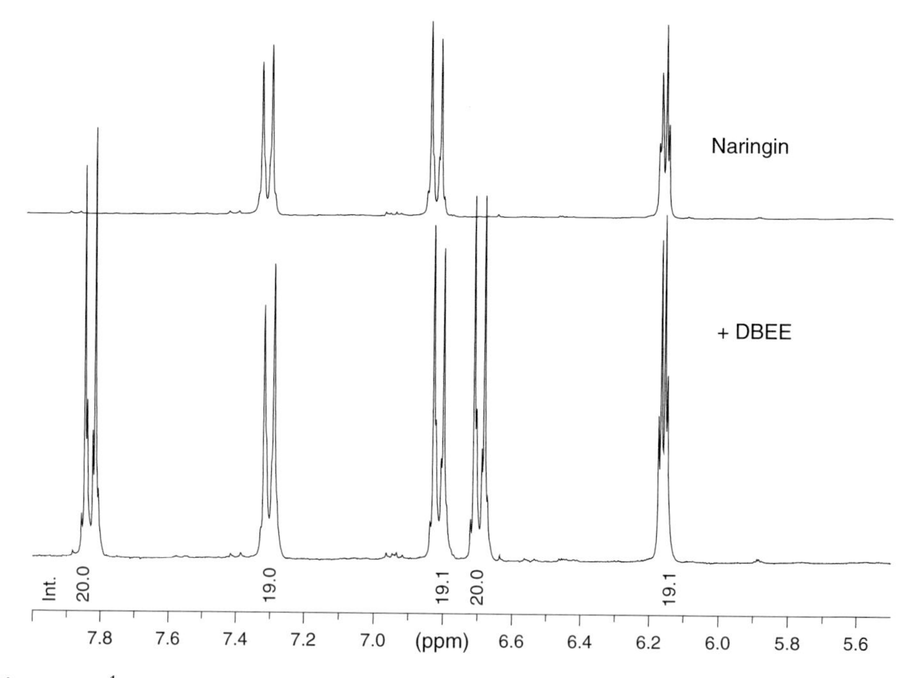

Figure 12 ^{1}H NMR spectrum of reference item of naringin with (bottom) and without (top) internal standard DBEE.

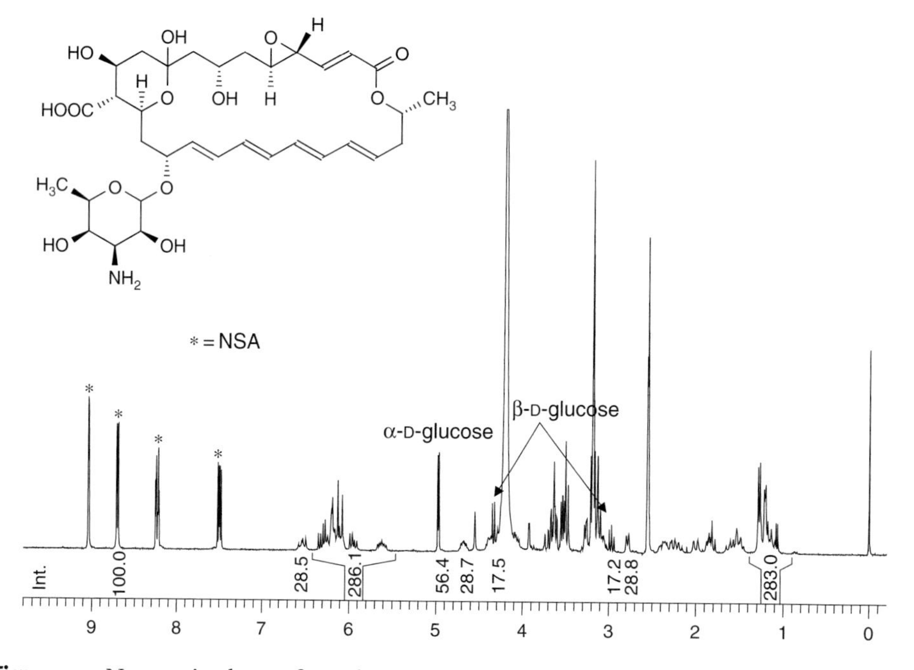

Figure 13 Natamycin glucose formulation in DMSO-d_6/MeOD, internal standard NA.

method allows a complete validation of the qNMR (quantitative nuclear magnetic resonance) method for this preservative formulation.

5. Multi-Component Analysis of Complex Natural Substance Mixtures and Phytopharmaceutics

5.1. Extracts from medical plants

Multi-component analysis by NMR spectroscopy in phytopharmaceutics is a further development. The analysis of extracts from *Aloe vera* (*barbadensis* Miller) will demonstrate the procedure. As many others *A. vera* extracts consist of a mixture of natural substances classes, e.g. amino acids, fruit acids (derived from the citric acid cycle), mono and polysaccharides, lipids and flavanoids.

First the NMR analysis is used to test the origin of an extract and to distinguish between different plant sources. The origin test is based on the presence and the amount of the biopolymer aloverose, a partly acetylated polymannose (see Chapter 2, Figure 5). Besides the original test, it is necessary to find out whether a product was exclusively taken from the inner gel of an *A. vera* leaf or whether it is a product from the whole leaf (WL) – including substances from the green rind. The composition of fruit acids deriving from the citric acid cycle can be used as marker to distinguish between these raw materials. Only in the green part of an *A. vera* leaf citric and isocitric acid are produced (Figures 14–16), the latter is partly

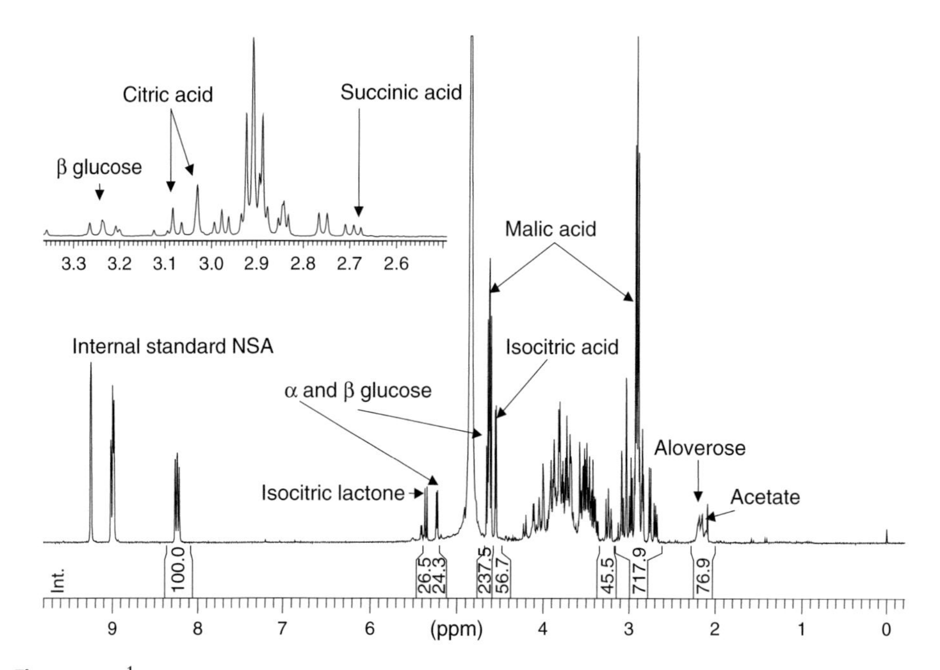

Figure 14 ^{1}H NMR spectrum of *A. vera* whole leaf + NSA as internal standard.

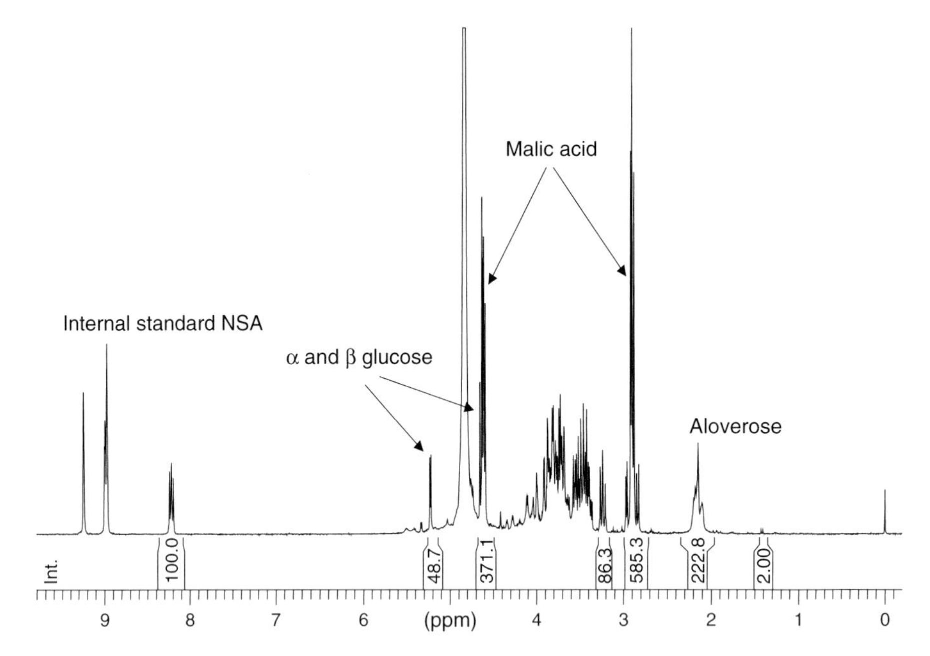

Figure 15 ^{1}H NMR spectrum of fresh *A. vera* gel + NSA as internal standard.

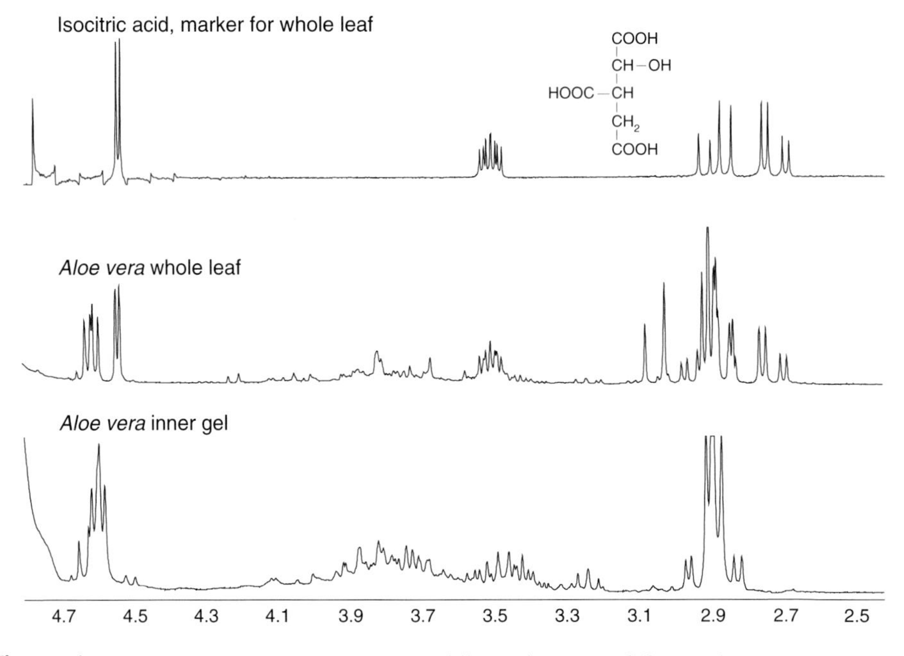

Figure 16 Isocitric acid (top), WL *A. vera* (middle) and *A. vera* gel (bottom).

converted to the corresponding lactone. If both the molecules can be detected in an *A. vera* product, the sample is therefore either a WL material or it is contaminated with parts of the rind.

Using the internal standard method, the characterisation of *A. vera* samples can be done quantitatively or at least semi-quantitatively. Malic acid as well as glucose and the polymer aloverose are main constituents of fresh *A. vera* gel and are quantified routinely by ^{1}H NMR. Additional signals originated from WL material are quantified in the same way (Figure 14).

For validation purposes the following procedure was performed. An *A. vera* leaf was split into two parts. From one part the inner filet was taken. A juice was prepared by using a blender (GEL). The second part of the leaf was equally processed without removing the rind to give a WL product. Both samples were analysed separately as well as in mixtures of different ratios by ^{1}H NMR. The increasing signals of the WL markers (isocitric acid and its lactone) correspond to the amount of *A. vera* WL material used.

To complete the quality control of *A. vera* products, besides their main constituents, chemical and bacterial degradation products are also detected as well as additives and adulterations, e.g. preservatives or maltodextrin. The detection is independent of the chemical structures (mono or polymer) and does not need an UV-active chromophore as been necessary in HPLC/UV analyses. Also, volatility, a condition for gas chromatographical analysis, is not required (Table 1).

5.2. Lipids

NMR spectroscopy has become a universal method in lipid analysis. It allows structure elucidation, qualitative and quantitative analysis of defined molecules and even complex mixtures. Not all nuclei are accessible to the NMR experiment.

Table 1 NMR chemical shifts of compounds in *A. vera* products

Compound	δ (ppm)[a,b]	No. of protons
Formic acid	**8.4 (CH)**	1
Fumaric acid	**6.5 (2 × CH)**	2
Benzoic acid	8.0, 7.6, **7.5 (CH)**	2
Sorbic acid	7.0, 6.2, 5.8, **1.8 (CH_3)**	3
Glucose	**5.3(α-CH)**, 4.6 and **3.2 (β-CH)**, 3.2–4.0	1[c]
Malic acid	**4.3 (CH)**, 2.4–2.8 **(CH)**	1
Succinic acid	**2.5 (2 × CH_2)**	4
Citric acid	**2.4–2.8 (2 × CH_2)**	4[d]
Acetic acid	**1.9 (CH_3)**	3
Lactic acid	4.1, **1.3 (CH_3)**	3
Isocitric acid	**4.55 (CH)**, 3.5 2.7–3.0	1
Isocitric lactone	**5.35 (CH)**, 3.5 2.9–3.2	1

[a] Chemical shifts are pH dependent and may change slightly.
[b] Bolded signals normally are used for calculation.
[c] Mix of anomeric forms. The integral value of the α-type has to be added to the β-type.
[d] Integration is complicated by interference with malic acid signals.

But those nuclei that are important in lipid chemistry like ^{1}H, ^{13}C and ^{31}P are recordable as a matter of routine with modern instruments equipped with a multi-nuclear probe.

Compared to common chromatographic analysis, e.g. gas chromatography (GC) and HPLC, NMR spectroscopy is a non-destructive ab initio method. Analysis of oils and lecithin of different origin demonstrates the usefulness of multi-nuclear high-resolution NMR spectroscopy.[9] Besides the composition of the main components, e.g. fatty acid and phospholipid composition, secondary substances such as sterols or terpenoids can be determined within the same analysis.

5.3. Lecithin

The 100% amount of the ^{31}P isotope in nature, the selectivity and the sensitivity make ^{31}P NMR spectroscopy the reference method of the ILPS (International Lecithin and Phospholipid Society). Artificial standards, e.g. triphenyl phosphate (TPP) or distearoyl phosphatidyl glycerol (DSPG), are used to quantify all phospholipids from complex mixtures of any origin. The quantification of phospholipids is a valid method. Method precision (reproducibility) and instrument precision for main components show a standard deviation <1%, components near the limit of quantification below 5%.

In the case of liposome lyophilisate preparation, the extraction recovery of the phospholipids is 100%. The analysis method is robust. It is almost insensitive to variation in sample preparation and parameters of measurement. Prepared samples can be stored up to 5 days at room temperature (typically <24 h) until they are measured without negative influence on the results. The internal standard method had been fully validated concerning selectivity, recovery, reproducibility, instrument precision and robustness. A lyophilised drug formulation on the basis of egg yolk phosphatidyl choline (PC) was used. The results of all experiments are summarised in Table 2.[10]

For further validation studies, a round robin test involving more than 23 NMR labs had been done for approving the ^{31}P NMR as an official method.[11] The measurements were performed on different field strength between 200 and 700 MHz and by using 10 different samples. Five phospholipid mixtures for pharmaceutical approaches were tested – native soya bean lecithin, purified

Table 2 Survey on validation experiments

Test	*n*	PC by weight %
Recovery	7	33.3 ± 0.2
Instrument precision	6	33.4 ± 0.3
Variation of pH value	7	33.3 ± 0.2
Variation of solvent mix	5	33.2 ± 0.2
Variation of sample amount	7	33.4 ± 0.2

n = number of tests.

Table 3 Abbreviation of phospholipids

PC	Phosphatidyl choline	APE	*N*-Acylphosphatidyl ethanolamine
PI	Phosphatidyl inositol	DPG	Diphosphatidyl glycerol, cardiolipin
2-LPC	2-Lysophosphatidyl choline	PG	Phosphatidyl glycerol
PS	Phosphatidyl serine	PA	Phosphatidic acid
PE	Phosphatidyl ethanolamine	LPA	Lysophosphatidic acid
LPE	Lysophosphatidyl ethanolamine	SPH	Sphingomyelin

phosphatidyl serine, synthetic distearyl phosphatidyl choline, de-oiled egg yolk lecithin and a crude reaction mix of an enzymatically prepared phosphatidyl serine (based on soya lecithin). The results demonstrate the robustness of the NMR analysis, the deviation (mean of all five experiments) from the mean value is given in Figure 17. Only 3 of 29 experiments failed due to technical problems or by the use of irregular parameters.

Figure 18 shows the phospholipid profile of native soya bean lecithin. The NMR data evaluation of this kind of ^{31}P NMR spectra is comparable to usual chromatographic methods, but comprising the advantages of optimised selectivity, linearity and dynamics.

Figure 19 shows a medical application, the ^{31}P NMR of human blood plasma.

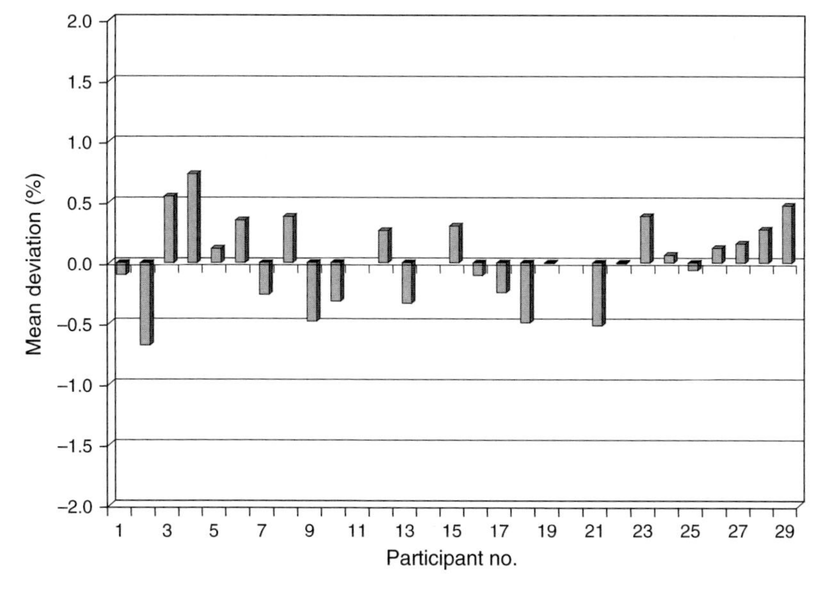

Figure 17 Results of the round robin test on ^{31}P NMR.

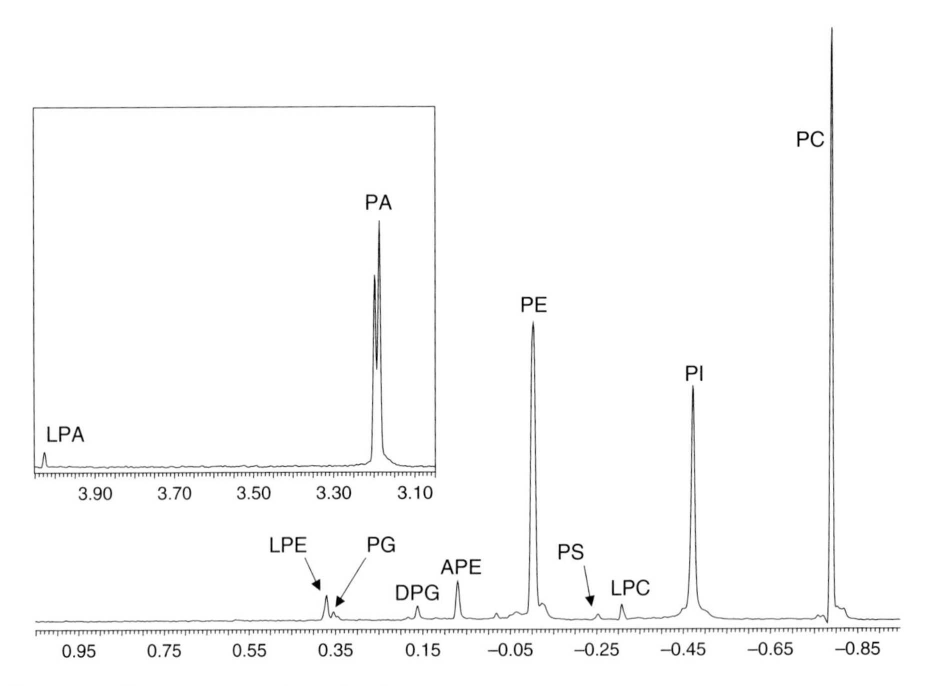

Figure 18 ^{31}P NMR of soya bean lecithin.

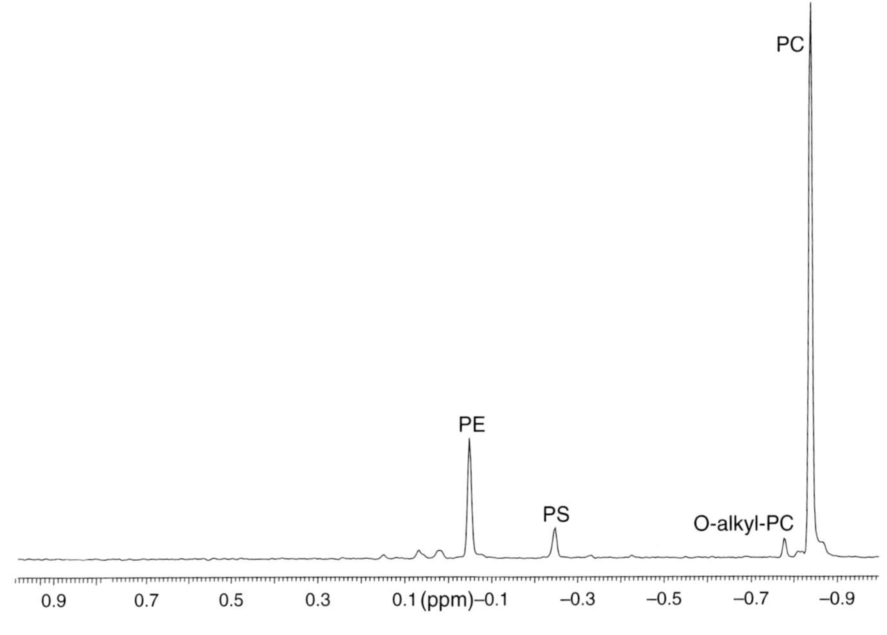

Figure 19 ^{31}P NMR human blood plasma.

5.4. Fats and oils

The determination of fatty acid composition in fats and oils can be done routinely by NMR spectroscopy. ^{1}H NMR enables analysis of double bonds as a sum parameter comparable to the iodine number,[12] a more detailed analysis is useful for the distinction of oils of different origin. So the relative amount of all ω-3 fatty acids can be calculated by using the integrals of the terminal methyl groups (Figure 20). Secondary constituents of plant and animal origin such as cholesterol or other sterols can be quantified using an internal standard,[8] lecithin, tocopherol, sesamol just to mention as well as degradation and oxidation products are other targets of ^{1}H NMR.

The higher spectral dispersion of ^{13}C NMR enables a more detailed look to the glyceride and fatty acid composition of fats and oils, which naturally are complex mixtures.[13] The carbonyl signal region is sensitive for position analysis of the different fatty acid types. Distinction of SN1/3 and SN2 are possible even for phospholipids and glycolipids.[14] The carbonyl signal is sensitive to the double-bond position in the responding fatty acid. A signal separation depending on the double-bond distance up to 11 atoms in the chain is given. This fact allows the individual fatty acid composition especially for lipids of marine origin with its high amounts of polyunsaturated fatty acids. For example, the ^{13}C NMR spectrum of the carbonyl region of a fish oil is shown in Figure 21.

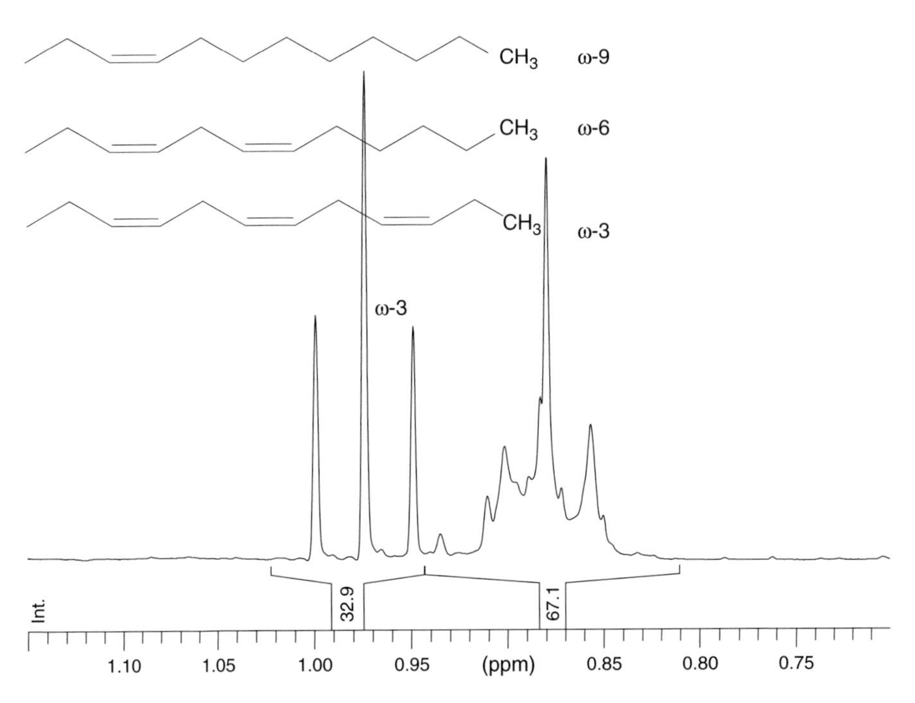

Figure 20 ^{1}H NMR spectrum, methyl region of a vegetable oil.

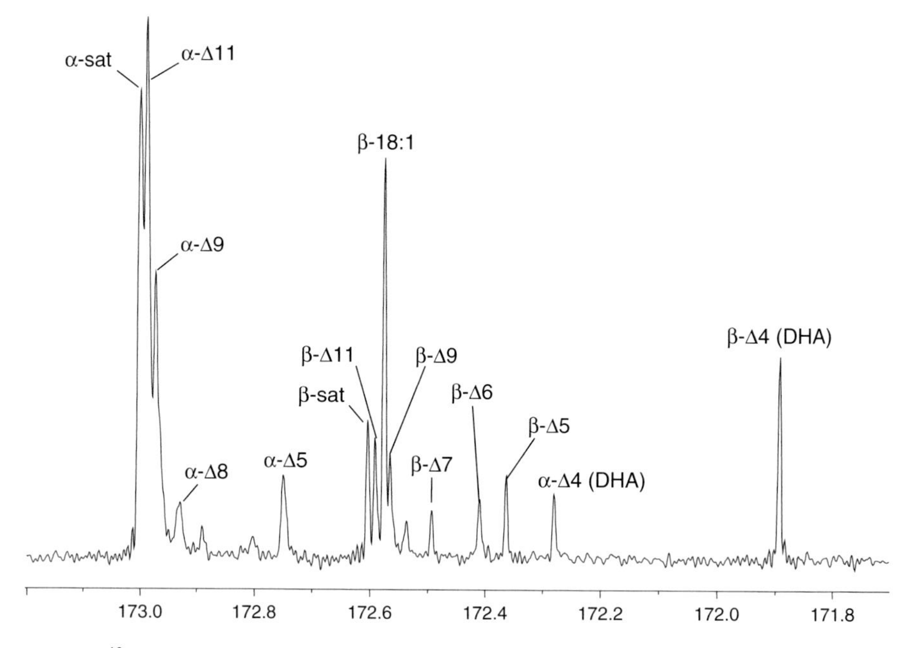

Figure 21 ^{13}C NMR spectrum, carbonyl region of fish oil.

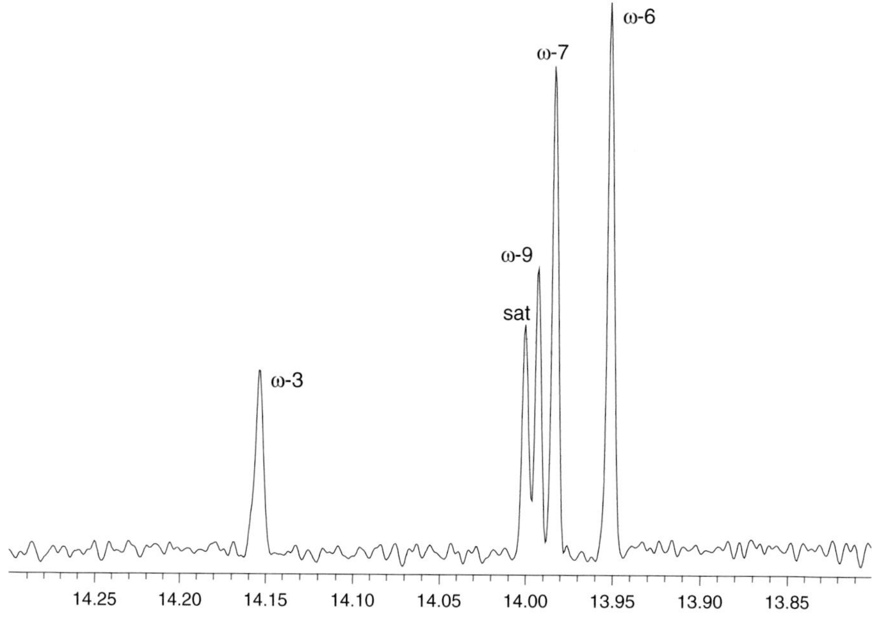

Figure 22 ^{13}C NMR spectrum, methyl region of fatty acid methyl ester mixture.

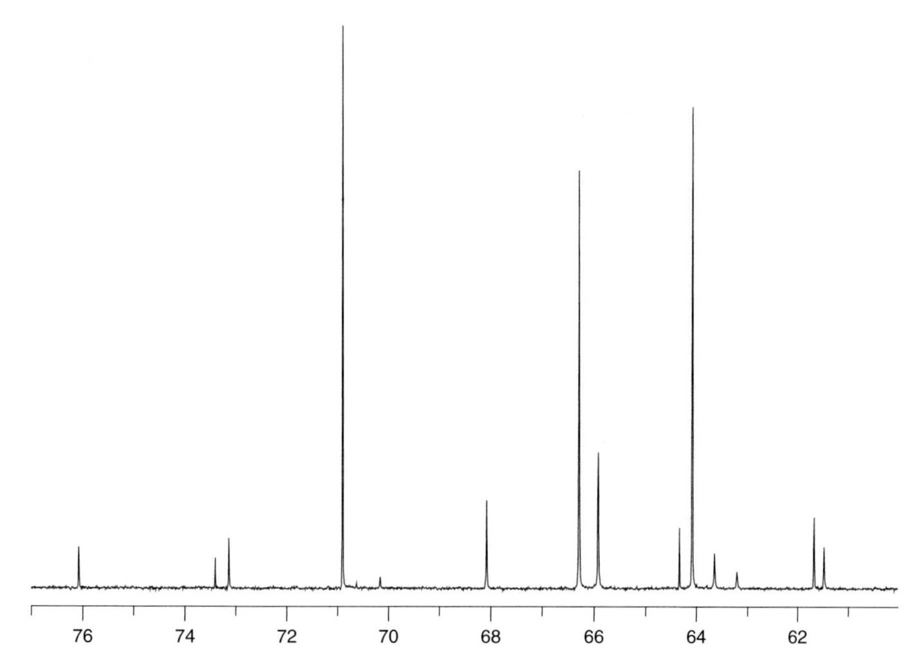

Figure 23 ^{13}C NMR glycerol region of lipids.

The individual position of the fatty acids within the glycerol backbone is reflected in the double-bond signals, too. Using the signals of the methyl groups, which are representing the chain ends, a quantitative analysis of the fatty acid profile concerning the ω-*n* and saturated types is possible (Figure 22).

In this chapter, the glycerol part of a mixture of glycerides is shown in Figure 23. The signal assignment is given in Table 4. The amount of a special glyceride (e.g. 2-acylmonoglucoside) used in an emulsifier mixture is quantified using a defined triglyceride (e.g. trinonadecanoate) as an internal standard. The chemical similarity of the atoms used for quantification avoids the problems normally caused by differences in ^{13}C NMR relaxation time and the nuclear Overhauser effect (NOE).

Table 4 Signal assignment of glyceride C-atoms

	2 M	Gly	1,2 D	1 M	T	1,3 D	1 M
ppm	76.1	73.5	73.1	70.9	70.2	68.1	66.4
	1,3 D	Gly	1 M	1,2 D	T	2 M	1,2 D
ppm	65.9	64.3	64.1	63.6	63.2	61.7	61.5

Gly = glycerol, T = triacylglycerol, M = monoacylglycerol, D = diacylglycerol.

6. Experimental Data

All NMR spectra given in this chapter were performed using a Bruker 300 MHz Avance 1 spectrometer equipped with a BBI probe. Data evaluation was performed using WIN-NMR.

REFERENCES

1. G.F. Pauli, B.U. Jaki, D.C. Lankin, Quantitative ^{1}H-NMR: Developement and Potential of a Method for Natural Products Analysis, J. Nat. Prod. 68 (2005), 133–149.
2. AiF Project, data not published.
3. M.C. Yimdjo, A.G. Azebaze, A.E. Nkengfack, A.M. Meyer, B. Bodo, Z.T. Fomum, Phytochemistry 65 (2004), 2789–2795.
4. Y.C. Shen, M.-C. Hung, L.T. Wang, C.Y. Chen, Inocalophyllins A, B and Their Methyl Esters from the Seeds of *Calophyllum inophyllum* Chem. Pharm. Bull. 51(7)(2003), 802–806.
5. U. Holzgrabe, I. Wawer, B. Diehl, NMR Spectroscopy in Drug Developement and Analysis, Wiley-VCH, Weinheim, (1999) 49–50.
6. P.K. Agrawal, (Ed.), Studies in Organic Chemistry 39, Carbon–13 NMR of Flavonoids, Elsevier, Amsterdam (1989).
7. G.F. Pauli, qNMR – A versatile Concept for the Validation of Natural Product Reference Compounds Phyt. Anal. 12 (2001), 28–42.
8. Spectral Service, internal studies.
9. B.W.K. Diehl, Lipid Analysis of Oils and Fats. Chapter 4: In Multinuclear High Resolution NMR Spectroscopy, edited by Prof. R.J. Hamilton, Blackie Academic & Professional, London, (1997), 87–135, ISBN 07514 0414 4.
10. B.W.K. Diehl, W. Ockels, Quantitative Analysis of Phospholipids. In Proceeding of the 6th International Colloquium Phospholipids. Characterization, Metabolism and Novel Biological Applications, edited by G. Cevc & F. Paltauf, Champaign: AOCS Press (1995), 29–32.
11. Round robin test on ^{31}P-NMR spectroscopy on phospholipids, in Press.
12. K. Manz, Schneider: Bestimmung der Jodzahl in epoxidiertem Sojabohnenöl mit automatisierter ^{1}H-NMR Spektroskopie. GIT Fachzeitschrift für Laboratorien, 3 (1995), 197–199.
13. B.W.K. Diehl, W. Ockels, Fatty Acid Distribution by ^{13}C-NMR-Spectroscopy, Fat Scientific Technol. 97(3) (1995), 115–118.
14. B.W.K. Diehl, H. Herling, E.H. Heinz, I. Riedl, ^{13}C-NMR analysis of the positional distribution of fatty acids in plant glycolipids. Chem. Phys. Lip. 77 (1995) 147–153.

CHAPTER 4

Solid-State Measurements of Drugs and Drug Formulations

I. Wawer

Contents

Abstract

Solid-state nuclear magnetic resonance (NMR) spectroscopy is a non-destructive and fast analytical technique. Powdered samples are retained in their original form, tablet samples need only to be crushed, and substances isolated from plants can be used for further assays. ^{13}C cross-polarization magic-angle spinning (CPMAS) NMR measurements are used to identify a drug or excipient, to evaluate their quality, to differentiate crystalline and amorphous state, as well as for studying structure and conformation in the solid phase. Solid-state NMR data, supported by theoretical calculations of shielding constants, may have practical value in treating solids that do not form single crystals. In the international pharmacopoeias, there is no example of application of solid-state NMR, although this technique is especially suitable for characterization of pharmaceutical polymorphs and may be valuable tool for solving regulatory problems (patent issues).

Keywords: ephedrine derivatives, erythromycin, paracetamol, procaine, rifampicin, ketoconazol, paclitaxel, steroids, troglitazone, nifedipine, ibuprofen, sildenafil, bupivacaine, warfarin, alkaloids

1. Solid-State NMR in Pharmacy

Although magic-angle spinning (MAS) was initiated 50 years ago,[1] the beginning of solid-state ^{13}C NMR can be dated to the 1970s. The first experiments using cross-polarization (CP) and high power decoupling were performed mainly on synthetic polymers or zeolites although, since the work of Schaefer and Stejskal,[2] the advantages of these techniques in the studies of organic solids became obvious. The first paper on polymorph of pharmaceutical significance (*tamoxifen citrate*) appeared in 1981.[3] The studies published by the groups of Byrn and of Harris in the 1980s illustrated the potential of this method but pharmaceutical companies did not fully appreciate it. In the review of Bugay[4] published in 1995 the author states, "It seems ironic that although 90% of the pharmaceutical products on the market exist in the solid form, solid state NMR is still in its infancy as applied to pharmaceutical problem solving."

NMR spectroscopy has traditionally been limited in its applications due to high costs of the instrumentation. However, after many years of my experience with NMR spectrometers, I can notice also "a psychological factor." High-field super-conducting magnets have to be located in stable conditions, in a separate room, and there is no free access to the spectrometer. Therefore, the whole machinery is not familiar to students and other users, although it is "user friendly," and 99% of the operator's jobs are performed at the computer desk. Solid-state NMR equipment has an additional "magic" feature because the spinning at "magic" angle removes line broadening in solid samples. Practically, this means that the stator is tilt of 54°. Fast spinning (up to 15 kHz in a 4 mm rotor and up to 35 kHz in a 2.5 mm rotor) is realized by two gas bearings and the method is not very difficult even for pharmacists.

Solid-state ^{13}C CPMAS NMR spectra of tablets or capsules of *prednisolone, enalapril maleate, lovastatin, simvastatin, ibuprofen, flurbiprofen, mefenamic acid, indomethacin, diflunisal, sulindac,* and *piroxicam* were obtained[5] in 1993. The technique permitted discrimination between two prednisolone polymorphs present in tablets obtained from various manufacturers even though the tablets contain only approximately 5% (w/w) of the drug. These results encouraged other research groups to perform solid-state measurements of drugs and related substances.

The advantages of solid-state NMR in the studies on conformations of various biomolecules and drugs, as well as the studies on drug's polymorphism, were discussed in a previous book[6] which appeared in 1999. In 2003 the authors[7] while reviewing the use of solid-state NMR for the characterization of pharmaceutical solids already noticed that it is a fast developing method. Besides polymorphism, solid-state NMR techniques were applied to study phase transition, drug–excipients interactions, drug stability, and reactions in the solid phase.

Fortunately, nowadays solid-state NMR has become an integral technique in the field of pharmaceutical sciences. The method is useful for studying structure and conformation, analyzing molecular motions (relaxation and exchange spectroscopy), and measuring internuclear distances. Assignment of resonances is more reliable using spectral editing and two-dimensional correlation spectroscopy. MAS NMR is making important contribution to our understanding of polymorphism and solvate formation, as illustrated by Harris[8,9] on a wide range of studies performed in the recent past.

Frequently, the spectra are only used, in association with other techniques, to identify the existence of polymorphs and act routinely as fingerprints. Generally, when there are two or more molecules in the asymmetric unit, there will be observable splittings in the spectra.

Chemical shifts are highly sensitive to short range structure within 0.1–0.4 nm and small changes in crystallography. The ability of solid-state NMR to provide detailed information about molecular geometry has become apparent; the spectra may also provide data on intermolecular geometry.

In recent years, there is a growing area which can be referred to as NMR crystallography.[10–12] Structural information is provided by dipolar coupling constants and chemical shifts (full tensor). The use of slow MAS yielding the sidebands can sometimes suffice, although the principal methods involve the use of sophisticated pulse sequences. For measurement of dipolar coupling, samples containing simple spin systems, preferably spin pairs isolated in the crystal structure, are required. The usual technique is to selectively enrich the sample with ^{13}C (or ^{15}N) at a given site. The simplest aspect of NMR crystallography provided by MAS spectra is the number of independent molecules in the unit cell.

Computation of chemical shifts is also an expanding area. The aim of such studies is to enable the prediction of solid-state structure. A comparison can be made of the shifts calculated for each computer-generated structure with those measured experimentally to find a best match. However, computing chemical shifts of crystalline solids has a number of controversial features. In effect, the computations are done at 0 K, but using molecular geometries derived from diffraction measurements at ambient temperature (or above 77 K). Calculations are carried out on isolated molecules and therefore no intermolecular interactions are taken into account. Nevertheless, a comparison of the theoretically calculated with the experimental principal components of the chemical shift tensor was performed with success; numerous examples include antibiotics *penicillin V* and *ampicillin*,[13] testosterone, or paclitaxel (see below).

As opposed to solution NMR, solid-state NMR is a less destructive technique. Powdered samples are retained in their original form, tablet samples need only to be crushed, and capsule samples merely require the powder and a container to be separated. This feature is especially valuable in the studies of compounds isolated in small amounts from plant material. These samples are frequently subjected to the pharmacological testing; therefore, there is a remarkable trend to avoid destructive analytical methods. Spectroscopic techniques, such as solid-state NMR, can be applied for identification of the constituents and structural studies (e.g., detecting sugar units in glycosides).

2. SOLID-STATE NMR MEASUREMENTS OF DRUGS

2.1. Amphetamine and ephedrine hydrochlorides and related compounds

As a part of forensic investigations on controlled drugs, the ^{13}C solid-state NMR spectra of the amphetamines, ephedrines, and their mixtures with lactose were recorded.[14] The observed chemical shifts of *R*,*S*-MDMA (*3,4-methylenedioxy-N-methylamphetamine*) hydrochloride changed when lactose monohydrate was added. In the presence of lactose, both carbons that are β to the nitrogen exhibit a significant chemical shift change (5.5–6.2 ppm). Thus, the spectra of *R*,*S*-MDMA·HCl in "Ecstasy" tablets are different from those of pure crystals, and the interaction with lactose appears to be specific to this compound.

Solid-state ^{13}C CPMAS NMR spectra of *(3S,2S)-(+)-pseudoephedrine·HCl* and of *(3R,3S)-(–)-ephedrine·HCl* (Scheme 1) contain eight peaks (identical to the number obtained in solution) but the chemical shifts for both diastereoisomers differ. The spectra are readily assignable, apart from the resonances of aromatic carbons C7 and C8.[14]

The crystal structure of (3*S*,2*S*)-pseudo-ephedrine·HCl is almost identical to 3*R*,2*S*-ephedrine·HCl, except for the hydrogen bonds distances. Both the amine hydrogens are able to form hydrogen bond, other hydrogen bonds involve Cl. Thus, solid-state chemical shifts differences between diastereoisomers are determined by stereochemistry and also by lattice packing. The differences are large for C1 and C3 carbons enabling fast identification of ephedrine and pseudoephedrine, even in a mixture of both the compounds.

The calculation of NMR shielding constants was helpful in the assignment of resonances. The DFT (density functional theory) calculations were performed[15] for (3*R*,3*S*)-(−)-ephedrine and (3*S*,2*S*)-(+)-pseudoephedrine with

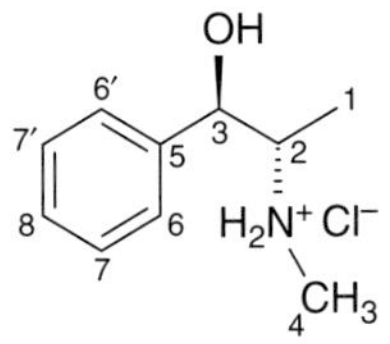

Scheme 1 (3*R*,3*S*)-(−)-ephedrine.

Table 1 Solid-state ^{13}C NMR peak assignments (δ_{CPMAS}, ppm) for (3*R*,3*S*)-(−)-ephedrine and (3*S*,2*S*)-(+)-pseudoephedrine hydrochlorides and the calculated shielding constants ($\sigma_{GIAO\ DFT}$, ppm)

C	Ephedrine			Pseudoephedrine		
	Lee et al.[14]	Wolniak et al.[15]		Lee et al.[14]	Wolniak et al.[15]	
	δ_{CPMAS}	δ_{CPMAS}	$\sigma_{GIAO\ DFT}$	δ_{CPMAS}	δ_{CPMAS}	$\sigma_{GIAO\ DFT}$
1	6.8	6.8	166.44	12.7	12.7	167.7
2	63.3	62.3	119.71	62.6	62.6	116.8
3	72.6	72.6	96.93	76.5	76.5	96.2
4	33.9	33.9	147.55	33.6	33.6	147.9
5	139.8	139.8	34.13	140.0	140	33.2
6	124.2	128.2	48.07	126.7	129.2	49.1
6′		128.2	45.71		129.2	49.8
7	128.2 or 127.7	124.2	50.92	131.2 or 129.2	126.7	50.96
7′		124.2	49.59		126.7	49.4
8	128.2 or 127.7	127.7	49.39	131.2 or 129.2	131.2	50.4
		$R^2 = 0.9941$			$R^2 = 0.9962$	

the GAUSSIAN03 program package at the 6-311++G** and B3LYP level. For the calculations of shielding constants, the gauge-independent atomic orbital (GIAO) DFT approach was used. Calculations were carried out on isolated molecules and therefore no intermolecular interactions in the solid state were taken into account. It was assumed that the measured δ_{CPMAS} follow the hierarchy of calculated shielding constants. The relationship $\delta_{CPMAS} = f(\sigma_{GIAO\ DFT})$ is linear, and the correlation coefficient $R^2 > 0.99$. The ^{13}C chemical shifts and the calculated (GIAO DFT) shielding constants are summarized in Table 1.

2.2. Erythromycin A – formation of a hemiketal

The macrolide *erythromycin A* is considered to be effective and one of the safest antibiotics. Recently, there has been much interest in the solid phase because of numerous polymorphic forms of erythromycin A and their intriguing hydrogyroscopic properties. The primary structural features of erythromycin A are a 14-member lactone ring, an amino sugar at C5, and a sugar attached to the lactone at C3 (Scheme 2). Ring-chain tautomerism occurs in solution, erythromycin A exists as a mixture of the 6,9-cyclic hemiketal, 9,12-cyclic hemiketal (with the C6—O—C9 and C9—O—C12 bridges, respectively), as well as the 9-ketone.

The question is whether the tautomeric forms exist in the solid phase. Marketed erythromycin A, the dihydrate, is stable and the structure has been characterized as the 9-ketone by X-ray diffraction. Previous solid-state NMR studies were focused on polymorphic and hydroscopic forms.[16] The erythromycin A was used as a

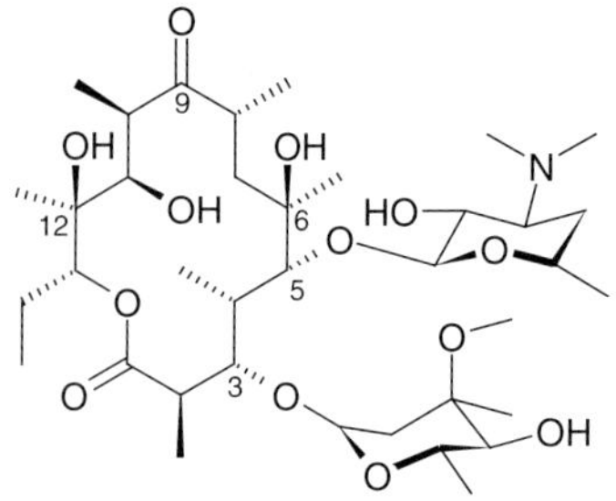

Scheme 2 Erythromycin.

model compound to demonstrate novel MAS NMR techniques. The homonuclear correlation method[17] provides a means to assign ^{13}C isotropic chemical shifts via recoupling of the homonuclear dipolar interaction under fast MAS. The magic-angle turning method[18] enables the measurement of chemical shift principal values of complex systems: 35 out of 37 ^{13}C principal values of erythromycin A have been measured. Unambiguous assignment of the tensor data to the nuclear positions of molecule requires uniform isotopic enrichment that hinders its general use.

In the ^{13}C CP/MAS spectrum of erythromycin A dihydrate nearly all 37 carbon lines are resolved. The use of chemical shift modeling with ab initio calculations (DFT/B3LYP method) of the chemical shielding tensor made possible the assignments of the experimental data.[19] A new resonance at 110.8 ppm was observed in the ^{13}C CP/MAS spectrum upon the application of heat or sample desiccation. With the use of the dipolar-dephasing spectral editing technique, this resonance was identified as a hemiketal carbon. Hence, the ketone–hemiketal tautomeric reaction has been shown to occur in the solid state and the formation of a cyclic hemiketal has been proven by solid-state NMR. The principal components of the ^{13}C chemical-shift tensor corresponding to this hemiketal are reported for the first time. The experimental principal values were correlated with theoretical shielding values and showed an rms deviation of 3.6 ppm.

2.3. Paracetamol – characterization of polymorphic forms

Paracetamol, a popular drug (Scheme 3), is easy to identify and quantify in its formulations because all the signals fall outside the spectral region of typical excipients. The signals of carbohydrates are in the range 60–100 ppm and alkyl carbons of polyethylene glycol or stearate within 20–40 ppm. In the ^{13}C MAS spectrum of a tablet,[6] aromatic carbons of paracetamol molecule C2 and C6 as well as C3 and C5 show separate resonances, due to the frozen conformation of —NH(CO)CH_3 substituent at C1 and the —OH group at C4 (Scheme 3).

The metastable orthorhombic form of *paracetamol* was prepared[20] from the melt of the commercially available monoclinic form. A ^{13}C CP/MAS solid-state NMR spectrum of the monoclinic form and a spectrum of a mixture of forms were obtained. Subtraction of the spectrum of the monoclinic form from that of the mixture yielded a spectrum of the orthorhombic form. These spectra show distinct

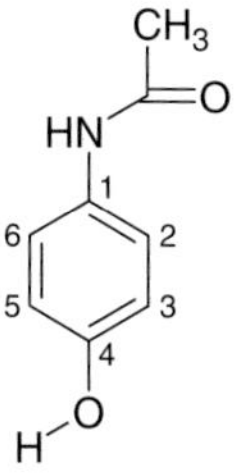

Scheme 3 Paracetamol.

differences, especially for the resonances arising from the aromatic carbons, i.e., between 115 and 125 ppm. The ^{13}C CP/MAS solid-state NMR spectrum of commercial paracetamol is in agreement with that reported previously, but it is more resolved. The resonances arising from those carbons bound to the nitrogen (132.5 and 169.3 ppm) are broader or appear as asymmetric doublets due to ^{13}C–^{14}N dipole–dipole interactions which are not completely averaged out by MAS. The effect is helpful by assignment.

2.4. Procaine and procainamide derivatives

Potential Class III antiarrhythmic agents of structure *p*-HOOC—R—CO—NH—C_6H_4—CO—X—C_2H_5—$N(C_2H_5)_2$ were obtained[21] as the amide and ester derivatives: succinylprocainamide I, succinylprocaine II, maleylprocainamide III, and maleylprocaine IV. The changes in their chemical structures produced different solubilities of the crystalline solids. The solution NMR spectra were unambiguously assigned to the zwitterionic structure; in the solid-state procainamide derivatives I and III crystallized as zwitterionic but procaine derivatives II and IV as neutral forms. The main support came from the chemical shifts of C1 in the CPMAS spectra ($\delta = 172.3$ ppm) that correspond to the anionic carboxylate group, several ppm shifted from the neutral species resonance (calculated chemical shift for zwitterionic form: $\delta = 174.4$ ppm, whereas for neutral form $\delta = 166.1$ ppm).

2.5. Rifampicin – molecular structure of polymorphs

Variable bioavailability of rifampicin from its formulations has been noticed as a problem in successful treatment of tuberculosis. *Rifampicin* (Scheme 4) has a complex structure, in commercial bulk samples[22] it exists as various combinations of form I, form II, and amorphous. Solid-state ^{13}C NMR spectra of rifampicin crystalline forms showed sharp peaks, whereas the neat spectrum for amorphous form could not be obtained because of broad and diffused peaks.

Molecular modeling (AM1 method) was employed for estimating hydrogen bonding interactions. The conformation of rifampicin is characterized by six intramolecular hydrogen bonds, including the C23—OH...O=C35 and C1—OH...O=C15 interactions. Form II shows four hydrogen bonds; and the C23—OH...O=C35 interaction is missing in form I. The differences in ^{13}C NMR

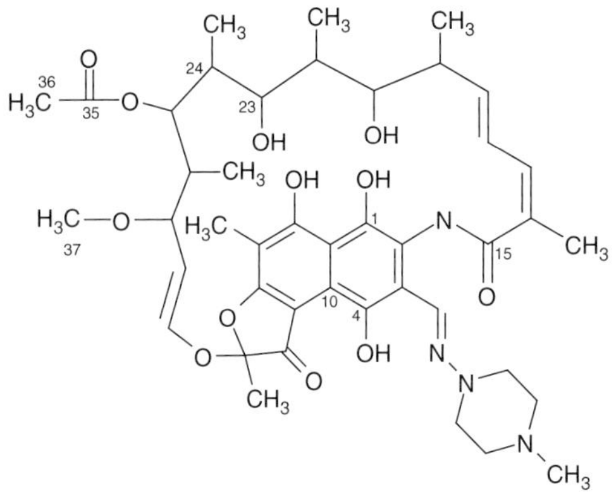

Scheme 4 Rifampicin.

spectra of two polymorphs were observed for: C15, C23, C24, C36, and C37 (singlet or doublet in forms I or II); other observed peaks in form I have different chemical shifts than their counterparts in form II whereas aromatic carbons C1–C10 show a similar pattern. As revealed by solid-state NMR and FTIR, in form I, C4—OH makes a strong H-bond with furanone thus forming a seven-membered ring. In addition, C1—OH is hydrogen bonded to amide carbonyl, and this intramolecular bonding increases the solubility of form I in acidic pH. On the other hand, in form II there is H-bonding between C23—OH and acetyl group, and the seven-membered ring is not formed. The differences in H-bonding in the solid state that change acidity/basicity of rifampicin may explain a lower dissolution rate of the metastable form, as compared with the stable form. The observed differences in solid-state NMR spectra of two polymorphic forms may be utilized for quantitative estimation of polymorphic purity.

2.6. Ketoconazole – amorphous form required

Ketoconazole is an imidazole antifungal agent that is administered both orally and topically. Its solubility in aqueous media is insufficient for the whole dose to be dissolved in the gastrointestinal fluids under normal conditions. Amorphous drugs are characterized by enhanced dissolution; however, they are not stable, and eventual devitrification is inevitable. Therefore, the glass-forming properties of *amorphous ketoconazole* and the stabilization of the amorphous drug using polyvinylpyrrolidone (PVP) as a carrier in binary solid dispersions have recently been studied.[23] ^{13}C CPMAS NMR spectra of a solid dispersion containing 30% (w/w) of ketoconazole and of a physical mixture consisting of 30% of amorphous ketoconazole and PVP were obtained. Although the peaks are very broad, it was clear that peak maxima of the carbon atoms of the drug in the spectrum of solid dispersion did not shift, as compared with those in the spectrum of the pure amorphous drug. Thus, the absence of strong and specific drug–polymer interactions was confirmed.

2.7. Paclitaxel (Taxol) – molecular structure and interactions

Paclitaxel is a naturally occurring substance that was originally extracted from the pacific yew tree *Taxus brevifolia*. It is a structurally complex, lipophilic molecule (Scheme 5) which became important in biology and medicine. The drug shows good antitumor activity against ovarian and breast cancer and a wide spectrum of carcinomas of the lung, colon, prostate, and brain.

Grant and coworkers[24,25] applied solid-state NMR to study *baccatin* and *paclitaxel polymorphs*. In 2006, ^{13}C CPMAS NMR was used to characterize the structure of two paclitaxel forms.[26] The ^{13}C spectrum exhibited a chemical shift range of over 200 ppm; the signals from aromatic side-chain residues in the range of 127–138 ppm were heavily overlapped whereas the signals of nonaromatic residues were much better resolved. Most of the signals arising from the taxane ring residues showed well-resolved doublets, with a splitting that ranged from 50 to 150 Hz. The resonances of C15, C16, and C17 from the center of the taxane ring remained as singlets indicating that the central part of the ring remained structurally similar. The doublets (e.g., for C4, C5, C9, C5′) originated from two slightly different molecular conformations of the taxane ring. Doublet splittings can serve as a direct measure of the structural difference between the two conformations, which could possibly correlate with the antitumor activity of Taxol.

New paclitaxel/polystyrene–isobutylene–styrene (SIBS) formulation was characterized by solid-state NMR.[27] Paclitaxel was incorporated within the polymeric matrix in solution and was then deposited as a coating onto a drug-eluting stent (TAXUS™). The ^{1}H and ^{13}C NMR spectra showed that most of the paclitaxel signals retained unchanged chemical shifts. The interactions between paclitaxel and SIBS are limited to nonbonded interactions occurring in solution. In particular, the doublet splittings in ^{13}C solid-state spectra indicated a rigid crystalline packing for paclitaxel. When paclitaxel is mixed with SIBS, the paclitaxel molecule displayed increased mobility, as indicated by proton and carbon spin-lattice relaxation times in the rotating frame.

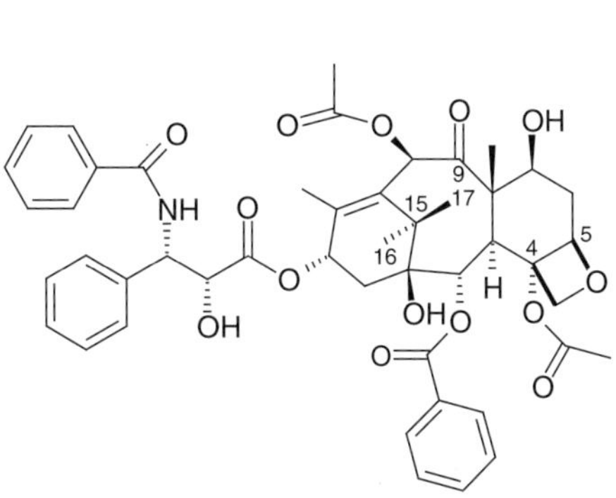

Scheme 5 Paclitaxel.

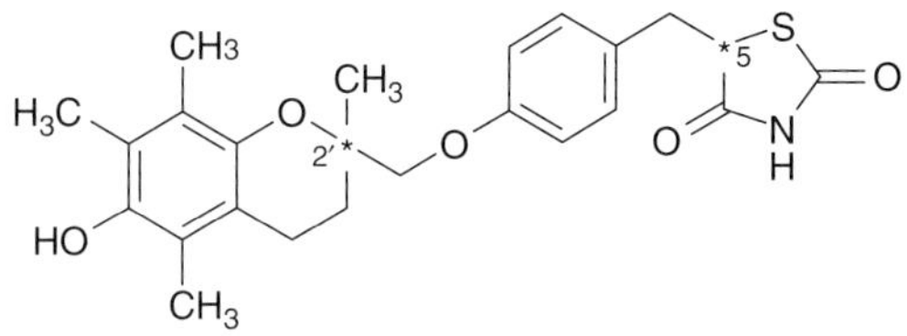

Scheme 6 Troglitazone.

2.8. Troglitazone – hydration of diastereomers

Troglitazone is a novel oral antidiabetic drug (Scheme 6) that improves insulin sensitivity and responsiveness. It has two asymmetric carbons, C2'/such carbon no. can be seen in Scheme 6 and C5, and is produced as a mixture of equal amounts of four optical isomers.

The solid-state form of troglitazone drug substance is a simple physical mixture of two diastereomers. The *RR/SS* diastereomer adsorbs water easily as a monohydrate; therefore, making troglitazone drug substance amorphous increases the solubility of the stereoisomers. Troglitazone has a hydrophilic function only at both edges of its molecule, and the other groups are hydrophobic. In the solid, the molecules are connected by head-to-tail hydrogen bonds, and the hydrated water breaks these chains. ^{13}C MAS NMR spectra of *troglitazone* showed[28] the differences during the hydration; the rebuilding of weaker hydrogen-bonding interactions changed the shielding, which is best seen as the shift of carbonyl resonance. It is worth noting that solid-state NMR could distinguish the hydrated and the nonhydrated *RR/SS* forms more clearly than powder X-ray diffractometry. Troglitazone was proved to exist in its amorphous form in tablets, and kept its solid-state form amorphous against heat and humidity during the 6-month storage period.

2.9. Delavirdine mesylate tablets – hydration and solubility

^{13}C CPMAS NMR was used to identify and quantify delavirdine form changes in tablets.[29] Dissolution extent of delavirdine mesylate tablets was substantially decreased after exposure to high humidity. This effect is related to the amount of water present in the tablet matrix. ^{13}C CPMAS NMR detected about 30% conversion from the *mesylate salt of delavirdine* to its free base form in the tablet matrix. The limiting solubility of delavirdine free base and a change in interparticle bonding can explain the decrease in the extent of dissolution.

2.10. MAS NMR of steroids: finasteride, estradiol, testosterone

The ^{13}C CPMAS NMR spectra of steroids usually exhibited clearly resolved narrow signals for the majority of carbon atoms of the molecule. These signals were assigned mainly on the basis of liquid-state chemical shifts. The skeleton of steroids contains numerous aliphatic protonated carbons which give resonances in a narrow spectral range, their assignment is thus tentative. It should be noted that in some of the assignments resonances separated by 0.5 ppm or less were assigned to particular carbons, although a reversed assignment could be equally valid.

Solid-state NMR studies of steroids[30,31] showed that the spectra of remarkably good quality could be obtained in only just over half an hour each. Additionally, the majority of solid-state peaks have chemical shifts almost the same as their liquid-state counterparts.

Azasteroids have received much attention, especially the compounds with 5α-reductase inhibitory activity and of androgen receptor binding. A group of biologically active 4-azasteroids was studied by ^{13}C NMR spectroscopy in solution and in the solid phase.[32] The ^{13}C CPMAS spectra were obtained for five *4-azasteroids* including finasteride. The number of resonances in the solid-state spectra of two 4-azasteroids exceeds the number of carbons in the molecule, which indicates that these compounds exhibited polymorphism (or pseudopolymorphism, i.e., cocrystallization with the solvent). In the crystals of finasteride, the lactam group is involved in different hydrogen-bond interactions. The comparison of chemical shifts for three forms confirms the conclusion that the steroid skeletons are similar because the chemical shift values for carbons of B, C, and D rings are close. The differences, which appear for A-ring carbons, are related to the formation of intermolecular hydrogen bonds. The form that included dioxane and exhibited different chemical shifts was identified as a new molecular complex.

In 2007, solid-state NMR and XRD data were reported for further *new forms of finasteride*;[33] form III and a new anhydrous form X were found in mixtures of polymorphs and their ^{13}C NMR chemical shifts were obtained. The previously reported and new solvates were characterized and the solvates were found to have a finasteride:solvent molar ratio of 2:1. The solvate molecules are highly disordered and sited in channels in the structure. These solvates may be distinguished by the characteristic ^{13}C CPMAS signals from the solvent molecules, but the resonances of the host finasteride structures differ only marginally, and powder XRD patterns are almost indistinguishable. It is interesting to note that ^{1}H MAS spectra give sharp lines for the solvent peaks, confirming their high degree of mobility.

Four *crystal forms of estradiol* were obtained by recrystallization from various organic solvents such as methanol, ethanol, isopropanol, and acetone, and their physicochemical properties were characterized using different methods, including solid-state ^{13}C NMR.[34] Estradiol typically tends to crystallize in the form of its hemihydrate. The residual solvent incorporated into the crystal changed the local chemical environment, which in turn caused the changes in the observed ^{13}C chemical shifts. The results suggest that the thermal analysis and CPMAS NMR could be useful to characterize the solvate molecule within the crystals of estradiols.

Spectral assignment can be improved using the calculated chemical shifts and two-dimensional techniques, such as INADEQUATE. Chemical shift calculations could provide structural information and are especially valuable to account for shift differences between polymorphs, and also between different independent molecules in a given crystal.

The case study of *two forms of testosterone* (17β-hydroxyandrost-4-en-3-one) is an example.[35] Although testosterone has a number of forms, the α- and β-forms have known crystal structures. The α-form is anhydrous and has two molecules in the asymmetric unit. The molecules are linked alternately in a chain by hydrogen bonding between the hydroxyl group at C17 and the carbonyl group at C3. The β-form is a monohydrate, with a single molecule in the independent unit.

A ^{13}C CPMAS NMR experiment at high field (16.5 T) for the *α-form of testosterone* has produced significantly improved resolution, and the crystallographic splittings are clearly visible for most carbons. Initially, there were some ambiguities in the assignments of lines in the region between 28 and 39 ppm. Correlations shown by an INADEQUATE two-dimensional spectrum, recorded at 11.7 T, allowed the components of most of the doublet signals to be grouped into two sets (for the two crystallographically independent molecules). This experiment also made possible the assignment of the signals in the ambiguous region, revealing some necessary reassignments different from those published previously. The calculated shielding (DFT, GIPAW) was plotted against the newly obtained experimental values and a linear relationship was obtained (with an RMS error of 3.53 ppm for α-testosterone). Using a combined INADEQUATE NMR and computational approach, the authors were able to assign all the resonances in the crystalline α-form of testosterone to individual atoms in the two molecules in the asymmetric unit cell. This methodology has a high potential for future applications in pharmaceutical problem solving.

^{13}C labeling was used to enhance the sensitivity of ^{13}C solid-state NMR to study the effect of tableting on the polymorphism of a steroidal drug.[36] The steroidal drug Org OD 14 was ^{13}C labeled and formulated into tablets containing only 0.5–2.5% active ingredient. The crystalline form could be readily analyzed in tablets by solid-state ^{13}C CPMAS NMR. No change in the crystalline form was observed as a result of formulation or in subsequent stability studies. Solid-state NMR in combination with ^{13}C labeling can be used as a strategy to study the effect of formulation on the polymorphism of low-dose drugs.

2.11. Nifedipine and amlodipine

The calcium channel blocker *nifedipine* (Scheme 7) is extremely photolabile and its photodecomposition has been widely studied. Various methods, including ^{1}H NMR, were used to detect and quantify its decomposition products. Of special interest are the investigations into the stability of solid pharmaceutical preparations. In the case of nifedipine tablets, light between 400 and 420 nm causes degradation,[37] and the exact penetration depth could be determined due to the

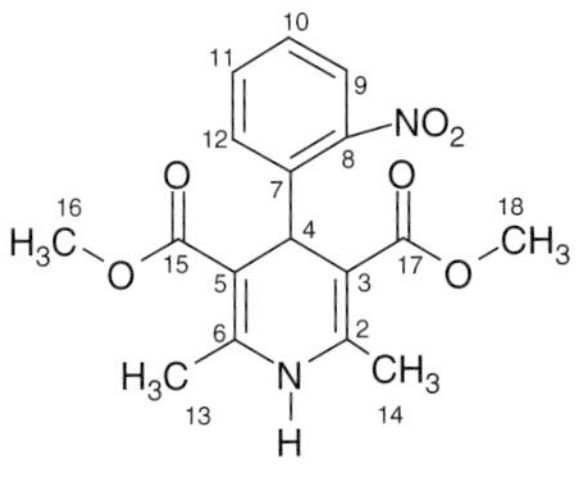

Scheme 7 Nifedipine.

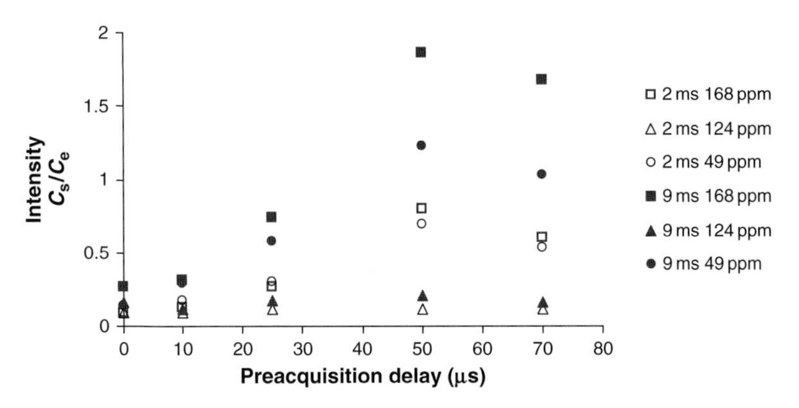

Figure 1 Optimization of intensity of nifedipine signals from tablets (C_s/C_e is a ratio of sample/excipient peak areas). Variable contact times (2 and 9 ms) and preacquisition delay 0–70 μs were used in the experiments.

discoloration of the drug substance upon irradiation.As the decomposition products of nifedipine act as photostabilizers by spectral overlay, light penetration and photodegradation in nifedipine tablets are limited.

An attempt has been made to detect the decomposition products using solid-state NMR.[38] To determine the optimal parameters for the measurements, i.e., to obtain the most intense resonances of the drug's signals, a series of spectra of nifedipine with contact times in the range 50 μs to 9 ms were recorded. Next, the spectra with two contact times of 2 and 9 ms and variable preacquisition delay time (dipolar dephased) were run. The results of these variable-contact CP experiments are illustrated in Figure 1.

The main criterion was the ratio of peak areas: substance (C_s)/excipient (C_e), calculated as the intensity of carbonyl (168 ppm), methoxy (49 ppm), and aromatic (124 ppm) carbon resonances with respect to the intensity of major CH (72 ppm) resonance.

It is evident that the optimal contact time for dihydropyridines should be longer than 1–2 ms used routinely, and should reach even ~8–9 ms. The spectra of tablets, recorded with $t_{cp} = 9$ ms and a preacquisition delay of 50 μs, benefit from fast relaxation of C—H carbons and the resonances of excipients (carbohydrates) are less intense. These parameters were applied in the measurements of powdered commercial tablets, containing nifedipine and *amlodipine* (Scheme 8).

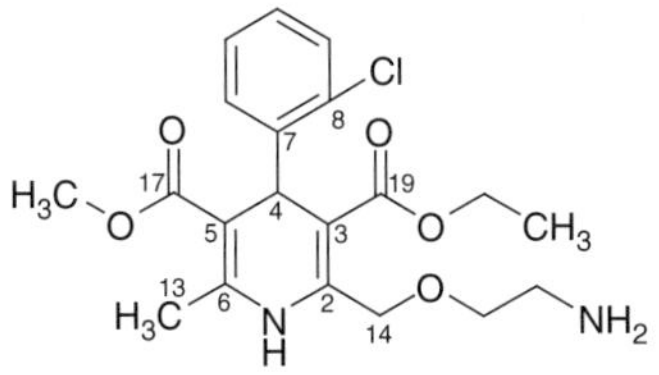

Scheme 8 Amlodipine.

The tablets of nifedipine (Cordafen, Cordipin) and amlodipine (Amlozek, Norvasc) were irradiated with UV light for 3 weeks. Then the tablets were powdered and subjected to CPMAS NMR measurements (Figures 2 and 3). The ^{13}C spectra recorded after irradiation did not differ from those of fresh material;

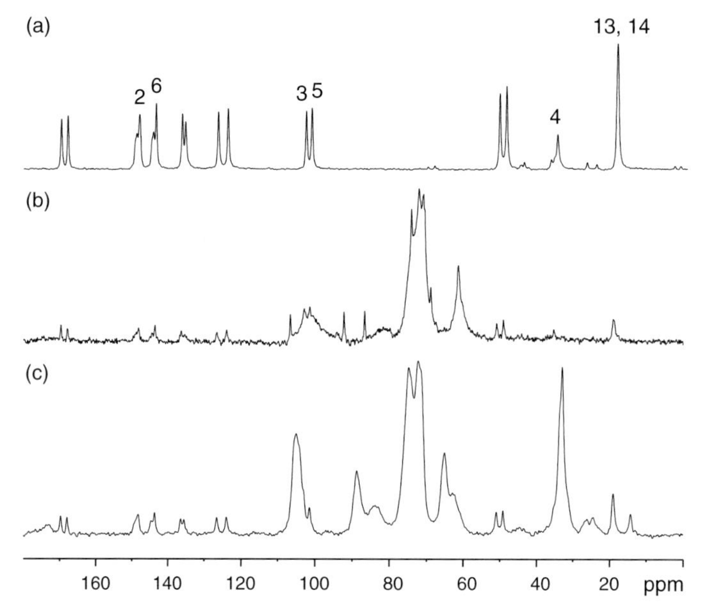

Figure 2 ^{13}C CPMAS NMR spectra of nifedipine (pure substance) (a) and its formulations: Cordafen 10 mg (b) and Cordipin 20 mg (c).

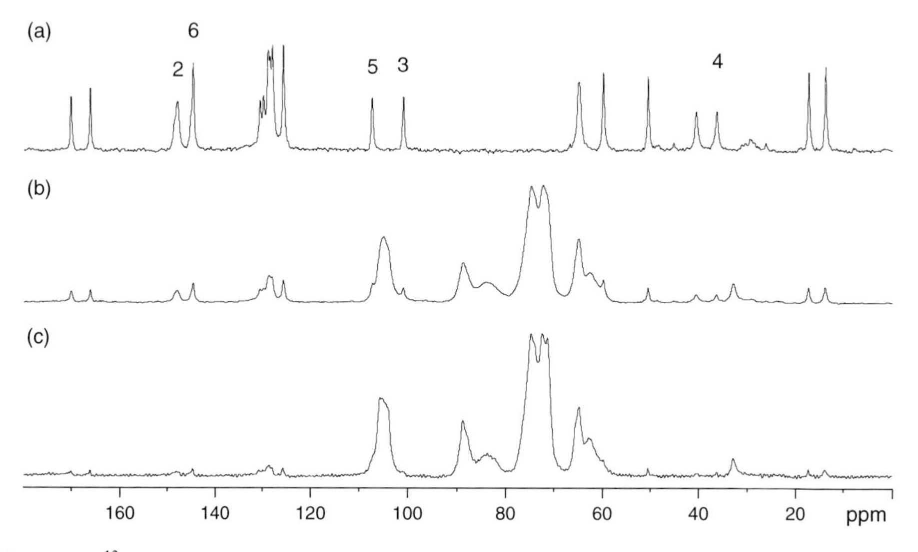

Figure 3 ^{13}C CPMAS NMR spectra of pure amlodipine (a) and its formulations: Amlozek 10 mg (b) and Norvasc 5 mg (c).

chemical shifts of active substances and of the excipients remained the same. It would be most interesting to monitor the signals of C4 (changing its position from $\delta = 35$ to 145 ppm), C3 (from 103 to 125 ppm), and C5 (from 102 to 126 ppm). However, expansion of the aromatic carbon region showed no signals which could be assigned to the nitroso-derivative. In conclusion, we found that solid pharmaceutical formulations containing nifedipine and amlodipine are resistant to UV light. It is worth noting that solid-state NMR can be used to study low-dose drugs, the signals of amlodipine can be observed in 5 mg tablet.

^{13}C CPMAS spectra of *amorphous nifedipine* (obtained from the melt) showed markedly broader resonances than those obtained for crystalline nifedipine.[39] Static ^{1}H and MAS ^{13}C spectra of the amorphous forms were examined as a function of temperature, monitoring the crystallization. Evidence for a metastable polymorph of nifedipine could be seen in the ^{13}C spectra and the authors supposed that this might help to explain the relative instability of amorphous nifedipine as compared with amorphous indomethacin. The value of T_1^H for amorphous nifedipine is significantly lower than that for the crystalline form, and this was also related to differences in stability.

Amorphous *nifedipine–PVP* and *phenobarbital–PVP* (PVP – polyvinylpyrrolidone) solid dispersions were prepared[40] with various drugs contents. Chemical shifts and spin-lattice relaxation times (T_1) of PVP, nifedipine, and phenobarbital carbons were determined by ^{13}C CPMAS NMR to elucidate drug–PVP interactions. The chemical shift of the PVP carbonyl carbon increased as the drug content increased, suggesting that there occurred hydrogen-bond interactions which decreased the mobility of the PVP carbonyl. T_1 of the drug carbons increased together with the increase of PVP content, and the results suggest that the localized motion of the PVP pyrrolidone ring and the drug molecules is reduced by hydrogen-bond interactions. A decrease in localized mobility appears to be one of the factors that stabilize the amorphous state of drugs.

2.12. Ibuprofen preparations

Solid-state NMR line widths were measured[41] for various *ibuprofen* preparations, including crystallization from different solvents (acetone, acetonitrile, methanol), melt-quenching, manual grinding, cryogrinding, compacting, and by blending with various excipients. Ibuprofen recrystallized from acetonitrile exhibited broader lines than ibuprofen recrystallized from either acetone or methanol. Manually ground ibuprofen had solid-state NMR line widths that were indistinguishable from the commercial sample, but cryoground ibuprofen had larger line widths than either. Physical mixtures with most excipients decreased the line widths. Only dilution in talc led to line width increases, which is attributed to the magnetic susceptibility anisotropy of the talc excipient. Therefore, line widths in solid-state spectra can be used to understand physical characteristics including particle size and morphology, degree of order in the materials, and physical environment.

2.13. Sildenafil citrate (Viagra) – structure and hydrogen bonding

Sildenafil citrate (SC) has been widely used for the treatment of erectile disorder. A detailed study concerning solid-state structure of this compound is very important for understanding enzyme (PDE5)-inhibitor (sildenafil) interaction. It is also of interest to determine sildenafil's protonation sites, as they may be responsible for its binding to the phosphodiesterase acidic amino acids.

Sildenafil citrate (Viagra) (see Scheme 3, Chapter 3, Part I) and *sildenafil base* in pure form were characterized by ^{1}H, ^{13}C, ^{15}N NMR spectroscopy in solution, solid-state, and pharmaceutical dosage forms.[42] The analysis of chemical shifts showed that: (i) N6-H forms intramolecular hydrogen bonds, (ii) N25 is protonated in the salt, and (iii) intermolecular OH...N hydrogen bonds involving N2 and N4 are present in the solid sildenafil citrate. The ^{13}C CPMAS spectra of the tablets containing different amounts of sildenafil citrate were recorded and showed that chemical shifts of sildenafil citrate in pure form and in pharmaceutical dosage forms are the same. SC is easily detected in the pharmaceutical dosage forms since only two of its carbon resonances (OCH_2 and quaternary carbon of the citrate anion) fall into carbohydrate-type region of the excipient.

Solid-state ^{13}C and ^{15}N MAS NMR have recently been used to investigate how water interacts with SC.[43] When the humidity is altered, the water concentration in the solid compound changes but does not reach a stoichiometric (e.g., 1:1) ratio to form a true hydrate. Only one set of ^{15}N and ^{13}C signals was observed for each humidity level indicating that water incorporated into the crystal lattice of SC is very mobile and exchanges rapidly between various sites. The ^{13}C data showed the formation of a hydrogen bond between water molecule and one carbonyl of the citrate anion. The spectra also show that the water content affects the conformation of the propyl group. Additionally, ^{15}N dipolar dephasing (DD) experiments confirmed that the sildenafil molecule is only protonated in the piperazine ring.

2.14. Oxybuprocaine hydrochloride – NMR crystallography

Oxybuprocaine hydrochloride acts as a local anesthetic and is used in eye drop formulations. Mod. II, the stable polymorphic form, contains two molecules with markedly different conformations (stretched and bent). The ^{13}C CPMAS NMR spectrum of this sample[12] showed crystallographic splittings arising from the fact that there are two molecules in the asymmetric unit. An INADEQUATE two-dimensional experiment was used to link signals for the same independent molecule. Of the four ethyl groups attached to NH^{+} nitrogens, one gives rise to unusually low chemical shifts, very different from those of the other three groups. This was attributed to gamma-gauche conformational effects, and confirmed by shielding computations. The oxybuprocaine system is too large for computations performed on the crystallographic unit and the intermolecular shielding effects have to be neglected. Nevertheless, the computed shifts do match the order of the experimental ones. The assignment of ^{13}C signals to specific carbons in the two crystallographically inequivalent molecules of *oxybuprocaine polymorph* showed the power of NMR crystallography.

2.15. Bupivacaine in protein matrix

Solid-state NMR spectroscopy is an effective technique for studying the state of both the drug and the matrix. Two systems have been studied[44] using solid-state NMR. The first system involved bupivacaine, a local anesthetic compound, which was incorporated into microspheres composed of tristearin and encapsulated using a solid protein matrix. Solid-state ^{13}C NMR was used to investigate the solid forms of bupivacaine in their bulk form or as incorporated into the tristearin/protein matrix. *Bupivacaine free base* and *bupivacaine-HCl* have very different solid-state NMR spectra, indicating that the molecules pack in different crystal forms. The capability of solid-state NMR to study interactions between the amino acid and the polymer matrix for synthetic peptides and peptidomimetics containing selective ^{13}C labeling was shown.

2.16. Warfarin – cyclic hemiketal

Warfarin and sintrom are effective coagulants usually named and depicted at the open tautomer. NMR data for solution provided the evidence[45] that warfarin exists in a dynamic equilibrium between the open form (the bioactive one) and two diastereoisomeric cyclic forms. ^{13}C CPMAS spectra showed the resonances in the range 25–165 ppm. The correlation between the experimental chemical shifts δ_{CPMAS} and the calculated carbon shieldings $\sigma_{GIAO\ DFT}$ was helpful in the assignment. The last signal, in the high-frequency range, appears at 161.5 ppm. There is no signal of carbonyl carbon from CH_2—CO—CH_3 fragment of the open form, which could be expected at 195–200 ppm. It confirms that there is no carbonyl C2=O, as in coumarins. The ^{13}C CPMAS spectra proved that both solid *coagulants* – *warfarin* and *sintrom* – appear as cyclic hemiketals. The chemical name of the solid-state form of warfarin should be 2-hydroxy-2-methyl-4-phenyl-3,4-dihydro-2*H*-pyrano {3,2-c}chromen-5-one (Scheme 9).

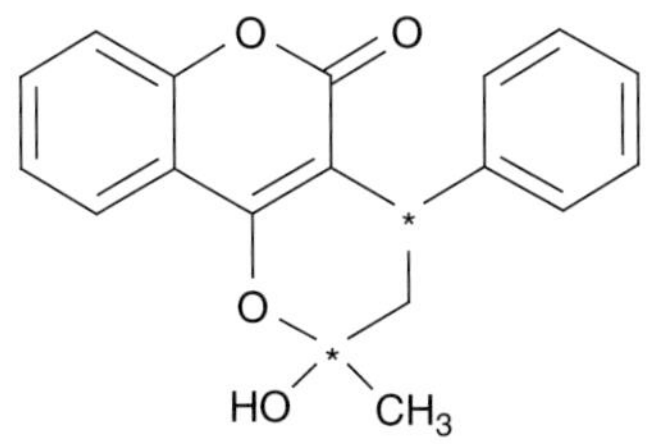

O O O HO CH_3

Scheme 9 Warfarin.

3. Medicinal Plants Extracts

Natural products, which have been known since ancient times, have contributed to the development of modern therapeutic drugs. The extracts containing alkaloids, flavonoids, terpenoids, or curcuminoids have been commonly used in folk medicine and have the efficacy against many diseases. The substances isolated from plants have to be investigated using modern analytical methods and subjected to in vitro and in vivo biological assays.

Solid-state NMR technique can be especially useful in the analysis and structural studies of insoluble materials and the substances that decompose in solution. Additionally, the analysis is nondestructive and the powder sample removed from the rotor can be used for further testing. Selected examples of such applications are presented.

3.1. Mixture of α-amyrin and β-amyrin

The extract from resins from Brazilian plants of Burseraceae (*Protioum heptaphyllum*, *Protium kleinii*) contains a mixture of two pentacyclic triterpenes *α-amyrin*, ursa-12-en-3β-ol (Scheme 10) and *β-amyrin*, oleana-12-en-3β-ol (Scheme 11). This mixture of triterpenes demonstrated immunostimulant, anti-inflammatory, as well as anxiolytic and antidepressant activities.

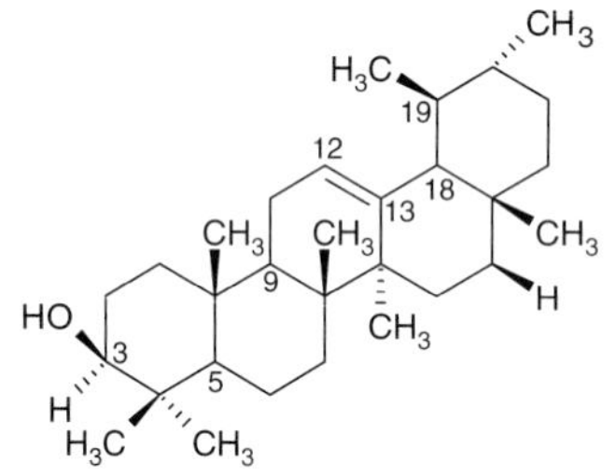

Scheme 10 α-Amyrin.

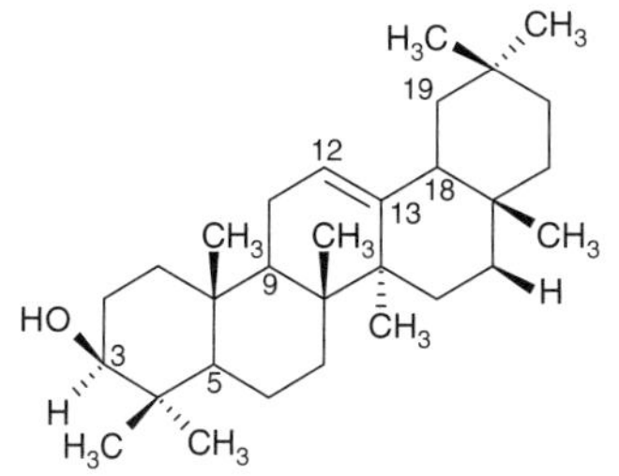

Scheme 11 β-Amyrin.

Table 2 ^{13}C CPMAS chemical shifts (ppm) for solid terpenoids: euphol, α- and β-amyrin

C	Euphol	α-Amyrin	β-Amyrin
1	35.2	40.0	40.0
2	28.9	27.6	27.6
3	79.7/78.4	78.2	78.2
4	40.0	37.6	37.6
5	51.0	55.5	55.5
6	19.8	20.8	20.8
7	27.7	32.8	32.8
8	134.3	40.7	39.1
9	134.3	48.0	48.0
10	37.9	37.6	37.6
11	22.2	23.7	23.7
12	32.2	125.2	122.4
13	44.4	139.8	145.6
14	50.8	41.6	41.6
15	30.0	28.7	27.6
16	28.9	27.6	27.6
17	50.8	34.0	34.0
18	14.9	59.9	48.0
19	19.8	40.0	48.0
20	35.9	39.1	31.1
21	19.8	31.1	34.0
22	35.9	42.7	37.6
23	26.1	28.7	28.7
24	125.3	16.0	16.0
25	130.5	16.0	16.0
26	17.8	17.6	17.6
27	26.1	24.0	27.6
28	28.9	28.7	28.7
29	16.3	17.6	32.8
30	26.1	20.8	23.7

Solid-state ^{13}C CPMAS NMR spectra of the mixture confirm the presence of these two compounds and the purity of the extract (lack of other signals) provided by Planta Analytica, USA.[46] ^{13}C MAS chemical shifts were assigned on the basis of solution data[47] and are summarized in Table 2. The α-amyrin dominates in the extract; it was evident by the inspection of resonances of aromatic carbons which appear in the region 120–150 ppm (Figure 4). The relative ratio of α-amyrin to β-amyrin can be determined from integrated areas of the C12 resonances ($\delta = 125.2/\delta = 122.4$ ppm) or of the C13 ones ($\delta = 139.8$ ppm/$\delta = 145.6$ ppm).

3.2. Euphol from *Euphorbia*

The euphane triterpenoids are isolated from the latexes of *Euphorbia* species. The chemical markers *euphol*, eupha-8,24-dien-3β-ol (Scheme 12) and nerifoliene were found to be common to both *E. antiquorum* and *E. nerifolia*, whereas euphol is the only marker for *E. tirucalli*. The latexes were highly valued in the Indian system

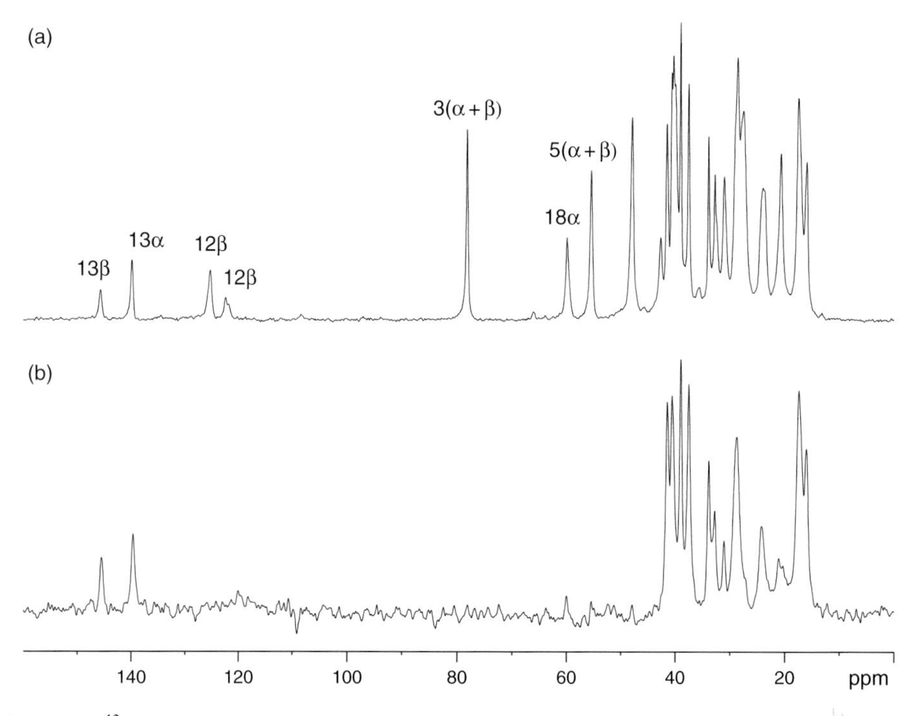

Figure 4 ^{13}C CPMAS NMR spectra of the mixture of α and β-amyrin: (a) standard and (b) dipolar dephased (50 μs preacquisition delay).

Scheme 12 Euphol.

of medicine. Now, the di- and triterpenes isolated from the resins have been the subject of intense studies because of their antiproliferative activity and the ability to inhibit multidrug resistance in cancer cells.

^{13}C NMR signals of euphol (provided by Planta Analytica, USA[46]) were assigned by 2D NMR techniques for $CDCl_3$ solution. The 2D correlations showed that this molecule has a skeleton with a $\Delta^{8,9}$ double bond, another double bond is located in the side chain. Solid-state ^{13}C CPMAS NMR spectra (Figure 5) exhibited the resonances for two double bonds (C23, C24, and C8, C9) in the region 120–140 ppm, resonances of eight methyl groups, and one doublet of hydroxylated carbon at 79 ppm (C3). The splitting of this signal suggested the presence of two molecules in the solid phase, with different conformations and/or intermolecular interactions. The

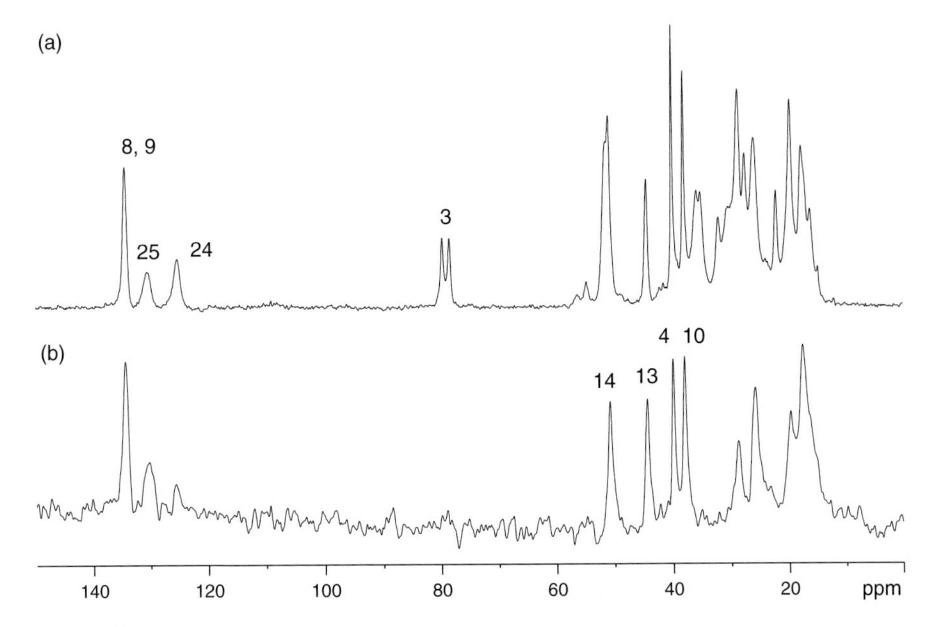

Figure 5 ^{13}C CPMAS NMR spectra of euphol: (a) standard and (b) dipolar dephased.

HPLC method was used to distinguish between the latexes; however, solid-state NMR can easily generate the fingerprinting patterns for the markers, such as euphol.

3.3. Oxindole alkaloids from *Uncaria tomentosa*

Uncaria tomentosa (Cat's claw, Una de Gato, Vilcacora) is a large, woody vine from the Amazon rainforest used medicinally by the native tribes for at least 2000 years.[48] Since the early 1990s, cat's claw has been used as an adjunctive treatment for diseases that target the immune system. Its bioactive components include oxindole alkaloids. A series of alkaloids were isolated[46] for the purpose of standardization of vilcacora products. Alkaloids may undergo isomerization in solution, therefore solid-state NMR technique was used to characterize the stereoisomers.[49] ^{13}C CPMAS NMR spectra for the most abundant alkaloids – *mitraphylline* (7*R*,20*R*), *pteropodine* or Uncarine C (7*R*,20*S*) (Scheme 13), *isomitraphylline* (7*S*,20*R*), and *isopteropodine* or Uncarine E (7*S*,20*S*) (Scheme 14) – were recorded; solid-state chemical shifts are listed in Table 3.

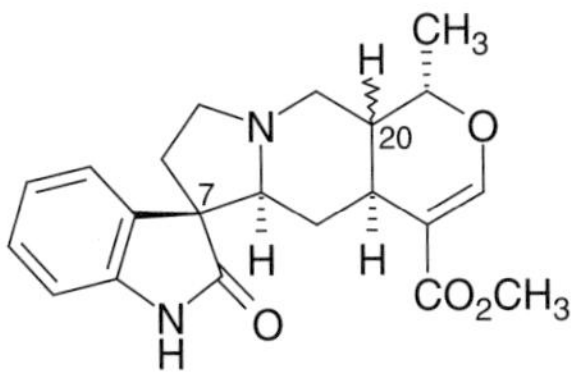

Scheme 13 Alkaloids: mitraphylline (7*R*,20*R*), pteropodine (7*R*,20*S*).

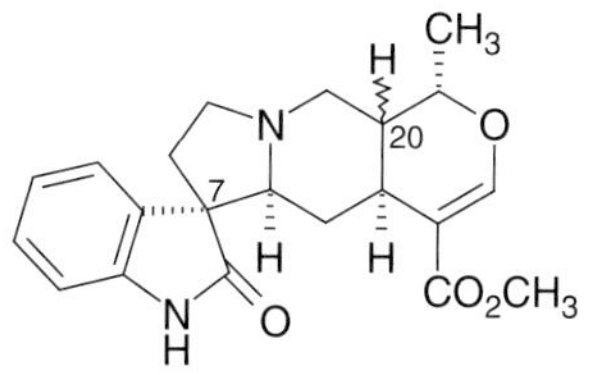

Scheme 14 Alkaloids: isomitraphylline (7*S*,20*R*), isopteropodine (7*S*,20*S*).

Table 3 ^{13}C CPMAS NMR chemical shifts δ (ppm) for solid alkaloids: mitraphylline, isomitraphylline, pteropodine, and isopteropodine (isolated by Planta Analytica, USA[46])

C	Mitraphylline	Isomitraphylline	Pteropodine	Isopteropodine
	7*R*, 20*R*	7*S*, 20*R*	7*R*, 20*S*	7*S*, 20*S*
2	181.7	181.1	182.0	180.9
3*	74.8	71.5	73.1	71.0/69.9
5	53.0	54.1	54.9	55.4/54.9
6	37.6	35.0	35.6	35.5
7*	55.1	56.7	56.1	56.9/56.3
8	133.2	134.2	132.8	133.9
9	123.3	124.3	123.1	123.5
10	121.8	122.0	121.5	121.8
11	128.3	127.7	128.5	128.6/128.1/127.6
12	111.5	110.0	111.0	111.9/110.7
13	142.7	141.5	142.8	142.1
14	29.8	30.4	29.7	30.0
15*	30.5	30.4	31.9	31.8
16	108.0	107.9	110.0	110.7/110.0
17	154.1	154.1	154.1	154.7/154.0
18	15.6	14.9	16.7	18.2/16.7
19*	74.8	74.0	73.1	72.7
20*	41.9	41.4	38.3	36.3
21	53.0	53.0	52.1	52.8
22	166.2	166.1	166.3	166.0
23	50.9	50.6	50.1	49.3/48.8/48.6

The resonances can be assigned directly by comparison with the solution data, only solid isopteropodine exhibited polymorphism. The identification of the quaternary carbons was made by means of DD experiments; the standard and DD spectra of pteropodine are illustrated in Figure 6.

Mitraphylline and isomitraphylline belong to the *normal*-type group, which is characterized by the *trans* relationships of the C/D and D/E ring junctions, the *allo*-type alkaloids pteropodine and isopteropodine are characterized by C/D *trans* and D/E *cis* relationships. In each type of compounds, the stereochemistry can be easily distinguished by solid-state chemical shifts of C3, C15, 1C8, C19, and C20 (Table 3). The 7*R* stereoisomers at the spiro C7 position are characterized by the C3 signals at $\delta = 73.1, 74.8$ ppm, whereas in the spectra of 7*S* the C3 signals appear at $\delta = 69.9\text{–}71.1$ ppm.

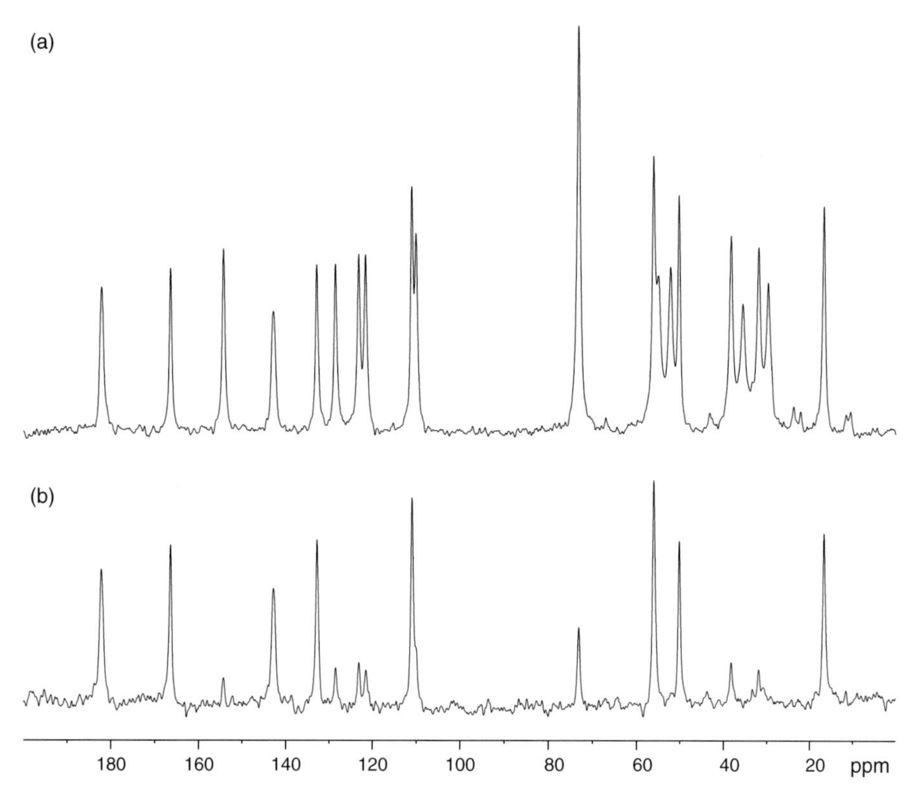

Figure 6 ^{13}C CPMAS NMR spectra of pteropodine (uncarine C): (a) standard with t_{cp} 2 ms and (b) dipolar dephased.

3.4. Lanatoside C

Digitalis lanata, the plant harvested in Brazil is the industrial source of digoxin and lanatoside C, used in the treatment of congestive heart failure. Agricultural production is the only economically feasible process to obtain these cardenolides and thus, investigations on methods of isolation and analysis are of interest. The raw material of lanatoside C for the preparation of the reference standard is usually examined by HPLC. However, prepurification procedures for extracts have a considerable effect on the chromatographic efficiency. Lanatosides can be easily identified by solution ^{1}H and ^{13}C NMR. And the solid-state NMR?

^{13}C CPMAS NMR spectrum[50] of solid *lanatoside C* exhibits 37 narrow signals of 49 carbons of this molecule (Figure 7). In the spectral range of carbohydrates (60–90 ppm) appear the signals of three units of digitoxose (one acetylated at C3′″) and one glucose. Overlapped resonances of aliphatic carbons at 20–40 ppm are typical of steroids, and the skeleton of lanatoside contains nine methylene groups. The characteristic resonances are those of C—O (180.6, 172.5 ppm), olefinic carbons C20, C22 (178.7, 115.7 ppm), and of C1″, C3, C21, C14, between 90 and 110 ppm. The purity and authenticity of lanatoside C can also be attested by solid-state NMR.

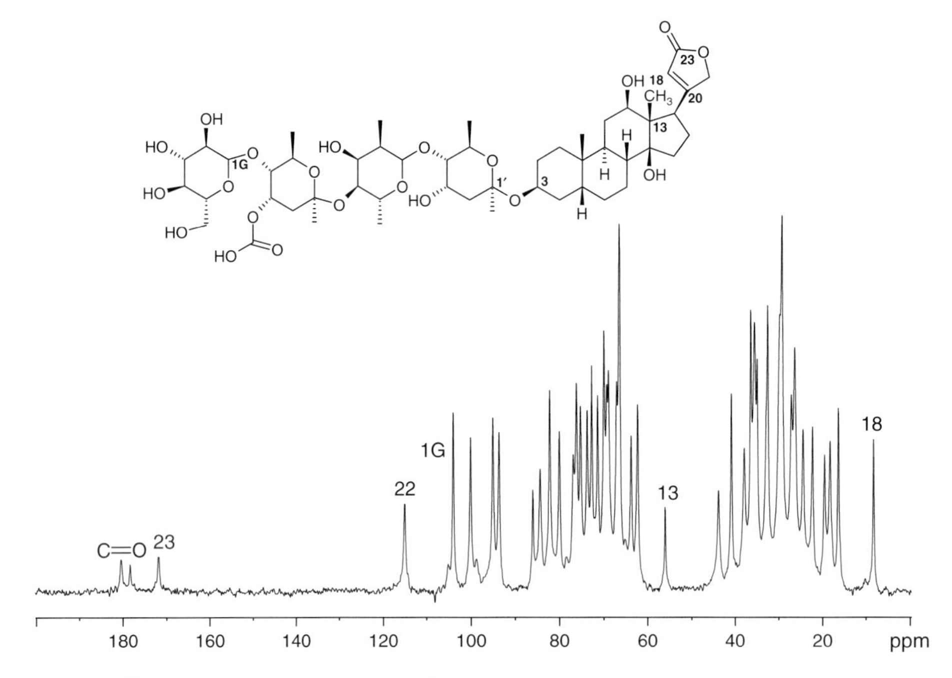

Figure 7 ^{13}C CPMAS NMR spectrum of lanatoside C.

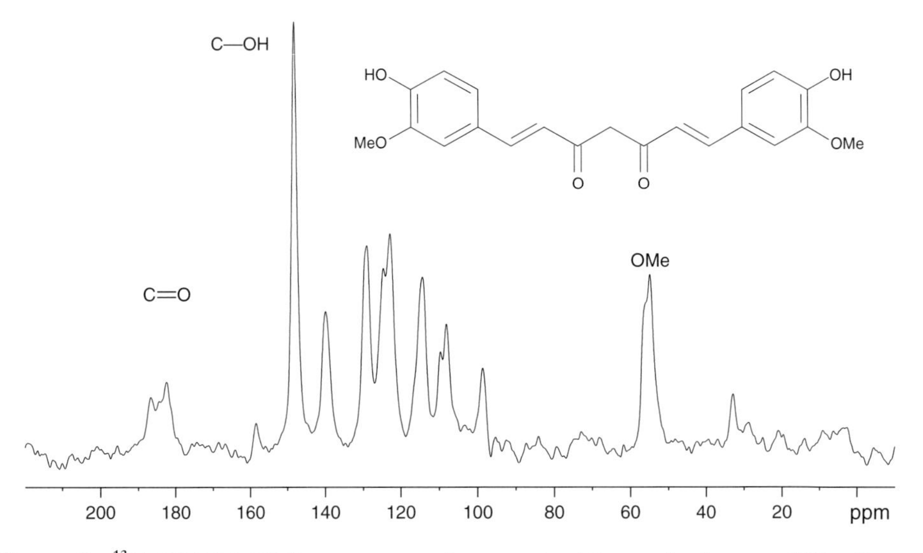

Figure 8 ^{13}C CPMAS NMR spectrum of curcumin (Curcumin forte – 95%, dietary supplement, DiSunhed.Net, Denmark).

3.5. Curcumin

Curcumin, a dietary spice from turmeric, was found to be anti-inflammatory, anticarcinogenic, and antithrombotic. Molecular mechanisms underlying its chemopreventive activities are widely studied. The dried ground rhizome of the perennial herb *Curcuma longa* has been used in Asian medicine since the second millenium BC. The most active constituent is a polyphenol, curcumin (diferuloylmethane). Commercial grade curcumin contains a mixture of curcuminoids (desmethoxycurcumin 10–20%, bisdesmethoxycurcumin 5% or less) whereas the capsules, sold as a dietary supplement, frequently contain amino acids (phenylalanine) or carbohydrate additives. There is considerable interest in fast identification of curcumin, which is relatively insoluble in water, and solid-state MAS NMR techniques can be useful.

The ^{13}C CPMAS NMR spectrum of *curcumin*, taken from a capsule of dietary supplement, is illustrated in Figure 8. Besides the signals of the methoxy groups, the spectrum shows only the signals of aromatic and carbonyl carbons. Intense resonances in the region 90–150 ppm indicate that the sample is 95% pure curcumin.

4. ^{13}C CPMAS NMR of Excipients

Pharmaceutical products are dominantly administered by solid dosage forms, tablets, and capsules. It implies the development of new techniques for formulation and processing of powdered solids. Usually, the powders enter the pharmaceutical development process in the most stable crystalline form and the changes of their structural state should be avoided. One of the most usual processes applied to the crystalline powders is the particle size reduction by micronization or milling. It improves the solubility and the dissolution rate of the drug and subsequently may improve therapeutic efficiency.

However, the drugs as well as the excipients may undergo amorphization or polymorphic transformations. Structural transformations can induce the solubility, the dissolution rate or the bioavailability, and may also modify some macroscopic properties of the powder like the tableting ability or the flowability. It is thus essential to define efficient characterization methods of solids reduced to the micro/nano scale and to understand the effects of the different size reduction treatments, to predict and control the state transformations they induce.

Solid-state NMR is widely used for characterization of crystalline powders but also of amorphous samples and polymers. The investigations included *amyloses* and *starches*,[51,52] hydrous and anhydrous *lactose*,[53] and also the *cyclodextrins* – easily available natural host compounds, increasingly popular for encapsulation of drugs.[54]

Starch was subjected to chemical treatments, such as cross-linking or hydroxypropylation, to meet the material requirements for food uses or controlled release in the pharmaceutical industries. Two types of cross-linking formulations have been employed to prepare high amylose starch for use as an excipient for sustained drug release.[55] Structural differences and chain dynamics of the modified starches in the dry and hydrated states have been compared by the use of variable contact time ^{13}C CPMAS NMR spectroscopy.

The effects of moisture on the polymorphic transition of crystalline *mannitol* were investigated.[56] Mannitol has three polymorphic forms: α, β, and δ. The water uptake of δ form was greater than that of the β form when each crystalline form was stored at 97% humidity. The different powder X-ray diffraction patterns confirmed that moisture induced polymorphic transition from the δ to β form. Thus it was suggested that the observed higher hygroscopicity of the newly formed β form arose from the gradual increase in the surface area with the polymorphic transition. The results from molecular modeling, solid-state CPMAS NMR spectra and scanning electron micrographs suggest that water molecules act as a molecular loosener to facilitate conversion from the δ form to the β form as a result of multinucleation.

The MAS NMR technique has been used to determine the structure and properties of *cellulose* and cellulosic products. In native cellulose fibers, cellulose chains are aggregated to form fibrils. As technologically important physical properties of cellulose are related to the supermolecular structure of the polymers, the molecular structure, mobility, and ordering is of interest. A method for quantifying the states of order/disorder is based on ^{13}C CPMAS NMR in combination with spectral editing. Typical ^{13}C MAS NMR spectra from cellulose I are made up of six signals from the anhydroglucose unit split into fine structure clusters due to the supramolecular structure. To obtain quantitative information, a processing of the overlapped spectra is necessary. The most informative region is a signal cluster between 80 and 92 ppm. This region contains the signals corresponding to C4 carbons situated in crystalline cellulose Iα and Iβ domains together with paracrystalline cellulose. The C4 carbons of more disordered regions are distributed in a broad band ranging from 80 to 86 ppm.[57] Cellulose and noncellulose polysaccharides can be distinguished on the basis of spin-lattice relaxation times in the laboratory and rotating frames; cellulose is always associated with the long components of T_1 and $T_{1\rho}$, whereas pectin and hemicellulose are associated with the short ones. Solid-state ^{13}C CPMAS NMR spectra of plant cell walls are complex owing to the superposition of resonances from different polysaccharides. The NMR signals in the C4 region 80–92 ppm were assigned[58,59] to cellulose at accessible fibril surfaces, cellulose at inaccessible fibril surfaces, and hemicellulose.

The particle properties and solid-state characteristics of two celluloses, Avicel PH101 and cellulose obtained from the alga *Cladophora* sp., were evaluated[60] and related to the compaction behavior and the properties of the tablets made from them. The cellulose fibril surface area was best reflected by solid-state ^{13}C MAS NMR. The difference in fibril dimension and, thereby, the surface area of the two celluloses was shown to be the primary factor in determining their properties and behavior. Properties such as the crystallinity and tablet disintegration could be related to the fibril dimensions. The *Cladophora* cellulose resulted in rather strong compacts that still disintegrated rapidly. The irregular surface morphology of the particles and the fragmenting behavior probably contributed to the strength of the tablets.

So, the question arises: to use or not to use solid-state NMR for the studies of tablets? An argument in favor of the positive decision may be found when considering chemical shifts of the drug in comparison with those arising from excipient. While the physical properties of pharmaceutical excipients have been well characterized, the spectral data, including solid-state NMR, are difficult to find.

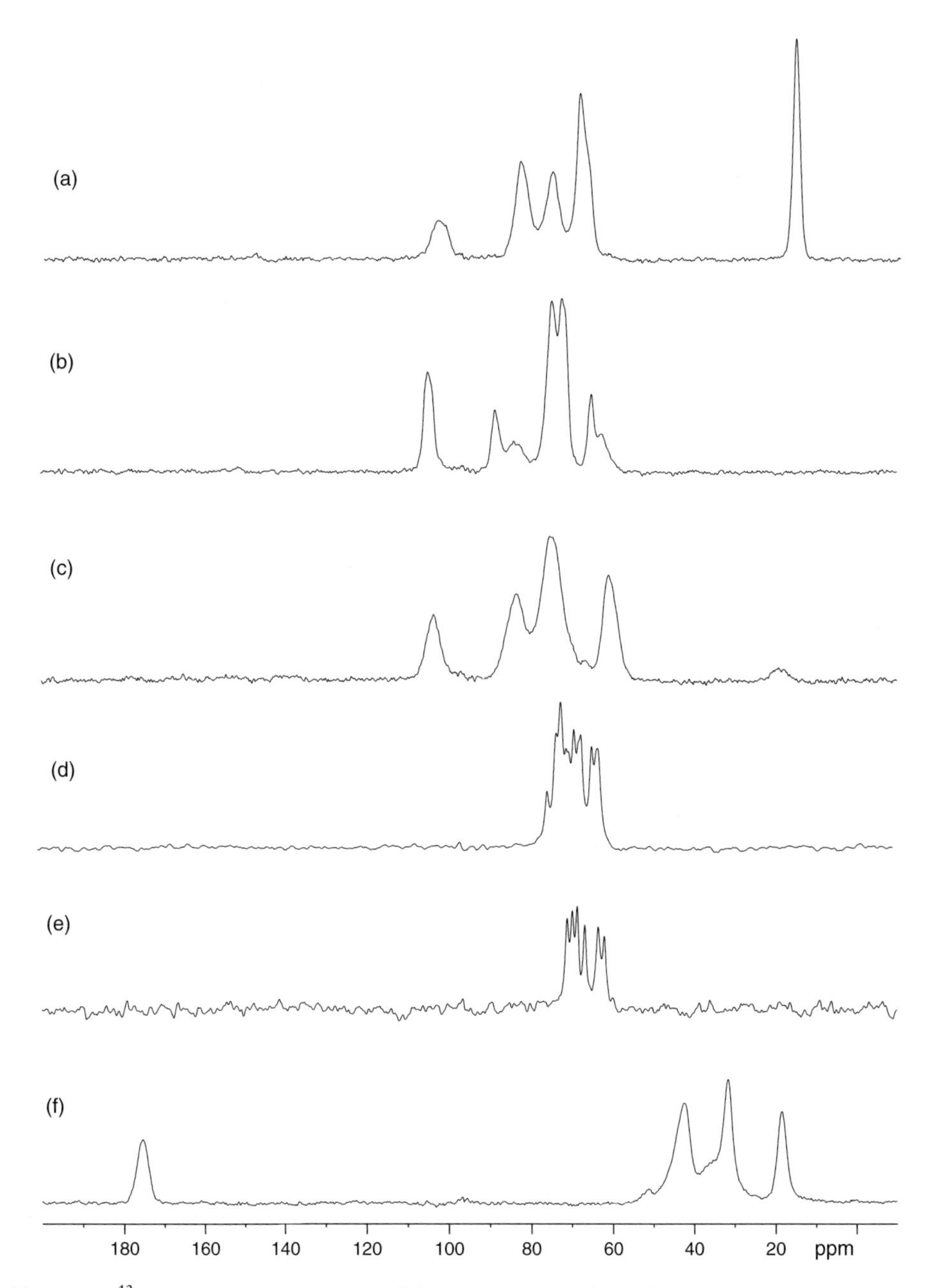

Figure 9 ^{13}C CPMAS NMR spectra of the most commonly used excipients: (a) AQUALON N-22 (ethy cellulose), ser. 41252-Hercules, (b) AVICEL (microcrystalline cellulose), PH-105, ser. 5527C, (c) BENECEL (hydroxypropylmethy cellulose HMPC), ser. VK 8494-Hercules, (d) SORBITOL (corion instant), ser. M261140-MERCK, (e) MANNITOL PARTECH 200-MERCK, and (f) COLLIDONE CL (crospovidone PVP), ser. 89-1609-BASF.

Characteristic signals of the pharmacologically active compound should be outside the spectral region covered by excipients. A set of the ^{13}C CPMAS spectra for the most popular excipients is illustrated in Figure 9.

For comparison of their ^{13}C MAS spectral regions, different *excipients* were selected: Avicel PH-1112, Avicel PH-105 and Avicel PH-102, Benecel HPMC (Hercules), Aqualon N-22, Collidon CL(PVP) (BASF), Mannitol Powder (Merck), Mannitol Partec (Merck), and Sorbitol (Merck).

The most commonly used excipient is microcrystalline cellulose (for instance, the three Avicel's samples). It gives a stable mixture in the solid-state formulation and is neutral to many compounds in pharmaceutical formulations. The ^{13}C signals of *cellulose*, like other carbohydrates (mannitol, sorbitol, starch), appear in the central part of the spectrum, between 60 and 110 ppm leaving free space below and above, i.e., the resonances of aromatic and aliphatic carbons of the drug can be observed undisturbed.

Additionally, microcrystalline cellulose is a good source of dietary fiber, with negligible calories; it will not interfere or interact with other compounds added to the drugs for fortification, such as vitamins.

Recently, biodegradable drug-delivery systems attract a lot of attention, they can be formulated to release drug for hours to years and have been used for the controlled release of medications. An important contribution in developing a drug-delivery matrix is the knowledge of the long-term stability of the form (drug + matrix) after formulation and any changes that might occur to the drug throughout the delivery process. These changes may be followed using solid-state NMR technique.

5. Drugs in Membranes

Biological membranes are complex assemblies in which constituents such as cholesterol and proteins experience slower and reduced dynamics as compared with phospholipids. *Cholesterol* dynamics and the interaction with phospholipids have been widely studied[61–63] using ^{1}H and ^{13}C MAS NMR. In a membrane, cholesterol inserts normal to the plane of the bilayer, with its hydroxyl group in close vicinity to the ester carbonyl of glycerophospholipids and its alkyl side-chain extending towards the bilayer center. The changes of its location and mobility are reflected in chemical shifts (^{1}H, ^{2}H, and^{13}C NMR), line broadening, and relaxation parameters. The effects of incorporating other compounds (peptides, steroids, canabinoids) can be likewise studied.

Chlorpromazine-HCl (CPZ), an amphipathic cationic psychotropic drug of the phenothiazine group, is known to partition into lipid bilayer membranes of liposomes with partition coefficients depending on the acyl chain length. It can alter the bilayer structure in a manner depending on the phospholipid headgroups. The effects of adding CPZ to the membranes were studied by differential scanning calorimetry (DSC) and ^{13}C CPMAS NMR.[64] MAS NMR spectra of the dipalmitoyl phosphatidylcholine (DPPC) liposomes show that CPZ has low or no interaction with the phospholipids of bilayer. This CPZ interaction causes about 30% of the acyl chains to enter the gauche conformation with low or no CPZ interdigitation among the acyl chains at 25°C. The DPPC bilayer at 37°C does show CPZ interdigitation among the phospholipids, as the phospholipid acyl chains carbon resonances shift downfield by 5–15 ppm.

Physicochemical methods were used[65] to study the thermal and dynamic changes caused by *losartan* in the membrane bilayers, and molecular modeling was implemented to explore its topography in membranes. Its incorporation resulted in modification of the thermal profile of DPPC bilayers in a concentration-dependent way up to 20 mol% as it is depicted from the combination of DSC and MAS NMR data. In particular, the presence of losartan caused the lowering of the phase transition temperature. The use of a combination of biophysical methods along with docking experiments brought out a possible two-step mechanism that involves incorporation of losartan at the interface of membrane bilayers and diffusion in the upper parts of AT(1) receptor helices IV–VII.

Several applications of both wide-line and solid-state MAS NMR of bicelles in which are embedded fragments of a tyrosine kinase receptor or enkephalins[66] were presented. The magnetically orientable bicelle membranes are of particular interest for studying the functional properties of lipids and proteins in a state that is close to their natural environment. Shielding interactions can be used to determine minute alterations of internal membrane dynamics and the orientation of peptides with respect to the membrane plane.

To understand various beneficial effects of flavonoids on health, the oxidative damages of lipoprotein in membranes have to be studied. As antioxidants, flavonoids have an ability to scavenge free radicals, including the lipid peroxy radicals. The mechanism of preventing oxidation suggests the propensity of flavonoids toward the lipid membrane and some intermolecular interactions. The interaction of different *flavonoids* of varying polarity with monounsaturated lipid model membranes was investigated using ^{1}H MAS NMR.[67] From ^{1}H ring current-induced chemical shifts and NOESY cross-relaxation rates, the localization and orientation of flavonoids in the lipid membrane could be determined. The largest flavonoid-induced shifts are observed for flavone, which suggests a distribution of the flavonoids according to their polarity. The most apolar flavone is located deeper in the membrane while an increasing number of hydroxyl groups provided a higher propensity toward the aqueous phase. No significant alterations of the bilayer structure have been found by static ^{31}P and ^{2}H NMR measurements, indicating that the bilayers can easily accommodate high concentrations of flavonoids. It is interesting that in the glycosylated form, flavonoids are still distributed over the entire membrane to protect any given double bonds from oxidation. As efficient inhibitors of lipid peroxidation. flavonoids are promising molecules for further pharmaceutical developments.

REFERENCES

1. I.J. Lowe, R.E. Norbert, Phys. Rev. 107 (1957) 46; E.R. Andrew, A. Bradbury, R.G. Eades, Nature 183 (1959) 1802.
2. J. Schaefer, E.O. Stejskal, J. Am. Chem. Soc. 98 (1976) 1031–1032.
3. G.E. Balimann, C.J. Groombridge, R.K. Harris, K.J. Packer, B.J. Say, S.F. Tanner, Philos. Trans. R. Soc. Lond. A 299 (1981) 643.

4. D.E. Bugay, Magnetic Resonance Spectrometry, pp. 94–125, in: H.G. Brittain (ed.) Physical Characterization of Pharmaceutical Solids, Marcel Dekker Inc., New York, 1995.
5. P.J. Saindon, N.S. Cauchon, P.A. Sutton, C.J. Chang, G.E. Peck, S.R. Byrn, Pharm. Res. 10(2) (1993) 197–203.
6. I. Wawer, Solid state NMR in drug analysis, pp. 231–256, in: U. Holzgrabe, I. Wawer, B. Diehl (eds) NMR spectroscopy in Drug Development and Analysis, Wiley-VCH, Weinheim, New York, Chichester, Brisbane, Singapore, Toronto, 1999.
7. P.A. Tishmack, D.E. Bugay, S.R. Byrn, J. Pharm. Sci. 92 (2003) 441–474.
8. R.K. Harris, Analyst 131 (2006) 351–373.
9. R.K. Harris, J. Pharm. Pharmacol. 59 (2007) 225–239.
10. R.K. Harris, Solid State Sci. 6 (2004) 1025–1037.
11. J.K. Harper, D.M. Grant, Cryst. Growth Des. 6 (2006) 2315–2331.
12. R.K. Harris, S. Cadars, L. Emsley, J.R. Yates, Ch.J. Pickard, R.K.R. Jetti, U.J. Griesser, Phys. Chem. Chem. Phys. 9 (2007) 360–368.
13. O.N. Anzutkin, Molecular structure determination: application in biology, pp. 280–384, in: M.J. Duer (ed.) Solid State NMR Spectroscopy, Principles and Applications, Blackwell Science, Cambridge, 2002.
14. G.S.H. Lee, R.C. Taylor, M. Dawson, G.S.K. Kannangara, M.A. Wilson, Solid State NMR 16 (2000) 225–237.
15. M. Wolniak, K. Paradowska, K. Zawada, I. Wawer, unpublished results.
16. G.A. Stephenson, J.G. Stowell, P.H. Toma, R.R. Pfeiffer, S.R. Byrn, J. Pharm. Sci. 86 (1997) 1239–1244.
17. E. Mary, L. Mueller, B. Sun, S.W. Fesik, R.G. Griffin, J. Am. Chem. Soc. 120 (1998) 10602–10612.
18. D.W. Alderman, G. McGeorge, J.Z. Hu, R.J. Pugmire, D.M. Grant, Mol. Phys. 95 (1998) 1113–1126.
19. R.J. Iuliucci, J. Clawson, J. Zhi Hu, M.S. Solum, D. Barich, D.M. Grant, C.M.V. Taylor, Solid State NMR 24 (2003) 23–38.
20. H.A. Moynihan, I.P. O'Hare, Int. J. Pharm. 247 (2002) 179–185.
21. M.E. Olivera, M.V. Ramirez Rigo, A.K. Chattah, P.R. Levstein, M. Baschini, R.H. Manzo, Eur. J. Pharm. Sci. 18 (2003) 337–348.
22. S. Agrawal, Y. Ashokraj, P.V. Bharatam, O. Pillai, R. Panchagnula, Eur. J. Pharm. Sci. 22 (2004) 127–144.
23. G. Van den Mooter, M. Wuyts, N. Blaton, R. Busson, P. Grobet, P. Augustijns, R. Kinget, Eur. J. Pharm. Sci. 12 (2001) 261–269.
24. J.K. Harper, J.C. Facelli, D.H. Barich, G. McGeorge, A.E. Mulgrew, D.M. Grant, J. Am. Chem. Soc. 124 (2002) 10589–10595.
25. J.K. Harper, D.H. Barich, E.M. Heider, D.M. Grant, R.R. Franke, J.H. Johnson, Y. Zhang, P.L. Lee, R.B. Von Dreele, Cryst. Growth Des. 5 (2005) 1737–1742.
26. Y. Ho, D.-L.M. Tzou, F.-I. Chu, Magn. Reson. Chem. 44 (2006) 581–585.
27. J.-Z. Chen, S.V. Ranade, X.-Q. Xie, Int. J. Pharm. 305 (2005) 129–144.
28. N. Suzuki, T. Kawasami, J. Pharm. Biomed. Anal. 37 (2005) 177–181.
29. B.R. Rohrs, T.J. Thamann, P. Gao, D.J. Stelzer, M.S. Bergren, R.S. Chao, Pharm. Res. 16 (1999) 1850–1856.
30. I. Wawer, J. Nartowska, A. Cichowlas, Solid State NMR 20 (2001) 35–45.
31. J.W. Morzycki, K. Paradowska, K. Dąbrowska-Balcerzak, I. Jastrzębska, L. Siergiejczyk, I. Wawer, J. Mol. Struct. 744–747 (2005) 447–455.
32. J.W. Morzycki, I. Wawer, A. Gryszkiewicz, J. Maj, L. Siergiejczyk, A. Zaworska, Steroids 67 (2002) 621–626.
33. A. Othman, J.S. Evans, I.R. Evans, R.K. Harris, P. Hodgkinson, J. Pharm. Sci. 96 (2007) 1380–1397.
34. J.S. Park, H.W. Kang, S.J. Park, C.K. Kim, Eur. J. Pharm. Biopharm. 60 (2005) 407–412.
35. R.K. Harris, S.A. Joyce, Ch.J. Pickard, S. Cadarsc, L. Emsley, Phys. Chem. Chem. Phys. 8 (2006) 137–143.
36. K.J. Booy, P. Wiegerinck, J. Vader, F. Kaspersen, D. Lambregts, H. Vromans, E. Kellenbach, J. Pharm. Sci. 94 (2005) 458–463.

37. W. Aman, K. Thoma, Pharmazie 58 (2003) 645–650.
38. M. Zielińska, M. Pisklak, I. Wawer, unpublished results.
39. D.C. Apperley, A.H. Forster, R. Fournier, R.K. Harris, P. Hodgkinson, R.W. Lancaster, T. Rades, Characterisation of indomethacin and nifedipine using variable-temperature solid-state NMR, Magn. Reson. Chem. 43 (2005) 881–892.
40. Y. Aso, S. Yoshioka, J. Pharm. Sci. 95 (2006) 318–325.
41. D.H. Barich, J.M. Davis, L.J. Schieber, M.T. Zell, E.J. Munson, J. Pharm. Sci. 95 (2006) 1586–1594.
42. I. Wawer, M. Pisklak, Z. Chilmończyk, J. Pharm. Biomed. Anal. 38 (2005) 865–870.
43. D.C. Apperley, P.A. Basford, C.I. Dallman, R.K. Harris, M. Kinns, P.V. Marshall, A.G. Swanson, J. Pharm. Sci. 94 (2005) 516–523.
44. J.W. Lubach, B.E. Padden, S.L. Winslow, J.S. Salsbury, D.B. Masters, E.M. Topp, E.J. Munson, Anal. Bioanal. Chem. 378 (2004) 1504–1510.
45. M. Pisklak, D. Maciejewska, F. Herold, I. Wawer, J. Mol. Struct. 649 (2003) 169–176.
46. J. A. Gliński, M. H. Davey, Planta Analytica, LLC, Brookfield, CT, 2004.
47. R.R. Miranda, G.D. Silva, L.P. Duarte, I.C. Fortes, S.A. Filho, Magn. Reson. Chem. 44 (2006) 127–131.
48. K. Keplinger, G. Laus, M. Wurm, M.P. Dierich, H. Teppner, J. Ethnopharmacol. 64 (1999) 23–34.
49. K. Paradowska, M. Pisklak, M. Wolniak, J. A. Gliński, M. H. Davey, I. Wawer, unpublished results.
50. K. Paradowska, J. Nartowska, I. Wawer, unpublished results.
51. K.R. Morgan, R.H. Furneaux, R.A. Stanley, Carbohydr. Res. 235 (1992) 15–22.
52. H.R. Tang, J. Godward, B. Hills, Carbohydr. Polym. 43 (2000) 375–387.
53. H.G. Brittain, S.J. Bogdanowich, D.E. Bugay, J. DeVincentis, G. Lewen, A.W. Newman, Pharm. Res. 8 (1991) 963–973.
54. S.J. Heyes, N.J. Claydenand, Ch.M. Dobson, Carbohydr. Res. 233 (1992) 1–14.
55. H. Thérien-Aubin, F. Janvier, W.E. Baille, X.X. Zhu, R.H. Marchessault, Carbohydr. Res. 342 (2007) 1525–1529.
56. T. Yoshinari, R.T. Forbes, P. York, Y. Kawashima, Int. J. Pharm. 247 (2001) 69–77.
57. H.R. Tang, Y.L. Wang, P.S. Belton, Solid State NMR 15 (2000) 239–248.
58. K. Wickholm, P.T. Larsson, T. Iversen, Carbohydr. Res. 312 (1998) 123–129.
59. P.T. Larsson, E.L. Hult, K. Wickholm, E. Pettersson, T. Iversen, Solid State NMR 15 (1999) 31–40.
60. C. Gustafsson, H. Lennholm, T. Iversen, C. Nyström, Drug. Dev. Ind. Pharm. 29 (2003) 1095–1107.
61. O. Soubias, V. Reat, O. Saurel, A. Milon, J. Magn. Reson. 158 (2002) 143–148.
62. G.P. Holland, T.M. Alam, J. Magn. Reson. 181 (2006) 316–326.
63. R.M. Epand, A.D. Bain, B.G. Sayer, D. Bach, E. Wachtel, Biophys. J. 83 (2002) 2053–2063.
64. W. Nerdal, S.A. Gundersen, V. Thorsen, H. Høiland, H. Holmsen, Biochim. Biophys. Acta 1464 (2000) 165–175.
65. P. Zoumpoulakis, I. Daliani, M. Zervou, I. Kyrikou, E. Siapi, G. Lamprinidis, E. Mikros, T. Mavromoustakos, Chem. Phys. Lipids 125 (2003) 13–25.
66. C. Sizun, F. Aussenac, A. Grelard, E.J. Dufourc, Magn. Reson. Chem. 42 (2004) 180–186.
67. H.A. Scheidt, A. Pampel, L. Nissler, R. Gebhardt, D. Huster, Biochim. Biophys. Acta 1663 (2004) 97–107.

CHAPTER 5

Metabolic Profiling

K.A. Kaiser, C.E. Merrywell, F. Fang, *and* C.K. Larive

Contents

Abstract

Metabolic profiling experiments have been of great use to researchers in the field of drug analysis both in the search for clinical biomarkers of disease or toxic insult and to assess the impact of a drug on natural metabolic processes. The inherent quantitative nature of nuclear magnetic resonance (NMR) and its ability to analyze biofluids and tissues, often with little or no sample pretreatment, makes it an ideal analytical technique for metabolic profiling. This chapter will discuss the NMR approach to metabolic profiling, primarily addressing experimental design, data acquisition, and data processing. The quality of NMR metabolic profiling results depends on the care with which data is collected and whether the results are subjected to robust statistical treatment and modeled onto biochemical pathways enabling the generation of meaningful conclusions.

Keywords: metabolomics, bacteria, fungi, plant, animals, biofluid, tissues, data acquisition, spectral processing, validation

1. INTRODUCTION

1.1. Definition of metabolic profiling

Metabolic profiling involves the identification and quantification of specific endogenous metabolites. A single metabolic profile can be thought of as a snapshot of the metabolic state of an organism at a given moment. Comparison of metabolic profiles can distinguish the metabolic impacts of factors such as gender, genotype, age, and disease state. Metabolic profiling often focuses on single classes of compounds (such as polar lipids, isoprenoids, carbohydrates, amino acids) or members of a particular metabolic pathway. Because of its quantitative and selective nature, metabolic profiling can be useful in a clinical diagnostic setting.[1]

A related field is metabolomics, sometimes called metabonomics (for clarification of this terminology, see a review by Robertson[2]), which strives for the identification and relative quantification of all endogenous metabolites present in an organism or biofluid sample. Metabolomic investigations are particularly useful when the effect of a genetic transformation, disease state, or administration of a drug or toxicant is unknown or affects multiple biochemical pathways. The goal of metabolomic investigations is typically a general understanding of the organism's response to a perturbation. Therefore these experiments typically do not focus on a specific pathway or metabolite class. Metabolomic data is sometimes merely qualitative, in these cases, it is termed metabolic fingerprinting, and the data set is used for the unsupervised classification of samples or the generation of predictive models to describe the difference between several states of an organism.[3] This chapter focuses on metabolic profiling and only experiments with the aim of a quantitative outcome will be discussed.

1.2. Goals of metabolic profiling experiments

Quantitative spectroscopic experiments can be carried out in two distinct modes: absolute and relative. An experiment aimed at absolute quantitation is considerably more challenging, owing to the fact that for most spectroscopic techniques, analytes with equal concentrations give different detector responses. This is not the case for nuclear magnetic resonance (NMR), where the signal is directly proportional to the number of nuclei giving rise to a resonance, assuming the data has been acquired with appropriate care, making NMR an excellent tool for quantitative comparisons of metabolite levels. Absolute quantitation by NMR is discussed in detail in

Chapter 5, section 3.1, but more commonly the objective of metabolic profiling experiments is the discovery of one, or more, measured variables whose values can be used to distinguish between two populations, classically deemed "case" and "control." This experimental design fits the binary class discrimination type, which has been employed in mainstream biology and medicine for many years.[4] In this type of analysis absolute quantitation is not required and the signals in each case can be reported relative to the signals in each control either as a difference (case − control) or as a ratio (case/control) as long as the two spectra are acquired under identical conditions.[5]

1.3. Dynamic range of NMR measurements

Although metabolic pathways are largely conserved across organisms, it is difficult to find published values for average in vivo metabolite concentrations even for biologically common model systems such as *Escherichia coli* (bacterium), *Saccharomyces cerevisiae* (fungus), *Caenorhabditis elegans* (invertebrate), *Drosophila melanogaster* (insect), and *Rattus norvegicus* (vertebrate). A report by Soga et al. provides concentration data for 63 metabolites in *Bacillus subtilis* determined using mass spectrometry (MS)[6] (Figure 1). Note that the concentration range spans five orders of magnitude, but the majority of metabolites (~83%) measured span only two orders of magnitude. Quantitative proton NMR has demonstrated an achievable dynamic range of greater than 300:1, which would allow a majority of metabolites

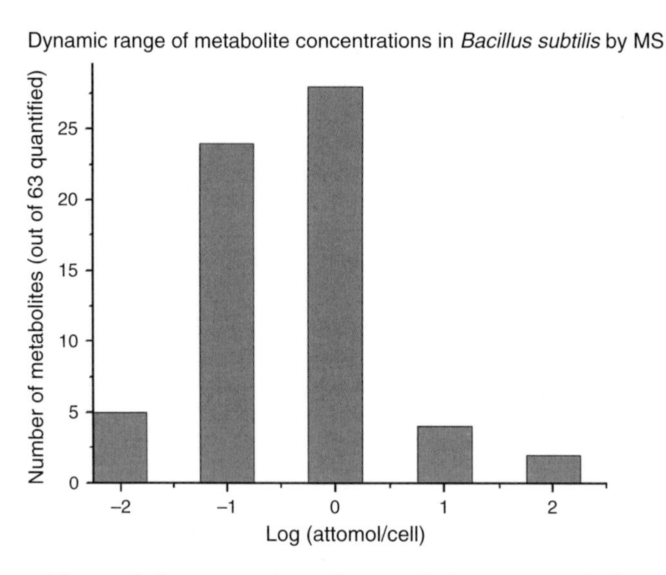

Figure 1 Logarithm-scaled concentrations of 63 metabolites quantitatively measured by MS. Data adapted from Table 1, T. Soga, Y. Ohashi, Y. Ueno, H. Naraoka, M. Tomita, T. Nishioka, J. Proteome Res. 2 (2003) 488–494. Note that the concentration range spans five orders of magnitude, but the majority of metabolites measured (~83%) spanned only two orders of magnitude.

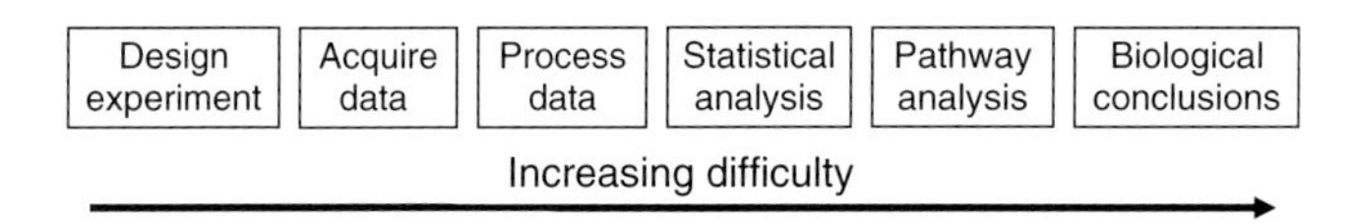

Figure 2 Workflow of metabolic profiling experiments.

to be sampled quantitatively by this technique in a single experiment.[7] This is demonstrated by Moing et al.,[8] who report concentrations of 17 metabolites in extracts of *Arabidopsis thaliana* (a model plant) over a wide range of concentrations ($1–10^4$). NMR was shown to be linear and correlated well with other standard methods of quantitation, including enzymatic assay and high-performance liquid chromatography (HPLC) determinations for selected metabolites.

1.4. Scope of chapter

Metabolic profiling was carried out as early as 1969 by MS[9] and issues relating to analysis of biofluids by NMR[10] were reported in 1978 leading to studies of intact cell metabolism[11] by 1984. A useful discussion of the merits and drawbacks to the primary analytical platforms for metabolic profiling (NMR and MS) is provided by Robertson,[2] who states in conclusion "most groups who are seriously pursuing metabonomics are currently or soon will be using both platforms." A review of recent literature shows this statement to be true.[12,13] Despite the impending convergence of these technologies, this chapter will cover only the NMR approach to metabolic profiling. Figure 2 shows the work flow by which a metabolic profiling experiment can lead to biological conclusions. This chapter will primarily address the first four steps in Figure 2: defining goals, experimental design, data acquisition, and data processing. The robust statistical treatment of data, projection of data onto biochemical pathways, and generation of biological conclusions are beyond the realm of this chapter, although these are the ultimate aims and often the most challenging aspects of a metabolic profiling experiment.

2. Sample Preparation

2.1. General considerations

There are many considerations that must be taken into account to ensure meaningful outcomes of metabolic profiling experiments. Depending on the organism under investigation, parameters that can be controlled include genotype, ecotype, phenotype, gender, age, diet, and environment. However, in large-scale human studies, there is much less control over these variables.[14] A recent study has determined that even factors such as intensity and duration of exercise, frequency of tobacco and alcohol intake, and the use of over-the-counter medications have significant impacts on the metabolic profiles of healthy human subjects of the same gender and age.[15] For this reason, careful design at the outset of an experiment will help to produce outcomes that are both statistically significant and biologically meaningful. This section will focus on the factors which can be brought under experimental control, and give examples of the data output if such recommendations are followed.

2.1.1. Obtaining a valid, representative sample

Whether the sample is derived from an animal, plant, or microorganism, care needs to be taken in deciding what constitutes the sample. For example, in plant studies, a sample may consist of one leaf, many leaves from one plant, a single whole plant, or a homogenous pool of leaves from many plants. Diet or growth media will have a metabolic consequence, as well as the feeding schedules and access to water or humidity. Natural diurnal and seasonal variations in metabolic profiles have been observed across organisms of all kingdoms: bacteria, fungi, plants,[16] and animals.[17] These factors can be somewhat controlled by careful experimental design.[18] For example, in plant studies, planting and harvesting at regular time intervals can average out variations in the growth environment if the resulting tissues are pooled.[19]

2.1.2. Replicates and variability

The term "replicate" is often used in the literature without indication of what type of replicate was conducted in the experiment. A machine replicate means that a single identical sample was analyzed multiple times by the same instrument. For NMR experiments, there is little variability (on the order of 1% using a high-purity standard[20]) between machine replicates and for this reason machine replicates are not typically conducted. In an interlaboratory comparison of identical urine samples, errors arose from factors such as efficiency of water suppression and it was discovered that different acquisition parameters were used.[21] When identical acquisition parameters were used, the differences fell to 1%.

An analytical replicate is a single homogenous sample subjected to the complete analytical protocol, including the sample preparation procedure.[22] In particular, the efficiency and reproducibility of extraction must be determined for each set of metabolic profiling experiments when evaluating samples that require extraction steps.[23] It is difficult to estimate the magnitude of variation caused by sample preparation because it depends on the protocol, the sample under investigation, and the care taken by the user who prepares the sample for analysis. Variation between analytical replicates has been minimized through the use of automated sample handling.[20] An advantage of NMR analysis of biological fluid samples is that minimal sample preparation is required, limiting the error contributed by sample manipulations.

Biological replicates are samples derived from separate organisms in which an attempt has been made to control all possible sources of biological variability such as genotype, age, diet, growth environment, and even the phase of an organism's diurnal cycle at the time of sampling. To evaluate the robustness of the metabolic profiling experimental design, biological replicates can be split into analytical replicates and assayed several times to constitute machine replicates (Figure 3). It is essential that the variation between analytical replicates is less than between

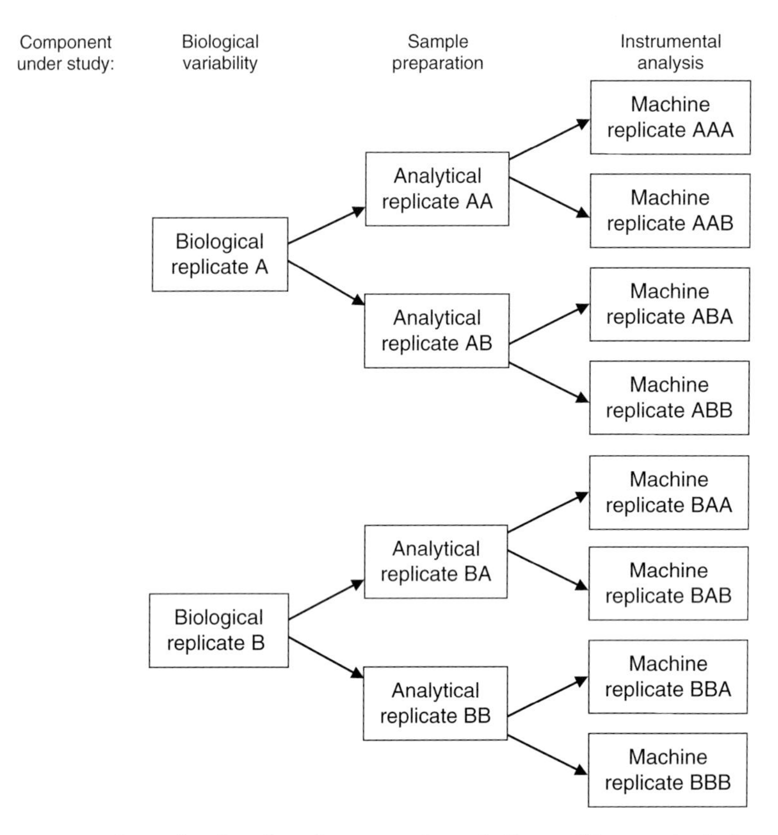

Figure 3 Strategy for evaluating the robustness of metabolic profiling methods through the use of replicates.

biological replicates.[24] For example, Fiehn et al.,[25] estimated that the analytical variation was ~8% using gas chromatography/mass spectrometry (GC/MS) for metabolic profiling of plant extracts whereas the biological variance was estimated at 26–56%. As with analytical variability, the biological variability that depends on many factors should be established for each experiment.

The question of sample size has been addressed rigorously by Dobbin and Simon,[26,27] who conclude that there is no "rule of thumb" for how many replicates are necessary. Instead, the authors propose the use of formal calculations based on several parameters: the goals of the experiment, whether pooling is employed in the sampling procedure, and the relative magnitudes of different sources of variability. Although the calculations are applied to data from microarray experiments, the data from microarrays and metabolic fingerprints is comparable in the degree of complexity and high dimensionality. In plant studies, samples from several plants are often combined to average out inherent biological variability.[28,29] In one such study, it was found that up to 30% of the major metabolites varied significantly in tomato fruit from plants grown under identical conditions.[30] In metabolic profiling studies of *Drosophila melanogaster* (fruit fly) and *Anisomorpha buprestoides* (walking stick) it was found that individual insects have widely divergent metabolic profiles and pooling will result in a loss of important information.[31,32]

2.1.3. Stability

Stability of samples is an important consideration in any biological experiment. In a comparison study, the storage of tissues, biofluids, or tissue extracts at 4°C, −20°C, and −80°C produced widely divergent metabolic fingerprints.[33] While sample classification may still be possible with samples stored at higher temperature, storage at −80°C is necessary for accurate quantitation.[34] Sodium azide is often added to samples of biofluids and tissue extracts to prevent microbial degradation of sample components.

2.1.4. pH and temperature

At constant pH and temperature, chemical shifts are highly invariant, making automated integration techniques possible. However, for nuclei near protonation sites, resonance chemical shifts can vary significantly even with small changes in pH near the pK_a of the ionizable group. For example, the pH dependence of resonances in the ^{1}H NMR spectra of corn root extracts is shown in Figure 4.[35,36] The simplest way to control pH is through the use of a buffer, although care must be taken that the sample components do not overwhelm the buffer capacity by measuring the final pH of the sample prior to analysis.

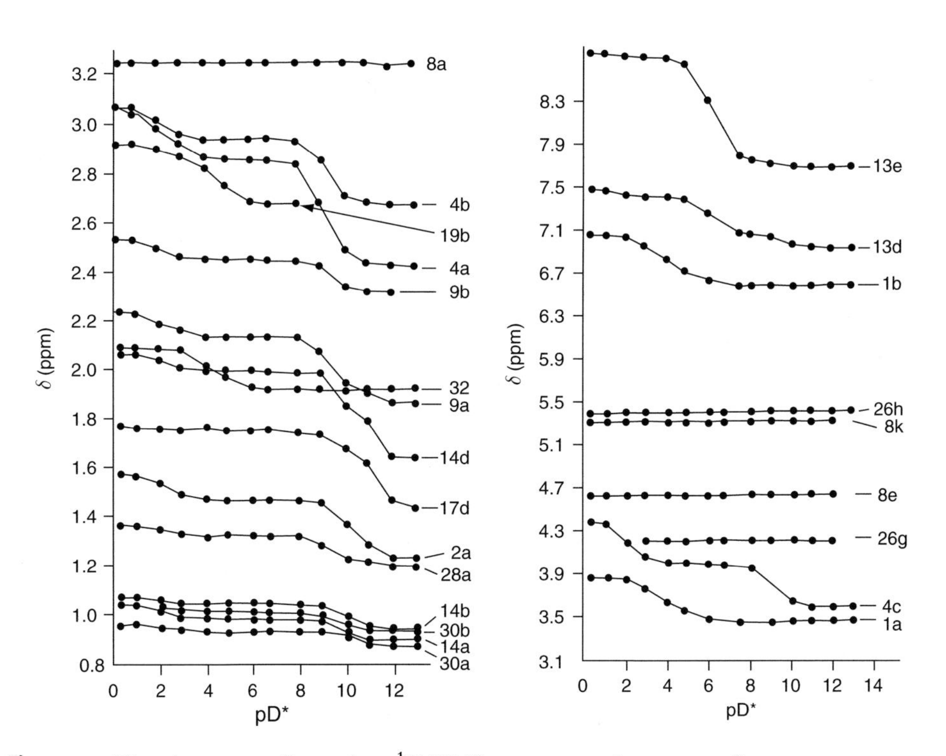

Figure 4 Titration curves for various ^{1}H NMR resonances in spectra of corn root extracts. Reprinted from T.W.-M. Fan, R.M. Higashi, A.N. Lane, O. Jardetzky, Combined use of ^{1}H NMR and GC-MS for Metabolite Monitoring and in vivo ^{1}H NMR Assignments, Biochem. Biophys. Acta 882 (1986) 154–167. Copyright (1986), with permission from Elsevier.

Temperature variation can also have an effect on chemical shifts, particularly those of solvent resonances, leading to inefficient solvent suppression and baseline problems. For these reasons, it is important that all samples in a data set be at the same pH and temperature at the time of data acquisition.[37]

2.1.5. Chemical shift and quantitation standards

Internal or external standards are used in NMR experiments to provide a chemical shift reference and for purposes of absolute quantitation. As discussed in Section 3.2, care is required in measurements where the goal is quantitation, especially as the resonances of most standards are due to isolated nuclei with long relaxation times. In theory, any compound can be used as an NMR standard providing it is chemically stable, does not interact with sample components, and produces a resonance that does not overlap with those of the sample. Because of the complexity of spectra obtained for biological samples, the latter requirement practically limits the choice of reference compounds. For ^{1}H NMR experiments sodium trimethlysilyl propionate-d_4 (TSP-d_4) or sodium 2,2-dimethyl-2-silapentane-5-sulfonate-d_6 (DSS-d_6) are common standards since they produce a single well-resolved resonance at a frequency well removed from the signals produced by most organic compounds.[37–40]

When TSP is to be used as a concentration reference, care needs to be taken to ensure that atmospheric moisture is not absorbed by the TSP solution, as this will decrease the true concentration of the TSP, thus causing calculated metabolite concentrations to be higher than their actual values. One strategy to avoid such accuracy errors is to use commercially available TSP in D_2O stored in small volume ampules that can be utilized immediately after opening.[41] TSP has also been shown to adsorb to glass surfaces over time, resulting in significant losses of its effective solution concentration.[42] The use of TSP as an internal reference in blood plasma and serum samples is not recommended as it has been shown to bind to serum albumin.[43]

To prevent interactions between the sample and the standard compound, external standards can be employed. This approach typically uses a capillary tube or other device to isolate the standard. In high-resolution-magic angle spinning (HR-MAS) experiments, Kel-F, a chemically resistant fluoropolymer, is often used in the fabrication of rotor inserts. This material gives a signal at 0.06 ppm which can be used as both a chemical shift and concentration reference, eliminating the need for an internal standard such as TSP.[44] Another useful external standard method is ERETIC (Electronic REference To access In vivo Concentrations) in which a signal is generated electronically and detected along with the sample resonances.[45] This ERETIC signal is used to generate a chemical shift and concentration reference at a desired frequency in the spectrum, eliminating the need to add an internal standard. For example, this method has been used to monitor levels of several important metabolites in the media used for cofermentation of glucose and citric acid by lactic acid bacteria.[46]

2.2. Sample specific considerations

The most commonly analyzed biofluids are urine and plasma, as sample acquisition is facile and minimally invasive and because their metabolic profiles report on the overall status of the organism.[47] Other biofluids sampled for metabolic profiling studies include bile,[48] saliva and sweat,[49] seminal fluid,[1,39] cerebrospinal fluid,[50] and breast milk,[51] which although more difficult to obtain, have the advantage of providing metabolic profiles unique to specific tissues. Tissue-specific metabolic information can also be obtained by analysis of tissue extracts or microdialysis samples, as well as by the use of HR-MAS NMR for analysis of intact tissue samples.

2.2.1. Biofluids

Fresh biofluid samples can be contaminated with particles that must be removed by filtration or centrifugation prior to NMR analysis. To provide a field frequency lock, D_2O can be added directly to biofluid samples. Typical deuterium content ranges from 10% for samples that are analyzed directly to 100% for samples that are dried and reconstituted in deuterated solvent.[14,47,52]

As discussed in Section 2.1.4, it is important that the pH remain constant for all samples in a data set, which can be achieved by the addition of a buffer (often in D_2O) to the biofluid. Because some samples can overwhelm the buffer capacity as they are more concentrated or have an unusual initial pH, it is important to measure sample pH after buffer addition and as needed titrate samples to the desired value. Typical buffers range from simple acetate and phosphate buffers to more complex systems such as synthetic urine buffer.[19,28,37,40] In cases where paramagnetic ions are present to the extent that they affect spectral quality, EDTA may be added as a chelator.[53]

2.2.2. Blood plasma and serum

Although blood is a clinically important biofluid, the sample preparation and analysis methods for metabolic profiling experiments using blood plasma or serum are more complex than those described in Section 2.2.1 and have been thoroughly discussed by Deprez et al.[54] In addition to small molecule components such as sugars, organic acids, lipids, and hormones, blood plasma contains many important proteins including serum albumin, globulins, and fibrinogen. Blood plasma is obtained by collecting whole blood into vacutainers coated with an anticlotting agent like heparin or EDTA. Blood plasma is subsequently separated from erythrocytes by centrifugation. Blood serum, the supernatant remaining after allowing blood to clot, is less commonly used for metabolic profiling. The two basic approaches to plasma analysis by NMR are discussed in the following sections. The first uses spectral editing approaches to analyze the intact biofluid. An alternative approach separates the proteins from the small molecule metabolites by ultracentrifugation or precipitation.

2.2.2.1. Relaxation editing

^1H NMR experiments focused on small molecule metabolites must overcome the intense spectral background produced by plasma proteins. Because small molecules typically have longer T_2 relaxation times than proteins, they can be selectively detected by experiments that incorporate a T_2 filter, such as the Carr–Purcell–Meiboom–Gill (CPMG) pulse sequence.[11,55,56] Substitution of a ROESY spin lock for the CPMG pulse train allows for editing based on differences in $T_{1\rho}$ relaxation. One disadvantage of this approach is that small molecules involved in protein binding equilibria may have T_2 relaxation times that are shorter than would be expected based on their molecular weight and will also be attenuated.

Diffusion-edited experiments can be used alone or in combination with T_1 or T_2 editing to selectively detect plasma components (Figure 5).[57] For example, as shown in Figure 5(b), resonances due to lipoproteins that have relatively long T_2 relaxation times and diffuse slowly can be selectively detected by appropriate choice of gradient

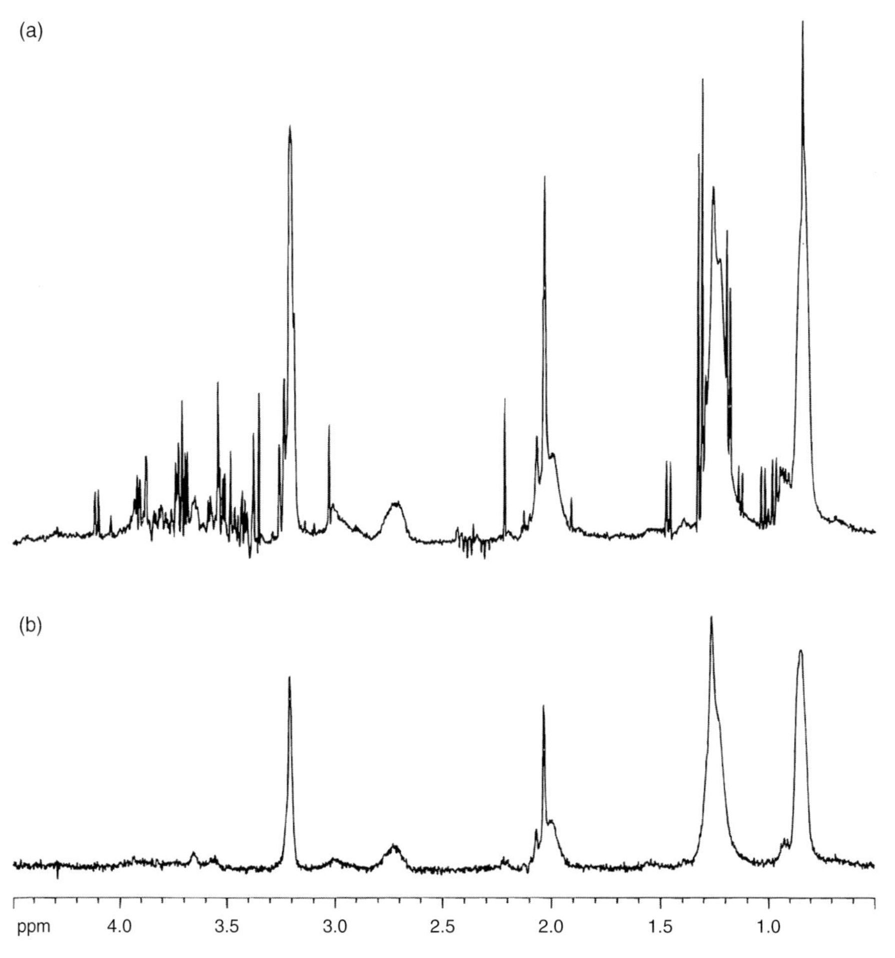

Figure 5 400 MHz ^{1}H NMR spectra of control human blood plasma edited on the basis of differences in both diffusion coefficients and relaxation times. (a) The use of a low-amplitude gradient and a 30 ms spin-echo period to remove resonances from molecules with short T_2 relaxation times. (b) The combined use of strong gradient pulses and a 30 ms spin-echo period to remove the resonances of small, fast-diffusing molecules and molecules with short T_2 values and selectively detect the resonances of lipoproteins. Reprinted with permission from M.L. Liu, J.K. Nicholson, J.C. Lindon, Anal. Chem. 68 (1996) 3370–3376. Copyright (1996) American Chemical Society.

amplitude and T_2 relaxation time. In all spectral editing methods, caution must be used in the quantitative interpretation of metabolite levels. Relaxation editing using the CPMG pulse sequence and stimulated echo (STE) diffusion-based gradient filtering were compared by Lucas et al. to investigate which method produced spectra with superior baseline characteristics.[58] Although T_2 filtering with the CPMG experiment is more common in metabolic profiling experiments, the STE data were found to be more amenable to automatic baseline correction, reducing the variability between machine replicates. Additionally, the STE experiment was found to be more favorable for biofluid analysis because it does not utilize presaturation-based water suppression, which has been found to attenuate nearby resonances, adversely affecting the potential for quantitative analysis.[59]

2.2.2.2. *Centrifugal ultrafiltration and protein precipitation*

Centrifugal ultrafiltration is a simple deproteination method for plasma samples that relies on a mechanical separation to remove proteins above a certain physical size. Limitations of this method are the loss of protein-bound metabolites and clogging of the filter micropores leading to low ultrafiltration efficiency and long centrifugation times. An additional disadvantage for NMR studies is that ultrafiltration devices require extensive presoaking and rinsing to remove glycerol and avoid sample contamination.

An alternative method for removing proteins from blood plasma is by precipitation using an organic solvent such as acetonitrile or methanol. In metabolic profiling experiments, it is important that the protein precipitation process does not cause excessive loss of small molecule metabolites. Loss of metabolites can be minimized by optimizing the organic solvent–plasma ratio as well as rate of solvent addition, with the best results obtained when the organic solvent is added slowly over several hours to allow the metabolite–plasma protein binding equilibria time to adjust.[60]

2.3. Tissues

While biofluids such as urine and blood plasma are easier to acquire and require minimum sample preparation, it can be desirable to profile metabolites at the tissue level, particularly in cases where a specific tissue type is very different from the rest of the organism such as in tumor cells. Biopsies from tissues such as muscle, fat, and organs are also common analytical targets particularly in the search for biomarkers of health or disease.[61] Whole plants or localized regions of a specific tissue type are employed in plant genetics.[13] Cell-based studies[62] and imaging[63,64] have been applied to metabolic investigations. However, these techniques are beyond the scope of this section, which will focus on tissue analysis, whether as extracts or as intact tissues via HR-MAS.

2.3.1. Tissue extracts

Metabolic profiling studies of specific tissues are often performed using tissue extracts rather than whole tissues because extracts provide much of the same information as whole tissues, but yield much better resolved NMR spectra than can be obtained by

HR-MAS. There are many methods available for extracting metabolites from tissues and preparing them for NMR analysis, but perchloric acid extractions are the most widely utilized.[65,66] Although most water-soluble metabolites are recovered by this extraction method, hydrophobic metabolites such as lipids are lost. In cases where these hydrophobic metabolites are of interest, an organic solvent such as chloroform can be used to extract the tissue pellet remaining after perchloric acid extraction. Combining the data from these two extraction methods has been shown to represent a more complete picture of an organism's metabolic profile.[65] One important disadvantage of perchloric acid extraction is that this strong oxidizing acid reacts only with select metabolites, leading to altered and potentially misleading metabolic profiles. As an alternative, tissues may be extracted using an aqueous buffer or a mixed solvent such as water/acetonitrile. As it is very difficult to establish quantitative recoveries, most metabolic profiling studies that use tissue extracts rely on relative quantitation in comparing sample groups.

2.3.2. HR-MAS analysis of intact tissues

Acquiring data for whole tissue samples ensures that metabolites are neither lost nor altered in the extraction process. Direct analysis of intact tissue samples using static NMR measurements is limited because of the broad resonances obtained as a result of irregular sample shape and inhomogeneities present in tissue samples, as well as contributions from chemical shift anisotropy and dipolar couplings inherent in solid samples. HR-MAS allows the acquisition of high resolution NMR spectra on small pieces of intact tissues with no sample pretreatment. Spinning a sample at 54.7° relative to the applied magnetic field serves to average the magnetic field inhomogeneities and remove the effects of chemical shift anisotropy and dipolar coupling.[67] Although spectral resolution can be improved by HR-MAS experiments, tissues often produce broad spectra due to proteins and other macromolecules. Therefore, spectral editing approaches based on T_2 or $T_{1\rho}$ relaxation are routinely used in the acquisition of HR-MAS spectra for purposes of metabolic profiling.

HR-MAS has been used to study a range of biological tissues such as liver,[68,69] brain,[70] prostate,[41,71] and tumors.[72,73] After excision, samples are usually washed with D_2O or a D_2O saline solution to remove blood and other unnecessary components.[68,74] Unless analysis is to be performed immediately, tissue samples are snap-frozen and stored at −80°C until analysis. Tissues are then weighed into the MAS rotor and a small amount of D_2O containing a chemical shift reference, often TSP, is added.[68,70] HR-MAS NMR has been used to distinguish between healthy and diseased tissue even in cases where this cannot be achieved through conventional methods such as histological examination.[75–77]

Although HR-MAS of intact tissue produces spectra with much higher quality than could be obtained from static spectra, the resolution is not typically as good as can be obtained by static spectra of tissue extracts. Unlike liquid-state NMR spectra that are free of spinning side bands, MAS creates spinning side bands that could be mistaken for key peaks or hinder quantification.[78] The high spinning speeds used, typically 3–6 kHz, can cause sample heating which can facilitate degradation of the tissue over the course of longer experiments.

2.3.3. Microdialysis samples

Another approach for obtaining tissue-specific metabolic profiles is through microdialysis, a sampling technique in which a probe containing a semipermeable

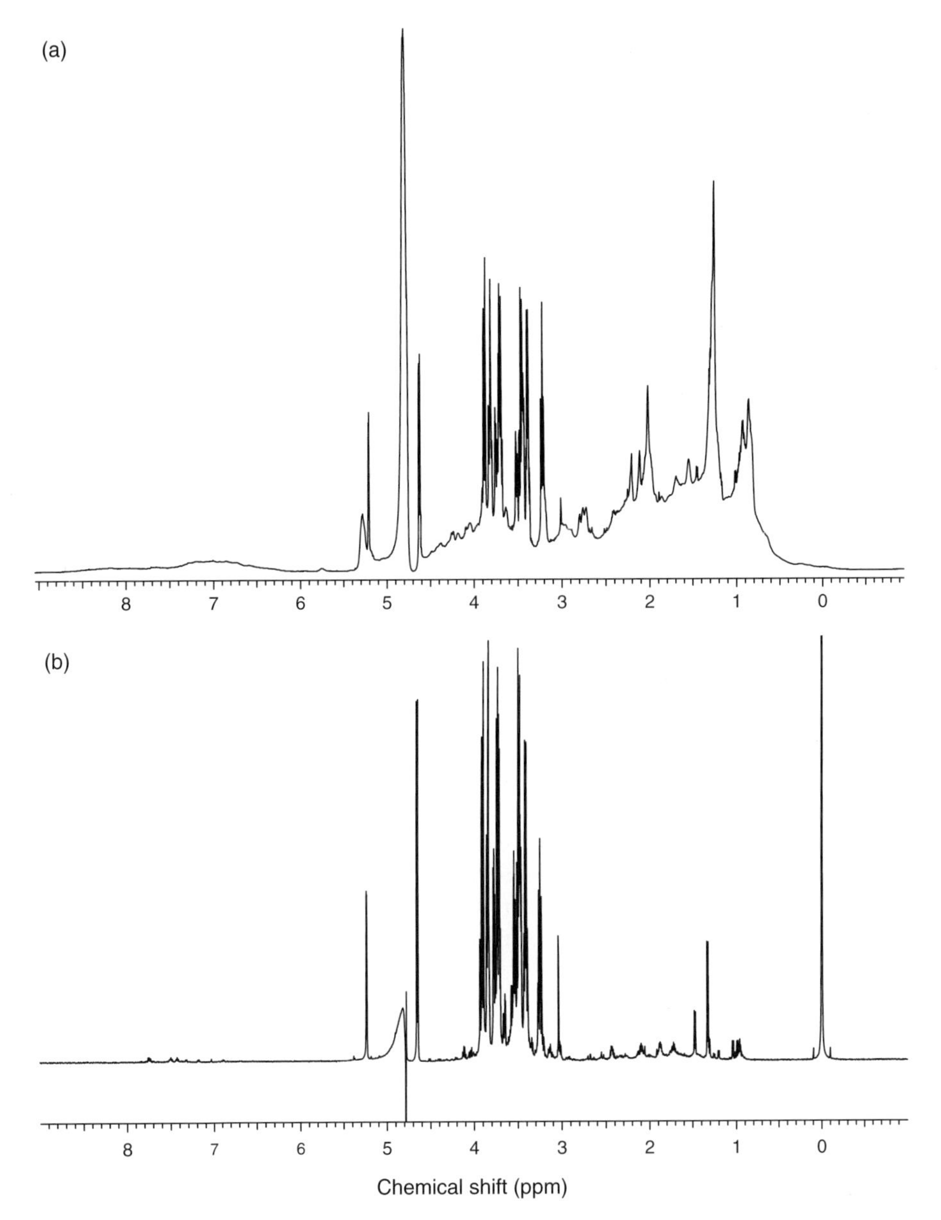

Figure 6 Comparison of ^{1}H NMR spectra of (a) rat plasma and (b) rat plasma dialysate collected by microdialysis sampling in the jugular vein. Reprinted from K.E. Price, C.E. Lunte, C.K. Larive, Development of Tissue-targeted Metabonomics: Part 1. Analytical Considerations, J. Pharm. Biomed. Anal. (2008) in press, Copyright (2008), with permission from Elsevier.

membrane is implanted in a tissue.[79] An isotonic solution is perfused through the probe, and molecules in the surrounding extracellular fluid diffuse into the membrane lumen along their concentration gradient. Samples of these molecules are collected at the probe outlet and can be analyzed by NMR to provide a metabolic profile of the tissue extracellular matrix.[47,80,81] Because flow rates on the order of μl/min are used, sample volumes obtained are small and samples are best analyzed using 1 mm capillary tubes or small volume microcoil probes. In addition to site-specific sampling, the dialysis membrane selectively excludes macromolecules that can give rise to broad NMR resonances that interfere with small molecule quantitation, as can be seen by comparison of the plasma spectra shown in Figure 6.[82]

3. DATA ACQUISITION

Careful selection of NMR data acquisition parameters is required to produce the best possible data for the system being studied and it is essential that all data sets are acquired with the same parameter set if any direct quantitative comparison is to be made.[37,52,83] Important parameters to consider in optimizing experiments are the excitation pulse width, receiver gain, acquisition and relaxation times, and solvent suppression parameters, which are addressed in greater detail in the following section. Because sample ionic strength can vary significantly among biological samples, it advisable to tune the probe for each new sample. Similarly, it is common practice to optimize field homogeneity by shimming on each sample to achieve a minimum standard established by the laboratory for quality, often specified by the linewidth at half-height of a prominent singlet resonance, typically that of the internal standard. Equally important is the linewidth at the height of the ^{13}C satellite peaks, indicative of peak resolution.

If several magnets are available, the magnet with the highest field strength should be employed for metabolic profiling analyses. Bertram et al.[84] recorded the metabolic profile of urine on four magnets (250, 400, 500, and 800 MHz) and reported the expected increase in spectral resolution and signal to noise as field strength increases. High-field magnets can facilitate resolution of resonances that would be unresolved at lower field strengths. In addition, higher field instruments can reduce the experimental time or increase the signal to noise.

While processing methods such as baseline and phase correction can be applied postacquisition, there are parameters in the acquisition setup that can affect the quality of the spectra produced as well as the ease and efficacy of postacquisition processing. For example, choosing a spectral width much larger than required by the sample resonance frequencies to produce regions of baseline several ppm wide on both sides of the spectrum can be helpful in making baseline adjustments to the processed spectra. In addition, the use of digital filters and oversampling in data acquisition, standard options for modern spectrometers, produce much flatter spectral baselines than could be obtained with older instruments equipped with analog filters. Oversampling increases the spectral width by sampling the signal at a frequency higher than required, providing flatter baselines. Oversampling has the additional benefit of improving the signal to noise (S/N) ratio because the noise is spread over a larger region as shown in Figure 7.

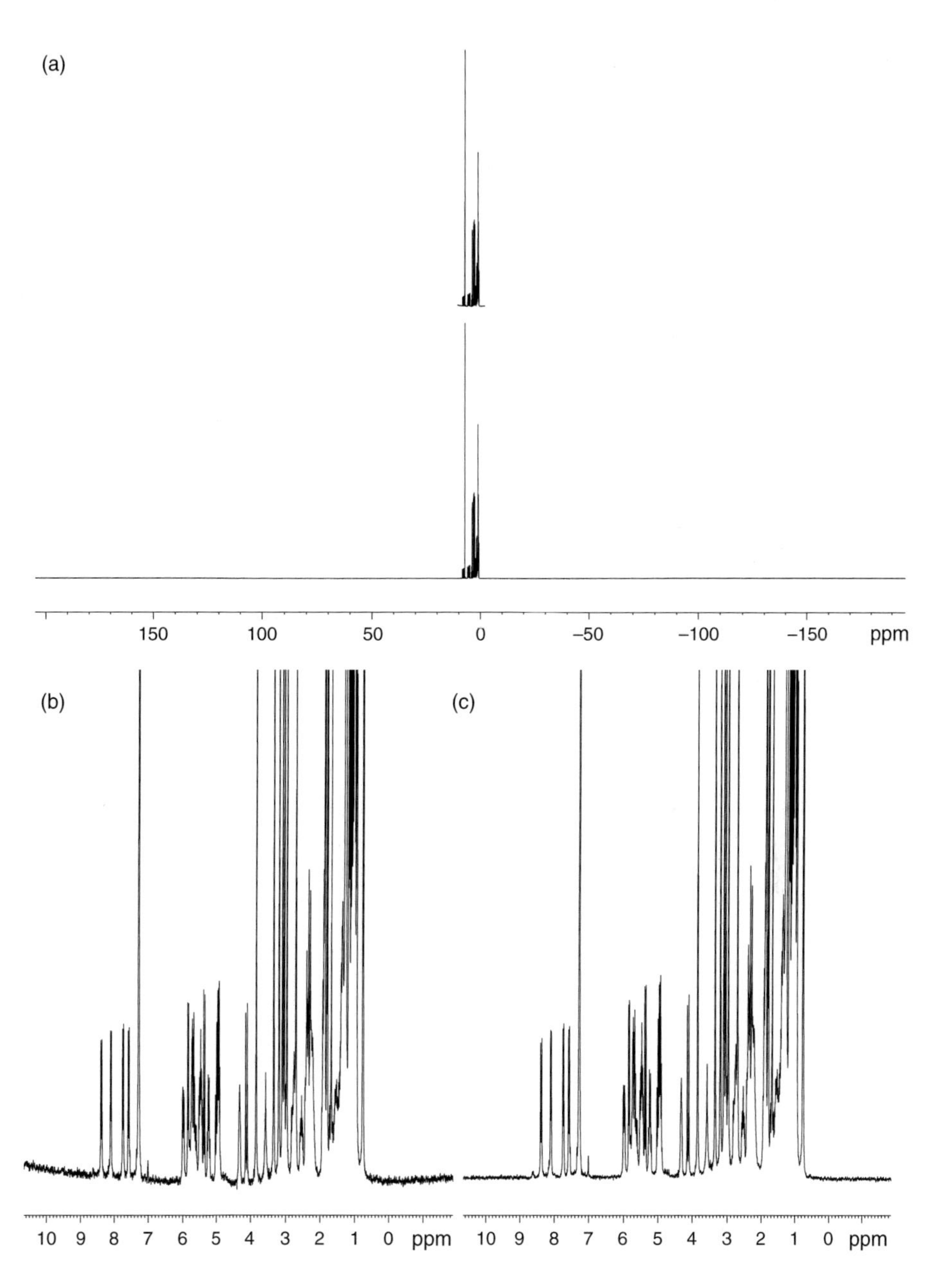

Figure 7 The principal of oversampling is illustrated with 300 MHz proton NMR spectra of cyclosporine A. (a) Two spectra are shown, where the spectrum on bottom was recorded with 32-fold oversampling (spectral widths are 120, 482, and 3765 Hz). (b, c) The full spectrum recorded with and without oversampling. The advantages of oversampling (reduced noise and flat baseline) are clearly visible in (c). Reprinted from D.D. Traficante, M. Rajabzadeh, Concepts Magn. Reson. A 12 (2000) 83–101. Copyright (2000), reprinted with permission of Wiley-Liss, Inc., a subsidiary of John Wiley & Sons, Inc.

3.1. Requirements for quantitative NMR measurements

Relative quantitation is often sufficient in metabolic profiling studies, though there are cases in which the absolute concentration of a metabolite is desired. Quantitative analysis requires spectra measured with adequate digital resolution, a repetition time sufficient for relaxation of the signals of interest, and with a high S/N ratio. For example, to be 99% certain that the measured integral falls within ±1% of the true value, a S/N ratio of 250 is required.[37] Proper postacquisition processing of the NMR spectrum can also affect the reliability of quantitative data. The spectrum must be appropriately apodized, properly phased and baseline corrected, and integration limits carefully selected.[85] However, the major challenge in quantitative metabolic profiling is ensuring that the resonances of the metabolite and the quantitation standard are fully relaxed. This can be accomplished by setting the recycling time to at least 5 times the longest T_1 relaxation time of the resonances of interest. This is not always a practical approach since metabolites are typically present at low concentration and lengthy data acquisitions may be required to obtain adequate S/N. In such cases, a shorter repetition time can be used and a T_1 correction factor applied to the resulting integrals.[37]

3.2. Water suppression

Most biologically derived samples require some method of water suppression in order to obtain high-quality ^{1}H NMR spectra. NMR spectrometers have analog-to-digital converters with a limited dynamic range, making it difficult to digitize and detect metabolite resonances in the presence of an intense solvent signal. Incomplete water suppression can also produce baseline roll and phase distortions that hinder quantitative measurements. The simplest methods of solvent suppression are those that rely on selective saturation of the solvent resonance during the recycle delay. More sophisticated techniques include WET (water suppression through enhanced T_1 effects) that utilizes selective excitation of the solvent followed by dephasing gradients and WATERGATE that uses gradient tailored excitation.[28,37,39]

Arabínar et al.[38] compared five water suppression techniques for robustness, repeatability, sensitivity, and practicality in the analysis of urine samples. The pulse sequences evaluated can be separated into two groups, those that leave the intrinsic water signal intact: WET and ES (excitation sculpting using a double-pulsed field gradient echo), and those that incorporate selective saturation of the solvent resonance: NOEPR (NOESY pulse train with presaturation during the recycle delay and mixing time), ESCW (ES with continuous wave on-resonance saturation), and ESWGL (ES with adiabatic frequency modulation). The spectra shown in Figure 8 illustrate the effectiveness of these water suppression methods. WET was the least effective at suppressing the water resonance and produced a baseline distortion that covered a fairly broad chemical shift range. ES and NOEPR were more effective than WET at suppressing the solvent resonance but produced baseline distortions in the spectral region around the water chemical shift (4.5–5.3 ppm). The best results were achieved when excitation sculpting was combined with selective saturation using either CW or adiabatic frequency

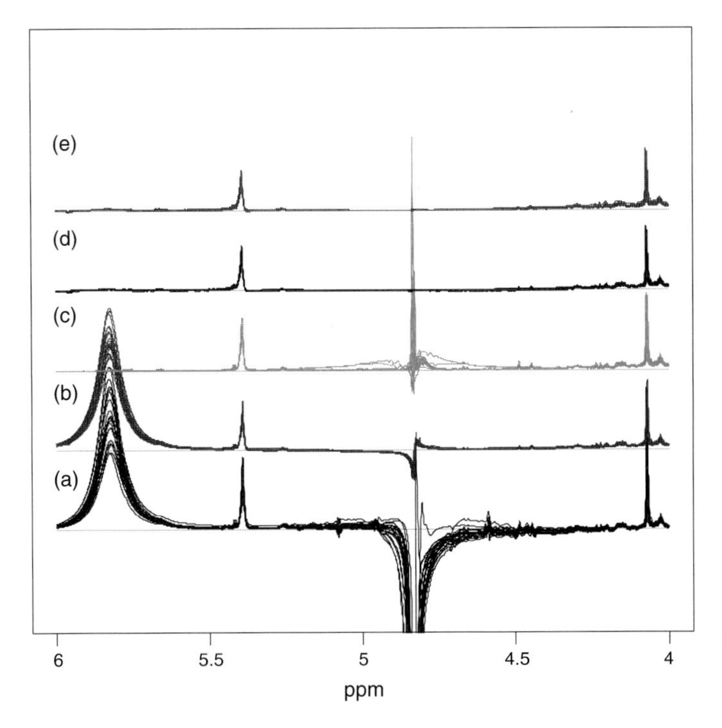

Figure 8 An expansion showing the region around the water resonance in urine spectra measured with (a) WET, (b) ES, (c) NOEPR, (d) ESCW, and (e) ESWGL pulse sequences. Reprinted from N. Arabinar, K.-H. Ott, V. Roongta, L. Mueller, Metabolomic Analysis using Optimized NMR and Statistical Methods, Anal. Biochem. 355 (2006) 62–70. Copyright (2006), with permission from Elsevier.

modulation. The spectra produced by these methods showed excellent water suppression, flat baselines, and high reproducibility and were amenable to automated spectral processing.[38]

3.3. Experiments utilizing nuclei other than ^{1}H

Because protons have a high natural abundance and inherent sensitivity, ^{1}H NMR is most commonly used for metabolic profiling experiments. Although ^{1}H NMR is a nearly universal method for detection of organic compounds, it suffers from problems with specificity in cases where the resonances of the metabolite of interest are obscured due to overlap from signals of chemically similar species. This problem can be circumvented by measuring spectra for nuclei with a larger chemical shift range or for which the number of metabolites containing the nucleus is more limited, e.g., ^{31}P. Other NMR-active nuclei, such as ^{13}C, ^{15}N, ^{19}F, and ^{31}P, can be used in the analysis of biological samples. ^{19}F and ^{31}P are nearly 100% abundant, facilitating their detection. In biological mixtures, ^{31}P detection is used in the analysis of purine metabolism and for phosphorous-containing metabolites such as ATP and phosphocreatine (PCr).[86–88] Fluorine-19 NMR is not generally used for

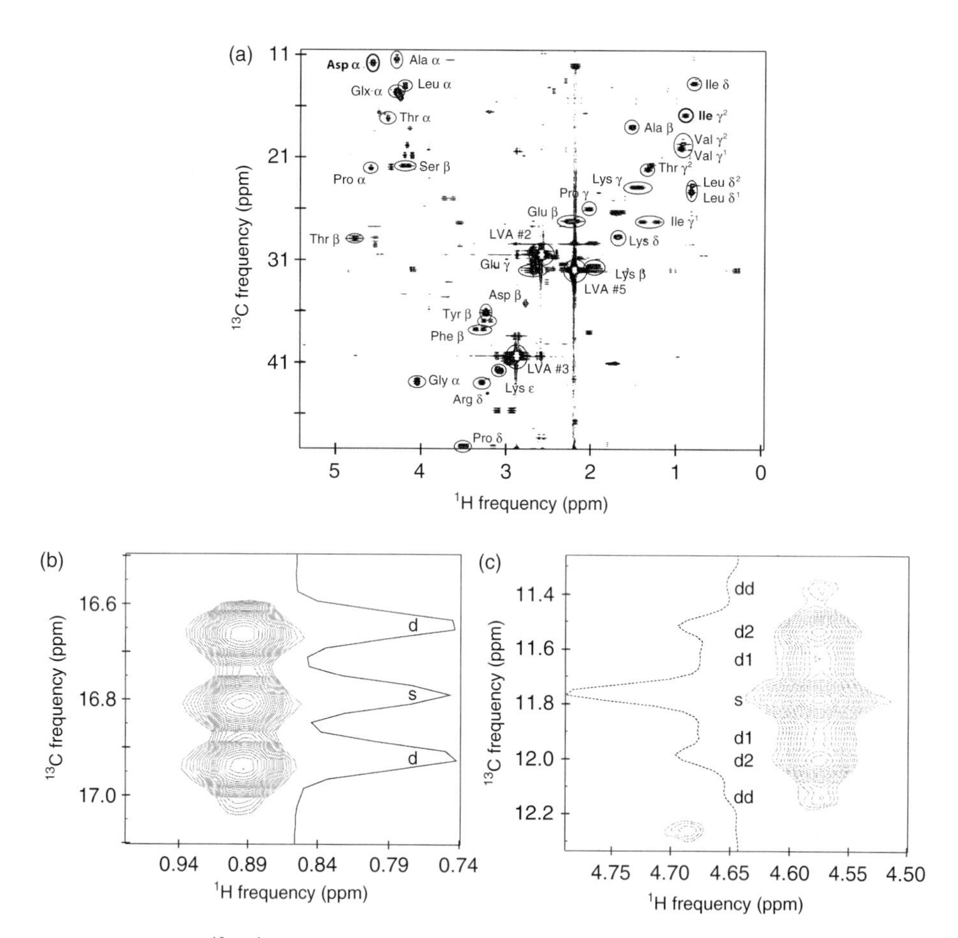

Figure 9 (a) 2-D [^{13}C, ^{1}H] HSQC spectrum of hydrolyzed aqueous extract of *C. roseus* hairy roots grown on (5% (w/w) U−^{13}C) sucrose. Cross-peaks represent carbon atoms of hydrolysate constituents (proteinogenic amino acids, levulinic acid (LVA), and hyroxyacetone (HyA)). The names of some amino acid nuclei are omitted for clarity. Expanded views of (b) Ile γ^2, and (c) Asp α cross-peaks. One-dimensional slices are shown alongside. The multiplet peaks are s: singlet; d, d1, d2: doublet; dd: double doublet. Reprinted from G. Sriram, D.B. Fulton, J.V. Shanks, Flux Quantification in Central Carbon Metabolism of *Catharanthus roseus* Hairy Roots by ^{13}C labeling and Comprehensive Bondomer Balancing, Phytochemistry 68 (2007) 2243–2257. Copyright (2007), with permission from Elsevier.

the detection of endogenous metabolites, but rather for the detection of fluorinated drug compounds and their metabolites. As an example, the metabolic products of 5-fluorouracil, a drug used in the treatment of gastrointestinal and breast cancers, have been analyzed using ^{19}F NMR.[89] A novel use of ^{19}F NMR is as an indirect probe of the intracellular oxygen concentration in live tissue, as demonstrated by Gross et al.[90]

Although ^{13}C and ^{15}N are only 1.1% and 0.37% abundant, respectively, stable isotope labeling can be employed to compensate for the poor sensitivity of these nuclei.[91] Uniform enrichment of organisms in ^{13}C and/or ^{15}N can be achieved by growing plants, algae, and bacteria using uniformly labeled carbon (i.e., [^{13}C]$_6$-glucose or [^{13}C]-carbon dioxide) or nitrogen ([^{15}N]-ammonium chloride) sources.[92] These uniform labeling techniques are highly efficient, facilitating the analysis of nuclei other than ^{1}H and enabling information-rich NMR experiments, such as HSQC, HMBC, and HCCH COSY, and TOCSY for more confident resonance assignments.

An alternative to uniform labeling involves the introduction of a single labeled compound while monitoring the fate and transport of the label as it is incorporated into other molecules by metabolic networks.[83,93–95] Some commonly fed substrates include [^{13}C]-carbon dioxide, [^{15}N]-ammonia, and labeled amino acids. In a study by Massou et al.,[96] *Escherichia coli* (bacteria) were grown on [^{13}C]-glucose media and the fractional enrichment of ^{13}C in various metabolites was detected by ^{1}H–^{1}H, ^{1}H–^{13}C, and ^{13}C–^{13}C NMR experiments. The ^{1}H–^{13}C HSQC spectrum in Figure 9 shows the relative levels of ^{13}C enrichment in extracts from *Catharanthus roseus* hairy roots grown on (5% (w/w) U–^{13}C, 95% (w/w) naturally abundant) sucrose.[97] Not only can this approach provide information on positional enrichment within a metabolite indicating the metabolic pathway through which the labeled precursor traveled, it can also give insights into previously unknown metabolic pathways if the position or incorporation of the stable isotope in the metabolite is unexpected.

4. Spectral Processing

After careful collection of NMR spectral data sets, there are many processing steps including apodization, phasing, baseline correction, alignment, integration, scaling, and normalization, that can be applied to aid in the extraction of quantitative and metabolically relevant information. Processing routines should be employed in a context-dependent manner and there is no single method that can be used for all metabolic profiling and metabonomics experiments.[98] Although spectral processing software packages often include these processing steps into automated routines, to achieve the best quantitative data possible it is useful to consider the effect of each step on the integrity of the data set. As discussed in earlier sections, the NMR spectrum of most biological samples is complex, containing many overlapping peaks and/or broad resonances. If sampling and acquisition strategies for improving spectral resolution have not been employed, or have been less successful than anticipated, some degree of spectral deconvolution may be applied post-acquisition.

4.1. Apodization

Theoretically, an FID has an infinite duration in the time domain, although in practice signals are collected for a finite time. Often the acquisition time is set to allow the signal to decay completely into the noise (roughly $3 \times T_2^*$, where T_2^* is the time constant for loss of transverse magnetization in an inhomogeneous magnetic field). However, there are circumstances in which acquisition times shorter than $3 \times T_2^*$ are used; for example, in two-dimensional NMR experiments or when the goal is to enhance the spectral S/N. In such cases, apodization of the FID by multiplication with a function (often called a window function) can be used to force the FID intensity to zero and eliminate distortions that would be produced by the Fourier transformation of a truncated FID.

One of the most commonly used apodization functions is an exponential decay that can enhance S/N at the cost of increased line width and poorer resolution. This can become a problem in NMR spectra of complex biological samples, consisting of thousands of resonances, where overlapping peaks are already evident. In a study in which the NMR acquisition and processing parameters including pulse flip angle, number of scans, and amount of exponential apodization were optimized for the metabolomic analysis of rat urine, it was found that the best compromise between increased S/N and broadened resonances was achieved by minimizing the use of apodization and maximizing the number of scans.[85]

Traficante and Rajabzadeh[99] discuss the properties of the line broadening exponential in detail and propose a novel approach that treats the real and imaginary parts of the spectrum with different functions, with recombination to yield a spectrum which has an increased S/N with no appreciable loss of resolution. The authors note that this approach works perfectly only in special cases, but it could be applied to biological data sets to determine its usefulness in metabolic profiling experiments.

4.2. Phasing and baseline correction

Zero- and first-order phase errors are introduced into the NMR spectrum by mismatch between the receiver and the reference signal phases and by precession of magnetization during the short delay required for pulse ring down prior to the start of acquisition. These errors are corrected by a process called phasing which can be carried out manually by visual inspection of the results of the phase correction or using an automated routine. Although automated phase correction is preferred for analysis of large data sets and may be more reproducible than manual adjustment, the results obtained should be carefully evaluated especially for spectra that do not have flat baselines or contain a significant residual solvent resonance, as these features can hamper automated phasing routines.

Baseline correction can also be applied manually by inspection of the baseline in a region free of resonances or using automated routines. The goal of this correction is a flat spectral baseline with a zero offset so that resonance integrals reflect the true peak intensity without distortion from integration of the underlying baseline. The digital signal processors and digital filters in most modern NMR spectrometers avoid the baseline distortions created by analog filters and have diminished, but not completely eliminated, the need for postacquisition baseline correction.[100]

Another important consideration specific to metabolic profiling is that perceived baseline distortions may actually result from residual protein resonances present in the biological sample,[101] as discussed in Sections 2.2 and 2.3. One classical solution to this problem is to subtract a spectrum measured for a sample containing only protein from the spectrum of solution containing the protein and the analytes of interest.[102] Although it is not really practical to prepare an identical protein-only blank for most biological samples, de Graaf and Behar effectively accomplished this using a process called "diffusion-sensitized" ^{1}H NMR spectroscopy.[103] In this approach, spectra of rat plasma are measured using an NMR diffusion pulse sequence with gradient amplitudes large enough to eliminate the fast diffusing small molecules and generate a reference spectrum containing only the resonances of the macromolecules present in the sample. This spectrum was subtracted from the standard single pulse NMR spectrum measured for the same plasma sample, yielding a baseline-corrected spectrum which could subsequently be used for quantitation,[103] as shown in Figure 10. In this study the results of the

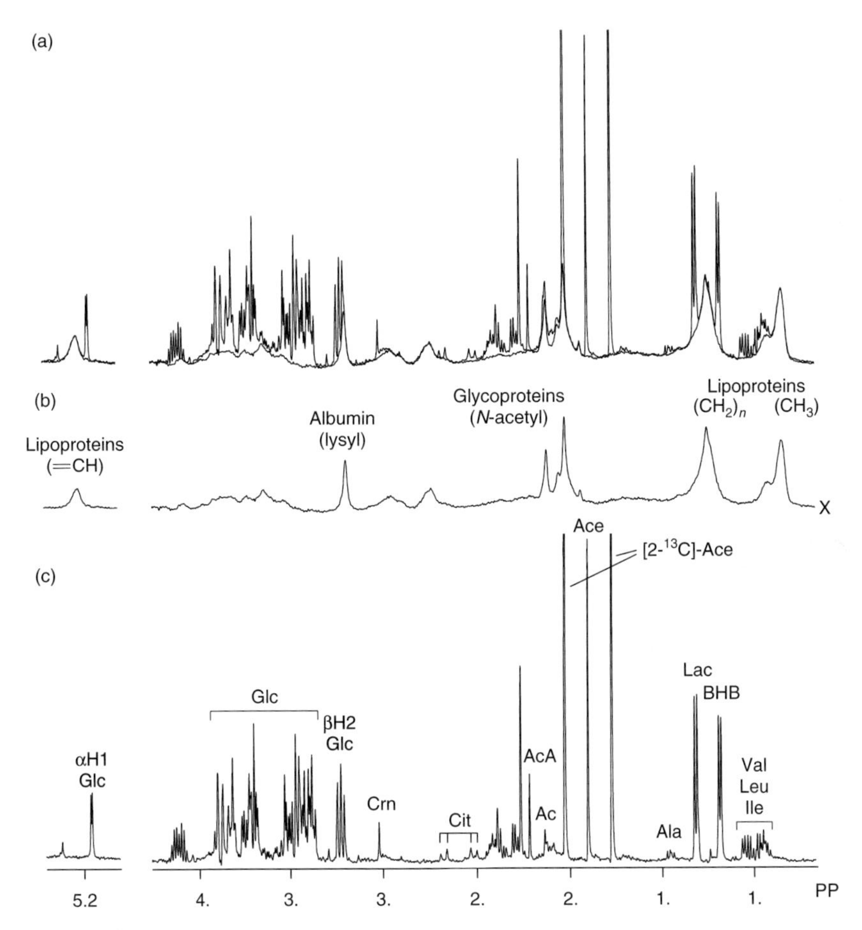

Figure 10 (a) ^{1}H NMR spectrum of blood plasma acquired with a low sensitivity toward diffusion ($b = 4.1\ s/mm^2$), overlaid with a spectrum obtained with high diffusion sensitivity ($b = 10{,}000\ s/mm^2$). (b) Proton NMR spectrum of blood plasma acquired obtained with high diffusion sensitivity ($b = 10{,}000\ s/mm^2$). (c) The difference spectrum obtained by subtracting spectrum (b) from (a). Abbreviations are given for acetate (Ace), acetoacetate (AcA), acetone (Ac), alanine (Ala), β-hydroxybutyrate (BHB), citrate (Cit), creatinine (Crn), glucose (Glc), isoleucine (Ile), lactate (Lac), leucine (Leu), and valine (Val). Reprinted with permission from R. A. de Graaf, K. L. Behar, Anal. Chem. 75 (2003) 2100–2104. Copyright (2003) American Chemical Society.

diffusion-edited NMR method were compared with the single-pulse spectra obtained for samples treated by ultracentrifugation. Excellent correlation was obtained between the two methods for most of the major metabolites including acetate, alanine, creatinine, β-hydroxybutyrate, glucose, lactate, and valine. Regression analysis of the results produced by both methods yielded a straight line with a correlation coefficient of 0.9989, a slope of 0.978, and an intercept of 0.054 mM.

A purported advantage of NMR-based metabolomic experiments is that there is little interlaboratory variability.[2] The combined effect of apodization, phasing, and baseline correction is a determining factor of whether this statement is true. On behalf of the German Federal Institute of Materials Research and Testing, Malz and Jancke[85] attempted interlaboratory validations of quantitative NMR data. In a 1999 trial, it was found that the results differed by up to 100% between labs when analyzing a simple five-component reference mixture. The poor agreement was found to be caused by differences in the acquisition and processing procedures used in each laboratory. In the 2005 trial, the cooperative mandated the use of a 30° pulse, collection of 32k data points, and a relaxation delay of 20 s. The data was then zero filled to 64k points and apodized by line broadening equivalent to 0.3 Hz. Phase and baseline corrections were performed manually, as was the definition of integral regions. In the second trial, the maximum uncertainty was found to be 1.5%, corresponding to a confidence interval of 95%. A similar interlaboratory comparison was conducted[24] using rat urine, although the goal of the experiment was a more qualitative identification of biomarkers for hydrazine-induced toxicity. Interestingly, it was found in this study that the biomarkers identified were not different whether baseline correction and phasing were conducted manually or automatically.

4.3. Selection of integral regions

It is well known that the accuracy of integrals depends on digital resolution as well as on the appropriate selection of integration limits.[104] It is relatively simple to achieve adequate digital resolution either by acquisition of sufficient data points or by extending the number of points in the FID through zero-filling or linear prediction. Selection of integral regions can be more of an art. As NMR peaks have Lorentzian line shape, much larger integral regions are required than for data sets with Gaussian signals. For a Lorentzian peak with a width at half-height of 0.5 Hz, integration regions set at 3.2 or 16 Hz on either side of the resonance would include approximately 95% or 99% of the peak area, respectively. Note that this analysis does not include the ^{13}C satellites which account for an additional 1.1% of the intensity of carbon-bound protons in samples containing ^{13}C at natural abundance. Use of a fixed integration interval is impractical since line widths vary from resonance to resonance and selection of integral regions sufficiently wide to capture >95% of the resonance intensity is not feasible in most metabolic profiling experiments because of the complexity of the spectra produced by biological samples.

One approach to the problem of resonance overlap is to use curve fitting to obtain peak areas.[105] Methods for automated peak recognition and metabolite identification have been developed by industry: AstraZeneca,[106] GlaxoSmithKline,[107] Merck,[101] Hoffman La Roche,[108] Bristol-Myers Squibb,[38] Chemomx,[37,109] Bio-Rad,[110] Varian,[111] and Bruker.[13] Calculation of peak integrals by fitting procedures has been said to solve overlap problems; however, some claim that the precision of the resulting integral rapidly decreases as peak separation becomes smaller.[112]

In a study conducted by Martin,[112] results determined by unsupervised peak fitting showed superior precision to integration by summation if line width was optimized in the peak-fitting integration. An alternative is to use supervised peak fitting, as described by Weljie et al.[109] who employed a predictive metabolite library to generate a Lorenzian peak-shape model of each reference, shown in Figure 11. This approach, called targeted profiling, is expected to become an increasingly common method for spectral deconvolution in data sets acquired for purposes of metabolic profiling and it is likely that a complete set of reference standards comprising all available characterized metabolites will be incorporated into the analysis at the outset of an experiment.[107] Other attempts have been made to use mathematical approaches to decompose NMR spectra into individual components, called independent component analysis[113] and molecular factor analysis, but these methods have so far failed to produce biochemically relevant results, as the biomarkers identified rarely match known metabolites.[107]

Many metabolomic experiments employ equidistant binning, also known as bucketing, in which the NMR spectrum is divided into many small integral regions of an arbitrary width, 0.05 ppm for example. Binning does not necessarily yield direct quantitative information about specific metabolites because bins may include intensity from more than one resonance or because a single peak may be divided into two or more integral segments. An advantage to bucketing lies in the observation that some resonances display changes in their chemical shift as a result of small differences in the sample pH or ionic strength. The use of relatively wide bins

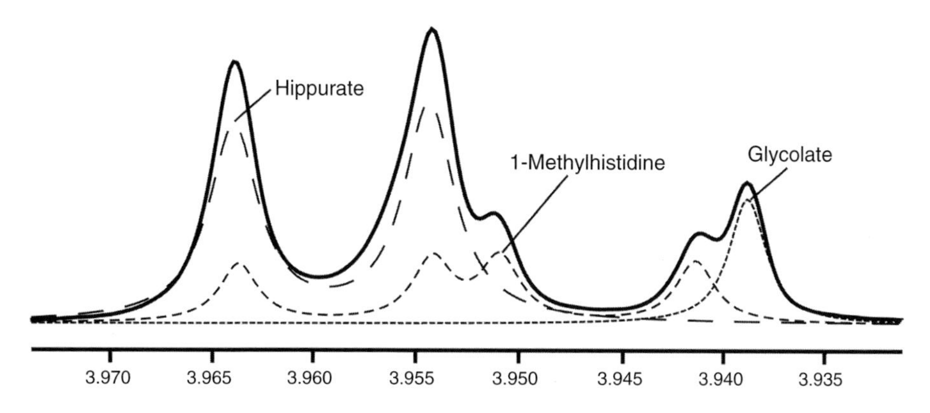

Figure 11 This figure illustrates how overlapping resonances may be deconvoluted by targeted profiling. The spectral line is shown in black, and the subspectra for 1-methylhistidine, hippurate, and glycolate are shown as dashed lines. Reprinted with permission from A.M. Weljie, J. Newton, P. Mercier, E. Carlson, C.M. Slupsky, Anal. Chem. 78 (2006) 4430–4442. Copyright (2006) American Chemical Society.

allows a resonance to have some deviation in its chemical shift and still lie within the same integration region. Although this strategy can simplify the statistical treatment of the data set, the use of larger bins translates into a reduction of the effective spectral resolution. Dieterle et al.[108] describe a nonequidistant binning approach that takes into account resonance linewidths. In this method an average spectrum is calculated using all of the measured spectra. The first derivative of the average spectrum is calculated and the borders of the bins defined based on regions of maximum slope. The application of this binning algorithm resulted in more than 3 times the number of integral regions than the previously used equidistant binning, with a corresponding increase in effective spectral resolution.

4.4. Peak alignment

^{1}H NMR spectra are typically aligned through the use of a chemical shift reference signal, as discussed in Section 2.1.5. However, deviations may arise due to variations in the sample matrix and instrumental instabilities that can produce erroneous integrals if automated integration routines are employed. For example, Figure 12 shows the chemical shift variance of one of the aromatic resonances of 1-methylhistidine in six different samples with very similar ionic strengths and solution pH values within 0.1 pH unit.[109]

Peak alignment has been employed by Forshed et al.[106] using two recently developed genetic algorithms, segment-wise peak alignment (SWA) and peak alignment by reduced set mapping (PARS), which were compared to classical bucketing of unaligned spectra using principal components analysis (PCA) and partial least

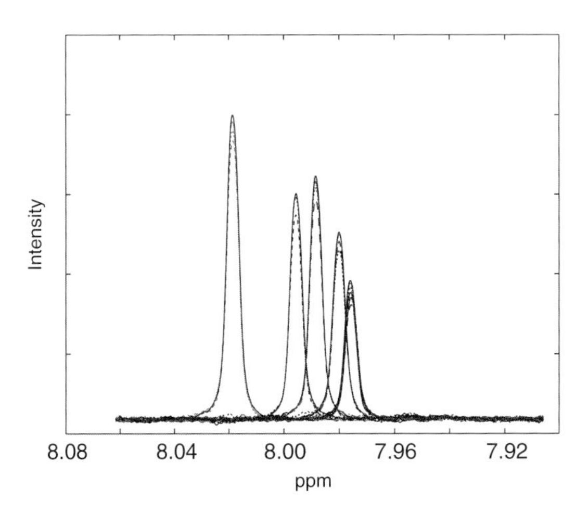

Figure 12 The variance in chemical shift of the downfield 1-methylhistidine aromatic proton in the NMR spectra of six synthetic urine samples. For each sample eight analytical replicates were collected. This experiment illustrates the large variation in chemical shift that can result because of slight changes in ionic strength and pH. All samples were titrated to a pH meter reading of 7.00 ± 0.05 prior to analysis. A.M. Weljie, J. Newton, P. Mercier, E. Carlson, C.M. Slupsky, Anal. Chem. 78 (2006) 4430–4442. Copyright (2006) American Chemical Society.

squares-discriminant analysis (PLS-DA) regression analysis.[114] The two dedicated peak alignment methods produced better class separation in the PCA or PLS-DA scores than could be obtained with the bucketed data sets. The incorporation of a peak alignment procedure was also shown to improve the outcomes of a metabolite fingerprinting experiment carried out on a set of green tea samples of different type, grade, and geographical origin.[115] The effect of these peak alignment methods on the quantitative accuracy of metabolic profiling data sets has not been shown, but it is thought that errors would be introduced into peak integrals if the integration is achieved by summation since these alignment methods are known to reduce spectral resolution.[114]

4.5. Normalization

After apodization, phasing, baseline correction, integration, and possibly alignment have been carried out, the experimenter is left with a table in which each row represents the data from a given sample and each column corresponds to the integrated area of a single resonance or a spectral bin, as discussed in Section 4.3. To be able to perform quantitative analysis on even a relative basis, most data sets must be subjected to normalization.[98] Normalization allows data to be directly compared in cases where the biological samples under comparison are either of unknown/nonuniform amount or of unknown/nonuniform concentration. The most commonly used method of normalization, applied in cases where there is no a priori knowledge of the biological sample mass or volume, is normalization to a constant sum.[12] In this method, the sum of each row is calculated (all integrals from a sample) and each integral is then expressed as a fraction of the total spectral integral. This normalization approach is often used in conjunction with binning integration and is not recommended for metabolic profiling of selected, specific integral regions, as the sum of all integrals is likely to have more variation when less of the spectrum is sampled.

In cases where the grams of wet or dry weight of solid[13] or volume of fluid constituting each sample is known, the normalization factor can lend more biological relevance to the assay. It has been pointed out that this approach is less biologically relevant for metabolic profiling of urine due to the variable nature of urinary volume, and it may be more clinically useful to normalize to the time interval between urinary excretions and body weight of the organism under study.[98]

Some normalization techniques employed recently in metabolic profiling have been influenced by the state-of-the-art quantitative biological techniques such as mRNA microarrays, perhaps because metabolic profiling data sets are now being combined with data sets generated from these types of experiments.[116,117] In this approach, the integrated area of each chosen metabolite resonance is referenced to the integrated area of a "housekeeping" metabolite.[98] This method has been applied in gel electrophoresis for protein quantitation since the early 1990s[118] and more recently in reverse transcription-polymerase chain reaction (RT-PCR) and gene chip array technologies for quantitation of mRNA. Alternatively, the concentration of a specific metabolite can be determined by an independent means, for example, by enzymatic assay,[119] which can provide a scaling factor for the entire spectrum.

Normalization to a "housekeeping" metabolite is not without pitfalls however, because it can be difficult to identify a representative metabolite to use for this purpose. Creatinine is a commonly used reference compound for the normalization of urine clinical parameters and is thought to be a good indicator of the glomerular filtration rate.[120] Normalization of metabolite peak intensities in urine ^{1}H NMR spectra as ratios relative to creatinine has been suggested as a method to compensate for urinary volume. However, mounting evidence in neonates,[121] children, and the elderly,[98] and even healthy adults[122] suggests that the variation in excreted creatinine levels among individuals is significant and results from differences in metabolism and not urine volume effects.

It has been noted multiple times that nonweighted linear multivariate pattern-recognition techniques, such as principal components analysis, favor more intense resonances, and fail to identify biologically important compounds that are present in relatively low quantities.[107,123] Gipson et al.[107] propose a non-uniform weighting scheme to tease out significant biomarkers which may be present at low levels or which have resonances deeply buried in crowded spectral regions. Nonuniform weighting involves a different normalization factor to be used in each column and in each row of the table of data as described earlier. It has been stated that nonuniform weighting can aid in binary classification of samples based on differences in low-abundance metabolites; however, the potential for absolute quantitation is lost, therefore, it is not recommended for metabolic profiling data sets.

4.6. Centering, scaling, and transformation

Centering and scaling are performed on columns of data, representing in this case the integrated area of one resonance or one spectral bin across all samples. Centering, in general, adjusts for differences in the offset between high- and low-abundance metabolites. Scaling can adjust for differences in the magnitude of the metabolite fold change by converting the data into differences relative to a scaling factor. Transformations, in general, are nonlinear conversions of the data which aim to correct for nonlinear variance or to make skewed distributions more symmetric.[124] Mean centering is achieved by subtracting the column mean from each value in the column, giving each column of the table a mean of zero. This is done so that all components found by multivariate analyses have their origin at the centroid of the data.

The most commonly applied scaling technique goes by many names including variance scaling, unit scaling, unit variance scaling, and autoscaling. Variance scaling is achieved by dividing all the values for a certain variable by the standard deviation for that variable. After variance scaling, all metabolites have a standard deviation of one and therefore the data are analyzed on the basis of correlations instead of covariances, as is the case with centered data. Many times, mean centering and variance scaling are performed on the same data set. Pareto scaling[124] uses the square root of the standard deviation as the scaling factor. This has advantages over variance scaling because large-fold changes are decreased more than small-fold changes. Also, the data does not become dimensionless as after mean centering and variance scaling.

Two types of transformations in common use are the log transformation and the power transformation. Both of these techniques reduce large values more than small values, resulting in a pseudoscaling effect. It can be useful to perform a transformation prior to one of the aforementioned scaling methods because these transformations do not use a scaling factor with any biological or analytical relevance. Unfortunately, it is not clear how transformation and scaling methods influence each other when applied to complex metabolic profiling data sets.[124] Parsons et al.[125] compared variance scaling, Pareto scaling, log transformation, and mean centering using statistical approaches to quantify the effects of the various scaling, centering, and transformations. The data compared resulted from metabolic profiling of canine urine, mussel muscle extracts, and fish liver extracts analyzed by one- and two-dimensional ^{1}H NMR. Multivariate statistical methods (PCA and ANOVA) were utilized to evaluate the effects of scaling techniques upon each data set as a whole instead of treating each resonance or spectral bucket individually. It was found that logarithm transformation improved the binary classification accuracy for the mussel and fish tissue extracts, but the use of the log transformation and variance scaling had approximately equal classification accuracies for the canine urine samples.[125]

5. Illustration

An interesting demonstration of the power of quantitative NMR measurements for metabolic profiling is the study by Serkova et al. to examine the effect of circannual hibernation cycles on liver metabolites of 13-lined ground squirrels.[126] These ground squirrels hibernate in winter during which time they experience long periods of topor, a state characterized by low body temperature and reduced metabolic, respiratory, and heart rates. These multiday periods of topor are interrupted by periods of arousal during which the animal's body temperature returns to normal. Dramatic changes in metabolite status are expected for animals in the homeothermic state (SA, summer active) and the distinct heterothermic (winter) state of topor. In this study animals were examined in two stages of torpor; late torpor (LT) and animals entering topor (Ent). Animals reentering torpor were selected for measurement rather than those in aroused state, because aroused animals could be in various stages of reestablishing the metabolic components that allow torpor, whereas those animals just entering topor must have completed that process.

The authors used quantitative ^{1}H and ^{31}P NMR of perchloric acid extracts from animals in the SA, LT, and Ent metabolic stages to quantify metabolite concentrations in the liver. Representative 500 MHz ^{1}H NMR spectra for the three metabolic states are shown in Figure 13. Concentrations of 26 metabolites were determined from the quantitative ^{1}H NMR spectra relative to an external reference of TSP-d_4. Results were obtained for six individuals in each grouping, with significance determined by ANOVA. Although these spectra are complex, the spectral differences were sufficient to allow unsupervised PCA to distinguish the three sample groupings based primarily on six metabolites: glutamine, betaine, glutathione (total and reduced), succinate, glucose, and lactate. To a lesser extent, alanine and the ketone body β-hydroxybutyrate also contributed to the group

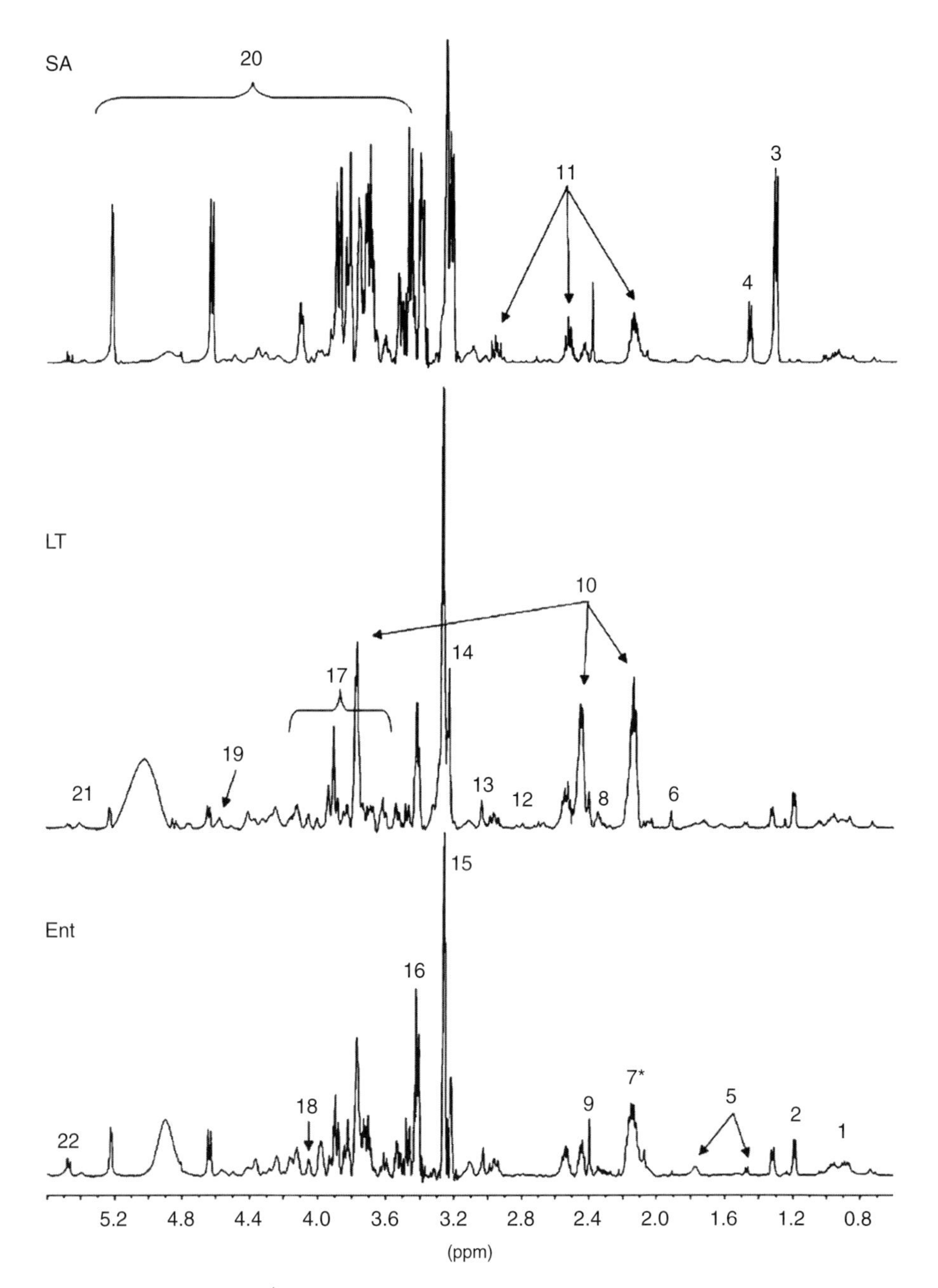

Figure 13 Representative ^{1}H NMR spectra from hydrophilic extracts of ground squirrel livers. NMR peak assignment: 1, valine + leucine + isoleucine; 2, β-hydroxybutyrate; 3, lactate; 4, alanine; 5, lysine + arginine; 6, acetate; 7, CH_3-acetyl groups (*glutamine, glutamate, glutathione); 8, glutamate; 9, succinate; 10, glutamine; 11, total glutathione (including reduced glutathione (GSH) at 2.98 ppm); 12, aspartate; 13, creatine; 14, cholines; 15, betaine; 16, taurine; 17, polyols + sugars; 18, myo-inositol; 19, nucleotides; 20, glucose; 21, glycogen; and 22, other sugars. N.L. Serkova, J.C. Rose, L.E. Epperson, H.V. Carey, S.L. Martin, Physiol. Genomics 31 (2007) 15–24. Copyright (2007), with permission of the American Physiological Society.

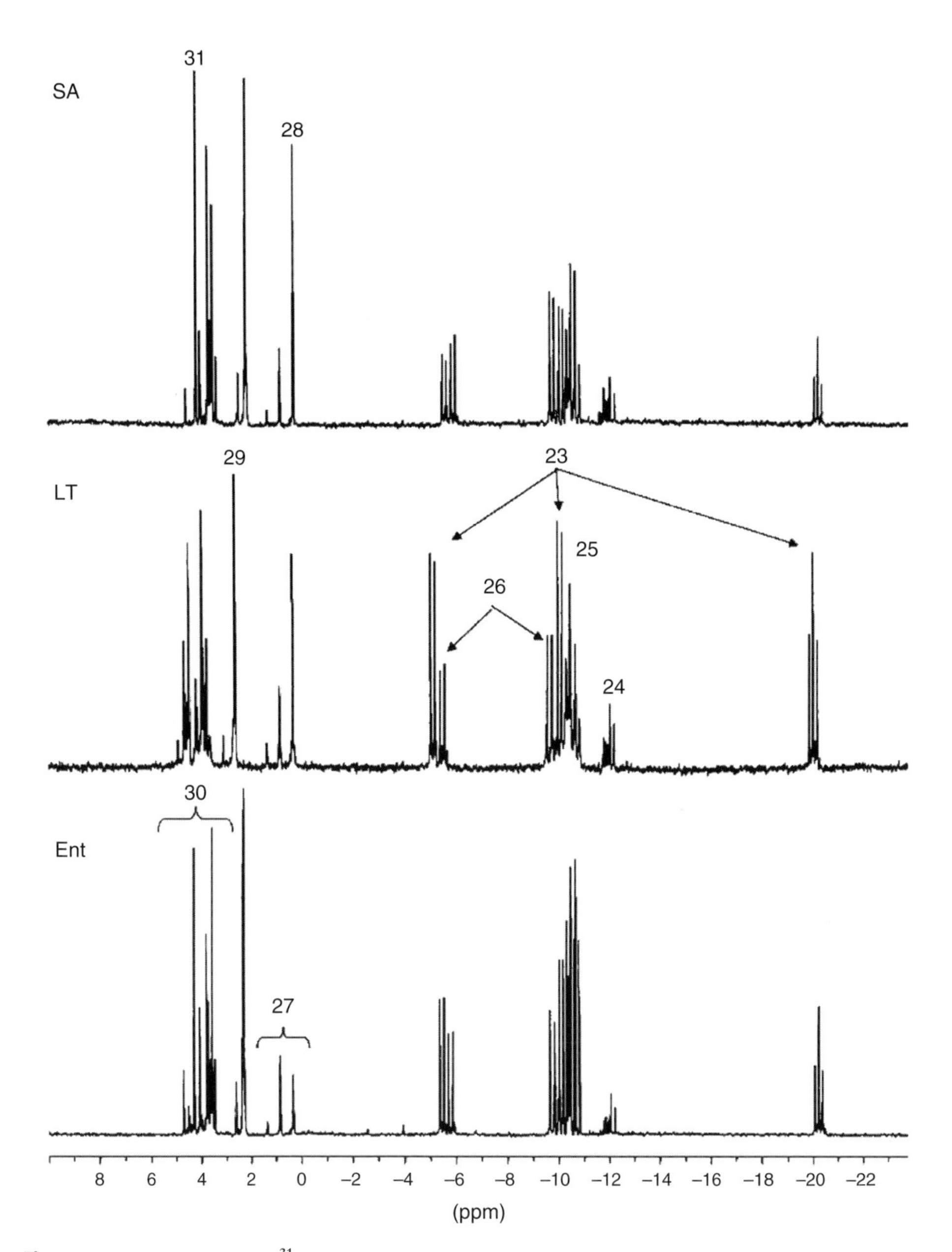

Figure 14 Representative ^{31}P NMR spectra from hydrophilic extracts of ground squirrel livers. NMR peak assignment: 23, adenosine triphosphates (α-, β-, and γ-ATP), 24, sugar phosphates; 25, nicotinamide adenine dinucleotide (NAD^{+}); 26, adenosine diphosphate (α- and β-ADP); 27, total PDE; 28, GPC; 29, inorganic phosphate (endogenous and exogenous); 30, total PME; 31, PC. N.L. Serkova, J.C. Rose, L.E. Epperson, H.V. Carey, S.L. Martin, Physiol. Genomics 31(2007) 15–24. Copyright (2007), with permission of the American Physiological Society.

segregation. Levels of glucose, lactate, alanine, succinate, and reduced glutathione were significantly lower in the winter states, LT and Ent, than in SA animals. In contrast, β-hydroxybutyrate was elevated in the livers of hibernating animals. The hibernation states LT and Ent were largely distinguished by differences in the levels of the osmolytes betaine and glutamine. Both metabolites were elevated in LT relative to Ent animals. These results are consistent with carbohydrate catabolism in the summer and a shift to lipid catabolism during hibernation. Because ground squirrels do not eat for long periods during hibernation, the amino acid building blocks for protein synthesis are recycled by protein catabolism leading to a buildup of glutamine as an ammonia sink during LT.

Representative ^{31}P NMR spectra of the perchloric acid liver extracts from animals in the SA, LT, and Ent metabolic states are shown in Figure 14. The spectra were measured using a 300 MHz spectrometer (^{31}P frequency of 121.5 MHz) using a 5 mm QNP probe. Addition of 100 mM EDTA was used to chelate divalent metal ions bound to ATP. The ^{31}P spectra were used to quantify the levels of nine phosphorous-containing compounds in the hydrophilic extracts relative to an external standard of methylene diphosphonic acid. Results were obtained for six individuals in each grouping, with significance determined by ANOVA. Somewhat surprisingly, elevated levels of ATP were found in the LT animals. The phospholipid precursor phosphocholine (PC) along with the total levels of phosphomonoesters (PME) were significantly decreased in LT compared with SA, although there was no significant difference between Ent hibernators and SA animals. The degradation products of membrane phospholipds, glycerophosphocholine (GC), and other phosphodiesters (PDE) were significantly depleted in the Ent liver extracts compared with the other two groups. The ratio of PME to PDE, which represents a balance between lipid biosynthesis and breakdown, increased 16.3-fold in Ent compared with LT animals. The likely source of increased betaine in LT is the breakdown and conversion of membrane phospholipids during topor using the liberated choline to produce betaine, an osmolyte.

Quantitative NMR profiling of liver metabolites in this study reveals unique metabolic signatures that distinguish the various stages of circannual hibernation cycle at a biochemical level and provides interesting insights into possible mechanisms of tissue protection and metabolic control that could ultimately benefit humans.

6. Conclusions and Future Perspectives

Metabolic profiling experiments have been shown to be of great use to researchers in the field of drug characterization,[117,127,128] in the search for clinical biomarkers of disease or toxic insult,[14,72,116,129] and in the impact of genetic modifications on metabolism.[13,19,93] The inherent quantitative nature of NMR and its ability to deal with relatively unadulterated biofluids and tissues makes it an ideal analytical technique for metabolic profiling. Many in the field have labored to investigate the effects of acquisition and processing parameters, as shown in this chapter, and such studies have been validated using biological samples. However, the future use of NMR for metabolic profiling and metabolomic experiments will depend on the degree to which the results can be modeled onto biochemical pathways, and the generation of meaningful biological conclusions that are ultimately supported by the NMR data.

One barrier to the use of metabolic profiling data to draw biological conclusions is that often there is a lack of thorough *metadata* reporting, meaning data about data, such as details from the experimental design, acquisition, and processing parameters. A large team of researchers with experience in the field of metabolic profiling[130] provided a full list of recommendations in an attempt to standardize both the degree of disclosure and the terminology used. Another helpful resource offers kingdom-specific reporting guidelines (i.e., mammals, plants, microbes, etc.) as well as specific standards for organisms studied as they exist freely in nature.[131] If metadata is included with metabolic profiling experiments when they are published, as some journals are beginning to require, there is a greater likelihood that the results will become part of the larger "omic" field including genomics, transcriptomics, and proteomics. Several studies have already begun to combine data sets from different omics approaches,[116,117] which allow more confident biological conclusions to be reached.

REFERENCES

1. E.E. Kline, E.G. Treat, T.A. Averna, M.S. Davis, A.Y. Smith, L.O. Sillerud, J. Urol. 176 (2006) 2274–2279.
2. D.G. Robertson, Toxicol. Sci. 85 (2005) 809–822.
3. O. Fiehn, Plant Mol. Biol. 48 (2002) 155–171.
4. D.I. Broadhurst, D.B. Kell, Metabolomics 2 (2006) 171–196.
5. S.C. Connor, R.A. Gray, M.P. Hodson, N.M. Clayton, J.N. Haselden, I.P. Chessell, C. Bountra, Metabolomics 3 (2007) 29–39.
6. T. Soga, Y. Ohashi, Y. Ueno, H. Naraoka, M. Tomita, T. Nishioka, J. Proteome Res. 2 (2003) 488–494.
7. G.F. Pauli, B.U. Jaki, D.C. Lankin, J. Nat. Prod. 70 (2007) 589–595.
8. A. Moing, M. Maucourt, C. Renaud, M. Gaudillere, R. Brouquisse, B. Lebouteiller, A. Gousset-Dupont, J. Vidal, D. Granot, B. Denoyes-Rothan, E. Lerceteau-Kohler, D. Rolin, Funct, Plant Biol. 31 (2004) 889–902.
9. D.C. Dejongh, T. Radford, J.D. Hribar, S. Hanessia, M. Bieber, G. Dawson, C.C. Sweeley, J. Am. Chem. Soc. 91 (1969) 1728–1740.
10. F.F. Brown, I.D. Campbell, P.W. Kuchel, D.L. Rabenstein, FEBS Lett. 82 (1977) 12–16.
11. D.L. Rabenstein, J. Biochem. Biophys. Methods 9 (1984) 277–306.
12. F.P.J. Martin, M.E. Dumas, Y.L. Wang, C. Legido-Quigley, I.K.S. Yap, H.R. Tang, S. Zirah, G.M. Murphy, O. Cloarec, J.C. Lindon, N. Sprenger, L.B. Fay, S. Kochhar, P. van Bladeren, E. Holmes, J.K. Nicholson, Mol. Syst. Biol. 3 (2007) 112.
13. F. Mounet, M. Lemaire-Chamley, M. Maucourt, C. Cabasson, J.-L. Giraudel, C. Deborde, R. Lessire, P. Gallusci, A. Bertrand, M. Gaudillère, C. Rothan, D. Rolin, A. Moing, Metabolomics 3 (2007) 273–288.
14. E.M. Lenz, J. Bright, I.D. Wilson, A. Hughes, J. Morrisson, H. Lindberg, A. Lockton, J. Pharm. Biomed. Anal. 36 (2004) 841–849.
15. S. Rezzi, Z. Ramadan, L.B. Fay, S. Kochhar, J. Proteome Res. 6 (2007) 513–525.
16. J.T.P. Albrechtova, M. Vervliet-Scheebaum, J. Normann, J. Veit, E. Wagner, Biol. Rhythm Res. 37 (2006) 381–389.
17. C. Touma, R. Palme, Ann. N. Y. Acad. Sci. 1046 (2005) 54–74.

18. M.E. Bollard, E. Holmes, J.C. Lindon, S.C. Mitchell, D. Branstetter, W. Zhang, J.K. Nicholson, Anal. Biochem. 295 (2001) 194–202.
19. H.K. Choi, Y.H. Choi, M. Verberne, A.W.M. Lefeber, C. Erkelens, R. Verpoorte, Phytochemistry 65 (2004) 857–864.
20. G. Maniara, K. Rajamoorthi, S. Rajan, G.W. Stockton, Anal. Chem. 70 (1998) 4921–4928.
21. M.E. Dumas, E.C. Maibaum, C. Teague, H. Ueshima, B.F. Zhou, J.C. Lindon, J.K. Nicholson, J. Stamler, P. Elliott, Q. Chan, E. Holmes, Anal. Chem. 78 (2006) 2199–2208.
22. V. Govindaraju, K. Young, A.A. Maudsley, NMR Biomed. 13 (2000) 129–153.
23. H. Rischer, K.M. Oksman-Caldentey, Trends Biotechnol. 24 (2006) 102–104.
24. H.C. Keun, T.M.D. Ebbels, H. Antti, M.E. Bollard, O. Beckonert, G. Schlotterbeck, H. Senn, U. Niederhauser, E. Holmes, J.C. Lindon, J.K. Nicholson, Chem. Res. Toxicol. 15 (2002) 1380–1386.
25. O. Fiehn, J. Kopka, P. Dormann, T. Altmann, R.N. Trethewey, L. Willmitzer, Nat. Biotechnol. 18 (2000) 1157–1161.
26. K. Dobbin, R. Simon, Biostatistics 6 (2005) 27–38.
27. K.K. Dobbin, R.M. Simon, Biostatistics 8 (2007) 101–117.
28. O. Hendrawati, Q.Q. Yao, H.K. Kim, H.J.M. Linthorst, C. Erkelens, A.W.M. Lefeber, Y.H. Choi, R. Verpoorte, Plant Sci. 170 (2006) 1118–1124.
29. J.L. Ward, C. Harris, J. Lewis, M.H. Beale, Phytochemistry 62 (2003) 949–957.
30. P. Krishnan, N.J. Kruger, R.G. Ratcliffe, J. Exp. Bot. 56 (2005) 255–265.
31. A.T. Dossey, S.S. Walse, J.R. Rocca, A.S. Edison, Chem. Biol. 1 (2006) 511–514.
32. P.R. Powell, T.L. Paxon, K.A. Han, A.G. Ewing, Anal. Chem. 77 (2005) 6902–6908.
33. A.D. Maher, S.F.M. Zirah, E. Holmes, J.K. Nicholson, Anal. Chem. 79 (2007) 5204–5211.
34. E.J. Saude, B.D. Sykes, Metabolomics 3 (2007) 19–27.
35. T.W.-M. Fan, R.M. Higashi, A.N. Lane, O. Jardetzky, Biochem. Biophys. Acta 882 (1986) 154–167.
36. T.W.-M. Fan, Prog. Nucl. Magn. Reson. Spectrosc. 28 (1996) 161–219.
37. E.J. Saude, C.M. Slupsky, B.D. Sykes, Metabolomics 2 (2006) 113–123.
38. N. Arabínar, K.-H. Ott, V. Roongta, L. Mueller, Anal. Biochem. 355 (2006) 62–70.
39. T.A. Averna, E.E. Kline, A.Y. Smith, L.O. Sillerud, J. Urol. 173 (2005) 433–438.
40. D.J. Crockford, E. Holmes, J.C. Lindon, R.S. Plumb, S. Zirah, S.J. Bruce, P. Rainville, C.L. Stumpf, J.K. Nicholson, Anal. Chem. 78 (2006) 363–371.
41. M.G. Swanson, A.S. Zektzer, Z.L. Tabatabai, J. Simko, S. Jarso, K.R. Keshari, L. Schmitt, P.R. Carroll, K. Shinohara, D.B. Vigneron, J. Kurhanewicz, Magn. Reson. Med. 55 (2006) 1257–1264.
42. C.K. Larive, D. Jayawickrama, L. Orfi, Appl. Spectrosc. 51 (1997) 1531–1536.
43. J.M. Rydzewski, T. Schleich, Biophys. J. 70 (1996) 1472–1484.
44. K.W. Jordan, W. He, E.F. Halpern, C.-L. Wu, L.L. Cheng, Biomark. Insights 3 (2007) 147–154.
45. S. Akoka, L. Barantin, M. Trierweiler, Anal. Chem. 71 (1999) 2554–2557.
46. V. Silvestre, S. Goupry, M. Trierweiler, R. Robins, S. Akoka, Anal. Chem. 73 (2001) 1862–1868.
47. K.E. Price, S.S. Vandaveer, C.E. Lunte, C.K. Larive, J. Pharm. Biomed. Anal. 38 (2005) 904–909.
48. E.M. Lenz, R.A. D'Souza, A.C. Jordan, C.D. King, S.M. Smith, P.J. Phillips, A.D. McCormick, D.W. Roberts, J. Pharm. Biomed. Anal. 43 (2007) 1065–1077.
49. D.J. Penn, E. Oberzaucher, K. Grammer, G. Fischer, H.A. Soini, D. Wiesler, M.V. Novotny, S.J. Dixon, Y. Xu, R.G. Brereton, J. R. Soc. Interface 4 (2007) 331–340.
50. E. Holmes, T.M. Tsang, J.T.J. Huang, F.M. Leweke, D. Koethe, C.W. Gerth, B.M. Nolden, S. Gross, D. Schreiber, J.K. Nicholson, S. Bahn, PLoS Med. 3 (2006) e327.
51. H.C. Holmes, G. Snodgrass, R.A. Iles, Eur. J. Pediatr. 159 (2000) 198–204.
52. H.R. Tang, Y.L. Wang, J.K. Nicholson, J.C. Lindon, Anal. Biochem. 325 (2004) 260–272.
53. J.C. Lindon, J.K. Nicholson, E. Holmes, J.R. Everett, Concepts Magn. Reson. A 12 (2000) 289–320.
54. S. Deprez, B.C. Sweatman, S.C. Connor, J.N. Haselden, C.J. Waterfield, J. Pharm. Biomed. Anal. 30 (2002) 1297–1310.

55. Q.N. Van, G.N. Chmurny, T.D. Veenstra, Biochem. Biophys. Res. Commun. 301 (2003) 952–959.
56. D.L. Rabenstein, K.K. Millis, E.J. Strauss, Anal. Chem. 60 (1988) A1380–A1391.
57. M.L. Liu, J.K. Nicholson, J.C. Lindon, Anal. Chem. 68 (1996) 3370–3376.
58. L.H. Lucas, C.K. Larive, P.S. Wilkinson, S. Huhn, J. Pharm. Biomed. Anal. 39 (2005) 156–163.
59. B.C.M. Potts, A.J. Deese, G.J. Stevens, M.D. Reily, D.G. Robertson, J. Theiss, J. Pharm. Biomed. Anal. 26 (2001) 463–476.
60. K.O. Boernsen, S. Gatzek, G. Imbert, Anal. Chem. 77 (2005) 7255–7264.
61. U. Sharma, S. Atri, M.C. Sharma, C. Sarkar, N.R. Jagannathan, NMR Biomed. 16 (2003) 213–223.
62. A. Miccheli, A. Tomassini, C. Puccetti, M. Valerio, G. Peluso, F. Tuccillo, M. Calvani, C. Manetti, F. Conti, Biochimie 88 (2006) 437–448.
63. K. Wolf, A. van der Toorn, K. Hartmann, L. Schreiber, W. Schwab, A. Haase, G. Bringmann, J. Exp. Bot. 51 (2000) 2109–2117.
64. W. Kockenberger, J. Exp. Bot. 52 (2001) 641–652.
65. J.E. Le Belle, N.G. Harris, S.R. Williams, K.K. Bhakoo, NMR Biomed. 15 (2002) 37–44.
66. J.-P. Usenius, P. Vainio, J. Hernesniemi, R.A. Kauppinen, J. Neurochem. 63 (1994) 1538–1543.
67. L.H. Lucas, S.F. Wilson, C.E. Lunte, C.K. Larive, Anal. Chem. 77 (2005) 2978–2984.
68. N.J. Waters, E. Holmes, C.J. Waterfield, R.D. Farrant, J.K. Nicholson, Biochem. Pharmacol. 64 (2002) 67–77.
69. J.L. Griffin, O. Corcoran, Magn. Reson. Mater. Phys. 18 (2005) 51–56.
70. J.L. Griffin, C. Blenkiron, P.K. Valonen, C. Caldas, R.A. Kauppinen, Anal. Chem. 78 (2006) 1546–1552.
71. K.W. Jordan, L.L. Cheng, Expert Rev. Proteomics. 4 (2007) 389–400.
72. Y.X. Yang, C.L. Li, X. Nie, X.S. Feng, W.X. Chen, Y. Yue, H.R. Tang, F. Deng, J. Proteome Res. 6 (2007) 2605–2614.
73. G.S. Payne, H. Troy, S.J. Vaidya, J.R. Griffiths, M.O. Leach, Y.L. Chung, NMR Biomed. 19 (2006) 593–598.
74. F.P.J. Martin, E.F. Verdu, Y.L. Wang, M.E. Dumas, I.K.S. Yap, O. Cloarec, G.E. Bergonzelli, I. Corthesy-Theulaz, S. Kochhar, E. Holmes, J.C. Lindon, S.M. Collins, J.K. Nicholson, J. Proteome Res. 5 (2006) 2185–2193.
75. L. Schenetti, A. Mucci, F. Parenti, R. Cagnoli, V. Righi, M.R. Tosi, V. Tugnoli, Concepts Magn. Reson. A 28 (2006) 430–443.
76. O. Saether, O. Risa, J. Cejkova, J. Krane, A. Midelfart, Graefes Arch. Clin. Exp. Ophthalmol. 242 (2004) 1000–1007.
77. M.G. Swanson, D.B. Vigneron, Z.L. Tabatabai, R.G. Males, L. Schmitt, P.R. Carroll, J.K. James, R.E. Hurd, J. Kurhanewicz, Magn. Reson. Med. 50 (2003) 944–954.
78. M.A. Burns, J.L. Taylor, C.-L. Wu, A.G. Zepeda, A. Bielecki, D. Cory, L.L. Cheng, Magn. Reson. Med. 54 (2005) 34–42.
79. D.K. Hansen, M.I. Davies, S.M. Lunte, C.E. Lunte, J. Pharm. Sci. 88 (1999) 14–27.
80. P. Khandelwal, C.E. Beyer, Q. Lin, L.E. Schechter, A.C. Bach, Anal. Chem. 76 (2004) 4123–4127.
81. P. Khandelwal, C.E. Beyer, Q. Lin, P. McGonigle, L.E. Schechter, A.C. Bach, J. Neurosci. Methods 133 (2004) 181–189.
82. K. E. Price, C. E. Lunte, C. K. Larive, J. Pharm. Biomed. Anal. 46 (2008) 737–747.
83. F. Mesnard, R.G. Ratcliffe, Photosynth. Res. 83 (2005) 163–180.
84. H.C. Bertram, A. Malmendal, B.O. Petersen, J.C. Madsen, H. Pedersen, N.C. Nielsen, C. Hoppe, C. Molgaard, K.F. Michaelsen, J.O. Duus, Anal. Chem. 79 (2007) 7110–7115.
85. F. Malz, H. Jancke, J. Pharm. Biomed. Anal. 38 (2005) 813–823.
86. P.J. Marro, O.P. Mishra, M. Delivoria-Papadopoulos, Brain Res. 1073 (2006) 444–450.
87. M.R. Viant, C.A. Pincetich, D.E. Hinton, R.S. Tjeerdema, Aquat. Toxicol. 76 (2006) 329–342.
88. S.F.E. Praet, H.M.M. De Feyter, R.A.M. Jonkers, K. Nicolay, C. van Pul, H. Kuipers, L.J.C. van Loon, J.J. Prompers, Magn. Reson. Mater. Phys. 19 (2006) 321–331.

89. W.E. Hull, R.E. Port, R. Herrmann, B. Britsch, W. Kunz, Cancer Res. 48 (1988) 1680–1688.
90. J.D. Gross, R.C. Long, I. Constantinidis, A. Sambanis, Anal. Chem. 98 (2007) 261–270.
91. H.C. Keun, O. Beckonert, J.L. Griffin, C. Richter, D. Moskau, J.C. Lindon, J.K. Nicholson, Anal. Chem. 74 (2002) 4588–4593.
92. J. Kikuchi, K. Shinozaki, T. Hirayama, Plant Cell Physiol. 45 (2004) 1099–1104.
93. R.G. Ratcliffe, Y. Shachar-Hill, Plant J. 45 (2006) 490–511.
94. T.W.-M. Fan, A.N. Lane, R.A. Higashi, Russ. J. Plant Physiol. 50 (2003) 787–793.
95. G. Sriram, D.B. Fulton, V.V. Iyer, J.M. Peterson, R. Zhou, M.E. Westgate, M.H. Spalding, J.V. Shanks, Plant Physiol. 136 (2004) 3043–3057.
96. S. Massou, C. Nicolas, F. Letisse, J.C. Portais, Metab. Eng. 9 (2007) 252–257.
97. G. Sriram, D.B. Fulton, J.V. Shanks, Phytochemistry 68 (2007) 2243–2257.
98. A. Craig, O. Cloareo, E. Holmes, J.K. Nicholson, J.C. Lindon, Anal. Chem. 78 (2006) 2262–2267.
99. D.D. Traficante, M. Rajabzadeh, Concepts Magn. Reson. A 12 (2000) 83–101.
100. D. Moskau, Concepts Magn. Reson. B 15 (2002) 164–176.
101. Q.W. Xu, J.R. Sachs, T.C. Wang, W.H. Schaefer, Anal. Chem. 78 (2006) 7175–7185.
102. T.S. Derrick, E.F. McCord, C.K. Larive, J. Magn. Reson. 155 (2002) 217–225.
103. R.A. de Graaf, K.L. Behar, Anal. Chem. 75 (2003) 2100–2104.
104. J.C. Lindon, A.G. Ferrige, Prog. Nucl. Magn. Reson. Spectrosc. 14 (1980) 27–66.
105. L. Griffiths, Magn. Reson. Chem. 39 (2001) 194–202.
106. J. Forshed, R.J.O. Torgrip, K.M. Aberg, B. Karlberg, J. Lindberg, S.P. Jacobsson, J. Pharm. Biomed. Anal. 38 (2005) 824–832.
107. G.T. Gipson, K.S. Tatsuoka, B.C. Sweatman, S.C. Connor, J. Magn. Reson. 183 (2006) 269–277.
108. F. Dieterle, A. Ross, G. Schlotterbeck, H. Senn, Anal. Chem. 78 (2006) 3551–3561.
109. A.M. Weljie, J. Newton, P. Mercier, E. Carlson, C.M. Slupsky, Anal. Chem. 78 (2006) 4430–4442.
110. T. Abshear, G.M. Banik, M.L. D'Souza, K. Nedwed, C. Peng, SAR QSAR Environ. Res. 17 (2006) 311–321.
111. T. Saito, S. Nakaie, M. Kinoshita, T. Ihara, S. Kinugasa, A. Nomura, T. Maeda, Metrologia 41 (2004) 213–218.
112. Y.L. Martin, J. Magn. Reson. Ser A 111 (1994) 1–10.
113. M. Scholz, S. Gatzek, A. Sterling, O. Fiehn, J. Selbig, Bioinformatics 20 (2004) 2447–2454.
114. J. Forshed, I. Schuppe-Koistinen, S.P. Jacobsson, Anal. Chim. Acta 487 (2003) 189–199.
115. M. Defernez, I.J. Colquhoun, Phytochemistry 62 (2003) 1009–1017.
116. J. Roncalli, F. Smih, F. Desmoulin, N. Dumonteil, R. Harmancey, S. Hennig, L. Perez, A. Pathak, M. Galinier, P. Massabuau, M. Malet-Martino, J.M. Senard, P. Rouet, J. Mol. Cell. Cardiol. 42 (2006) 526–539.
117. T.W.-M. Fan, R.M. Higashi, A.N. Lane, Drug Metab. Rev. 38 (2006) 707–732.
118. O. Thellin, W. Zorzi, B. Lakaye, B. De Borman, B. Coumans, G. Hennen, T. Grisar, A. Igout, E. Heinen, J. Biotechnol. 75 (1999) 291–295.
119. K.P.R. Gartland, F.W. Bonner, J.K. Nicholson, Mol. Pharmacol. 35 (1989) 242–250.
120. W. Schoonen, C. Kloks, J. Ploemen, G.J. Horbach, M.J. Smit, P. Zandberg, J.R. Mellema, C.T.V. Zuylen, A.C. Tas, J.H.J. van Nesselrooij, J. Vogels, Toxicol. Sci. 98 (2007) 271–285.
121. S. Trump, S. Laudi, N. Unruh, R. Goelz, D. Leibfritz, Magn. Reson. Mater. Phys. 19 (2006) 305–312.
122. E.J. Saude, D. Adamko, B.H. Rowe, T. Marrie, B.D. Sykes, Metabolomics 3 (2007) 439–451.
123. E. Holmes, H. Antti, Analyst 127 (2002) 1549–1557.
124. R.A. van den Berg, H.C.J. Hoefsloot, J.A. Westerhuis, A.K. Smilde, M.J. van der Werf, BMC Genomics 7 (2006) 142.
125. H.M. Parsons, C. Ludwig, U.L. Gunther, M.R. Viant, BMC Bioinformatics 8 (2007) 234.
126. N.L. Serkova, J.C. Rose, L.E. Epperson, H.V. Carey, S.L. Martin, Physiol. Genomics 31 (2007) 15–24.
127. F. Dieterle, G.T. Schlotterbeck, A. Ross, U. Niederhauser, H. Senn, Chem. Res. Toxicol. 19 (2006) 1175–1181.

128. J. Tarning, Y. Bergqvist, N.P. Day, J. Bergquist, B. Arvidsson, N.J. White, M. Ashton, N. Lindegardh, Drug Metab. Dispos. 34 (2006) 2011–2019.
129. L.C.A. Amorim, Z.D.L. Cardeal, J. Chromatogr. B 853 (2007) 1–9.
130. J.C. Lindon, J.K. Nicholson, E. Holmes, H.C. Keun, A. Craig, J.T.M. Pearce, S.J. Bruce, N. Hardy, S.A. Sansone, H. Antti, P. Jonsson, C. Daykin, M. Navarange, R.D. Beger, E.R. Verheij, A. Amberg, D. Baunsgaard, G.H. Cantor, L. Lehman-McKeeman, M. Earll, S. Wold, E. Johansson, J.N. Haselden, K. Kramer, C. Thomas, J. Lindberg, I. Schuppe-Koistinen, I.D. Wilson, M.D. Reily, D.G. Robertson, H. Senn, A. Krotzky, S. Kochhar, J. Powell, F. van der Ouderaa, R. Plumb, H. Schaefer, M. Spraul, Nat. Biotechnol. 23 (2005) 833–838.
131. D. Robertson, Chair, Biological Sample Context Working Group, Metabolomics Standards Initiative, http://msi-workgroups.sourceforge.net/bio-metadata/, accessed November 17, 2007.

CHAPTER 6

DOSY NMR for Drug Analysis

V. Gilard, S. Trefi, S. Balayssac, M.-A. Delsuc, T. Gostan,
M. Malet-Martino, R. Martino, Y. Prigent, *and* F. Taulelle

Contents

Abstract

The DOSY method measures the translational self-diffusion of molecules in solution and allows a precise analysis of a complex mixture without any prior separation of the different components. In full analogy with the DOSY experiment, TOSY that is based on the relaxation time evolutions allows to approach the interactions between all the components of a mixture. After a part devoted to the theory of DOSY and TOSY, this chapter outlines the use of the DOSY or TOSY NMR method applied to drug analysis in solution or in solid form, respectively. To illustrate the interest of both methods for the analysis of drug formulations, examples are taken with pharmaceutical preparations of ciprofloxacin, fluoxetine, Cialis®, herbal Chinese formulation, and ibuprofen. Finally, the advantages and drawbacks of the methods are discussed.

Keywords: TOSY, DOSY, fluoxetine, tadalafil, ibuprofene, Chinese herbs

1. Introduction

The ability of nuclear magnetic resonance (NMR) to provide valuable information regarding mixture analysis has led to its widespread use across chemistry, biochemistry, biology, and medicine. As drugs can be considered as complex mixtures (composed of many different substances or including simultaneously large and very low quantities of compounds), NMR is therefore a great tool for studying such formulations. For a long time it has been preferred, when possible, to isolate each component of a mixture prior to its study by NMR rather than to analyze the mixture. However, there is indeed "a pure NMR" method that allows a precise analysis of a complex mixture without any prior separation of the different components: the Diffusion Ordered SpectroscopY (DOSY) method. Furthermore, DOSY experiments do not need heavy setups and the method can be easily standardized and automated. One should also keep in mind the nondestructive and noninvasive nature of NMR spectroscopy.

The DOSY method allows measuring the translational self-diffusion of molecules in solution. Based on the analysis of mono- and multiexponential decays, spectra of the components of a mixture can be separated depending on the value of their apparent diffusion coefficients. The measurement of diffusion by NMR and especially the DOSY-type experiments are thus powerful analytical tools, which have remained underemployed till now by most scientists.

In addition, one can notice that nowadays 80% of the pharmaceutical formulations are solids. It is therefore interesting to study these drugs in their original nonmodified form. Based on the relaxation time evolutions and in full analogy with the DOSY experiment, TOSY (raTe of relaxation Ordered SpectroscopY) is thus a relevant tool to obtain a fingerprint of a given formulation which also allows to approach the interactions between all the drug components.

Through a brief theoretical introduction and numerous examples, this chapter outlines the use of the DOSY NMR method applied to drug analysis in solution and points out the growing interest of studying the solid formulations by TOSY.

2. Theory

2.1. DOSY

2.1.1. Principle

The use of NMR for measuring the self-diffusion of molecules in solution was originally proposed in 1965 by Stejskal and Tanner[1] in an experiment close to the one presented in Figure 1(a). The principle of the measure lies in the fact that the phase of the magnetization is spatially encoded at the beginning of the experiment. Brownian motion in the liquid results in translational diffusion of the various solutes, and a mean molecule displacement is observed at the end of the delay Δ. This displacement has the effect of reducing the signal intensity with an exponential law:

$$I(q) = I_0 \exp\left(-D\Delta q^2\right) \text{with } q = \gamma g \delta \tag{1}$$

where D is the diffusion coefficient, γ the gyromagnetic ratio, and g and δ the intensity and the duration of the pulsed field gradient (PFG), respectively. The DOSY experiment relies on this evolution law to lead to a multidimensional

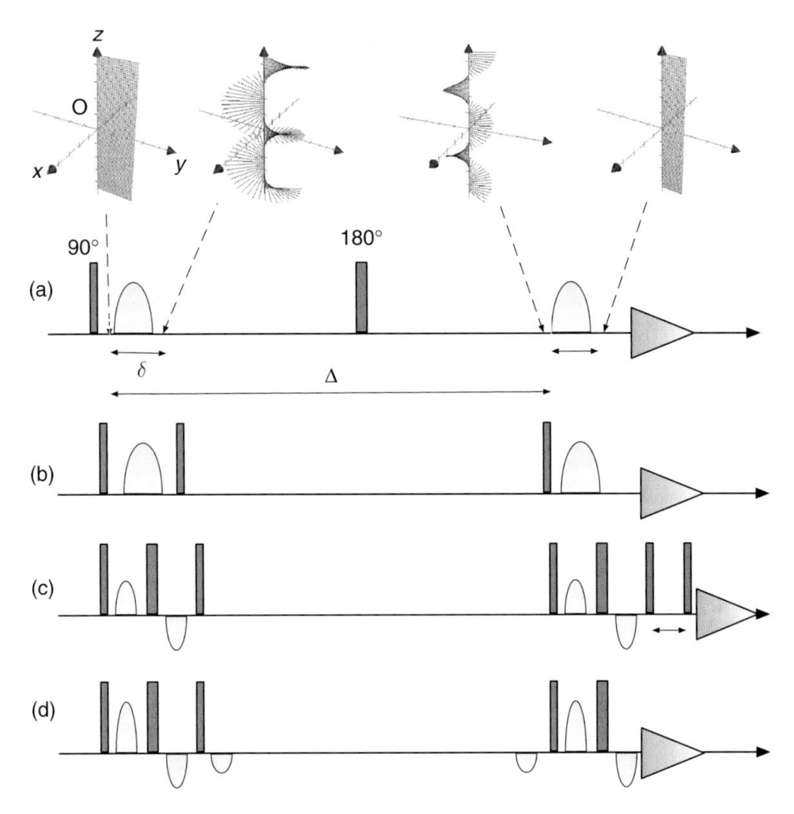

Figure 1 NMR experiments used in DOSY acquisition. (a) The original spin-echo (SE) experiment: This figure presents the simplest pulse sequence that can be used for diffusion measurement. After the first excitation 90° pulse, the magnetization is aligned on the xOy plane perpendicular to the applied static magnetic field. This static field is then briefly perturbed by the first PFG of length δ. During the PFG, the field intensity varies linearly along the main axis of the sample. This introduces a dephasing of the bulk NMR signal due to a spatial phase encoding which depends on the spin position along the z-axis. At the end of the PFG, a magnetization helix is thus observed. The final PFG applies an opposite dephasing to the magnetization, and the signal is normally restored to a state close to its initial one, the 180° in the center of the evolution creating a Hahn echo that removes any contribution of the chemical shift to the evolution. However, the diffusion takes place during the delay between the two PFG. If the mean molecular displacement is on the order of the pitch of the magnetization helix, this helix is blurred and the intensity of the signal is decreased. The measure is based on the analysis of the decay of the spin-echo signal intensity with increasing PFG intensities. (b) The STE experiment: In this experiment, the second 90° pulse applied directly after the coding gradient builds a spatial modulation of the transverse magnetization along the sample z-axis rather than a transverse modulation as in SE. Being on the z-axis, this modulation does not evolve any further, except for diffusion. The final 90° pulse restores along the y-axis the same modulation, which then produces an echo once refocused by the second PFG. (c) The bipolar pulse pair-stimulated echo-longitudinal eddy current delay (BPP-STE-LED) experiment: Here, the STE experiment is further optimized by applying two independent tricks: (i) the PFG are made bipolar, i.e., applied in two opposite pulses, sandwiching a 180° pulse; (ii) a final delay (LED) is added during which the magnetization is stored along the z-axis to let all the generated fluctuations die away before the acquisition. These two optimizations have the same purpose that is to reduce the intensity of the eddy current generated by the PFG and to minimize its impact on the observed signal. (d) The OneShot experiment brings further improvement to the standard STE pattern: It presents a better stability against the experimental fluctuations and allows a simplification of the measure. Less signal averaging is required permitting a faster measure when the S/N ratio is not a limiting factor.

spectrum on which the diffusion coefficient D is displayed along one of the spectral axes. This representation was first proposed by Morris and Johnson[2] in 1992, and several reviews cover this subject in the literature.[3–5]

The DOSY experiment differs from the usual modern NMR experiments in two important ways. First, while modern NMR experiments are based on the analysis of the time dependence of the signal, here all the delays are kept strictly constant to abstract the measure from the relaxation phenomenon. In contrast, only the intensities of the coding and decoding gradients are varying. Second, the time dependence of modern NMR experiments is usually analyzed by some spectral analysis methods, the Fourier transform being by far the most common one (the relaxation measurement being the main exception to this proposition, and we will see below that it shares some relationship with the DOSY analysis). In the case of DOSY, the signal variation follows an exponential law, and a Laplace analysis is to be applied.

2.1.2. Data acquisition

2.1.2.1. The stimulated echo experiment

While the principle is easily described by a Hahn echo experiment, the most common experiment used for drug analysis is the stimulated echo, usually called STE.[6] This experiment is described in Figure 1(b). Because of the way in which magnetization is handled, loss of signal due to T_2 relaxation is minimized. Also, because it is less sensitive to distortions due to homonuclear couplings, this experiment is usually preferred as it permits the use of longer diffusion delays and thus to sample slower diffusion processes.

This pulse sequence is reminiscent of other 2D experiments such as nuclear Overhauser spectroscopy (NOESY) or excitation spectroscopy (EXSY) and as such, it can be designed to produce either a regular echo, or the so-called "anti-echo" depending on the sign of the decoding gradient. It is particularly well suited for the analysis by high-resolution DOSY of complex mixtures where many signals are observed with a wide range of dynamic.

2.1.2.2. Experimental difficulties

Several difficulties are encountered when trying to push the technique to its limits. Obtaining high-resolution in the DOSY spectra needs increasing the resolution along both the chemical shift axis and the diffusion axis. This increase requires very good linearities and stabilities throughout the entire experimental setup. A comprehensive review has been devoted to this problem by Antalek.[4] However, we can pinpoint here some of the most important difficulties.

The linearity of the spectrometer, and particularly of the PFG amplitude, is important as any departure from a perfectly proportional response will be echoed in a shift or a spread of the signal along the diffusion axis. In particular, the spatial linearity of the probe gradient coil in the probe should be monitored. When strong nonlinearity is observed, some software corrections might be used.[7]

The overall stability of the spectrometer during the measurement is also paramount. Two principal difficulties are encountered: the presence of eddy currents and the temperature gradient. Eddy currents are transient currents that naturally appear in the conducting parts of the experimental apparatus and tend to oppose to the applied magnetic field variations. They are produced by the flux variation observed in the conducting parts, which then produces an electric current whose distribution obeys the Lenz law. Eddy currents particularly build up in the presence of the fast varying field gradients used in the DOSY experiments and can lead to strong distortion of the recorded signal. Several optimizations of the STE sequence have been proposed in the literature: use of BiPolar pulses, use of the Longitudinal Eddy current Delay (LED),[8] or the OneShot sequence.[9] These modifications are detailed in Figure 1(c) and (d).

A fine temperature regulation is always difficult in an NMR spectrometer and temperature gradients in the sample are not uncommon. The diffusion process is very sensitive to the temperature and in the presence of such gradients, the measured diffusion coefficient values will be imprecise. More importantly, temperature gradients may initiate convection movements in the NMR tube, completely destroying the ability to reliably measure any diffusion. So, care should be taken to control very accurately the temperature. If convection is suspected, one may resort to the double echo experiment proposed by Jerschow and Müller,[10] which is insensitive to convection perturbation but associated to a loss in sensitivity.

2.1.3. Data processing

The result of the experiment obtained while running one of the pulse sequences presented in Figure 1 is an evolution of the signal intensity following the law presented in Eq. (1). More precisely, Eq. (1) shows the evolution of the signal produced by a single monodisperse species. If several species are present in the solution or if polydispersity is present, a more general form should be used:

$$I(q) = \int_0^{\infty} A(D)\exp\left(-D\Delta q^2\right)\mathrm{d}D \tag{2}$$

This equation shows that the measure $I(q)$ is related to the Laplace transform of the diffusion distribution $A(D)$. The Laplace transform is formally related to the Fourier transform, and $A(D)$ can be seen as the Laplace spectrum of $I(q)$. This Laplace spectrum will be displayed along the diffusion axis of the DOSY spectrum. However, the difficulty is to estimate $A(D)$ from the measure of the signal decay since the inverse Laplace transform (ILT) is an operation which is much more difficult to implement numerically than the inverse Fourier transform. Two approaches are found in the literature. The first relies on a simplification of Eq. (2), reducing it either to a finite sum of a few exponentials or even to a single exponential as in Eq. (1). When this assumption is made, the spectrum $A(D)$ is much simplified and consists in one or a few nonnull positions along the diffusion axis. These points are then easily estimated by a nonlinear least-squares fit of Eq. (1), and the Laplace spectrum is then reconstructed from this parameter estimation. However, it is difficult to extract reliably more than one or two diffusion coefficients when using this approach.

Several improvements to this scheme have been proposed. It has been suggested to fit the whole spectrum with a small set of diffusion coefficients rather than fitting each spectral line independently.[11] This has the effect of somewhat stabilizing the exponential fit and permitting to separate more efficiently the various components of a mixture. In other approaches, the data set is handled as a whole, searching for signals which have common evolution. The signals are then isolated by mathematical methods close to multivariate analysis.[12–14] With this approach, a DOSY spectrum per se is not produced but rather a set of NMR subspectra that contain signals that all behave in the same manner. The DOSY spectrum is then reconstructed from this set of subspectra.

All the fitting techniques presented above assume that a finite (and usually small) number of pure components are present in the sample. This hypothesis might not be relevant in a complex mixture either because of strong spectral overlap, or because of the presence of intrinsically polydisperse species such as polymers (oligosaccharides for instance), or because of molecular aggregation for instance with surfactants or other detergents. In this case, which happens to be very common, the modeling of the signal decay with a few pure exponentials is to be abandoned, and one has to resort to a complete Laplace analysis of the signal. As it has been noted earlier, the ILT is not an easy numerical task, and a special algorithm has to be used.

Several techniques have been proposed in the literature based on an iterative solution of the inverse problem stated in Eq. (2). The approach consists in finding the distribution $A(D)$ that best explains the data $I(q)$. However, the problem is ill-posed and regularization is required. Several regularizations have been used in the literature but two techniques have known a large development: the CONTIN[15,16] or related approaches,[17] which rely on a Tikhonov regularization, and the Maximum Entropy (MaxEnt) approach[18] that tries to minimize the quantity of information present in $A(D)$. Both techniques permit to directly build a DOSY spectrum from the decay analysis. In these approaches, each decay is processed independently and a complete DOSY spectrum is built from the successive analysis of all the points along the 1D spectrum. One of the main advantages of these ILT approaches comes from the fact that very little assumption is made on the nature of the Laplace spectrum. In contrast to the fit techniques presented earlier, no hypothesis is made on the dispersity or the number of species present in the mixture. This insures a more robust result and more independence of the data to the processing details.

Finally, a new approach, called iRRT (regularized resolvent transform, the i denoting the adaptation of the method to evaluate the ILT), has recently been proposed based on a global inversion of both the Fourier (spectral) and the Laplace (diffusion) axes.[19] This approach relies on a particular sampling of the gradient values that permits to build a recursive expression for the diffusion axis similar to the recursive expression found along the spectral axis. It then uses an SVD (singular value decomposition)-based regularization to solve the problem in one step and generate a DOSY spectrum. This promising technique suffers nonetheless from the fact that a suboptimal sampling of the gradient series is to be employed.

It should be noted that, while the analysis of the diffusion equation is central in the processing of DOSY experiment, this is not the only required step. The quality of the final spectrum relies also on the quality of the preprocessing, such as baseline correction,[20] fluctuation correction,[21] and the processing should be conceived as a whole.[22]

2.2. TOSY

Relaxation has been considered since the beginning of NMR and is explicitly present in the first paper of Bloch.[23] Since then relaxation has received attention and treatments by so many contributors that a book would not suffice to quote them all. However, at least some contributions are of very special interest. The so-called BPP (Bloembergen-Pound and Purcell) theory leads to an operational exponential character for relaxation,[24] and a clear picture of this approach has been given by Goldman.[25] Redfield's contribution has laid the ground on which relaxation lies, though not developing every case that could be deduced from his theory.[26,27] The limits for computing relaxation expressions for a spin system are discussed in this latter contribution. It is quite clear that for isolated molecules with a limited number of spins, a complete theoretical treatment can be provided. For a larger number of spins as in biomolecules, one has to resort to drastic simplifications like the Lipari–Szabo model.[28] For molecular as well as inorganic solids, very few contributions have emerged due to the theoretical limits indicated by Redfield.[26,27] However, the practical operational approach that has been used by the vast majority of authors in this field is to consider for the three relaxation times, T_1, T_2, and $T_{1\rho}$, a single or multiexponential behavior of the magnetization decay with the evolution of the relaxation time.

The 2D NMR experiment TOSY, in full analogy with the DOSY experiment, displays the distribution of rates of relaxation in the indirect dimension and the traditional spectrum in the direct dimension. It is a generalization of the classical analysis of relaxation, replacing a discrete spectrum of relaxation rates, actually limited to one to few discrete values, by a continuous Laplace spectrum of relaxation rates.

2.2.1. Principle

Traditionally, the T_1 and T_2 relaxation times, introduced by Bloch, are measured with different but classical pulse sequences. T_2 is measured with a Hahn echo whose time between the echoes is incremented up to the point where magnetization vanishes. T_1 depending on the order of its magnitude is either measured with a saturation recovery experiment or with an inversion recovery sequence. In both cases, the time between the first event, saturation or inversion of the magnetization, and the observe pulse is progressively increased until the magnetization at equilibrium is recovered. $T_{1\rho}$ measurement is analogous to that of T_2 but the decay of the transverse magnetization is observed under radio frequency irradiation along the magnetization in the transverse plane. The Hahn echo is replaced by a locking period of increasing length. $T_{1\rho}$ is commonly measured indirectly in solid state by the transfer of magnetization on a different nucleus, routinely in cross-polarization experiments between ^1H and ^{13}C for organic or molecular compounds. The different expressions of relaxation are given as follows:

$$M = M_0 \exp\left(\frac{-t}{T_2}\right) \tag{3}$$

$$M = M_0\left(1 - \exp\left(\frac{-t}{T_1}\right)\right) \quad \text{or} \quad M - M_0 = M_0 \exp\left(\frac{-t}{T_1}\right) \tag{4}$$

$$M = M_0\left(1 - 2\exp\left(\frac{-t}{T_1}\right)\right) \quad \text{or} \quad \frac{M - M_0}{2} = M_0 \exp\left(\frac{-t}{T_1}\right) \tag{5}$$

$$M = M_0 \exp\left(\frac{-t}{T_{1\rho}}\right) \quad \text{or} \quad M - M_0 = M_0 \exp\left(\frac{-t}{T_{1\rho}}\right) \tag{6}$$

Each of these equations contains an exponential term if the relaxation expression can be described by a single exponential behavior. In a more general sense it is commonly accepted that any other description of the relaxation behavior is a superposition of exponentials. Therefore, the general expression of the time evolution signal can be given by the following expression:

$$G(t_i) = \sum_i {}_{1,n} A(R_j) \exp(-R_j t_i) \tag{7}$$

Though generally an integral and not a discrete sum, the discrete sum is the actual computed form and corresponds closely to the general procedure used by spectroscopists to analyze such decays by a single or a multiexponential model, limited usually to two to three exponentials. This equation can cover expressions of the relaxation that can be analytically derived from a theoretical approach as well as relaxation for complex materials for which the expression for a theoretical derivation might be just impossible to describe by lack of proper knowledge on the organization of the material. The distribution of rates of relaxation ($R_j = 1/T_j$) is the ILT of the magnetization decay.

2.2.2. Data acquisition

The acquisition schemes for relaxation are the saturation recovery or the inversion recovery schemes for T_1. Usually, the inversion recovery scheme is preferred for sensitivity reasons and for its robustness versus imperfections of pulses and radiofrequency inhomogeneity. However, for applying an ILT, the data set has to be reorganized to provide a decay of the magnetization. This leads to assuming that the infinite time has been reached in the experiment, a condition not always fulfilled, and that magnetization is initially properly inverted. The latter condition has received different types of solution with a quite robust one using an inversion composite pulse. But to allow a more direct processing of the ILT of T_1 measurement, a progressive saturation provides immediately a magnetization decay of the proper form.

T_2 measurements have been considered by using Hahn echo trains. This corresponds to a direct decay of magnetization and can be readily processed. However, the phase inhomogeneity may cause considerable problems for getting reliable T_2. Usually, restricting the sample to the coil region of good homogeneity is sufficient. This reduces the sensitivity and should be kept in mind.

Considering $T_{1\rho}$, two cases are commonly encountered, single nucleus measurement or indirect detection by double irradiation. In indirect detection with cross-polarization experiments between ^{1}H and ^{13}C usual in solid-state NMR, one should keep in mind that some ^{1}H may not cross-polarize ^{13}C and so a prior direct $T_{1\rho}$ is recommended. Though the indirect detection usually provides more resolution, comparing both directly and indirectly detected $T_{1\rho}$ allows combining complete enumeration of sites and increase in resolution. $T_{1\rho}$ data are usually properly conditioned and do not need to be preprocessed prior to ILT.

2.2.3. Data processing

The requirements to get a properly processed data set follow two directions. First, the data should have been optimally acquired: logarithmic increment for t_1 indirect dimension, optimal signal-to-noise (S/N) ratio for optimal resolution in relaxation rate dimension, and initial magnetization and the last part of relaxation evolution, the zero base line should have been particularly taken care of. Second, all the points risen in the DOSY part for ILT processing also apply to the TOSY case. At last, as the ILT is a linear transform and it applies on a linear combination of signals for the TOSY, ILT of TOSY is quantitative and integration in the indirect dimension is allowed as well as in the direct dimension.

3. Applications to the Analysis of Drug Formulations

3.1. DOSY ^{1}H NMR analysis of ciprofloxacin formulations

Ciprofloxacin (Figure 2) is an essential synthetic antibiotic belonging to the so-called "fluoroquinolones". It has a broad spectrum of antimicrobial activity and remains effective in a wide variety of indications. DOSY ^{1}H NMR spectra of solutions of two formulations of ciprofloxacin tablets, the brand formulation from BAYER (Ciflox®) and a copy from Syria (Sipro 250), along with their corresponding 1D spectra are presented in Figure 3. All the peaks of ciprofloxacin are lined up. Several excipients can be observed depending on the formulation. The two formulations contain the lubricant magnesium stearate (Mg st, △). They also contain a cellulose derivative that is a tablet binder, known to be hypromellose (hydroxypropylmethyl cellulose) for the brand formulation but unknown for the other formulation. Dibutyl phthalate (dbp, •), a plasticizer, and impurity D are detected in the Syrian formulation. The concentration of impurity D in this formulation was determined by comparing the signal areas of its protons 2 and 8 with that of H2 of ciprofloxacin. The level of impurity D reaches 1.2% well beyond the limit given in the ciprofloxacin monograph of the European Pharmacopeia, which authorizes a maximal content of 0.2% for impurity D.[29]

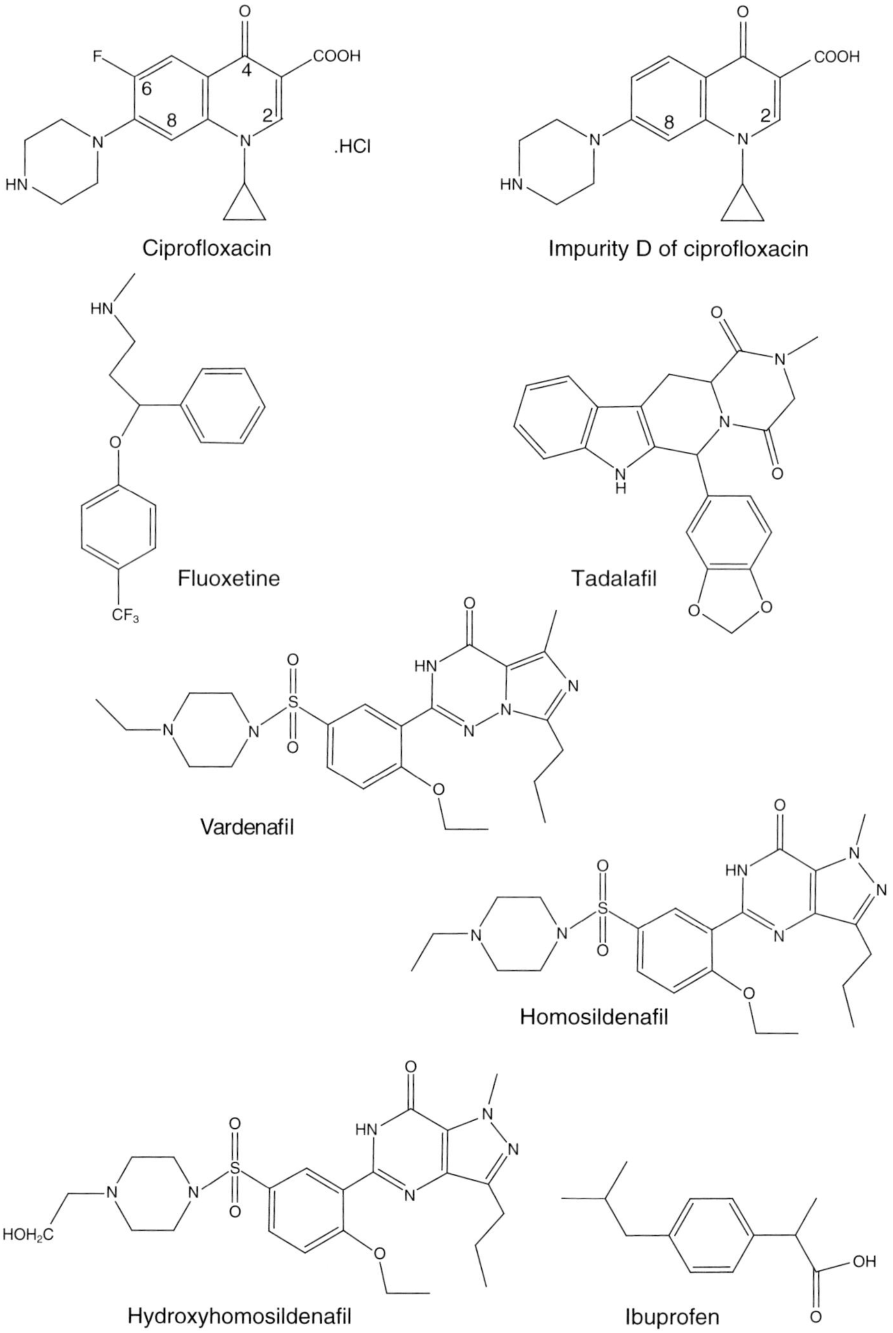

Figure 2 Chemical structures of drugs discussed in this chapter.

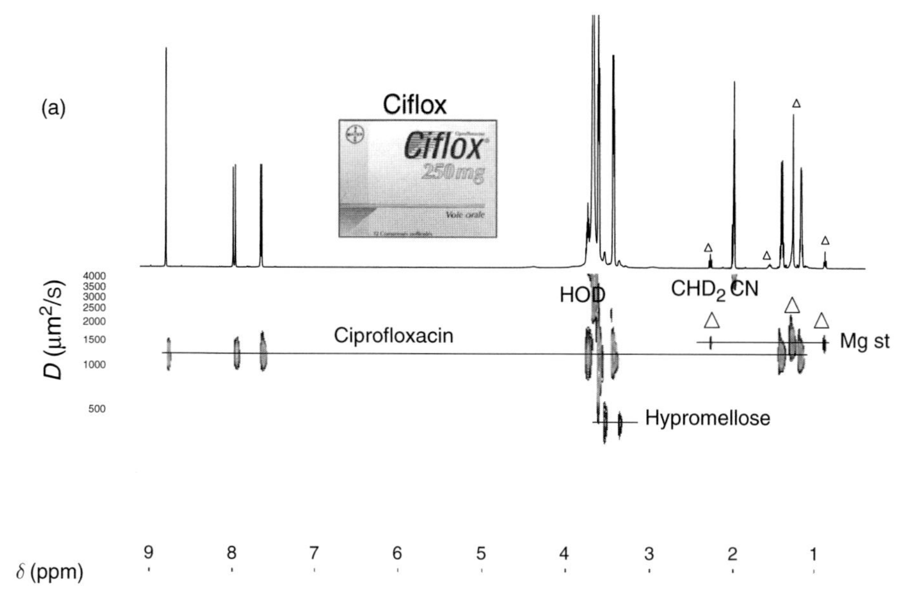

Figure 3 2D DOSY ^{1}H NMR spectra recorded at 500 MHz of commercial formulations of ciprofloxacin (solvent CD_3CN/D_2O, 80/20). (a) the brand formulation from Bayer (Ciflox®); (b) a copy from Syria (Sipro 250). Mg st (△): magnesium stearate; impurity D (□); dbp (•): dibutyl phthalate. Stimulated echo bipolar gradient pulse experiments were used with a pulse delay of 3 ms after each gradient, a pulse field gradient length of 1–1.7 ms and a diffusion delay of 80–120 ms. These sequence parameters were adapted to have the intensity of one NMR signal of the active pharmaceutical ingredient strongly decreased (at least divided by 50) at 95% of the full gradient strength. Forty experiments were recorded with gradient intensity linearly sampled from 5% to 95%. The gradient system had been calibrated to 46.25 G/cm at maximum intensity. All data were processed using Gifa 5.2 software with the ILT method using the MaxEnt algorithm. The processing parameters were 2048 points along the Laplace spectrum diffusion axis and 20,000 MaxEnt iterations. The ILT was computed only on the columns presenting a signal 32 times greater than the noise level of the experiment. DOSY spectra are presented with chemical shifts on the horizontal axis and diffusion coefficients expressed in $\mu m^2/s$ on the vertical axis.

3.2. DOSY ^{1}H NMR analysis of oral formulations of fluoxetine

Fluoxetine (FLX, see Figure 2) is belonging to a new generation of antidepressant drugs, the selective serotonin reuptake inhibitors (SSRIs), which are commercialized since the early 1990s. FLX commercially known as Prozac® is one of the most widely used SSRIs in therapy and is often the drug of choice in the treatment of severe depressive disorder.

Two oral solutions of FLX analyzed by DOSY ^{1}H NMR present different spectral patterns (Figure 4). The DOSY spectrum of the brand formulation Prozac® (Eli Lilly) shows the signals of the active ingredient FLX and the following excipients: glycerol, saccharose (sacc), glucose (gluc), ethanol (eth; from mint aroma), benzoic acid (benz), and menthol. The formulation from Arrow has a different spectral signature; it contains FLX, macrogol to adjust the viscosity, two sweetening agents sodium cyclamate (cy) and sodium saccharine (s), benzoic acid

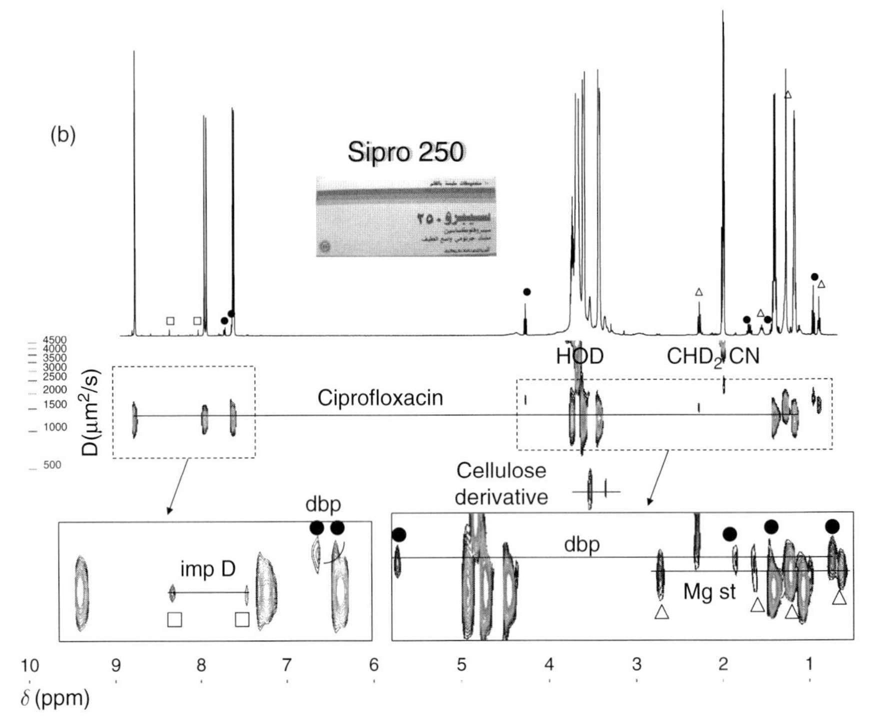

Figure 3 *(Continued)*

(benz), menthol, and citric acid (citric). In these two DOSY spectra, the self-diffusion coefficients of the various compounds are very low (<1000 μm^2/s) due to the high viscosity of the solutions.

3.3. DOSY ^{1}H NMR analysis of formulations of genuine Cialis® and a Chinese imitation

Nowadays, drugs for the treatment of erectile dysfunction that inhibit the phosphodiesterase type 5 enzyme (PDE-5), sildenafil citrate (Viagra®), tadalafil (Cialis®) or vardenafil hydrochloride (Levitra®, see Figure 2) are among the most counterfeited or imitated. Imitations come generally from Asia (India and China most often) which do not recognize the European and American patent laws so that products manufactured legally in such countries are illegal in Europe, USA and other countries. The original Cialis® tablet from Eli Lilly Laboratories and one copy commercialized in China were analyzed. The DOSY ^{1}H NMR spectra of their solutions are presented in Figure 5. It is obvious that the two formulations do not contain the same active ingredient. Their common excipients are the lubricant magnesium stearate (Mg st, △), the tablet binder hypromellose (#) and lactose (□) as a diluent. Moreover, the brand formulation is composed of another cellulose derivative with a higher molecular weight known to

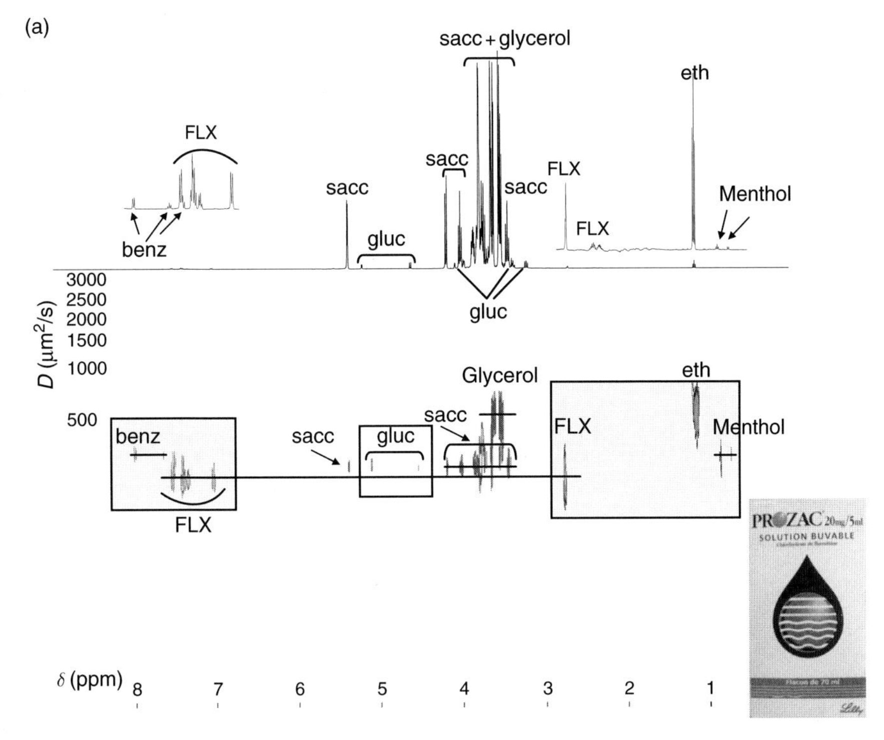

Figure 4 2D DOSY ^{1}H NMR spectra recorded at 500 MHz of liquid formulations of fluoxetine. (a) the brand formulation from Eli Lilly (Prozac®); (b) the generic formulation from Arrow. FLX, fluoxetine; benz, benzoic acid; sacc, saccharose; gluc, glucose; eth, ethanol; s, sodium saccharine; cy, sodium cyclamate; citric, citric acid. A deeper section of some signals is shown in the boxes. For the recording and processing conditions see legend of Figure 3.

be hydroxypropylcellulose (■). Signals of sodium lauryl sulphate (laur; ▲), a wetting agent, and triacetin (♦), a hydrophilic plasticizer, are also present.

The intensity of the peaks located in the aromatic region of the ^{1}H NMR spectrum demonstrates that the Chinese fake formulation of Cialis contains a mixture of two active pharmaceutical ingredients. These two compounds were purified by chromatography and then identified by NMR and LC–MS–MS (liquid chromatography–mass spectrometry–mass spectrometry) as vardenafil (V) and homosildenafil (H) (Figure 5). Their amounts were 34.3 ± 0.2 and 7.2 ± 0.1 mg per tablet, respectively.

Similar to tadalafil, vardenafil is a PDE-5 inhibitor, and indications and contraindications for this medicine are close to those of other PDE-5 inhibitors. However, the structural differences between tadalafil and vardenafil are reflected in dissimilarities of the clinical pharmacokinetics and pharmacodynamics.[30] The brand formulation of vardenafil (Levitra®) is available in 2.5, 5, 10, and 20 mg doses and the normal starting dose is 10 mg. The level of vardenafil found in the Chinese formulation might induce an overdosage in patients. In the case of vardenafil, there was no evidence of a dose-dependent improvement in efficacy beyond the 10 mg dose but problems of drug–drug interactions may occur. For example, vardenafil is contraindicated with nitrates as

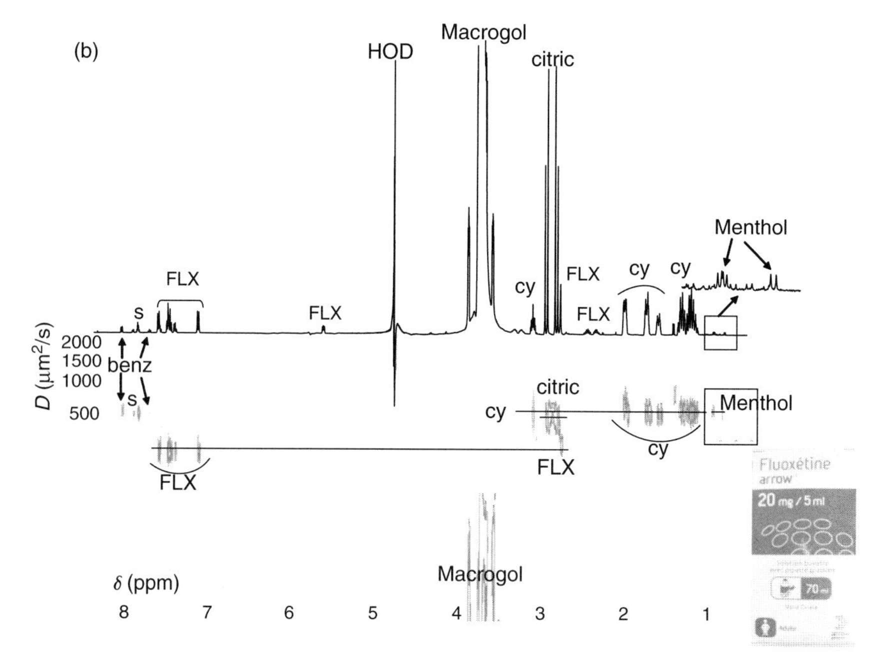

Figure 4 *(Continued)*

it potentiates their hypotensive effect. Also vardenafil (10 or 20 mg) when concomitantly administered with α-blockers (such as terazosin and tamsulosin) and with nifedidine to healthy volunteers resulted in some subjects experiencing hypotension.[30] The presence of homosildenafil (see Figure 2) has already been described in functional food marketed for penile erectile dysfunction[31] and in dietary supplements[32] but toxicological data are not known or not available. Thus, due to possible side effects or drug–drug interactions, a significant risk is faced by consumers who purchase drugs marketed for erectile dysfunction via the Internet.

3.4. DOSY ^{1}H NMR analysis of an herbal Chinese formulation

Worldwide traditional herbal medicines are gaining popularity as a source of complementary and alternative remedies. In contrast to conventional pharmaceuticals, herbal medicines are generally presumed as safe, harmless, and without side effects, because of their natural origin. The introduction of PDE-5 drugs has led to the promotion of herbal products as substitutes in enhancing male sexual function, often under the premise that natural products are safer. While there are many such herbal dietary supplements sold at local stores or over the Internet which claim to enhance sexual performance, there have been reports of adulteration of some "natural" products with known synthetic PDE-5 inhibitors.[33–37] It is clear that the adulteration of these products with undeclared, synthetic pharmaceuticals poses a serious health risk.

The DOSY ^{1}H NMR spectrum of a solution of an herbal drug commercialized in China as a capsule formulation called "Chui Hua San" (Junen Company, Taiwan) is shown in Figure 6. The DOSY spectrum shows the presence of the lubricant

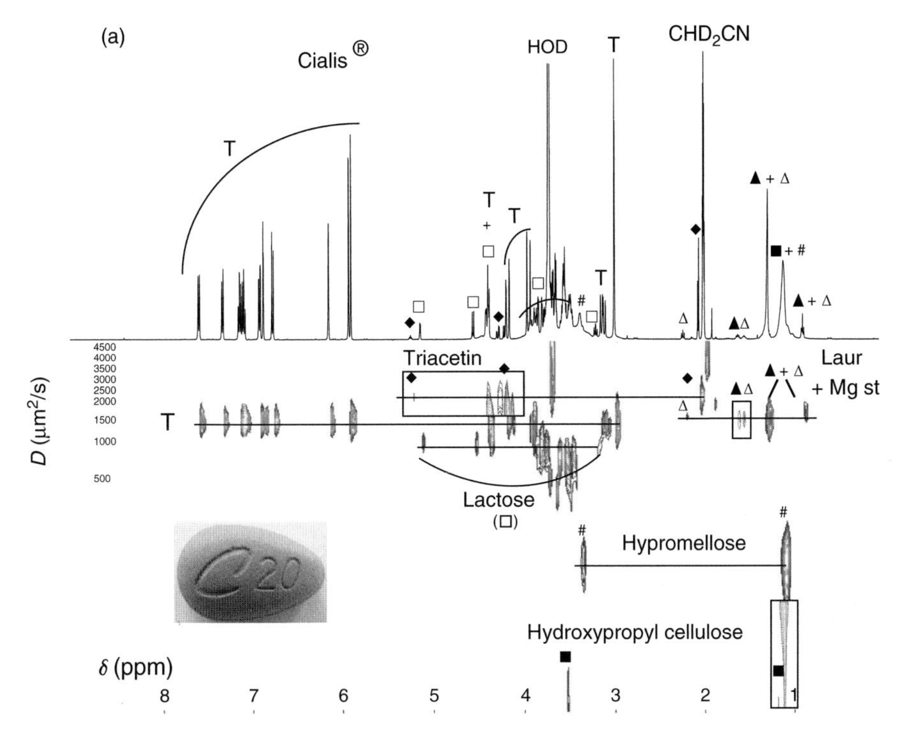

Figure 5 2D DOSY ^{1}H NMR spectra recorded at 500 MHz of Cialis formulations (solvent CD_3CN/D_2O, 80/20). (a) the brand formulation from Eli Lilly (Cialis®); (b) Chinese Cialis. T, tadalafil; V, vardenafil; H, homosildenafil; ■, hydroxypropylcellulose; #, hypromellose; □, lactose; Mg st (△), magnesium stearate; Laur (▲), sodium lauryl sulphate; ♦, triacetin; ◘, diethylphtalate; PEG, polyethylene glycol; ?, unknown. A deeper section of some signals is shown in the boxes. For the recording and processing conditions, see legend of Figure 3.

magnesium stearate (Mg st, △) and saccharose (sacc). Moreover, it contains two active pharmaceutical ingredients identified as tadalafil (T) and hydroxyhomosildenafil (HS) whose amounts are 31 and 48 mg per tablet, respectively. These values are superior to the normal dose for tadalafil as in the USA and Europe, the recommended dose is 10 mg once daily, the highest dose reaching 20 mg.[30,38] As for the other PDE-5 inhibitors, tadalafil should not be used without a previous approbation of treatment by medical instances due to potential risks when concomitantly administered with some antihypertensive medication (α-blockers). Hydroxyhomosildenafil has already been detected as adulterant in dietary supplements[32] but toxicological data are unknown or not available. The adulteration of "natural" herbal dietary supplements with erectile dysfunction drugs or analogues is a growing trend and poses a health threat to patients who unwittingly consume a synthetic drug that has been untested for safety and efficacy.

3.5. TOSY NMR analysis of ibuprofen

Tests have been run on different galenic formulations of ibuprofen (see Figure 2). The experiments shown in Figure 7 consist in a ^{1}H $T_{1\rho}$ measurement indirectly

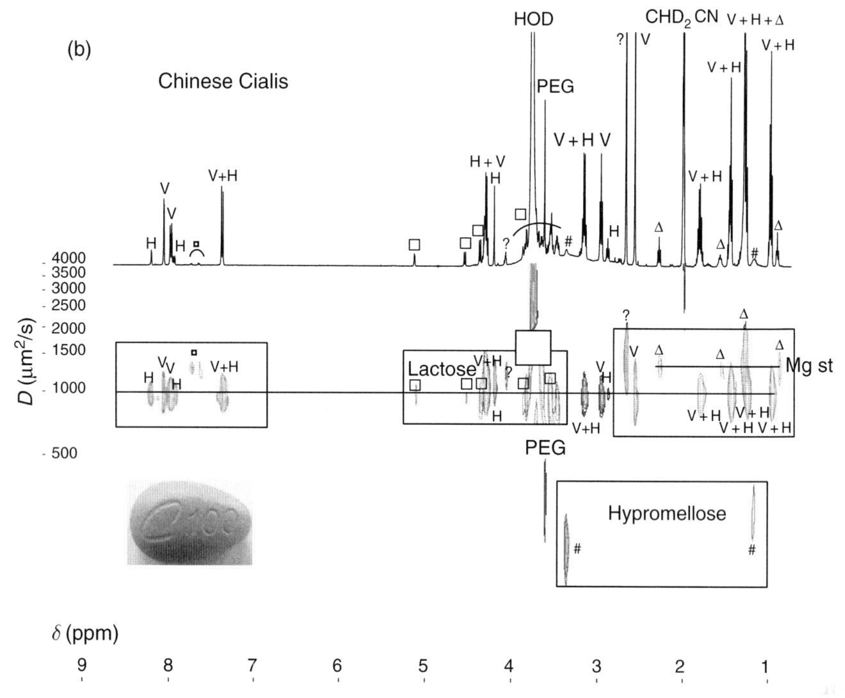

Figure 5 *(Continued)*

detected on ^{13}C. All ^{1}H are cross-polarizing in this situation. Though the crystalline state of ibuprofen is identical, the drug is well crystallized, the ^{13}C chemical shifts and the line widths are identical, one observes large differences in the indirect relaxation dimension as well as in the direct dimension for which variations in the composition of excipients appear. From the pure standard polycrystalline material to the three different classical tablets, one can immediately fingerprint the different origins of the formulated tablets. The relaxation dimension indicates most of the time a two relaxation rates behavior, one due to the core of ibuprofen grains almost independent of the galenic formulation, the other being very sensitive to the nature of the excipients or to the manufacturing process of the tablets. This is the basis for the recognitions of the different formulations. With such an experiment, reaching extremely reproducible formulations is accessible at almost no additional costs.

4. Advantages and Drawbacks of the Methods

DOSY and TOSY are two methods permitting correlations of microscopic chemical site identification with a physical aspect of the concerned chemical site. This correlation allows other correlations like the same indirectly sensed diffusion for a molecule in solution, or correlated relaxation rates for sites affected by the same fluctuations. The picture of the material provided by such techniques is characteristic

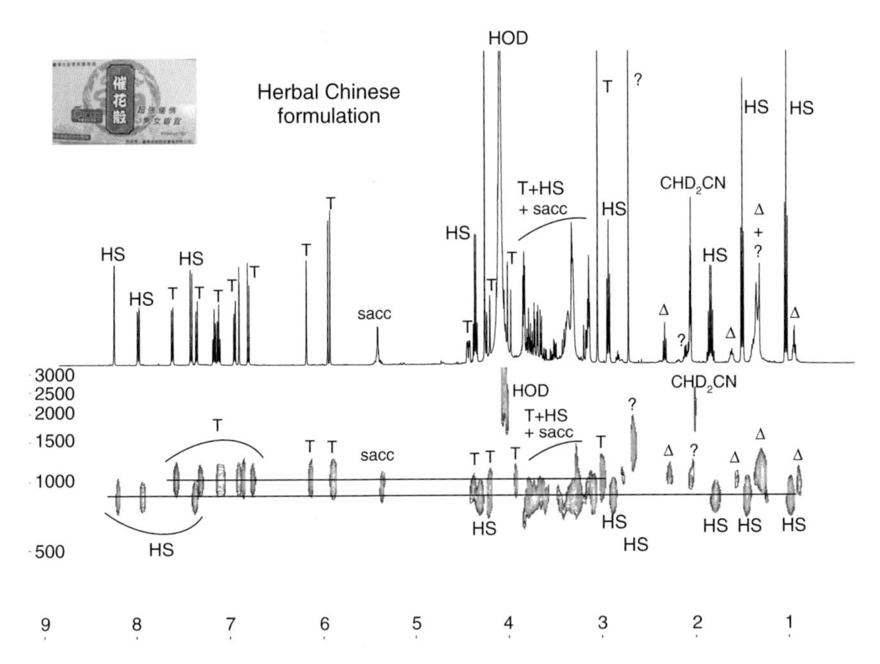

Figure 6 2D DOSY ^{1}H NMR spectrum recorded at 500 MHz of an herbal drug commercialized in China as a capsule formulation called "Chui Hua San" (Junen Company, Taiwan) (solvent CD_3CN/D_2O, 80/20). T, tadalafil; HS, hydroxyhomosildenafil; Mg st (△), magnesium stearate; sacc, saccharose; ?, unknown. For the recording and processing conditions, see legend of Fig. 3.

of a molecule and its surrounding environment. DOSY and TOSY characterize the properties of the materials separating chemical identification and the physical one without performing a mixture separation. This allows complex mixture characterization where not only the components are analyzed and separated but their interactions manifest themselves into the diffusion or relaxation parameters.

DOSY NMR spectra clearly show similarities and differences in the composition of pharmaceutical preparations, thus giving a chemical fingerprint of the formulation and a signature of the manufacturer. This spectral signature includes not only the active pharmaceutical ingredient(s) but also the excipients. The analytical methods that provide most information on excipient composition are solid-state methods such as Raman spectroscopy,[39–41] near-infrared spectroscopy,[42,43] or X-ray diffraction,[44] whereas chromatography allows an accurate identification of active pharmaceutical ingredients but does not give information on excipients. DOSY NMR is thus a holistic method that permits to consider the drug preparation as a whole.

Set against these advantages, however, are a number of disadvantages that need to be taken into account. There are several consequences of the acquisition schemes that may limit the use of both techniques. As presented here, the DOSY and TOSY spectroscopies both rely on the ILT implemented with the help of the MaxEnt approach. This approach presents some characteristics which should be detailed here. Whereas the implementation is iterative, it is an integral transform approach and as such ensures that

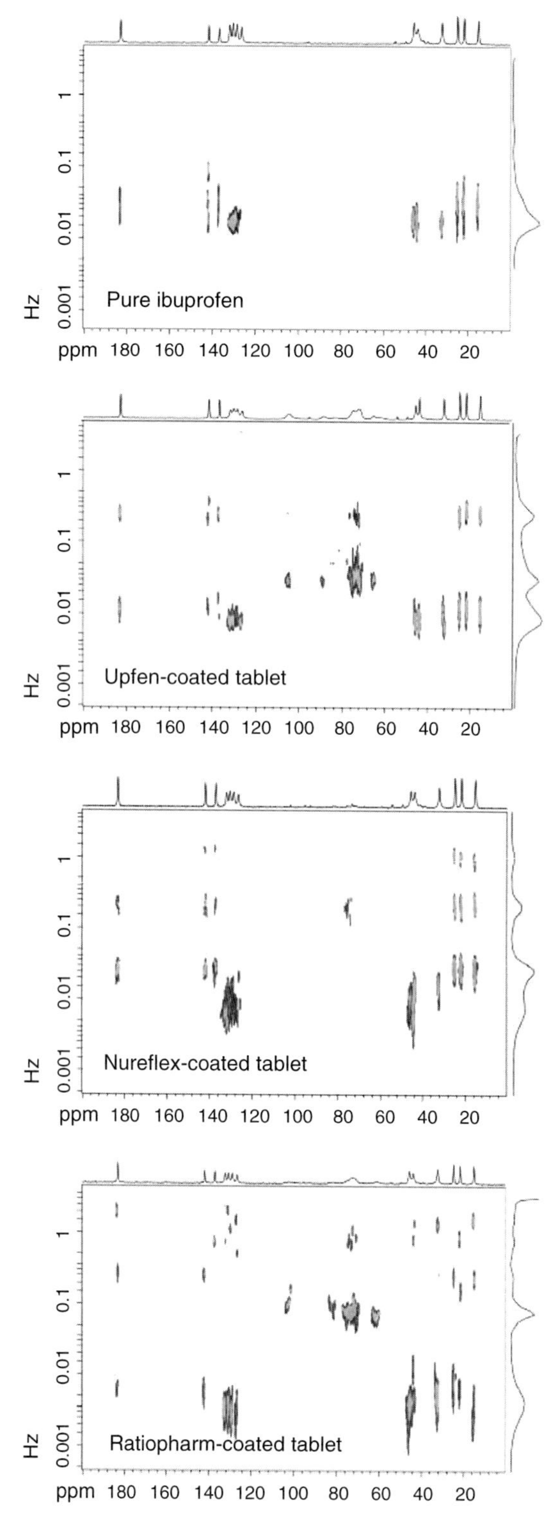

Figure 7 TOSY-$T_{1\rho}$ of (from top to bottom) pure standard ibuprofen, Upfen®-coated tablet, Nureflex®-coated tablet and Ratiopharm®-coated tablet.

the result is independent of any data model provided by the operator. A relative robustness against acquisition artifacts is thus assured. Thanks to this robustness, a feature observed in the ILT spectrum can reliably be associated to an actual feature in the measurement. This contrasts with the more classical modeling approaches which should thus be avoided or carefully controlled (see, for instance, Nilsson et al.[7]).

One drawback comes from the tendency to underweight or to shift small components relative to bigger ones;[45] this is the result of the relative weighting of the entropy and the χ^2 statistics performed during iterations. In extreme cases, this effect may completely wipe out a weak component hidden under a larger one. However, this will have an adversary effect only for species which appears at the same chemical shift, and will have no effect for isolated lines as each column of the spectra is computed in a locally optimized fashion. In this case, a weak compound, presenting lines isolated as well as hidden ones, will still be detected (see, for instance, aromatic lines of the dbp additive in Figure 3(b)).

Finally, whereas the Laplace transform (Eq. (2)) on which the DOSY and the TOSY experiments are based is intrinsically linear, and as such quantification of the spectra could be thought as straightforward, the inversion operation is not, and quantification should be very carefully performed in the case of large dynamic spectral superposition as mentioned above. For this reason, it is advised if possible not to realize quantifications directly on the ILT spectra.

The DOSY experiment itself is a very sensitive spectroscopy, and weak lines observed in the 1D spectrum will be readily analyzed by ILT for S/N on the order of 20–50 and above. Reliable spectra will easily be obtained as long as a few points are carefully taken care of (see Section 2 above). The linearity of the PFG system, both spatially and electronically, should have been checked on the system, and possible convection should particularly be scrutinized especially in low-viscosity solvent such as $CDCl_3$ as it will ruin the measurement if no counter measure is used.

For TOSY, it should be reminded that the accuracy of the determination of a relaxation parameter lies on the S/N ratio of the data sets analyzed. This point is usually underestimated in the literature. But to make it clear, a simple picture would suffice. To analyze a magnetization decay, for instance a one single exponential decay, by least-squares modeling or by ILT, if the data set is very noisy, one has an immediate perception of the fact that many relaxation rates will go through the noisy decay. As this implies to get reasonably high S/N ratio (a minimum of $\approx$50 for ILT-MaxEnt and generally above 200–300 for ILT without MaxEnt), TOSY can be considered as not being a very sensitive experiment. As a counterpart, it is quite reproducible due to the transform properties of ILT, and almost completely independent of the operator, which is far from being the case for relaxation modeling.

5. Conclusion

Nowadays, the drug market is subjected to profound transformations partly due to its globalization. The first one is the development of generic drugs that allow access to affordable treatment for people living in poor countries. Also, in wealthier countries, driven by healthcare insurance systems, generic drugs are substituted for more expensive brand-named drugs. Secondly, the Internet has revolutionized the way in which patients purchase medications. When prices of medicines are high and price differentials between identical products exist, there is a greater incentive for the consumer to seek medicines outside the normal supply system. Moreover, many consumers believe that online pharmacies are more convenient than traditional pharmacies and offer cost savings. However, if one buys a drug via the Internet, he never knows if the product has been manufactured in accordance with good manufacturing practices, contains the appropriate active ingredient, is free of potentially harmful contaminants or was properly stored and shipped. Thirdly, counterfeiting of medicines is rife as it is a hugely lucrative business due to high demand and low production costs. Finally, the growing development of alternative medicines leads to an increase in the trade of herbal products whose use and production remain largely unregulated.

Ensuring the quality of medicines becomes thus more important and challenging. This means that the pharmaceutical analyst needs more techniques to fulfill his duty. In addition to traditional techniques, DOSY and TOSY could now be part of the "analytical toolbox" of any spectroscopist dealing with complex mixture analyses in pharmaceuticals, especially analysts investigating quality control, counterfeit, and smuggling. Moreover, these techniques should be helpful in determining the relationships between different samples and so assist in the investigation of the sources of these drugs.

REFERENCES

1. E.O. Stejskal, J.E. Tanner, J. Chem. Phys. 42 (1965) 288–292.
2. K.F. Morris, C.S. Johnson, J. Am. Chem. Soc. 114 (1992) 3139–3141.
3. C.S. Johnson, Prog. Nucl. Magn. Reson. Spectrosc. 34 (1999) 203–256.
4. B. Antalek, Concepts Magn. Reson. 14 (2002) 225–258.
5. K.I. Momot, P.W. Kuchel, Concepts Magn. Reson. A 28A (2006) 249–269.
6. J.E. Tanner, J. Chem. Phys, 52 (1970) 2523–2526.
7. M. Nilsson, M.A. Connell, A.L. Davis, G.A. Morris, Anal. Chem. 78 (2006) 3040–3045.
8. D.H. Wu, A.D. Chen, C.S. Johnson, J. Magn. Reson. A 115 (1995) 260–264.
9. M.D. Pelta, G.A. Morris, M.J. Stchedroff, S.J. Hammond, Magn. Reson. Chem. 40 (2002) S147–S152.
10. A. Jerschow, N. Müller, J. Magn. Reson. 125 (1997) 372–375.
11. S.W. Provencher, R.H. Vogel, P. Deuflhard, E. Hirer, Numerical treatmet of inverse problems in differential and integral equations, pp. 304–319. Birkhäuser, Boston, 1983.
12. W. Windig, B. Antalek, Chemom. Intell. Lab. Syst. 37 (1997) 241–254.
13. L.C.M. Van Gorkom, T.M. Hancewicz, J. Magn. Reson. 130 (1998) 125–130.
14. R. Huo, C. Geurts, J. Brands, R. Wehrens, L.M.C. Buydens, Magn. Reson. Chem. 44 (2006) 110–117.
15. S.W. Provencher, Comput. Phys. Commun. 27 (1982) 213–227.
16. S.W. Provencher, Comput. Phys. Commun. 27 (1982) 229–242.
17. P.T. Callaghan, S. Godefroy, B.N. Ryland, J. Magn. Reson. 162 (2003) 320–327.
18. M.A. Delsuc, T.E. Malliavin, Anal. Chem. 70 (1998) 2146–2148.
19. G.S. Armstrong, N.M. Loening, J.E. Curtis, A.J. Shaka, V.A. Mandelshtam, J. Magn. Reson. 163 (2003) 139–148.
20. R. Huo, R. Wehrens, L.M.C. Buydens, J. Magn. Reson. 169 (2004) 257–269.

21. H. Barjat, G.A. Morris, S. Smart, A.G. Swanson, S.C.R. Williams, J. Magn. Reson. B, 108 (1995) 170–172.
22. D. Tramesel, V. Catherinot, M.A. Delsuc, J. Magn. Reson. 188 (2007) 56–67.
23. F. Bloch, Phys. Rev. 70 (1946) 460–474.
24. N. Bloembergen, E.M. Purcell, R.V. Pound, Phys. Rev. 73 (1948) 679–712.
25. M. Goldman, Quantum description of high-resolution NMR in liquids, Oxford University Press, Oxford, 1991.
26. A.G. Redfield, Phys. Rev. 98 (1955) 1787–1809.
27. A.G. Redfield, Adv. Magn. Reson. 1 (1966) 1–32.
28. G. Lipari, A. Szabo, Biochemistry, 20 (1981) 6250–6256.
29. European Pharmacopeia 5th ed (2005) 1300–1301.
30. N. Mehrotra, M. Gupta, A. Kovar, B. Meibohm, Int. J. Impot. Res. 19 (2007) 253–264.
31. M.H. Shin, M.K. Hong, W.S. Kim, Y.J. Lee, Y.C. Jeoung, Food Addit. Contam. 20 (2003) 793–796.
32. P. Zou, S.S.Y. Oh, P. Hou, M.Y. Low, H.L. Koh, J. Chromatogr. A 1104 (2006) 113–122.
33. J.C. Reepmeyer, J.T. Woodruff, J. Chromatogr. A, 1125 (2006) 67–75.
34. M.E. Abdel-Hamid, J. Liq. Chromatogr. Relat. Technol. 29 (2006) 591–603.
35. S.R. Gratz, C.L. Flurer, K.A. Wolnik, J. Pharm. Biomed. Anal. 36 (2004) 525–533.
36. Q. Liang, J. Qu, G. Luo, Y. Wang, J. Pharm. Biomed. Anal. 40 (2006) 305–311.
37. J.C. Reepmeyer, J.T. Woodruff, J. Pharm. Biomed. Anal. 44 (2007) 887–893.
38. D.O. Sussman, J. Am. Osteopath. Assoc. 104 (2004) S11–S15.
39. M. de Veij, P. Vandenabeele, K.A. Hall, F.M. Fernandez, M.D. Green, N.J. White, A.M. Dondorp, P.N. Newton, L. Moens, J. Raman Spectrosc. 38 (2007) 181–187.
40. K.A. Hall, P.N. Newton, M.D. Green, M. de Veij, P. Vandenabeele, D. Pizzanelli, M. Mayxay, A. Dondorp, F.M. Fernandez, Am. J. Trop. Med. Hyg. 75 (2006) 804–811.
41. M.R. Witkowski, Am. Pharm. Rev. 8 (2005) 56–60.
42. O.Y. Rodionova, L.P. Houmoller, A.L. Pomerantsev, P. Geladi, J. Burger, V.L. Dorofeyev, A.P. Arzamastsev, Anal. Chim. Acta, 549 (2005) 151–158.
43. M.J. Vredenbregt, L. Blok-Tip, R. Hoogerbrugge, D.M. Barends, D. de Kaste, J. Pharm. Biomed. Anal. 40 (2006) 840–849.
44. J.K. Maurin, F. Plucinski, A.P. Mazurek, Z. Fijalek, J. Pharm. Biomed. Anal. 43 (2007) 1514–1518.
45. T. Gostan, Diversification des champs d'application de la mesure de diffusion par RMN, Thèse de l'Université de Montpellier I, 2004.

CHAPTER 7

THE USE OF qNMR FOR THE ANALYSIS OF AGROCHEMICALS

R.J. Wells, J. Cheung, *and* (in part) J.M. Hook

Contents

Abstract

This review evaluates the chemical analysis of agrochemicals using quantitative nuclear magnetic resonance (qNMR) techniques. Experimental approaches to the quantification of both technical grade agrochemicals and impurities therein by qNMR using ^{1}H, ^{31}P or ^{19}F NMR data are discussed.

As a primary analytical method, qNMR permits the use of a standard reference material that is not directly related to the target analyte. Dimethylsulphone is a simple substance that fulfils all criteria necessary as a universal reference material to which other organic substances can be compared by qNMR analysis. Validated methods for the analysis of about 100 agrochemicals have been developed using a range of certified standard reference materials that have been certified relative to dimethylsulphone.

Keywords: reference materials, subtraction method, gentamicin, malathion, chlorpyriphos, glyphosate, ethephon, tebuconazol

1. Introduction

Most recent advances in the analysis of organic materials have involved improvements in selectivity and sensitivity at the trace level, thereby lowering the limits at which target analytes may be routinely and unambiguously quantified into the low nanogram and picogram range. Less progress is apparent in improvements in the quantification in the range 5–100%, that is required for the analysis of, for example, technical grade chemicals and formulations in the agrochemical industry.

Many of the validated, wet chemical methods currently available appear on closer examination to need improvement, while chromatographic techniques engender their own limitations. However, this is precisely the domain in which nuclear magnetic resonance (NMR) spectroscopy has come to the fore and excelled, because it may provide the analyst with both information-rich qualitative (structural) data as well as quantitative assessment of the analyte in a single measurement, and which is directly applicable to one of the hundreds of compounds registered for use in crop control and related activities.

This chapter discusses the application of NMR, particularly quantitative NMR (qNMR) for the analysis of agrochemicals, that is accurate, effective and straightforward, often using certified standard reference materials unrelated to the target analyte. This is not an exhaustive review of the topic but rather discusses some of the principles, comparisons and the practical uses of qNMR in the routine analysis of registration batches and quality control samples of agrochemicals. For the most part, agrochemicals can be readily analysed by ^{1}H qNMR, but there are specific cases in which the use other nuclei, such as ^{19}F and ^{31}P, are preferred for quantification.

2. Limitations of Present Methods for Analysis of Technical Grade Materials

Unlike qNMR, most other methods currently used for agrochemical analysis rely almost exclusively on the availability and use of a highly pure and certified sample of the chemical being analysed against which it can be compared or against which the detector used is calibrated.

2.1. Purity and supply of standard reference materials

When technical grade chemicals are analysed by chromatographic methods, the precisely known purity of certified standard reference material used is of critical importance. Often, technical grade agrochemicals are registered at levels of purity in the 95–99% range and a standard reference material that has a purity that deviates from its certified value by 0.5% has a profound influence of the results of analysis at such high levels of purity. Very few of the available certified reference materials are accompanied by uncertainty budgets and most standards that are supplied are not traceable to a substance that has been subjected to a primary analytical method such as titrimetry, gravimetry or differential scanning calorimetry or a combination of these. Furthermore, standard reference materials are usually certified only to $\pm 0.5\%$ or $\pm 1\%$. Therefore, it follows that the uncertainty of measurement must be greater than the uncertainty of the reference standard. This fact is seldom reflected in the relative standard deviation (RSD) quoted for any secondary chromatographic method using a certified reference material because the uncertainty about its purity is not included in statistical treatments.

There are many different suppliers of reference materials used as standards, which are undoubtedly acquired from a common source. Whether the certificate of analysis they provide with the standard is generated in-house or was provided by the source manufacturer is seldom made clear. Thus, there is as yet no commonly accepted method by which the same substance originating from different suppliers can be adequately compared to a single "primary" standard of that substance which has been independently certified by a specific primary analytical method.

In the vast majority of cases, the purity of a particular batch is *inferred* by lack of visible impurities using techniques such as chromatography and/or NMR spectroscopy. While these techniques may be generally acceptable for most simple substances, it becomes problematical for more complex substances, particularly those of biological origin,[1] where significant differences have been found between the purity claimed for a substance and that found by NMR measurements, such as found for the aminoglycoside antibiotics (see below).

Compounding these difficulties is the fact that many organic standard reference materials are often not stable over prolonged periods, particularly in solution, and must be discarded after a finite storage period. Furthermore, there is often no available method to link one batch of a particular standard with another from the same or a different supplier.

Finally, we have found that some agrochemical standards supplied by companies specialising in this area have not conformed to the Certificate of Analysis. Thus, the purity of metham sodium (1) from two separate suppliers was totally unacceptable. One, certified 95% pure as the trihydrate, was found to be less than 60% pure by NMR, while the other was totally insoluble in water.[2] Similarly, samples of clodinafop-propargyl (2) and haloxyfop-*R*-methyl (3), both chiral agrochemicals, were supplied without any information on optical purity and the supplier was unable to provide that information on request.[3]

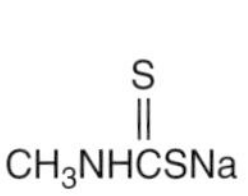

S
||
$CH_3NHCSNa$

(1)

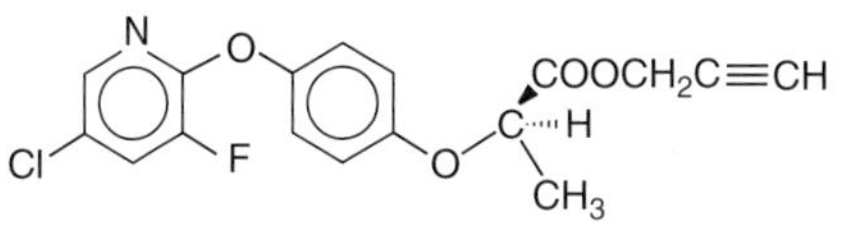

N O Cl F O $COOCH_2C{\equiv}CH$ C H CH_3

(2)

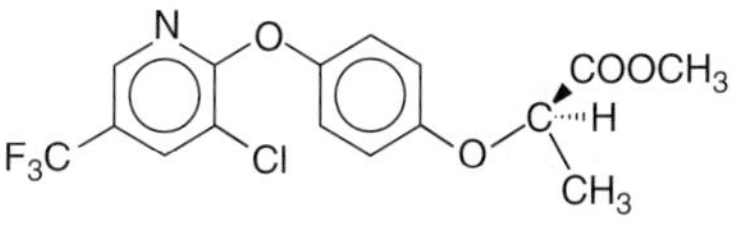

N O F_3C Cl O $COOCH_3$ C H CH_3

(3)

2.2. Limitations of chromatographic methods

Not withstanding the requirements for reference standards, sample preparation for chromatographic methods is often extensive. Even in the simplest case, a sample and an internal standard are weighed, followed by a series of dilutions to bring the analytical solution into the linear range of the detector or to accommodate the capacity factor of the separation column. Each chromatographic method will require a separate standard of the analyte and, if an internal standard is used, the response factor between the analyte and the internal standard will have to be determined. Many more steps are involved, all with attendant errors. As will be discussed later, this is a lesser problem for qNMR.

It is evident that ^{1}H NMR and chromatographic techniques differ greatly as quantitative methods. The quality of analytical data gathered by high-performance liquid chromatography (HPLC) and gas chromatography (GC) is more dependent on the physical properties of a pure substance than on its structural complexity. As long as the target analyte has good chromatographic properties and responds effectively to the detector, it matters little if the substance has a molecular weight of 80 or 1500. However, polar, low boiling and/or low-molecular-weight substances often pose significant analytical problems. They may be too volatile, they may lack convenient chromophores or be too polar to have favourable chromatographic properties. Such substances are not straightforward to analyse and, when accompanied by low-medium levels of impurities with similar properties, the quantitative analysis of the mixture becomes a significant challenge. By contrast, NMR spectra of low-molecular-weight or highly polar substances are usually simple and the detection and quantification of minor amounts of impurities are readily accomplished as with dalapon sodium.

2.3. Shortcomings of some official agrochemical analytical methods

There are practical problems with many agrochemical analytical methods over and above the problems associated with standard reference materials. A compendium of validated methods appears in the *Handbook of the Collaborative International Analytical Council* (CIPAC).[4] Many of these methods are very old and some are non-specific, such as methods for ethephon and metam-sodium, which are discussed in more detail later. Some of the older methods are wet chemical methods that are difficult to perform in a modern chemical laboratory which is heavily reliant on instrumental analysis. Also, independent corroboration of the purity of some substances, such as dithiocarbamates, determined by release of carbon disulphide, is difficult because of the cost and unreliability of the standards and the unavailability of alternative methods. Thus, in such cases where independent corroboration of the result of a method is impossible, the resultant purity is defined by the method.

3. Analysis of the Active Ingredient in Agrochemicals by qNMR

3.1. General considerations

^{1}H, ^{31}P and ^{13}C qNMR has been used sporadically for many years and is only recently gaining widespread acceptance as a reliable quantitative technique. The first complete critical evaluation of qNMR was published in 1998 in a benchmark paper by Maniara and co-workers.[5] ^{1}H and ^{31}P qNMR analysis of agrochemicals were used to systematically test experimental precision, accuracy, specificity, linearity, limits of detection and quantification as well as ruggedness. Standard reference materials used were National Institute of Standards and Technology (NIST)-certified benzoic acid and triphenyl phosphate and Cyanamid-certified, tris(hydroxymethylamino)methane.

Conclusions about these important analytical features and individual experimental NMR parameters are summarised below.

Precision: pooled RSD was calculated to be 0.25% for ^{1}H test substances and 0.53% for ^{31}P test substances, precision values that were comparable to those obtained for chromatographic methods.

Accuracy: determination of benzoic acid purity by ^{1}H NMR gave a value of 99.9% compared to 99.9958% by titration, giving a bias of 0.0958 and an accuracy of the qNMR method of 0.2. For ^{31}P NMR, the bias between NMR and gas chromatography was 0.3% and the accuracy of the NMR method was 0.7%.

Linearity: coefficient of determination for the linearity of detector response was $\geq$0.995, indicative of excellent linearity of detector response.

Specificity: specificity was found to be within the statistical variations expected. It should be noted here that determination of a certified test material of nicotine by both chromatographic and NMR methods showed the certified purity to be overstated by at least 3%.

Limits of detection and quantification: these are not applicable to major component analyses. However, in 500 MHz ^{1}H measurements, the LOQ was acceptable at concentrations greater than 10 mM while the LOD was three to five times lower.

Ruggedness: results obtained were not found to be significantly operator or instrument dependent.

The authors concluded that qNMR was suitable and valid for the characterisation of analytical standards and other chemical materials used in regulated industries such as agrochemicals and pharmaceuticals. They also pointed out that the qNMR method was equally applicable to both the analysis of active ingredients and of impurities above 0.1%.

In another key paper, Griffiths and Irving[6] also addressed the question of accuracy attainable from qNMR measurements. They came to the conclusion that qNMR is capable of quantification limits of $\pm$1%, markedly better than the limits generally accepted at the time of publication (1998), by employing the following procedure:

(a) Select a standard with as few resonances as possible.
(b) Weigh approximately 20 mg of sample and standard and ensure complete dissolution in approximately 1 ml of solution; vary the weight ratio between sample and standard in proportion to molecular mass if divergent.

(c) Select the highest possible magnetic field for specificity and sensitivity.
(d) Employ an ADC (analogue-to-digital converter) ≥16 bits.
(e) Maximise the RF power to the practical limit.
(f) Employ a tip angle of 45° if the sensitivity is adequate.
(g) Set the carrier frequency mid-way between the sample and standard resonances.
(h) Employ a repetition time of ≥60 s and acquire as many pulses as practical.
(i) Zero fill by a factor of two.
(j) Maximise vertical expansion during manual phase set.
(k) Integrate just inside ^{13}C satellites.
(l) If possible, employ a phase routine which maximises points used to determine slope and bias correction but which minimises those used to determine area.
(m) Avoid integrating resonances closer than 100 times their respective linewidths; if unavoidable, use a routine which sets the slope and bias tangentially.

Some of these conclusions are questionable and we do not adopt all of these recommendations in our work. For example, when considering conclusions (c) and (d), choosing a higher field magnet will not necessarily give better sensitivity if the probe is not of a modern design and all recent model NMR spectrometers have at least 16 bits ADC. In (f), the authors stipulate a tip angle of 45° but they employed a pulse width of 11.5 μs whereas it is far more usual to employ a pulse of less than 10 μs. A general rule is to use the shortest pulse duration possible at a moderate power level that does not damage the probe. We use a tip angle of 90°. The reason for setting T_1 at 60 s in (h) is not stated but presumably related to the requirement for complete relaxation. It is known that the repetition time should be at least five times the longest T_1 and we routinely adopt a value of 7 × the longest T_1. Also, the statement that accuracy will increase with the number of pulses is not correct. Indeed, the higher the number of pulses the lower the accuracy, even with a delay of >5 × T_1, because the spectrum will never be 100% relaxed. Therefore, with each additional scan the noise is magnified. The zero filling suggested in (i) will only improve the appearance of the peak shapes, it does not improve the accuracy of integration and it is more important to increase the number of data points used. The authors do not follow their own rule by integrating inside ^{13}C satellites. We routinely quantify using ^{13}C GARP decoupled spectra and integrate inside where the ^{13}C signal would have occurred. This procedure avoids errors arising from incomplete ^{13}C decoupling across the whole spectrum whilst allowing observation of any impurity resonances. We do not adopt recommendations (l) and (m), as integrals are not phased and only the baseline of the spectra is zeroed.

3.2. qNMR using the subtraction method

In the "subtraction" method, all organic impurities are quantified by NMR (relative to the active ingredient) and the sum of the organic and non-organic impurities, such as moisture and inorganic salts (determined separately), are subtracted from 100% to give the purity of the material (Schemes 2 and 4).

The "subtraction" method is a true primary ratio method in that the results are totally independent of the purity or weight any external agent. This has the practical

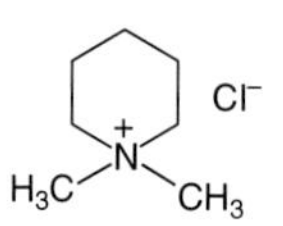

(4) 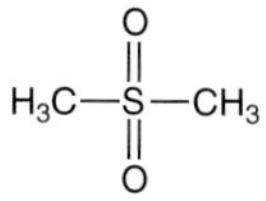 (5) 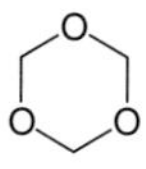 (6) 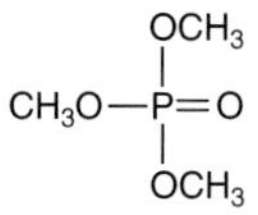 (7) 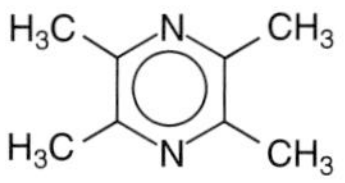 (8) 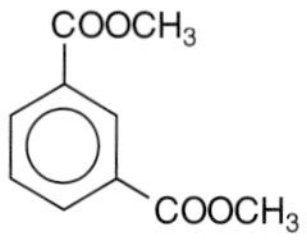 (9)

advantage that the active ingredient and associated organic impurities can be quantified without the use of reference materials, certified or otherwise, and also without the necessity to weigh the analyte. Implicit to the method, however, is that the chemical identity of every separate impurity is known and that peak area is optimised by proper attention to the experimental NMR spectroscopic parameters (see above).

It is generally an excellent method for relatively simple organic substances that have few separate resonances in their NMR spectra and is particularly valuable for the analysis of simple highly polar organic substances that are difficult to quantify by chromatographic methods because of poor chromatographic properties and/or the absence of a suitable chromophore, e.g. mepiquat chloride (4).

NMR is able to readily detect small amounts of impurities in organic substances provided that the chemical shift of at least one peak from the impurity is not coincident with those of either the target analyte or another impurity, which is seldom the case with simple substances. Examples of the use of the subtraction method for detection of organic impurities in commercially acquired dimethylsulphone (5), trioxane (6), trimethylphosphate (7), tetramethylpyrazine (8), sodium acetate and dimethyl isophthalate (9) are shown in Wells et al.[7] In none of these cases was any measurable impurity detected, apart from the residual solvent peak. The ^{13}C satellite peaks represent 0.55% of the main resonance and from these satellites it can be estimated that any impurity would be readily detected at 0.01% and accurately quantified at 0.03%.

It is possible to obtain excellent quantitative results using the "subtraction" method for organic materials of high purity (95% or higher) that contain few impurities greater than 0.1%. However, agreement of the "subtraction" method with the use of an internal standard often diverges as the complexity of the sample increases because it becomes more difficult to unambiguously identify and individually quantify all individual constituents.

3.3. qNMR using a standard of known purity

As discussed previously, one of the considerable advantages of qNMR is the ability to employ a reference material totally unrelated to the target analyte. For ^{1}H NMR, a wide range of standards is available and the only important criteria are that the standard must be of known purity, stable, non-hygroscopic, soluble in the same solvents as the target analyte and have a chemical shift that does not interfere with any of the analyte resonances. Clearly there are many substances that readily fulfil these criteria.

A large number of agrochemicals contain, in addition to hydrogen, NMR-sensitive nuclei such as phosphorus and fluorine, which may be also used in qNMR determinations. The choice of nucleus used then depends on individual circumstances. For instance, although ^{1}H NMR may be perfectly satisfactory for the determination of a substance in the technical grade active material, the same method may not be suitable for formulations of that active material because of interferences from other formulation constituents. In this case, the use of either ^{31}P or ^{19}F NMR would be preferable because few formulation constituents contain phosphorus or fluorine and those that do pose little risk of interfering with the active ingredient resonance.

In practice, there are many cases in which quantification of the active ingredient and associated impurities in technical grade chemicals by qNMR has been achieved with the use of a single standard reference material. This contrasts markedly with chromatographic methods for which a separate standard of each target analyte is necessary to accurately determine the composition of a mixture. Furthermore, the effort required to establish relative response factors for the impurities is considerable and can introduce significant errors in chemical analysis. This is not required in qNMR analysis, which measures directly the relative numbers of nuclei that are observed. Thus in qNMR, sample preparation using an internal standard comprises the weighing of sample and internal standard followed by dissolution of each in the NMR solvent. Since NMR spectra are routinely run on 10–60 mg of material, no dilution is required.

3.4. The concept of universal standard reference material

Wells and co-workers[7] have proposed the concept of a single universal reference material (URM) to which all other organic certified reference materials may be referenced to avoid some of the pitfalls mentioned earlier in Section 2. They have suggested dimethysulphone (Scheme 5) as an attractive candidate for such a standard.

The need for a URM for organic analysis arises from the fact that there is currently no generally accepted method that allows all standards to be referenced against a single reference of a high purity, which may be independently and simply established by a primary analytical method. By employing a single reference material for quantification of both the active ingredient and the impurities, there is no need to acquire certified reference materials for any of the target analytes, although it may be prudent to do so in some cases.

The ideal features of a URM should include the following properties:

- readily availability in a highly pure form
- inexpensive
- containing a small number of magnetically equivalent protons that resonate in a region of the proton NMR where interference with other substances is minimal
- very stable
- soluble in both aqueous and non-aqueous solvents
- low volatility
- non-hygroscopic
- chemically inert.

The NIST offers benzoic acid and naphthalene in very high purity, but these two compounds have limitations from the point of view of NMR: They both have multiple ^{1}H NMR peaks and are not readily soluble in both organic and aqueous solvents. Dimethylsulphone, on the other hand, possesses all of the attributes outlined above. It was purchased in a nominal purity of 99% but NMR and differential scanning calorimetric measurements established that two batches of this inexpensive substance from two separate suppliers each had a purity of >99.9% which, importantly, is not diminished over time. Furthermore, materials of high purity suitable as secondary standards could be readily certified against dimethylsulphone. The purity of a number of substances, tested as supplied, has been determined by qNMR using dimethylsulphone as the primary standard. All of these substances are very pure, suitable as qNMR standards and are supplied in up to 1 kg packages at low cost. In particular, trioxane is useful as a secondary standard and we have found that either dimethylsulphone or trioxane has covered the great majority of our internal standard requirements for analysis of more than 90 different agrochemicals.

3.5. Practical analysis of technical grade agrochemicals by qNMR

In this section we discuss specific examples of the quantification of technical grade agrochemicals by qNMR, Al-Deen and co-workers[8] have investigated the qNMR determination of the herbicide glyphosate (10) and the insecticide – miticide profenofos (11) by both ^{1}H and ^{31}P NMR techniques. The purity of seven independent samples of glyphosate from the same batch, determined in D_2O was 96.48 ($s=0.24$)% by ^{1}H NMR and 95.65 ($s=0.33$)% by ^{31}P NMR. Sodium acetate was used as the NMR standard for ^{1}H NMR measurements and sodium phosphate was used as the NMR standard for ^{31}P NMR measurements. The precision of each method was excellent but there was a disparity of almost 1% in the purity of the technical grade material analyses by the two methods.

Use of sodium phosphate as the ^{31}P NMR standard may lead a possible underestimation of purity, with concomitant uncertainty arising from the nuclear Overhauser effect (NOE) on the analyte peak. Sodium phosphate contains no protons and so cannot exhibit NOE whereas the peak from glyphosate has a strong CH_2—P coupling contributing to an NOE effect. Although the NOE was mimimised by using inverse-gated decoupling, complete elimination of coupling effects cannot be guaranteed and may be responsible for the disparity in measured values between the two techniques. For this reason, organic phosphate with similar NOE may be preferable as NMR standard.

By contrast, the purity of profenofos measured by each qNMR method was very similar. The NOE would be expected to be minimal because phosphorus and proton bearing carbon atoms are separated by an oxygen atom in this substance. The purity of seven independent samples, determined in deuterochloroform, was 94.63 ($s=0.14$)% by ^{1}H NMR and 94.61 ($s=0.33$)% by ^{31}P NMR. Dimethylsulphone was used as the NMR standard for ^{1}H NMR measurements and trimethyl phosphate (Scheme 7) was used as the NMR standard for ^{31}P NMR measurements. The precision of and agreement between each method was excellent.

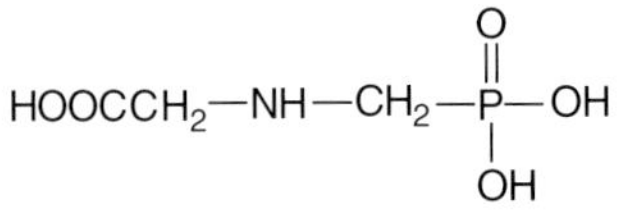

$HOOCCH_2{-}NH{-}CH_2{-}P(=O)(OH){-}OH$

(10)

Br, Cl, O, $P(=O)(OCH_3CH_3)(SCH_2CH_2CH_3)$

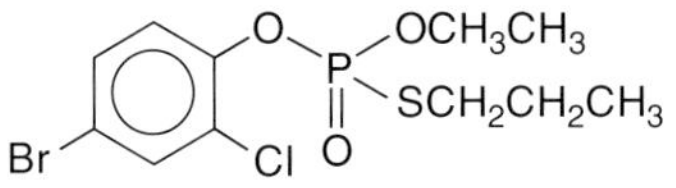

(11)

The spectra used to quantify the active ingredients in each of the agrochemicals above could also be used to identify and to quantify associated manufacturing impurities. The use of NMR in impurity profiling is dealt with in detail later in this review.

Al-Deen and co-workers[9] have also developed a protocol for assessing the uncertainty budget for the qNMR analysis of glyphosate (10) determined by both ^{1}H and ^{31}P NMR.

The analyses performed showed how it was possible to take intra-laboratory reproducibility uncertainty and combine other contributing effects. In the case of these qNMR determinations, the precision of the method was excellent, and the uncertainty of the purity of the NMR standard was a major effect, and even uncertainties in molar masses could not be neglected. The use of a stock solution of the NMR standard lowered the weighing uncertainty, even though it introduces an extra term because of the greater masses weighed out.

A large component in the uncertainties of both methods (^{1}H and ^{31}P) was the standard deviation of the replicate, independent determinations. An attempt to perform a bottom-up analysis estimating the effects that contribute to this term identified about half of those that contributed to the measured standard deviation. The greatest effects were from operator precision of weighing.

The uncertainty budget presented for the measurement of purity of the using ^{1}H and ^{31}P qNMR combined intra-laboratory precision from repeated independent measurements of a batch and other Type A factors. A calculation of the uncertainty for a single measurement gave expanded uncertainties of the purity by ^{1}H and ^{31}P qNMR as 0.66% and 0.82% (95% confidence interval), respectively.

Another class of compounds, produced by fermentation and used in the agricultural industry, are the aminoglycoside antibiotics represented by gentamicin (12). A detailed NMR study of gentamicin has been undertaken by Holzgrabe and

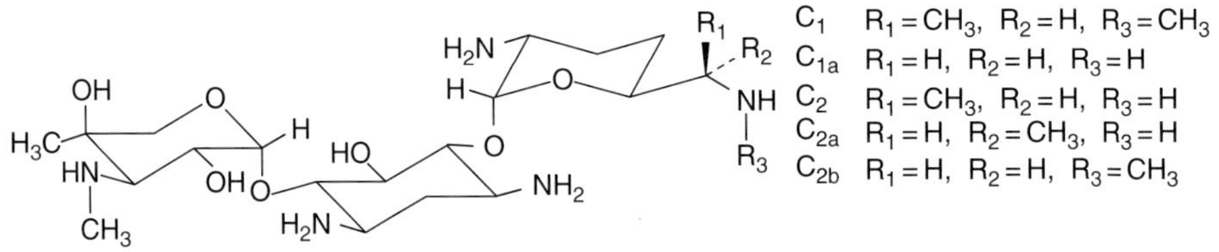

C_1 $R_1=CH_3$, $R_2=H$, $R_3=CH_3$
C_{1a} $R_1=H$, $R_2=H$, $R_3=H$
C_2 $R_1=CH_3$, $R_2=H$, $R_3=H$
C_{2a} $R_1=H$, $R_2=CH_3$, $R_3=H$
C_{2b} $R_1=H$, $R_2=H$, $R_3=CH_3$

(12)

co-workers[10] in which the relative amounts of the four of the five substances which comprise the gentamicin complex could be identified and measured. In a separate study with different objectives, quantification of gentamicin and related aminoglycosides by qNMR was explored with emphasis on the comparison of the absolute amounts and the relative distribution of gentamicin isomers in gentamicin from various sources.[11]

Commercially available batches of gentamicin sulphate were analysed in D_2O and in 25% pyridine–D_2O solutions by ^1H 600 MHz qNMR and a quantification method using potassium hydrogen maleate as the internal standard reference material was developed. The effects of temperature and of the addition of deuteropyridine have a significant effect on the separation of isomer resonances in the NMR spectrum. Methyl resonances in the garosamine portion of the gentamicin molecule are constant for all gentamicin isomers and allow quantification of total gentamicin. By contrast, resonances in the purpurosamine portion of the molecule are isomer dependent and allow constituent ratios to be determined. Figure 1 shows the ^1H 600 MHz NMR spectrum of a gentamicin batch and identifies the resonances suitable for use in quantification.

Commercial "USP" and lower grade gentamicin batches had a component ratio similar to that claimed by the Certificate of Analysis (CoA). However, the total

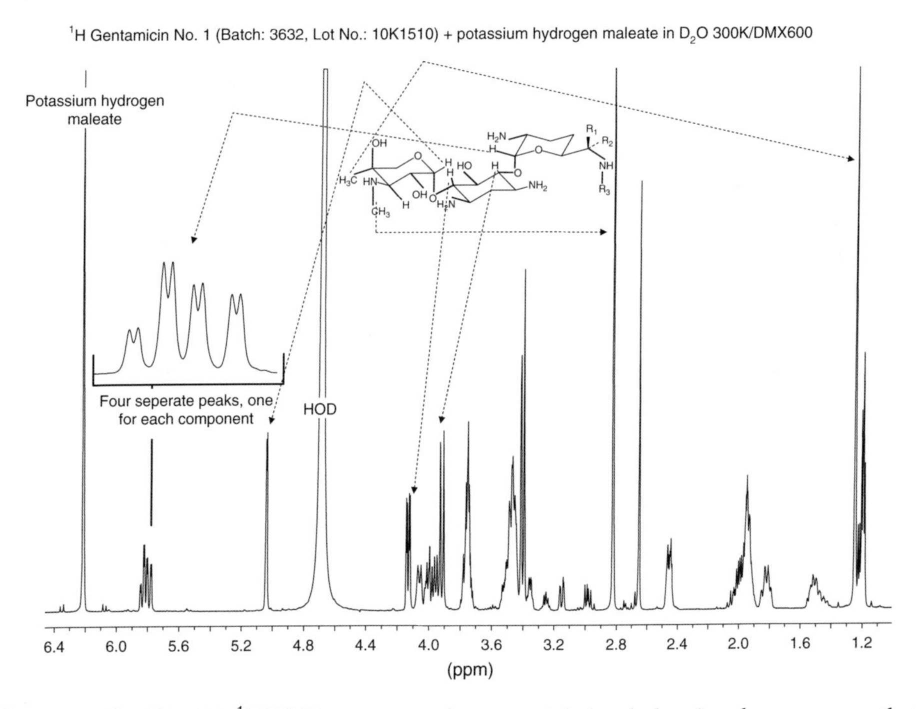

Figure 1 The 600 MHz ^1H NMR spectrum of a gentamicin batch showing the resonances that may be used for quantification.

Table 1 Quantification of gentamicin by 600 MHz qNMR using different gentamicin resonances

Resonance (ppm)	Multiplicity	Number of repeats	Operator 1 Weigh to 4 figures		Operator 2 Weigh to 5 figures	
			Mean %	%SD	Mean %	%SD
2.8	s	7	53.79	1.13	51.06	0.36
5.8	4 × bd	7	50.43	1.72	50.81	0.32
1.3	s	7	51.00	1.43	51.09	0.42
5.2	bd	7	51.10	1.40	51.36	0.30
4.15	dd	7	51.07	1.48	51.41	0.30

Table 2 Ratios of components of different gentamicin batches by 600 MHz qNMR

Gentamicin component	content relative to gentamicin C1 = 100%			
	C_{2a}	C_1	C_2	C_{1a}
Sigma 1 (2004)	40	100	82	73
USP 1 (2004)	47	100	97	62
Inject (2004)	25	100	83	56
Sigma 2 (2002)	41	100	82	67
USP 2 (2002)	52	100	108	69
Raffa (2002)	31	100	126	41

content of aminoglycoside antibiotic was less than 52%. Over a range of batches, gentamicin content was a minimum of 10% lower from the amount of gentamicin claimed in the CoA, even in USP-certified material. An injectable pharmaceutical grade gentamicin sulphate was also found to be 10% less than label (40 mg/ml) by the qNMR method and with a different ratio of C-1, C-1a, C2, C-2a, as W:X:Y:Z. Summaries of results are presented in Tables 1–3. The unavailability of a certified standard of gentamicin did not allow any firm conclusions to be drawn on the reasons for the difference in gentamicin content claimed in various Certificates of Analysis supplied with the batches and the quantitative results obtained by qNMR.

Other commercially available aminoglycosides are also amenable to this analytical approach by qNMR[12] and, in five out of six cases, the amount of aminoglycoside determined by qNMR result proved to be significantly lower than that claimed in the CoA (Table 3).

This study highlights the value of qNMR which readily yields both qualitative and quantitative information directly without the necessity of certified reference material standards for the gentamicin mixture or for each of the individual components. Table 1 demonstrates that not only a number of resonances are suitable for quantitative work but that the accuracy of sample weighing is the major contributor to the uncertainty in a qNMR method (12).

In a previous reports, we[2,7,13] and others[5] have shown that ^1H qNMR was as effective as HPLC for the analysis of a number of agrochemicals, and has distinct advantages over the chromatographic methods. The scope of that initial study has

Table 3 Quantification of various aminoglycosides by 600 MHz ^1H NMR with potassium hydrogen maleate as internal standard

Aminoglycoside	Content (from C of A)	Number of components (ratio)	NMR result (total %)	Peak used (ppm)	No. of repeats (%CV)
Streptomycin USP	774 IU/mg = 99.4% as sulphate	1	55.2	5.5	3 (0.15)
Dihydrostreptomycin	74.5% (by TLC)	1	68.7	5.2	3 (0.67)
Kanamycin	77.4%	1 + 1 minor	78.2	1.7	3 (0.55)
Tobramycin	68.2%	1 + 1 minor	64.7	5.7	3 (0.14)
Neomycin	73.2%	1 + 2 minor	58.7	5.2	3 (0.53)
Gentamicin	61.0%	4 (8:8:5:4)	50.5	3.1	7 (0.30)
Spectinomycin	62.5%	1 + >12% impurity	57.8	5.2	3 (0.39)
Spiramycin	Sum of isomers = 92.3% (I = 87.6% II = 0.8% III = 3.9%)	2 visible by NMR	82.9	5.2	3 (0.38)

been expanded and results are summarised in Table 4.[14] The structural diversity, polarity difference and variations of the chromatographic properties of the compounds studied here is large but they were all readily amenable to qNMR analysis. Satisfactory precision was achieved in all examples with RSDs of <1.0% for the determination of active ingredient in technical grade materials. These RSDs were obtained without any particular attention to sample homogenisation and used a stock solution of the chosen NMR standard in the NMR solvent, which was dispensed by volume rather than weight. The standard used for qNMR analysis was either dimethylsulphone or a substance that had been measured by qNMR directly against dimethylsulphone of >99.9% purity. Dimethylsulphone itself was obtained from two separate suppliers and certified by differential scanning calorimetry and GC–FID (flame ionisation detector) as being >99.9% pure in each case, and no trace of any organic impurity could be detected by ^{1}H NMR at 600 MHz. Although none of the methods listed in Table 4 has been formally published, all have been used to support registrations of agrochemical technical grade actives in Australia and have been peer-reviewed by Australian registration authorities.

The generic regimen that has been developed for this work allows the rapid validation of analytical methods for agrochemicals. It follows the following steps:

1. The NMR spectrum of the analyte is studied (in different solvents, at various pH and different temperatures if necessary) to gauge the correct standard reference material to use and to ensure that there are no interferences with the resonance chosen for quantification.
2. An inversion recovery experiment is then performed to determine the null points, T_D, for the peaks in the spectrum. The relaxation delay, D_1, used in subsequent experiments is calculated from the equation $D_1 = 7 \times T_D \times 1.44$, where T_D represents is taken to be the longest value found for either analyte or standard peaks found in the inversion recovery experiment.
3. To simplify spectra and to ensure that ^{13}C satellite peaks could not interfere with either analyte or standard peaks, the sample was ^{13}C decoupled during the acquisition process using GARP decoupling. However, to prevent excessive sample heating due to decoupling irradiation, particularly in aqueous media, inverse-gated decoupling was chosen.
4. Initially, the next step was to establish the linearity of the method by performing six repeat analyses on a mixture of analyte and standard at concentrations of x, $x/2$ and $x/4$, where x is a ratio of analyte to standard that is about double to that used for subsequent measurements. This step was inserted because method evaluation by primarily analytical chemists required a linearity step. All analyses detailed below had $r^2 > 0.999\%$ and, as a result of such precision, linearity data is now not required as part of validation procedures for registration of agrochemicals in Australia.
5. The precision of the method was established by performing seven individual analyses on one of the batch samples. The sample was weighed and dissolved in the chosen NMR solvent (typically 400 μl) and a specific volume (typically 300–500 μl) of standard dissolved in the same solvent was added. The amount of added standard was calculated to attain and approximately equal signal intensity from both the standard and the target analyte signal. The dispensing of standard

Table 4 Summary of some unpublished validated qNMR methods for technical grade agrochemicals

Agrochemical	Type	Nucleus	Standard used	Solvent	Resonance used
2,4-D	AI, SC	^{1}H	Trioxane	D_2O	4.26, $O{-}\mathbf{CH_2}{-}CO$
2-ethylhexyl 2,4-D	AI	^{1}H	Dimethylsulphone	Cf	4.1, $O{-}\mathbf{CH_2}{-}CO$
Abamectin B1	AI, EC	^{1}H	Dimethylsulphone	Cf	5.56, $\mathbf{H}{-}C{=}$
Acetamiprid	AI	^{1}H	Trioxane	Cf	8.16 + 8.10, $Ar{-}\mathbf{H}$
Alphacypermethrin	AI, EC	^{1}H	Dimethylsulphone	Cf	6.38, $\mathbf{H}{-}C{=}$
Amitraz	AI	^{1}H	Trioxane	Cf	3.77, $\mathbf{CH_3}{-}N$
Atrazine	AI, WDG	^{1}H	Dimethylsulphone	Cf	3.4. $=N\mathbf{CH_2}CH_3$
Bifenthrin[c]	AI, EC	^{1}H	Dimethylsulphone	Cf	7.00, $\mathbf{HC}{=}$
Brodifacoum	AI	^{1}H	Trioxane	Cf	4.88, 4.73, $\mathbf{CH_2}$ isomers
Bromoxynil octanoate	AI	^{1}H	Dimethylsulphone	Cf	7.87, $2 \times Ar{-}\mathbf{H}$
Captan	AI	^{1}H	Dimethylsulphone	Cf	3.3, $2 \times CH_2\mathbf{CH}CO$
Carbendazim	AI	^{1}H	Trioxane	DMSO	7.4, $2 \times Ar{-}H$
Chlormequat	AI, SC	^{1}H	Trioxane	D_2O	3.14, $\mathbf{CH_3}N^{+}\mathbf{CH_3}$
Chlorpyrifos	AI, EC	^{1}H	Dimethylsulphone	Cf	1.4, $\mathbf{CH_3}CH_2{-}OP$
Chlorthal dimethyl	AI	^{1}H	Dimethylsulphone	Cf	4.01, $2 \times \mathbf{CH_3}OOCAr$
Chlosulphfuron	AI, WDG	^{1}H	Dimethylsulphone	Py	2.47, triazine$-\mathbf{CH_3}$
Cinmethylin	AI	^{1}H	Dimethylsulphone	Cf	3.45, $-\mathbf{CH}OH$
Clethodim	AI, MC	^{1}H	Trioxane	Cf	6.3, $=\mathbf{C}{-}\mathbf{H}$
Clethodim	AI	^{1}H	Trioxane	Cf	4.7, $=C{-}\mathbf{CH_2}{-}O$
Clodinafop-propargyl	AI	^{1}H	Dimethylsulphone	Cf	2.5 $\mathbf{HC}{=}$
Clopyralid	AI	^{1}H	Dimethylsulphone	Ac/Cf	7.62, pyridine$-\mathbf{H}$
Cloquinocet-mexyl	AI	^{1}H	Dimethylsulphone	Cf	8.9 $\mathbf{HC}{=}$
Cyanazine	AI, WDG	^{1}H	Dimethylsulphone	Cf	3.5, $CH_3\mathbf{CH_2}N$
Cypermethrin	AI, EC	^{1}H	Dimethylsulphone	Cf	6.2, $\mathbf{HC}{=}$ isomers
Dalapon sodium[c]	MC	^{1}H	Dimethylsulphone	D_2O	2 1 m $\mathbf{CH_3}CCl_2CO$
Dazomet	AI	^{1}H	Trioxane	Cf	3.49, $\mathbf{CH_3}{-}N{-}$
Dicamba	AI, SC	^{1}H	Dimethylsulphone	Cf	3.97, $\mathbf{CH_3}OAr$

Table 4 *(Continued)*

Agrochemical	Type	Nucleus	Standard used	Solvent	Resonance used
Diclofop-methyl	AI, EC	^{1}H	Trioxane	Cf	4.68, O**CH**(CH$_3$)CO
Difenoconazole	AI	^{1}H	Trioxane	Cf	6.82, Ar—**H**
Diflufenican	AI	^{1}H	Dimethylsulphone	Cf	6.92, Ar—**H**
Dimethoate	AI	^{1}H	Trioxane	Cf	2.73 m **CH$_3$**—N
Diquat	AI, SC	^{1}H	Dimethylsulphone	D_2O	8.30, 2 × Ar—**H**
Diuron	AI,WDG	^{1}H	Trioxane	Cf	7.41, Ar—**H**
Epoxiconazole	AI	^{1}H	Trioxane	Cf	3.97, **CH**O or 7.89, Ar—**H**
Ethephon	AI. MC	31P	K_2HPO_4	D_2O	21.6, —**P**(O)(OH)$_2$
EPTC	AI	^{1}H	Trioxane	Cf	2.7, CH$_3$**CH$_2$**N
Fenoxaprop-ethyl	AI	^{1}H	Dimethylsulphone	Cf	4.74, O**CH**(CH$_3$)CO
Fipronil	AI	^{19}F	4,4′-DFBP	Cf	—SO**CF$_3$**, 73.44 + 84.97
Fluazifop-butyl	AI	^{1}H	Dimethylsulphone	Cf	4.7, O**CH**(CH$_3$)CO
Flumetsulam	AI	^{1}H	Trioxane	Ac	9.26, pyrimidine—**H**
Fluoxypyr-meptyl	AI	^{1}H	Dimethylsulphone	Cf	4.8, ArOC**H$_2$**CO
Flupropanate sodium	AI	^{1}H	Sodium acetate	D_2O	6.2, **H**CF$_2$CF$_2$
Fluquinconazole	AI	^{1}H	Dimethylsulphone	Cf	8.79, Ar—**H**
Flutriafol	AI, SC	^{1}H	Dimethylsulphone	Cf	7.73, triazole—**H**
Gibberellic acid	AI, SC	^{1}H	DMiP	MeOD	6.42, **HC**=
Glufosinate-ammonium	AI, SC	^{1}H	Dimethylsulphone	D_2O	1.47, **CH$_3$**—
Glyphosate	AI	^{1}H	Sodium acetate	Py/D_2O	2.77, N—**CH$_2$**—P
Glyphosate	AI, F	^{31}P	KH_2PO_4	D_2O	7.26, —**P**(O)(OH)$_2$ (St. 0)
Haloxyfop-methyl	AI	^{19}F	4,4′-DFBP	Cf	62.06, —**CF$_3$**, (St. 106.2)
Hydramethylnon	AI	^{1}H	Trioxane	Cf	3.02, 2 × **CH$_2$**—N
Imazapic	AI, SC	^{1}H	Dimethylsulphone	Ac/Cf	2.45, C**H$_3$**—pyridine
Imazapyr	AI	^{1}H	Trioxane	Cf	8.51, pyridine—**H**
Imazethapyr	AI	^{1}H	Trioxane	Cf	2.21, —**CH**(CH$_3$)$_2$
Imidacloprid	AI	^{1}H	Dimethylsulphone	Cf	7.35, pyridine—**H**

Iprodione	AI	^{1}H	Trioxane	Cf	4.0, N—$\mathbf{CH_2}$CO
Lambdacyhalothrin	AI, EC		Trioxane	Cf	6.4, **HC**=C
MCPA	AI, SC	^{1}H	Dimethylsulphone	Cf	6.76, Ar—**H**
Mepiquat chloride	AI, SC	^{1}H	Trioxane	D_2O	3.24, $\mathbf{CH_3}$—N—$\mathbf{CH_3}$
Metaldehyde	AI	^{1}H	Dimethylsulphone	Cf/Py	4.97, O—**CH**(CH3)—O
Methabenzthiazuron		^{1}H	Trioxane	Cf	3.50, N—$\mathbf{CH_3}$
Metham sodium	MC	^{1}H	Sodium acetate	D_2O	3.34, $\mathbf{CH_3}$NHC=S
Metolachlor	AI, EC	^{1}H	Trioxane	Cf	4.24, N—**CH**(CH_3)CH_2
Metribuzin	AI, WDG	^{1}H	Dimethylsulphone	Cf	2.61, S$\mathbf{CH_3}$
Metsulfuon-methyl	AI, WDG	^{1}H	Dimethylsulphone	Py	2.55, triazole—$\mathbf{CH_3}$
Oryzalin	AI	^{1}H	Trioxane	Cf	8.35, 2 × Ar—**H**
Oxyfluorfen	AI	^{1}H	Dimethylsulphone	Cf	7.91, Ar—**H**
Paclobutrazole	AI	^{1}H	Trioxane	Cf	8.30, triazole—**H**
Paraquat	AI, SC	^{1}H	Dimethylsulphone	D_2O	8.45, 2 × Ar—**H**
Parpargite	AI	^{1}H	Dimethylsulphone	Cf	2.53, **H—C**ΞC—
Pirimicarb	AI	^{1}H	Trioxane	Cf	3.08 $\mathbf{(CH_3)_2}$N—
Prochloraz	AI	^{1}H	Trioxane	Cf	7.90, imidazole—**H**
Procymidone	AI	^{1}H	Dimethylsulphone	Cf	1.50, $\mathbf{CH_3}$—C—$\mathbf{CH_3}$
Profenofos	AI	^{31}P	Me_3PO_4	Cf	
Propiconazole	AI, EC	^{1}H	Dimethylsulphone	Cf	7.89, triazole—**H**
Propineb	AI	^{1}H	Sodium acetate	Py	4.2, N—$\mathbf{CH_2}$
Quizalalofop-ethyl	AI	^{1}H	Dimethylsulphone	Cf	4.82, O**CH**(CH_3)CO
Sethoxydim	AI, MC	^{1}H	DMiP	Cf	4.06, $CH_3$$\mathbf{CH_2}$ON=
Simazine	AI, WDG	^{1}H	Dimethylsulphone	Cf	3.6. 2 × $CH_3$$\mathbf{CH_2}$N
Sulphometuron	AI, WDG	^{1}H	Trioxane	Py	2.26, 2 × $\mathbf{CH_3}$—pyrimidine
Sulphosulfuron	AI, WDG	^{1}H	Trioxane	Py	6.34, pyrimidine—**H**
Tebuconazole	AI, WDG	^{1}H	Trioxane	Cf	8.35, triazole—**H**
Thiabendazole	AI	^{1}H	Trioxane	DMSO	8.42, benzthiazole—**H**
Thidiazuron	AI, WDG	^{1}H	Trioxane	Cf	8.64, thidiazole—H

Table 4 *(Continued)*

Agrochemical	Type	Nucleus	Standard used	Solvent	Resonance used
Thiram	AI	^{1}H	TMP	Cf	3.53, $\mathbf{(CH_3)_2}NC{=}S$
Tolclofos-methyl	AI	^{1}H	Dimethylsulphone	Cf	3.97, $2 \times O\mathbf{CH_3}$
Tralkoxydim	AI	^{1}H	Trioxane	Cf	3.8, **HC**—
Triadimefon	AI, WP	^{1}H	Dimethylsulphone	Cf	0.88 + 1.05, $\mathbf{(CH_3)_3}C$
Triadimenol	AI, EC	^{1}H	Trioxane	Cf	8.35, triazole—**H**
Triallate	AI, EC	^{1}H	Trioxane	Cf	4.02, $S\mathbf{CH_2}$—C—Cl
Triasulphuron	AI, WDG	^{1}H	Dimethylsulphone	Py	8.53, Ar—**H**
Triclopyr-butoxyethyl	AI	^{1}H	Dimethylsulphone	Cf	4.93, O—$\mathbf{CH_2}$COO
Triflumuron	AI	^{1}H	Dimethylsulphone	Cf	7.64, Ar—**H**
Trifluralin[c]	AI, EC	^{1}H	Trioxane	Cf	8.07, 2 × Ar—**H**
Trinexapac-ethyl	AI	^{1}H	Trioxane	Cf	3.5, **CH**—C=O
Triticonazole	AI	^{1}H	Trioxane	Cf	7.83, triazole—**H**
Propineb	AI	^{1}H	Sodium acetate	Py	4.0, N—CH2

Abbreviations used: Cf = $CDCl_3$, Ac = acetone-d_6, DMSO = DMSO-d_6, Py = pyridine-d_5, MeOD = methanol-d_4, 4,4′-DFBP = 4,4′difluoro-benzophenone, TMP = tetramethylpyrazine DMIP = dimethylisophthalate Me_3PO_4 = trimethyl phosphate AI = active ingredient WDG = water dispersible granules, EC = emulsion concentrate, SC = solution concentrate (or suspension concentrate), MC = manufacturing concentrate * In the column labelled 'resonance used' the atoms shown in bold are the functional group resonance used for quantification.

by volume was found to be more convenient and equally reproducible as dispensing by weight. The %CV attained was usually less than 0.5%. If %CV exceeded 1%, the whole experiment was repeated. The most common reason for unacceptable CVs proved to be sample non-homogeneity resulting from insoluble foreign matter or from uneven sample drying.

4. qNMR Analysis of Proscribed Impurities in Agrochemicals

Most international registration authorities adopt standards for technical grade active ingredients that are identical to, or closely based on, the standards set by the Food and Agricultural Organisation of the UN (FAO). These standards not only stipulate the minimum permissible amount of active ingredient but also, on many occasions; stipulate the maximum allowable amount of toxic impurities associated

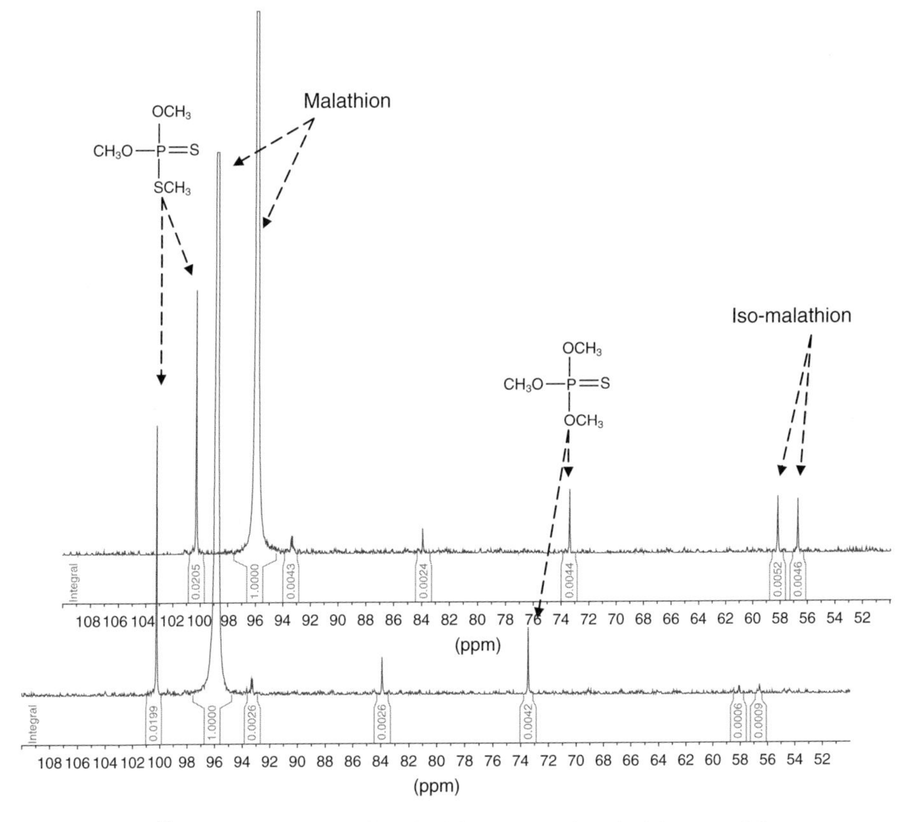

Figure 2 The ^{31}P NMR spectra of a freshly prepared malathion emulsion concentrate formulation standard (lower trace) and the same formulation stored for 14 days at 54°C (upper trace).

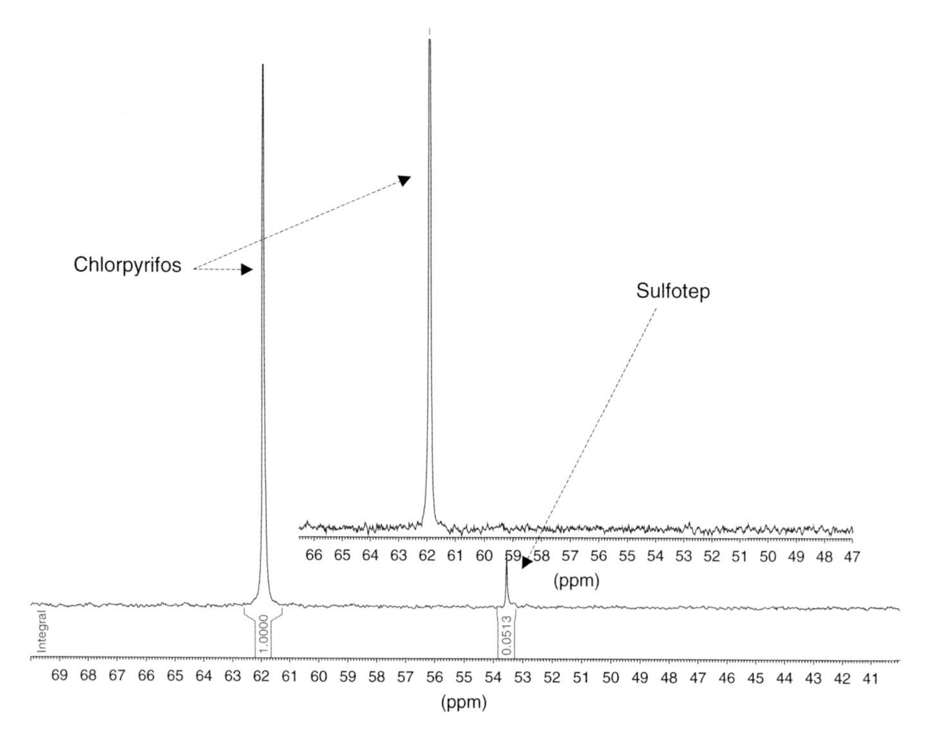

Figure 3 The ^{31}P NMR spectra of a chlorpyriphos standard containing 0.25% sulfotep (lower trace) and a technical grade batch samples of chlorpyriphos (upper trace).

with certain classes of agrochemical. Many of these proscribed impurities are conveniently measured by qNMR methods.

Two ^{31}P qNMR methods for the determination of proscribed impurities in organophosphorus pesticides appear in FAO specifications. The first of these methods was developed to access the decomposition of malathion (13) in various formulations during storage. Samples of technical malathion and various formulations were subjected to storage at 54°C (±2°C), in compliance with CIPAC MT 46.3.1, and were analysed for content of active ingredient and impurities. Figure 2 shows the ^{31}P spectra of an emulsifiable concentrate formulation of malathion immediately following formulation and after storage at 54°C for 14 days. The method allows quantification of *O,O,O*-trimethylphosphorothioate (14), *O,O,S*-trimethylphosphorodithioate (15), malaoxon (16) and iso-malathion (17) relative to the malathion concentration.[15]

The second method[16] was supplied by Syngenta and measures *O,O,O*-trimethylphosphorothioate (14), *O,O,S*-trimethylphosphorodithioate (15), *O,O,O*-trimethyl phosphoro chlorido-thioate (18), *O,O,S*-trimethylphosphorothioate (19), and iso-pirimiphos-methyl (20) relative to the pirimiphos-methyl (21) concentration.

qNMR may also conveniently be used for the simple and routine determination of other proscribed impurities such as free phenol in aryloxyacetic acid herbicides,

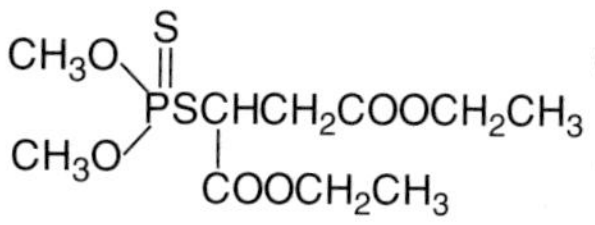

(13) 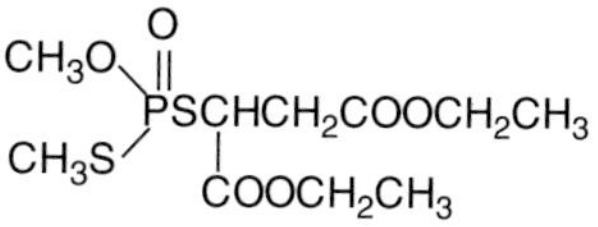 (16) 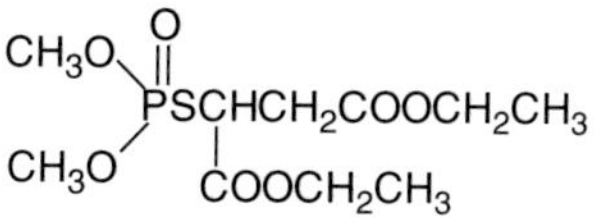 (17)

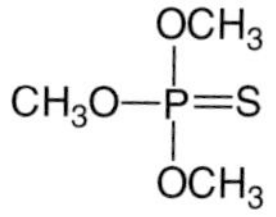

(14) 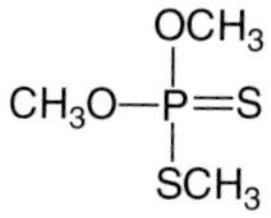 (15) 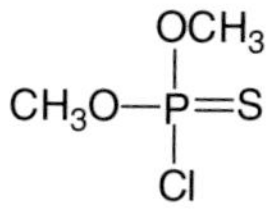 (18) 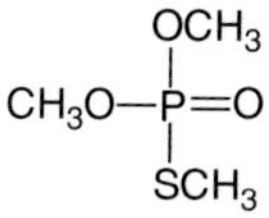 (19)

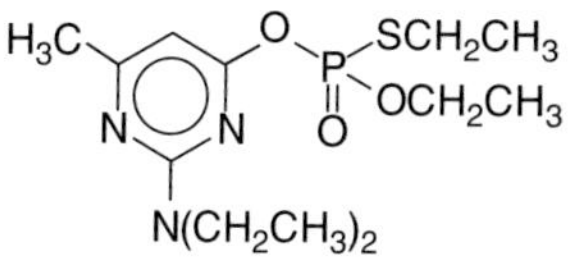

(20) 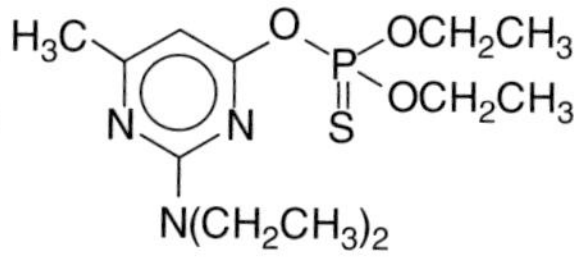 (21) 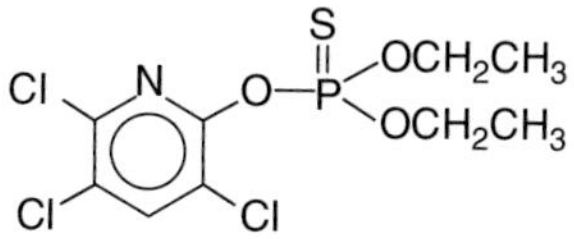 (24)

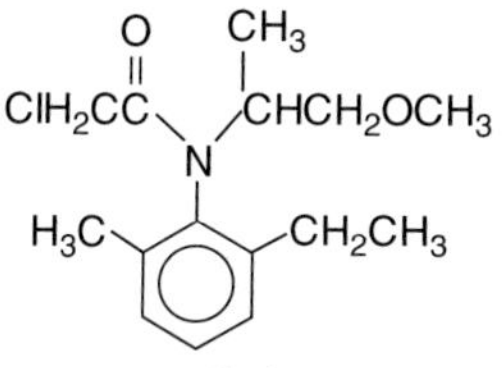

(22) 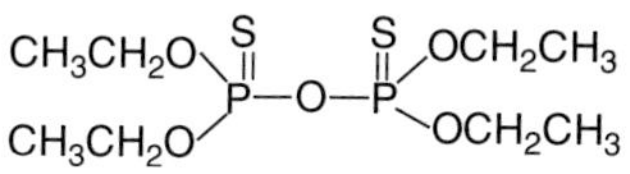 (23)

2-ethyl-6-methylaniline in metolachlor (22) and sulphotep (23) in chlorpyrifos (24).[14] Figure 3 shows the ^{31}P NMR spectra of chlorpyrifos standard plus 0.25% sulphotep and a technical grade batch of chlorpyrifos. A validated method for the determination of perchloromethylmercaptan, which lacks any protons, has also been developed. In this case, perchloromethylmercaptan is quantitatively converted to *N*,*N*-diethyl-2,2,2-trichloromethylthioamine before measurement by the addition of diethylamine.[14]

The cyclic impurities, ethylene thiourea and propylene thiourea found in zineb and propineb may be readily detected and quantified using sodium acetate in pyridine, the coordinating solvent facilitating dissolution of the normally insoluble zinc dithiocarbamate powders.[17]

5. qNMR Analysis of Manufacturing Impurities in Agrochemicals

The FAO standards for pesticides include a requirement to identify all manufacturing impurities that are found at 0.1% or higher in the technical grade material. European and US registration authorities require that these impurities be quantified by a validated analytical method under GLP and this

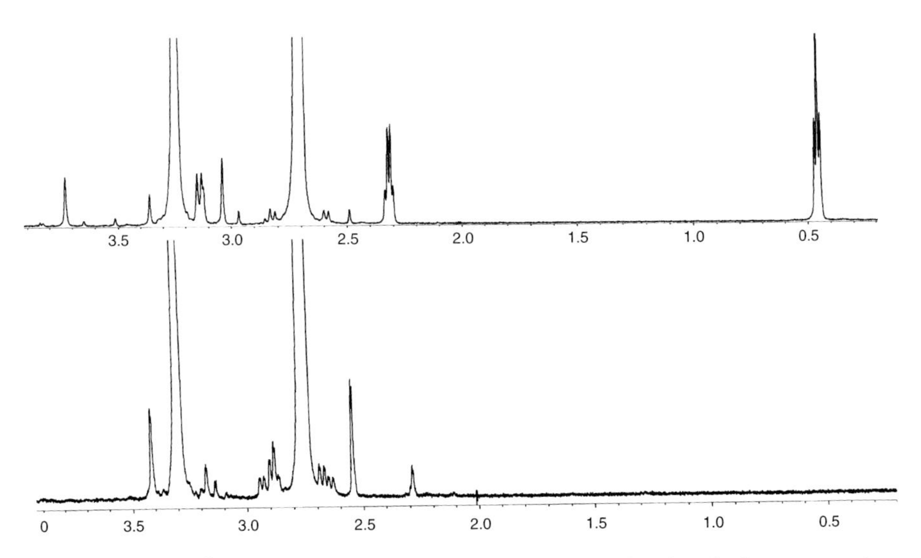

Figure 4 600 MHz ^{1}H NMR Spectra of glyphosate prepared by the glycine process (upper trace) and the iminodiacetic acid process (lower trace).

necessitates either the isolation or preparation of all impurities present. This is not a current requirement in Australia and many of these manufacturing impurities are conveniently measured by qNMR methods where an analytical standard of each impurity is unnecessary. Furthermore, the determination of impurities by NMR can yield valuable information about manufacturing process and manufacturing source.

Figure 4 shows the NMR spectra of two glyphosate batch samples prepared by completely different manufacturing processes. It can be seen that the NMR impurity fingerprint of each spectrum is completely different. Furthermore, the sample prepared by the glycine process has a fingerprint that is dissimilar from glyphosate prepared by the same process in other factories. The individual impurities in each glyphosate batch are readily quantified.

As discussed earlier, NMR has proved to be a simple analytical technique for the analysis of substances which are otherwise difficult to determine by chromatographic techniques. Ethephon (2-chloroethylphosphonic acid) is a good example: it is non-volatile, transparent to UV and has extremely poor chromatographic properties. The official CIPAC method,[18] published as recently as 1997, involves the thermal decomposition of the sodium salt of ethephon followed by titration with sodium hydroxide. Not only is the method cumbersome, but it is also non-specific and some manufacturing impurities are incorrectly included in the method. By contrast, the determination ethephon by ^{31}P NMR is simple using either potassium dihydrogen phosphate or hexamethylphosphoramide as standards. Furthermore, manufacturing impurities can be simultaneously identified and quantified.

Figure 5 shows the ^{31}P NMR spectrum of an ethephon formulation being sold in the Australian market. The figure also shows quantification details obtained from the spectrum which measures the manufacturing impurities at about 15.9%, well

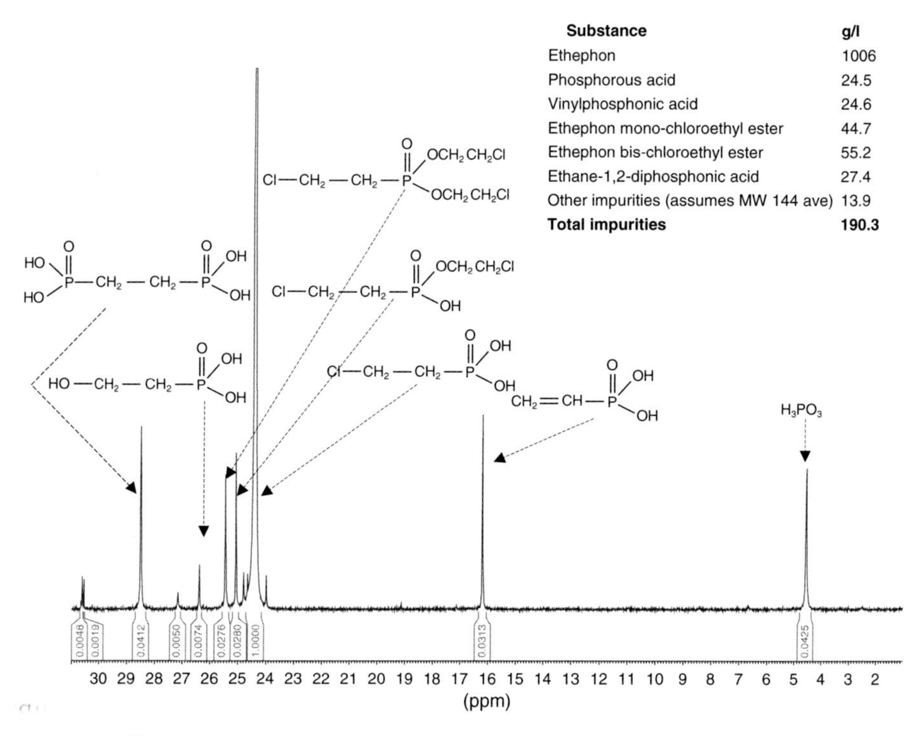

Substance	g/l
Ethephon	1006
Phosphorous acid	24.5
Vinylphosphonic acid	24.6
Ethephon mono-chloroethyl ester	44.7
Ethephon bis-chloroethyl ester	55.2
Ethane-1,2-diphosphonic acid	27.4
Other impurities (assumes MW 144 ave)	13.9
Total impurities	**190.3**

Figure 5 A ^{31}P NMR spectrum of an ethephon formulation containing 1000 g/l ethephon. Total impurities are about 15.5%.

above the 10% permissible maximum specified for total impurities in ethephon on an anhydrous basis.[19]

Other agrochemicals that are simple to analyse by qNMR but present a more significant challenge by other analytical techniques include the determination of the active ingredient and the impurity *N*-methylpiperidiine in mepiquat chloride (4)[14] and active ingredient and impurities in dalapon sodium (1).[13]

We have found that in virtually all analyses listed in Table 4, valuable qualitative information on impurities could be simultaneously obtained from the same NMR spectra used for the determination of the active ingredient. Thus, the triazole fungicides of which tebuconazole (25) is a typical example are always accompanied by the 4H-isomer (26). This isomer occurs as a two proton singlet in the NMR

(25) (26)

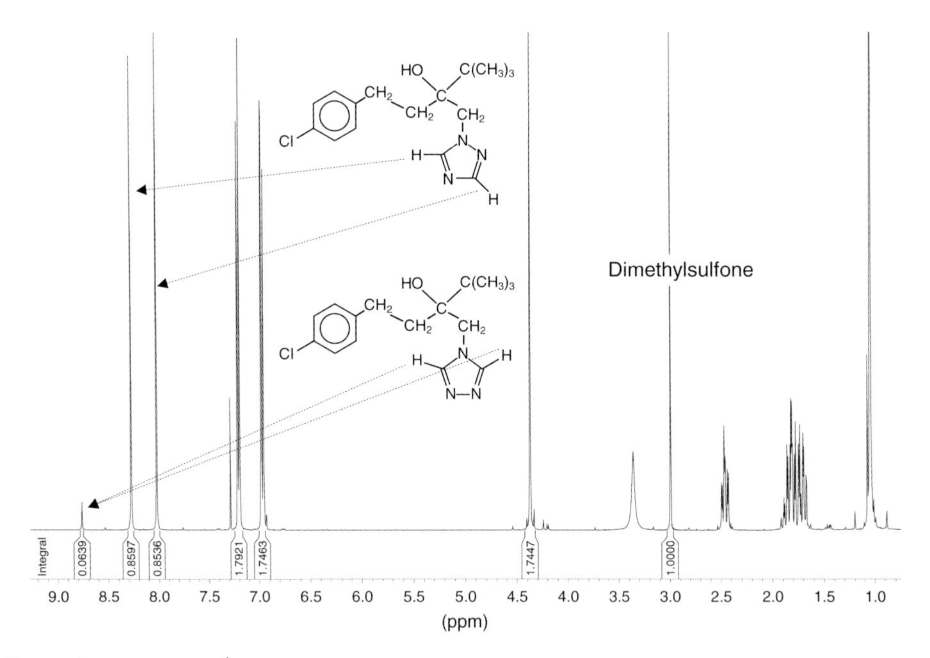

Figure 6 A 400 MHz ^{1}H NMR spectrum of a technical grade batch of tebuconazole.

which may be readily quantified. Figure 6 shows the NMR spectrum of the fungicide tebuconazole. The major impurity is the isomer (26) and this is readily quantified along with the active ingredient.[14]

We concur with Malz and Jancke[20] that a *S/N* ratio of 150:1 or greater is appropriate to give an integration uncertainty of <1%. This condition is not always achieved when quantifying minor constituents using the same data acquisition parameters employed or quantification of active ingredient. Under such conditions integration uncertainties of <5% may be often attained for impurities at the 0.1 – 1% level. However, greater accuracy is obtainable by suitable attention to experimental parameters.

6. Conclusion

The strength of the emerging analytical agrochemical methods based on qNMR rests on the general utility, readily availability and stability of a standard reference material for the determination of purity of organic chemicals. Dimethylsulphone and trioxane are simple inexpensive substances which are widely applicable. They have single ^{1}H NMR resonances in usually uncluttered portions of the ^{1}H spectrum and are available off the shelf in high quality.

qNMR is suitable not only for the analysis of the active ingredients in technical grade materials but may also be used in formulation stability studies and to quantify many toxic and manufacturing impurities. qNMR is also extremely suitable for routine quality control work.

It also rests on accessibility to medium field (300–600 MHz) NMR spectrometers suitably equipped and configured to perform analyses on the nucleus of choice, operated with the understanding of the parameters required for quantitative assessments.

ACKNOWLEDGEMENTS

We wish to thank Sean Ho of Genfarm Crop Protection Pty Ltd, Phil Patterson of 4 Farmers Pty Ltd and Andrew Hill of eChem Australia Pty Ltd for permission to use unpublished information in this review.

We also thank Dr Ian Luck, NMR Facility, University of Sydney for advice and discussions.

REFERENCES

1. G.F. Pauli, B.U. Jaki, D.C. Lankin, J. Nat. Prod. 68 (2005) 133–149 and references cited therein.
2. R.J. Wells, J. Cheung, The Chemistry Preprint Server CPS, analchem/0103002 2001.
3. R.J. Wells, unpublished observations.
4. CIPAC (Collaborative International Pesticide Analytical Council) Handbook, Blackbear Press, Cambridge, U.K. Volumes A-L.
5. G. Maniara, K. Rajamoorthi, S. Rajanand, G.W. Stockton, Anal. Chem. 70 (1998) 4921–4928 and references cited therein.
6. L. Griffiths, A.M. Irvine, Analyst 123 (1998) 1061–1068.
7. R.J. Wells, J. Cheung, J.M. Hook, Accred. Qual. Assur. 9 (2004) 450–456.
8. T.S. Al-Deen, D.B. Hibbert, J.M. Hook, R.J. Wells, Anal. Chim. Acta 474 (2002) 125–135.
9. T.S. Al-Deen, D.B. Hibbert, J.M. Hook, R.J. Wells, Accred. Qual. Assur. 9 (2004) 55–63.
10. R. Deubner, C. Schollmayer, F. Wienen, U. Holzgrabe, Magn. Reson. Chem. 41 (2003) 589–598.
11. R.J. Wells, J.M. Hook, N. Isoherranen, S. Soback, Second International Conference on Antimicrobial Agents in Veterinary Medicine (AAVM) Ottawa, Canada, June 13–17, 2004.
12. R.J. Wells, J.M. Hook, D.G. Lonnon, H. Stender, Second International Conference on Antimicrobial Agents in Veterinary Medicine (AAVM) Ottawa, Canada, June 13–17, 2004.
13. R.J. Wells, J.M. Hook, T. Tareq Al-Deen, D.B. Hibbert, J. Agric. Food Chem. 50 (2002) 3366–3374.
14. R.J. Wells, J. Cheung, unpublished data.
15. FAO Specifications and Evaluations for Agricultural Pesticides Malathion (S-1,2-bis(ethoxycarbonyl)ethyl O,O-dimethylphosphorodithioate) (Adapted from Cheminova analytical method VAM 203-01).
16. FAO Specifications and Evaluations for Agricultural Pesticides Pirimiphos-methyl (O-2-diethylamino-6-methylpyrimidin-4-yl-O,O-dimethyl phosphorothioate) (Syngenta method).
17. J.M. Hook, H. Stender, unpublished results.
18. CIPAC (Collaborative International Pesticide Analytical Council) Handbook, Blackbear Press, Cambridge, U.K. Volume H (1998) 168–169.
19. Australian Pesticides and Veterinary Medicines Authority, Standards for Active Consutuents.
20. F. Malz, H. Jancke, J. Pharm. Biomed. Anal. 38 (2005) 813–823.

PART III

SPECIAL APPLICATIONS

CHAPTER 1

NMR-Based Mixture Analysis on the Example of Fruit Juice Quality Control Using Statistics and Quantification

M. Spraul, E. Humpfer, H. Schäfer, B. Schütz, M. Mörtter, *and* P. Rinke

Contents

Abstract

Mixture Analysis is rapidly gaining importance for nuclear magnetic resonance (NMR) spectroscopy, mainly driven by the pharmaceutical applications of metabonomics. Similar requirements are existent for the quality control of food material. NMR is used to obtain statistical information and to quantify a predefined set of compounds for every sample investigated. Statistical data evaluation is only possible if strictly standardized conditions are used for sample preparation and measurement. Also important is the stability of the analytical device used. Here NMR can be considered as a system with extremely low system internal variance, allowing to observe smallest concentration ratios of many metabolites at the same time. For example, for the fruit juice quality control, possible results are presented for statistical data evaluation and quantification. Good quantification can only be achieved if the compounds are identified safely before quantification using fast 2D experiments that are also used to support the deconvolution of the mixture spectrum.

Keywords: ERETIC, ridge regression, apple and passion fruit juice, glucose, PCA

1. Introduction

For many years nuclear magnetic resonance (NMR) was seen as a tool to perform structure verification, elucidation and purity analysis. It was considered insensitive and slow compared to other analytical tools like mass spectroscopy (MS) or UV. In the previous years, NMR has seen a rapid expansion into the field of mixture analysis and screening applications. This was enabled by high-throughput sample changing technology, integrated sample preparation and the improved quality of digital spectrometers in general. IR spectroscopy has found major applications in the area of food and beverage quality control long before NMR and is well established integrated with statistical calculations.[1,2] Only with the advent of metabonomics has NMR found its applications in mixture analysis,[3–7] driven also by the pharmaceutical application of drug toxicity screening on animal urine.[8–15] Flow injection NMR[16] has enabled the throughput needed to measure and process thousands of samples in a rapid and highly reproducible way with no manual interaction. Today NMR is well established in the pharmaceutical field to investigate drug toxicity, drug efficacy as well as drug metabolites formed. Besides flow injection NMR, also the hyphenated methods of liquid chromatography (LC)-NMR and LC-NMR/MS in combination with post-column solid-phase extraction have especially enabled mixture analysis by isolating and identifying individual compounds.[17–21]

NMR has also found its way into clinical applications, currently in the research area mainly; however, lipoprotein subclass analysis is offered today as a clinical analysis tool.[22,23]

While the applications mentioned above are mostly performed using higher field instrumentation (~600 MHz) also with cryogenic probes, another application is currently coming into beverage quality control after some initial publications.[24–27,29,30] Here one requirement is the low-cost analysis. Therefore, 400 MHz is an ideal field for such an application.

Even so a 400 MHz NMR machine still is expensive compared to other analytical systems used like GC or LC, the cost per sample for NMR can be very low, depending on the turnover to be achieved. Another advantage of NMR is that the samples need minimal preparation before measurement. NMR also is known to be fully quantitative, if operated with suitable measurement parameters.

Site-specific natural isotope fractionation (SNIF)-NMR[31–33] has been established since many years, where, for example, the sugaring of wine before fermentation can be investigated or where compounds from natural sources with higher price can be clearly separated from chemically synthesized material like in vanillin.

In this chapter the quality control of fruit juices will be described in more detail as it is already developed into a push button tool and can stand as an example, how operation in medical and pharmaceutical area can be once statistical models are established and rules as well as knowledge bases established for quantification. The requirements for the analysis can be divided into two main groups, quantification of organic compounds in the juice and statistical analysis.

Fruit juices form a dominant part of non-alcoholic drinks and are produced in large amounts in all parts of the world and therefore have the need of massive transportation from the producer to the consumer. Therefore, fruit juices are mainly offered as re-diluted concentrates and to a much lesser extent as direct juice. The direct juice is the higher priced product. Compared to concentrates the volume for transportation is fivefold for the direct juices.

For customer products, in many countries the type of processing has to be on the label and therefore fraud can happen by, for example, mixing re-diluted concentrate into direct juice or even selling re-diluted concentrate as direct juice. Further quality aspects accessible by NMR are, for example, fruit type purity and control of geographical origin.

All investigations mentioned are possible by NMR with only one experiment, while in classical analysis, many different methods have to be applied. This forms another reason for the cost efficiency of NMR in this application.

Today the quality control of fruit juices is driven as a push button application without manual interaction from sample registration to the final report. Figure 1 shows the 1D spectra of apple and orange juice, the two major products on the market. On the left side, the spectra are scaled to the sugar signals, on the right side, a scaling of a factor of 32 has been applied.

The upper spectrum is from apple juice, the lower spectrum from orange juice.

It can readily be seen that there are some differences in the sugar distribution, the major difference in the left-hand part is due to the concentration of citric acid (low in apple, signals around 2.75–3 ppm) and malic acid (low in orange juice, signals around 4.5 and 2.75–3 ppm. It is obvious that the signals at 2.75–3 ppm show overlap. It will be described later how quantification works reliable despite the overlap. It can also be seen, that chemical shift of the two compounds mentioned is quite different in the two juices, even so both have been adjusted to a pH of 3. This is an additional problem for quantification.

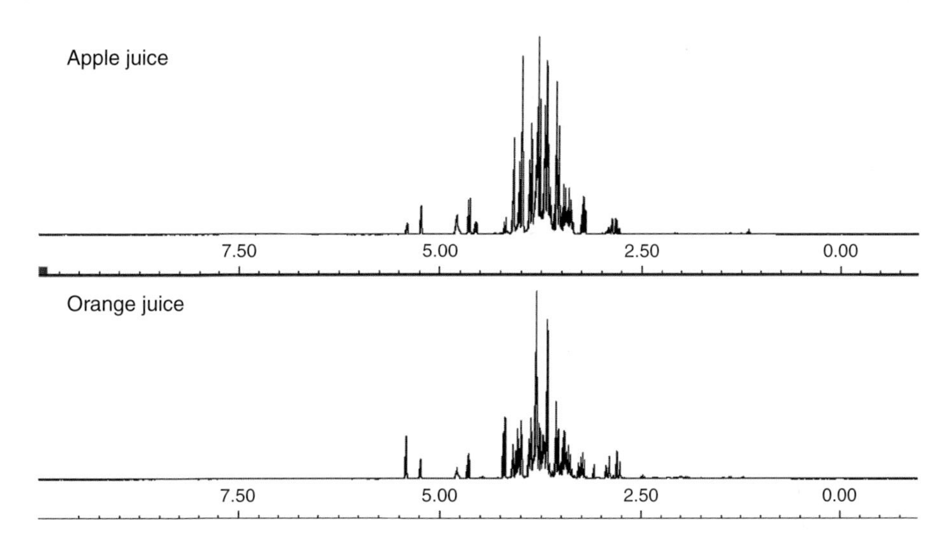

Figure 1 Comparison of orange and apple juice 1D NMR spectra at 400 MHz at different *Y*-scaling levels.

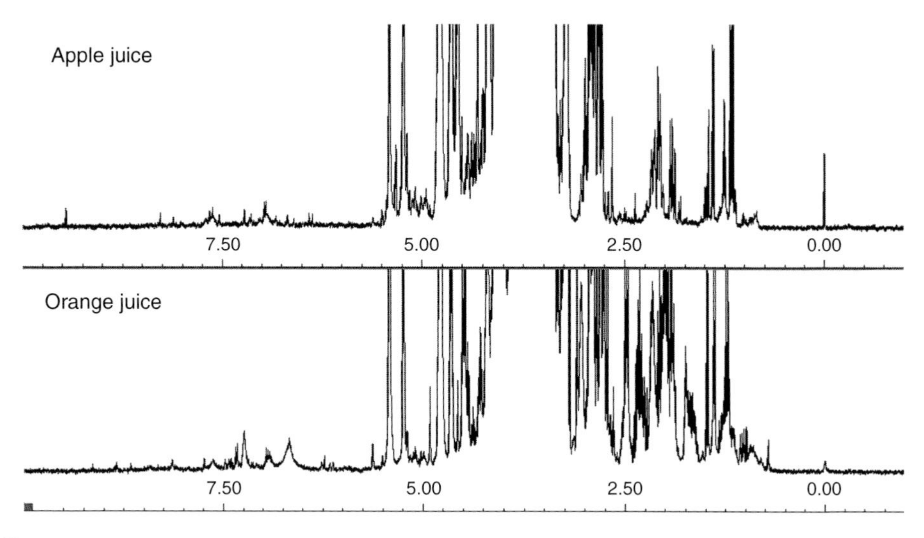

Figure 1 (*Continued*)

When comparing the right part of the figure, one can see the signals of amino acids, organic acids and phenolic compounds. It is clearly visible that there are substantial differences between the juices in many parts of the spectra. This already indicates that statistics have to be done preferably on the entire spectrum. Another difference on the Y-scaled spectra is on the standard signal of TSP. Even though the same amount of standard was added and the measurement conditions are also identical, in orange juice the signal is much smaller and broad. This is due to substantially higher protein concentration in orange juice (compared, for example, to apple juice), which can complex TSP. This observation shows the need for another standard for quantification. This problem is found in exactly the same way comparing, for example, urine and plasma NMR spectra. All tools used as part of the NMR measurement and analysis will be described and the results achievable are shown.

2. Instrumental Aspects of Fruit Juice Quality Control

Flow injection NMR is used in the juice application to assure high-throughput and low-cost operation. A typical 400 MHz system set-up is shown in Figure 2. Samples are positioned on a liquid handler, which can perform preparation and sample transfer to the NMR flowcell. Preparation consists of buffer addition and consequent mixing. Juices with solid particles like orange juice have to be centrifuged prior to buffer addition.

The buffer is prepared in D_2O to allow locking of the spectrometer for spectral quality reasons. The juice is mixed with the buffer in 90–10 aliquots. For sensitivity purposes, a 120 μl flowcell is used, where 120 μl is the active volume and 200 μl is

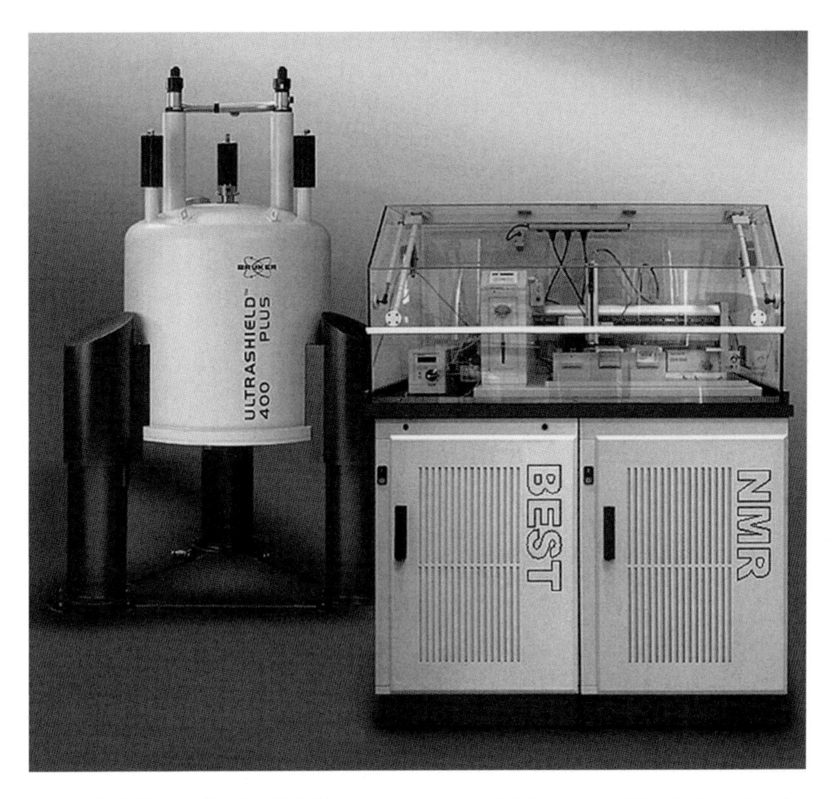

Figure 2 Typical flow injection NMR spectrometer with integrated liquid handler for sample preparation and transfer.

the total volume of the cell. To achieve rapid transfer, the cell is overfilled on sample transfer by injecting 300 μl of juice. Non-deuterated water is used for the transfer of the sample.

Thus the cost is minimized by not needing NMR tubes and having water as push solvent.

The flow probehead is equipped with a *Z*-gradient and temperature-controlled transfer line.

This allows to adjust the measurement temperature on transfer with minimum overshoot needed. Temperature equilibration for a typical 5 mm NMR tube can take 3–4 min, during which the instrument remains idle. This is completely obsolete in the flow solution with heated transfer line. To ensure correct transfer, a gradient image is acquired after the sample is positioned in the flow cell. Thus, transfer problems could be detected immediately without time loss. Such problems could be incomplete filling and air bubbles in the sample, as seen clearly in the image profile.

For correct quantification purposes, the proton 90° pulse is checked on every sample after locking and lock phase adjustment, having the probe tuned and matched as well as gradient shimmed automatically. The pulse determination can be done with a single scan acquisition in a few seconds.[34] The 1D version of the 2D NOESY[35] in a modified form is used for the 1D acquisition. This pulse sequence like many others is sensitive to errors in the 90° pulse in that the signal intensity is lost. The 1D NOESY sequence is using gradients and pre-saturation of the water signal. Owing to the use of gradients, the mixing time can be kept short (~10 ms) and signal loss due to t_2 relaxation is minimized. It is also assured, that the residual water is always in phase with the rest of the spectrum.

The importance of totally standardized conditions can be explained on the quantification of D-glucose, which is done co-adding the α- and β-anomeric proton signal integrals.

Figure 3 shows the effect of pre-saturation power on the integral changes of α- and β-anomeric signals. As the β-anomeric signal is closer to water it is affected much stronger.

A 25 Hz pre-saturation field is adequate to achieve good solvent suppression and no higher power should be applied. From the integral values obtained, it is obvious, that quantification can only be done, if the pre-saturation strength is calibrated. Because only then it is possible to use a correction factor for the β-anomeric integral value.

Modern instruments have digital receivers, allowing efficient filtering of the incoming signal.

Thus, it is possible to generate spectra with no baseline offset directly, making baseline correction obsolete. In addition, the sweep width and timing can be

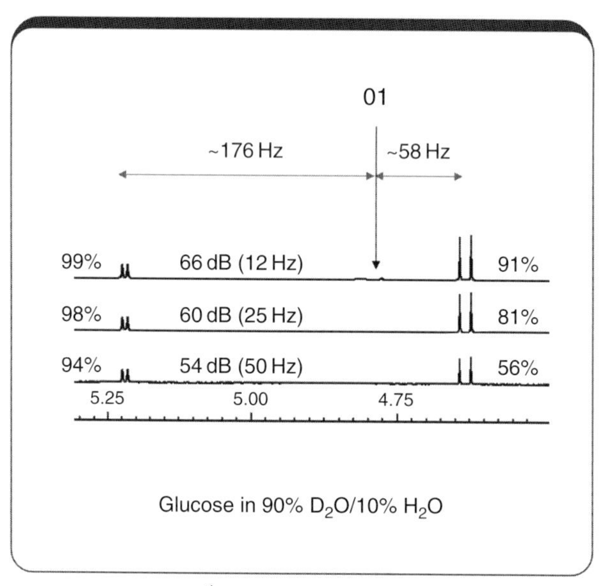

Figure 3 Effect of pre-saturation strength on the attenuation of α- and β-anomeric proton signals of D-Glucose at 400 MHz.

adjusted in a way that spectra are produced with 0° first-order phase correction, which together with a flat baseline allows fully automatic phase correction of highest quality.

Figure 4 shows 40 spectra obtained under the conditions described with automatic sample preparation and no manual interaction. It can be seen that phase and baseline are perfectly set in the left part and the signal positions are absolutely stable as seen in the expansion on the right side.

Together with precise temperature settings the system's internal variance of the instrument is at minimum. This is important for statistical analysis, where smallest effects in the samples have to be detected safely as well as for precise quantification.

As has been already discussed in the introduction, the internal standard is problematic in some fruit juices due to higher protein concentrations, especially among orange and pineapple juices. Using DSS 2,2-dimethyl-2-silapentane-5-sulfonate instead of TSP sodium-3-trimethylsilyl-2,2,3,3-d4-propionate leads to the same problem. Therefore an alternative is needed. Using an ERETIC (Electronic REference To access In vivo Concentrations) signal[36,37] overcomes this problem, as this is an artificial signal, that can be placed at every position in the spectrum. This signal has to be calibrated against a sample of known concentration and known number of protons. This calibration should be repeated in regular time intervals. Generation of the signal does not need a full frequency channel with amplifier, just the frequency generation is needed. Best performance is obtained if the signal is fed into the proton channel through a directional coupler, as this assures that tuning and matching of the proton channel have the same effect on ERETIC and real signals. Figure 5 shows an apple juice spectrum with ERETIC signal at 12 ppm and TSP at 0 ppm. It must also be mentioned that the phase of the ERETIC signal has to be pre-adjusted for correct quantification so that it is in phase with all other signals.

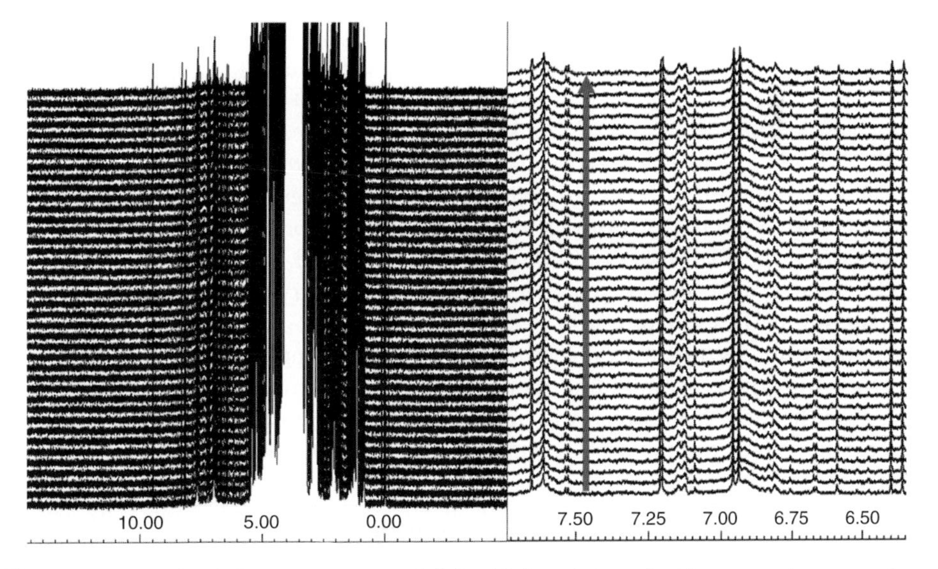

Figure 4 Reproducibility test on one apple juice 40 times injected with automatic preparation, measurement and processing. Left: whole spectrum, right: expansion of aromatic region.

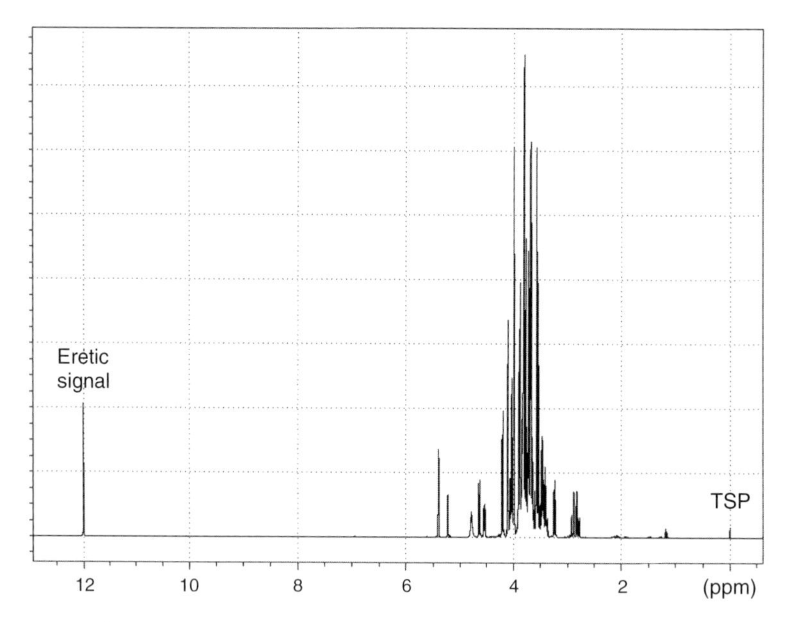

Figure 5 1D apple juice spectrum at 400 MHz with integrated ERETIC signal.

3. Identification of Compounds in the Mixture

Before being able to safely quantify compounds in the fruit juice spectrum, it is necessary to identify them correctly. Knowing that several compounds like citric acid and malic acid have strong chemical shift dependency from pH, ionic strength and general composition, it is often difficult in strongly overlapping signal regions to assign the signals correctly based on 1D spectra. Since fruit juice quality control is a high-throughput screening (HTS) assay, it is not possible to run extensive 2D data on each sample. Only 2D acquisition to a maximum of 3 or 4 min would be agreeable with the scenario.

Therefore the *J*-resolved experiment[38] was modified to include spoiler gradients and allow operation with one or two scans per increment. Such acquisition with one scan and 40 increments results in less than 2 min experiment time. To increase resolution in the F1 dimension, linear prediction is used.

The experiment allows to differentiate signals of varying multiplicities if they overlap in the 1D spectrum.

Figure 6 shows an expansion of a 1D passion fruit juice spectrum at 400 MHz together with the corresponding rapid *J*-resolved spectrum. It can clearly be seen that quantification would miss the singlet at 2.96 ppm completely as well as some other multiplets under the citrate signals; it is also obvious that the integration value achieved for citric acid would be too high, only using the 1D spectrum.

Identification of signals in the 2D spectrum is achieved using a reference compound database for biofluids and food materials.[39] This spectral database contains 1D and 2D spectra of more than 450 compounds at different pH values, for fruit juices the entry at pH = 3 is used.

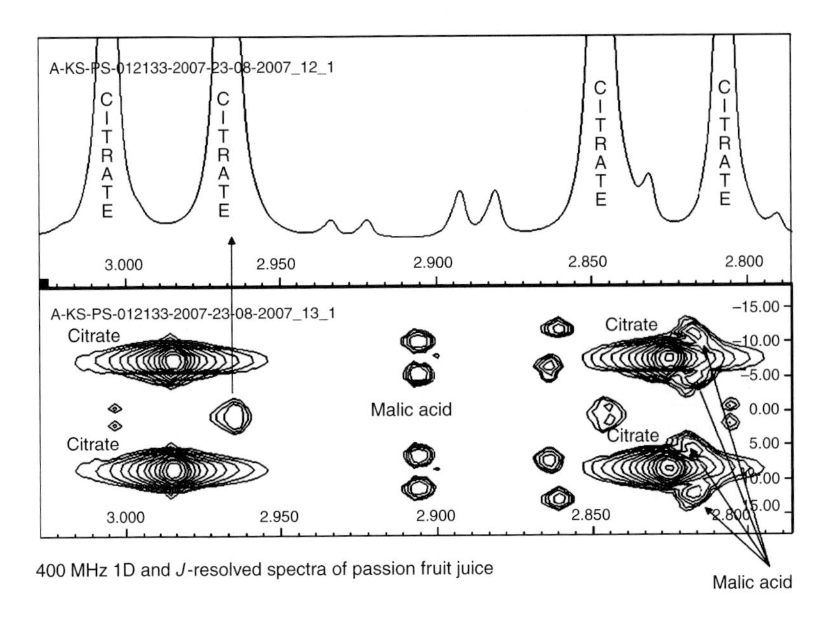

Figure 6 Resolving overlap in a 1D spectrum of passion fruit juice using a rapid 2D *J*-resolved spectrum.

As mentioned before, despite pH adjustment, signals can still shift. The amount of shift change can be identified by looking to a multitude of juices and taking the chemical shift range from there. It is important that only juices from the same fruit are used. This also means that depending on the fruit type, selective information is required. A software had to be developed that takes the shift range information and searches the corresponding coupling patterns taken from the spectral database in the juice 2D spectrum.[40]

4. Quantification of Compounds in Fruit Juice Spectra

Having identified the signals in the mixture spectrum, quantification can be enabled.

The additional information needed now is the number of protons represented by a signal or signal group (multiplet). As the fruit juice screening is run in high-throughput mode, it is not possible to wait for complete relaxation of all protons. This implies that correction factors have to be used, if a signal is not completely relaxed in the recycle time of the experiment.

Therefore, T_1 values of the compounds of interest have to be measured in the corresponding juices and correction factors have to be calculated.

The actual integration then works by deconvolving the spectrum, applying the corresponding correction factors for T_1 relaxation and pre-saturation reduction where needed.

The values found can be compared for example to the Code of Practice of the European Fruit Juice industry that contains concentration ranges for selected compounds so far accessible by conventional means (non-NMR) for several fruit types. A traffic light approach can be taken here to visualize the result. Red together with an up or down arrow indicates values that are above or below the allowed concentration range. A yellow traffic light can indicate if the value is just on the limit, again combined with an up or down arrow. Green indicates compliance with the defined concentration ranges. However, deviation from the reference guidelines should be evaluated and interpreted by experts. Refinement of guide values depending on geographical origin and/or production technique of the analysed juice is possible and necessary for quality control purposes. With NMR it is possible to quantify further compounds, after collection of a larger set of authentic samples for every fruit juice type, new standard values can be defined based on NMR results. Therefore, it is important that authentic samples had undergone a normal industrial process before analysis. Juices obtained from fresh fruits in a laboratory can show quantitative and qualitative differences.

Figure 7 shows very good correlation between concentration values determined by NMR compared to conventional analysis for lactic and malic acid in apple juices. No correction factors are applied.

A correction factor is needed in the quantification of D-glucose as already mentioned, due to the partial saturation of the beta-anomeric signal, if this is used for quantification. Figure 8 shows the correlation obtained for D-glucose (sum of α- and β-isomer). Correlations are aligned along a straight line, deviating from the diagonal. The NMR value is too small as expected. Applying a correction factor centres the points along the diagonal (not shown).

Such correlations of NMR and conventionally determined concentrations have to be investigated separately for every juice type, as the signal overlap situation is different for the various juices investigated.

Quantification also allows, for example, to detect addition of citric or malic acid, which sometimes happens if the juice is too sweet in taste.

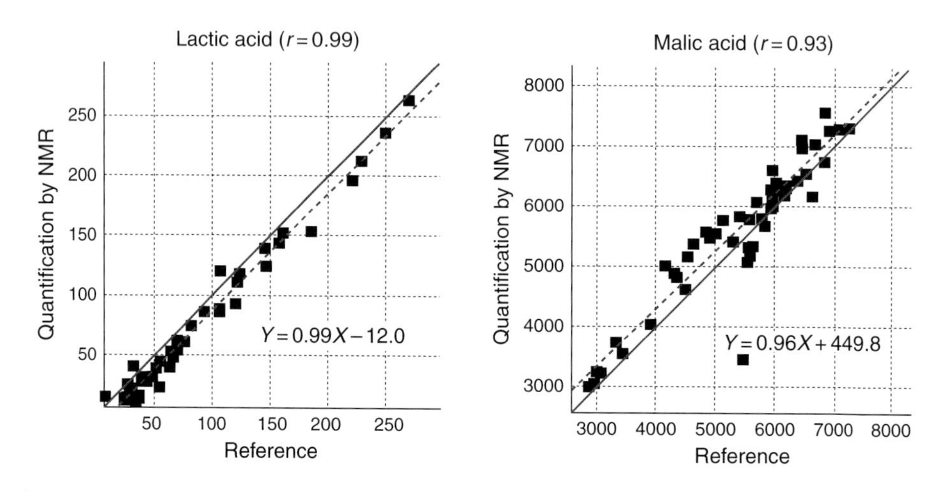

Figure 7 Correlation between NMR and conventionally determined concentrations of lactic (left, enzymatic) and malic acid (right, HPLC) in mg/l.

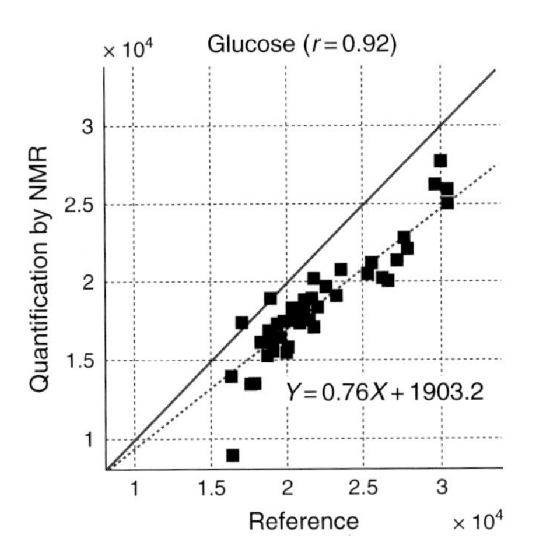

Figure 8 Correlation of NMR-based concentration of D-glucose and values determined with conventional means (enzymatic test kit).

5. Ridge-Regression in Fruit Juice Quality Control

Besides the deconvolution-based concentration determination, it is also possible to use ridge regression analysis.[41] For this purpose, a training set of data is needed, where the values for conventional analysis (enzymatic test kit) are available for every NMR juice spectrum obtained. As mentioned for the correlation based on deconvolution, this has to be done independently for every juice type. Figure 9 shows the result obtained for D-glucose in apple juice. Each juice was tested 5 times. The correlation of these five tests for all juices produces a measure of the regression quality. In the case shown, a value of 0.96 was obtained. The probability factor that the value was accidental is less than 1E − 10, clearly dismissing this option.

Ridge regression can be used to not only predict concentrations of single molecules, but it is also possible to predict, for example, the glucose/fructose ratio, a parameter used in fruit juice industry for quality assessment. Another parameter of rather historical type is the titration value to pH = 8.1, which is a measure for the total acidity of a fruit juice. The result is expressed as the corresponding amount of citric acid.

Figure 10 shows the correlations obtained for both quality parameters mentioned.

The glucose/fructose ratio is of interest, as fructose is present in the juice in the form of five isomers, where three are low in concentration and therefore difficult to access by NMR, the solution so far was to use a fixed factor to calculate from one isomer to the total fructose content. As glucose is precisely accessible by NMR-based ridge regression, fructose can be back-calculated, if the glucose/fructose ratio is precise as well.

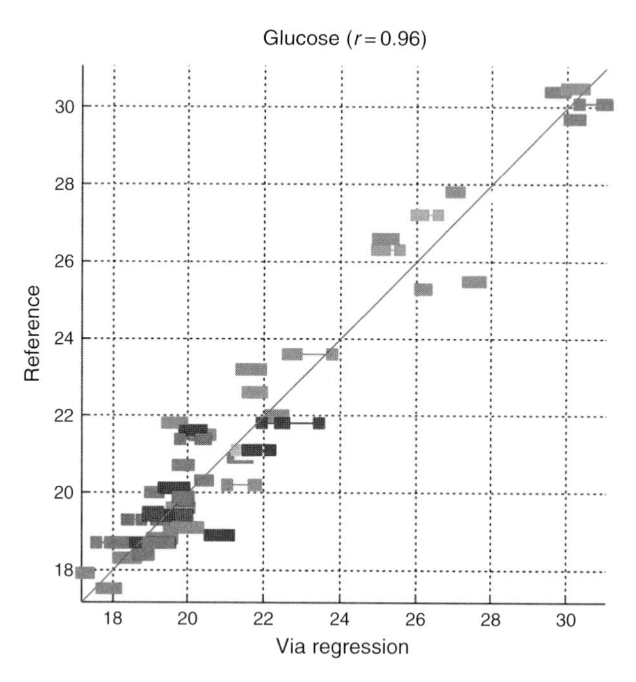

Figure 9 NMR concentrations of D-glucose in apple juice determined by ridge regression based on 400 MHz NMR spectra.

6. Statistical Analysis on Fruit Juices Other than Ridge Regression

There are several questions in fruit juice quality control besides the quantification and the ridge regression analysis. Control of juice properties described for the semi-finished product and the final juice available to the consumer is common practice in the European Community. Such properties are, for example, direct juice or re-diluted concentrate, fruit purity and geographical origin. It is also possible to detect so far unknown problems in the juices based on classification to the authentic models, as the statistics can define the region in the spectrum that is differing from the model. The region can be defined by the so-called "buckets" (for example, equidistant subfractions of the spectrum) or also individual data points of the spectrum. Likewise, it was possible to determine 5-hydroxymethylfurfural as a reason for spectra to be outliers, as the statistics directly pointed to the resonances of this molecule and a search in the reference compound database revealed the identity. This compound can be found in juices, if during treatment the normal temperature is exceeded or if sugar colour is added to the juice. In a similar way, it was possible to detect an excessive concentration of quinic acid in the whole acid profile, which indicates the use of unripe apples in juice production.

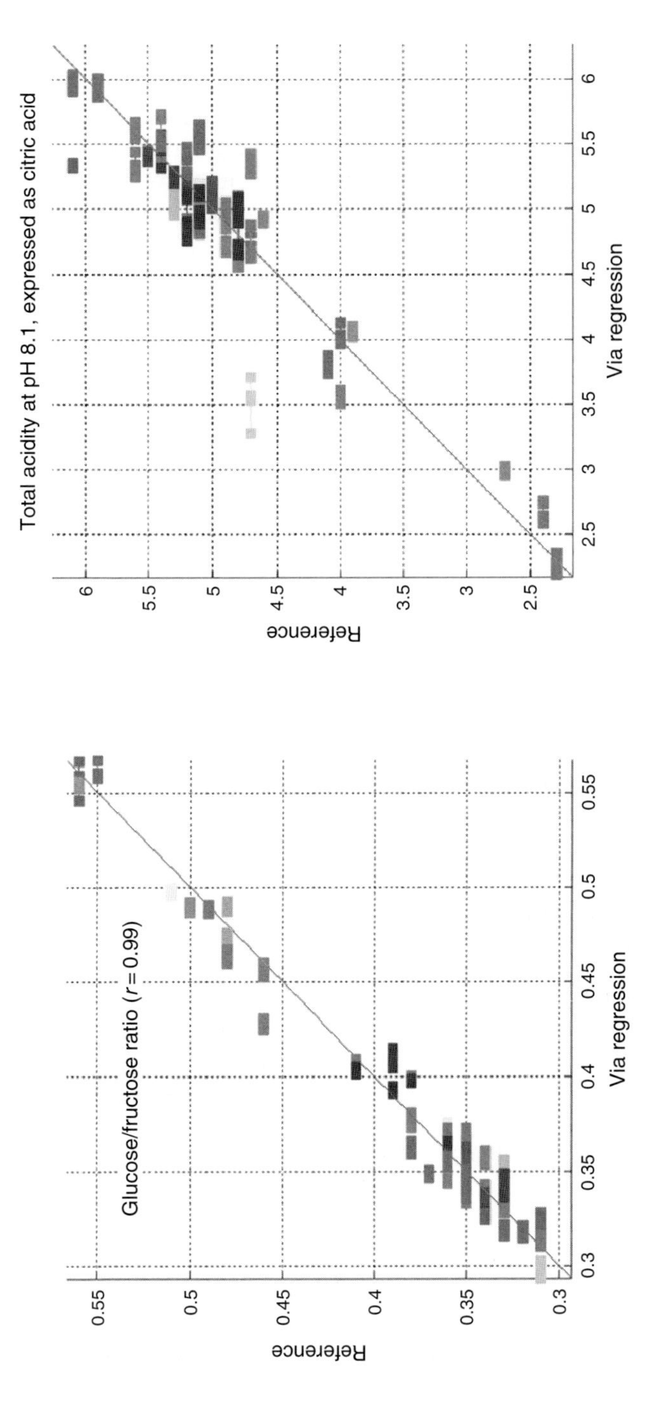

Figure 10 Ridge regression results for the glucose/fructose ratio (left) and the total acidity at pH 8.1 (right).

The first test mentioned is to check whether the juice is a direct juice, or whether it has been reconstituted from a concentrate by dilution with water. The direct juice is a higher price product. One reason for this is the transport volume. When orange juice is transported from Brazil as the largest producer to Europe or the US this is done to the major extent as concentrate, saving a factor of 5 in transport volume. Another product shipped in large amounts is apple juice from China, which is much sweeter than, for example, juice from central Europe. The final apple juice reaching the consumer therefore often is a mixture of juices from different origin.

This already forms the second analysis request for geographical origin purity. If the juice label defines a pure juice from one geographical origin, this has to be tested.

Coming back to the question direct juice or re-diluted concentrate, there are differences in the NMR spectra, which are due to the heat treatment during the concentration of the juice. This leads to minor changes in the compound composition, mainly on small molecules and therefore is suitable for NMR detection. To be able to perform statistical classification, a large enough number of authentic samples is needed to build a model, for example, of direct orange juice. Again these models have to be fruit-type specific to be meaningful.

This clearly defines that a large number of juices has to be measured, before different juice types can be classified with regard to the direct juice versus re-diluted concentrate question.

Only authentic juices should be used to build a model.

Figure 11 shows the result of testing a new orange juice sample against the models for direct juice and re-diluted concentrate. It has to be mentioned that the models in this case are not region specific, the analysis works independent of

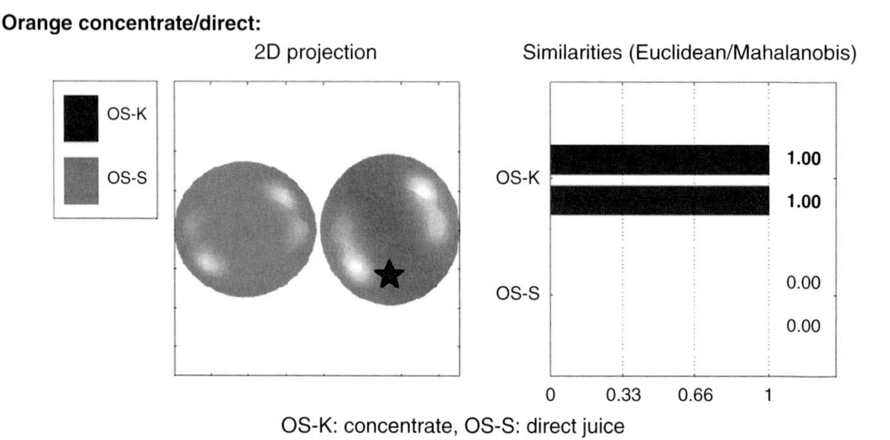

Figure 11 Recognition of direct orange juice versus re-diluted concentrate based on 1D NMR proton spectra at 400 MHz. The 99% confidence ellipsoids of 127 and 105 model samples are shown. The asterisk represents a test sample.

whether a juice comes from Brazil, Europe or the US. The asterisk in the model for re-diluted concentrate clearly indicates that the new orange juice is made from concentrate. Besides the graphical representation a similarity factor is shown that is calculated in two ways, using either Euclidian or Mahalanobis distance measurements.

When looking into the geographical origin, authentic samples have to be available from the main orange-producing countries in the world in sufficient amounts. One has to be aware that geographical origin in this case includes other parameters like the orange variety and the juice production method. Here direct juices and re-diluted concentrates are used in combination to build the geographical origin models. This is again a very good example to demonstrate how many different aspects can be investigated with one fully automatic measurement. It should be outlined that NMR has a unique strength in this respect.

Figure 12 shows the classification of a new orange juice sample towards the geographical origin. As always new samples are used, they have not taken part in the construction of the models. However, every new authentic sample coming in can take part in even stronger models, based on more samples, meaning that the system is learning and delivering better classification results. In the figure, six areas of origin are available as models. The samples of Belize (BH), Costa Rica (CR) and Mexico (MEX) form one model, this fact being necessitated by (a) the restricted number of authentic samples, (b) having a very different amino acid profile compared to the other areas and (c) coming from the tropical part of the American continent.

As more authentic samples are measured, other geographical areas can be implemented in the future like Italy or the North African countries like Morocco. For a country like Brazil, stretching over a very large area of land, it should be possible to obtain areas from dedicated regions within the country. A refined

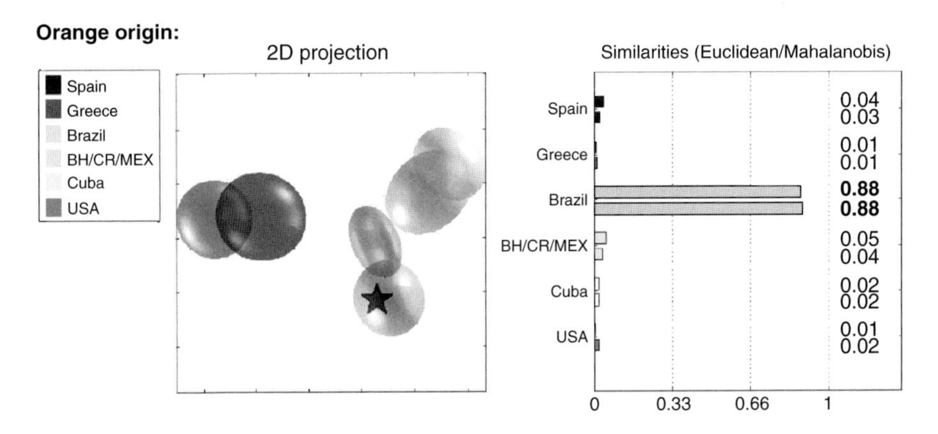

Figure 12 Classification of a new sample (asterisk) into the geographic origin models for orange juice based on 400 MHz 1D NMR spectra having on average 35 authentic samples per region for model building. The models are visualized by their 99% confidence ellipsoids.

recognition of geographical origin and process techniques contribute to the need of large food brands to standardize organoleptic properties to assure consumer satisfaction and corporate identity. This can often only be achieved by mixing juices of different origin and therefore it is likely that NMR can be trained towards a certain taste of the final product.

Staying with the orange juice, another problem is the addition of mandarin juice, which is not allowed in Europe as long as it is not stated on the label. Mandarin juice is typically the cheaper product. It can enhance colour of the juice and compensate missing sweetness. Therefore, in some cases fraud can be observed. Up to the addition of about 20% of mandarin juice, it is very difficult to taste the fraud, even by experienced inspectors.

This problem has more dimensions to it, as it cannot be said straight away that the fraud only happens where the fruits are pressed originally. If the same origin of *Citrus sinensis* (orange) and *Citrus reticula* (mandarin, tangerine and others) could be assumed, it would be easier to detect the fraud, as one could use regional models for both groups. However, it can also happen that addition takes place in the semi-finished product only, then the fruits co-added can stem from different geographical origins.

When looking to orange juice spectra from the tropical Middle-American region, they look more similar to mandarin juice spectra due to the very different amino acid profile. Despite that, there are other differences that can be used by the statistical methods.

Figure 13 shows a principal component analysis (PCA) evaluation of a set-up, where three different mandarin juices have been added to one orange juice in amounts of 10%, 20% and 30%, respectively. It is obvious, that the addition of 10%

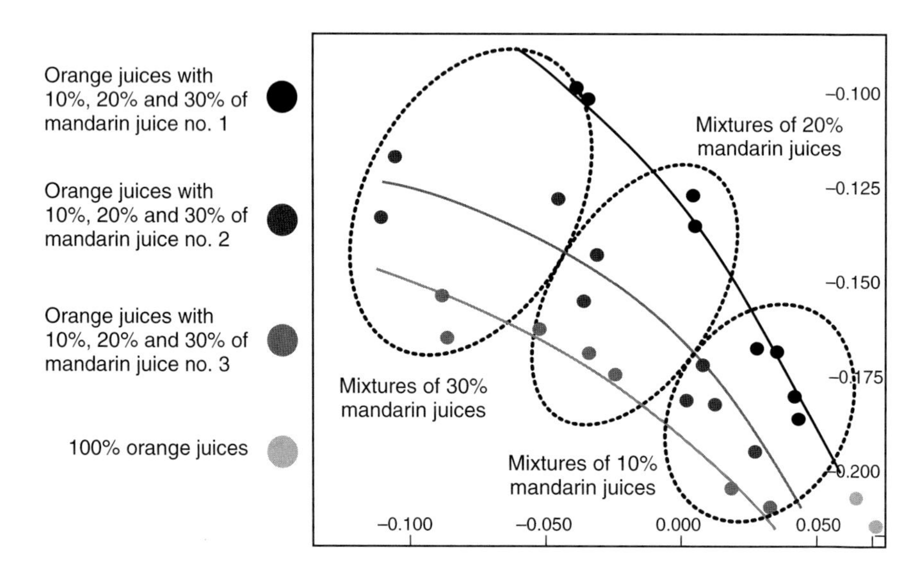

Figure 13 PCA analysis on a 400 MHz spectral test set, where three different mandarin juices are added to orange juice with amounts of 10%, 20% and 30%.

mandarin juice already leads to a substantial deviation from the pure orange juice positions in the 2D PCA map. Fruit juice industry would like to detect additions less than 10%; however, it is also clear that small amounts of mandarin fruits present in an orange harvest collection is hard to avoid, as often orange and mandarin trees grow on the same farm.

Therefore, it would be adequate to be able to detect smaller concentration than 10% too. Further work is in progress to improve the prediction.

Similar problems occur, for example, with passion fruit, where orange juice can be added to a certain content, without being able to taste it. In this example, orange juice is the low-cost juice.

To demonstrate the applicability of the routines described to pharmaceutical application, Figure 14 shows the quantification of vitamin C in various juices. Vitamin C is added to juices having low concentration per se for fortification or as anti-oxidative agent. This raises the value of the juice and represents a similar situation compared to vitamin C being added in drug formulations like headache

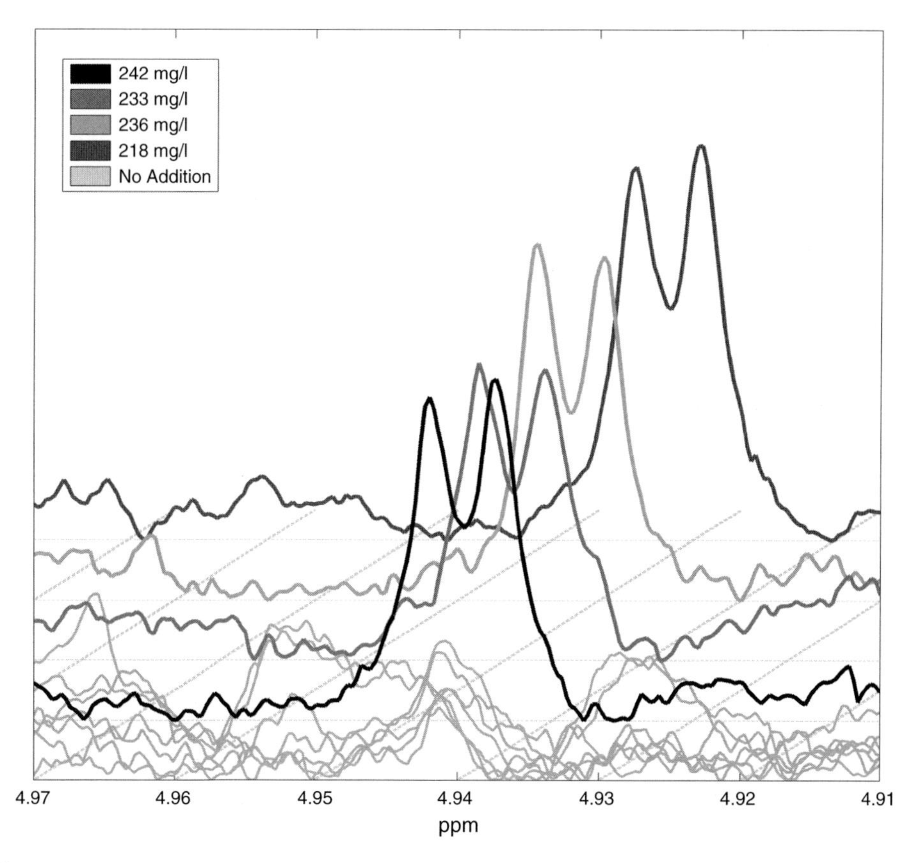

Figure 14 Automatic quantification of vitamin C added to fruit juices based on 400 MHz proton spectra.

or influenza medication. The identification and quantification of vitamin C is based on spectra in the reference compound database and a knowledgebase for juices at pH = 3.

7. Information on the Statistical Methods Applied

For the statistical analysis the amount of data of the spectra has first to be reduced (32 k data points per spectrum at a basis of about 1000 spectra can obviously yield overfitting). This is done via *bucketing*: the resolution of the spectra is reduced by the integration of equally sized regions (the so-called buckets). This data is now called a *bucket table*. Typical bucket widths are between 0.04 and 0.005 ppm. Further statistical analysis is always done on the basis of these bucket tables and not directly on the spectra.

The first step of the statistical analysis used is the recognition of the type of the sample. The challenge is here to safely differentiate between groups of juices and hence to predict meta information about the sample. This is done by cascading models: the first model can differentiate between some fruit types like orange, apple, banana, black currant and sour cherry. If this model detects, for example, the sample as "orange juice", all orange juice models are now applied to the data, predicting the origin, the product type (re-diluted concentrate/not from concentrate) or the addition of *Citrus reticula*.

These models for differentiation are based on discriminant analysis[42]. The most discriminating directions of the high-dimensional data space are estimated by a training set of spectra, which should describe most of the natural variance of all groups. To avoid overfitting, this building process is done in a Monte Carlo approach[43] with disjunct training and test sets. So the estimated accuracy can be calculated by applying the model to the test set for each run. The final classification of a sample is done by projecting its original bucket data into the calculated discrimination model. The nearest model group is the prediction result for this sample.

After the determination of the type of the sample, a second analysis is done for detecting any deviation with reference to the set of the model spectra. With the knowledge of the type, it is possible to test the sample against its most appropriate model, for example testing an orange juice from Brazil against the model "orange juices from South America". This methodology decreases the variances of the models and hence possible deviations can be detected easier.

The first step is to detect single compounds with significantly abnormal concentrations. This is done via a univariate analysis: for each bucket the normal range is calculated with respect to its distribution (defined by the model set). When applied to the bucket data of a test sample, this analysis directly gives information about deviating concentrations and abnormal spectral regions.

The univariate analysis cannot handle multivariate correlations, for example testing the glucose/fructose ratio, which often has to be constant. This check is done by SIMCA modelling[44] with calculation of in- and off-model distances and comparing them to limits obtained by a Monte Carlo analysis (which describes the "normal" behaviour of these values).

8. Conclusion

It could be shown that NMR spectroscopy can play an important role in mixture analysis on the example of fruit juice quality control. To achieve this, several facts had to come together:

(a) barcode operation of the preparation and measurement procedure
(b) reproducible high spectral quality in full automation
(c) flow injection NMR for highest throughput and lowest cost
(d) the existence of a spectral database of reference compounds at suitable pH value
(e) the use of a rapid 2D *J*-resolved experiment for improved compound identification
(f) new methods to assist deconvolution, based on *J*-resolved data
(g) ERETIC signal as internal standard
(h) existence of a spectral database of industrially processed fruit juices, certified as authentic and being in conformity with allowed production techniques
(i) robust statistical evaluation in automation
(j) automatic report generation

The main advantage of NMR is the fact that with a single NMR sample measurement (1D spectrum + rapid *J*-resolved) many questions can be answered, that by conventional means take several independent measurements or are not at all available. The cost per sample can as such be reduced substantially and that creates the possibility to test a much larger set of samples for improved consumer security. Another advantage of the NMR screening method is the ability to detect so far unknown problems. Before this NMR application becomes an official reference method itself, it should be wise to confirm NMR prediction by other established and officially recognized methods, in particular in case of deviation from the expectation and pronounced claims.

The methodology described for fruit juice quality control is only one example out of a new field of applications for statistical and quantitative analysis on mixture spectra and can be transferred to medical and pharmaceutical applications as the database of reference compounds is available also for pH = 7, where most biofluids are measured. Other statistical models are needed first, describing standard situations in biofluids of humans or animals.

Acknowledgement

The methodology described is co-developed by Bruker BioSpin GmbH in Rheinstetten (Germany) and SGF International e. V. in Nieder-Olm (Germany) under the acronym SGF-Profiling.

References

1. C.D. Patz, A. David, K. Thente, P. Kurbele, H. Dietrich, Wein Wiss. 5454 (1999) 80–87.
2. S.A. Kupina, A.J. Shrikhande, Am. J. Enol. Vitic. 54 (2003) 131–134.
3. J.C. Lindon, J.K. Nicholson, E. Holmes, J.R. Everett, Concepts Magn. Reson. 12 (2000) 289–320.

4. J.K. Nicholson, P.J.D. Foxall, M. Spraul, R.D. Farrant, J.C. Lindon, Anal. Chem. 67 (1995) 93–811.
5. C.L. Gavaghan, E. Holmes, E. Lenz, FEBS Lett. 484 (2000) 169–174.
6. J.R. Espina, J.P. Shockcor, W.J. Herron, B.D. Car, N.R. Contel, P.J. Ciaccio, J.C. Lindon, E. Holmes, J.K. Nicholson, Magn. Reson. Chem. 39 (2001) 559–565.
7. J.T. Brindle, H. Antti, E. Holmes, G. Tranter, J.K. Nicholson, H.W.L. Bethell, S. Clarke, P.M. Schofield, E. McKilligin, D.E. Mosedale, D.J. Grainger, Nat. Med. 8(2002) 1439–1444.
8. M. Anthony, K.P.R. Gartland, C.R. Bedell, J.C. Lindon, J.K. Nicholson, Arch. Toxicol. 66 (1992) 525–537.
9. J.K. Nicholson, J.C. Lindon, E. Holmes, Xenobiotica 29 (1999) 1181–1189.
10. J.C. Lindon, J.K. Nicholson, E. Holmes, H. Antti, M.E. Bollard, H. Keun, O. Beckonert, T.M. Ebbels, M.D. Reily, D. Robertson et al., Toxicol. Appl. Pharmacol. 187 (2003) 137–146.
11. J.K. Nicholson, J. Conelly, J.C. Lindon, Nat. Rev. Drug Discov. 1 (2002) 153–161.
12. E. Holmes, A.W. Nicholls, J.C. Lindon, Chem. Res. Toxicol. 13 (2000) 471–478.
13. M.L. Anthony, V.S. Rose, J.K. Nicholson, J.C. Lindon, J. Pharm. Biomed. Anal. 13 (1995) 205–211.
14. M.L. Anthony, B.C. Sweatman, C.R. Bedell, J.C. Lindon, J.K. Nicholson, Mol. Pharmacol. 46 (1994) 199–211.
15. H. Antti, M.E. Bollard, T. Ebbels, H. Keun, J.C. Lindon, J.K. Nicholson, E. Holmes, J. Chemom. 16 (2002) 461–468.
16. M. Spraul, M. Hofmann, M. Ackermann, A.W. Nicholls, S.J.P. Damment, J.N. Haselden, J.P. Shockcor, J.K. Nicholson, J.C. Lindon, Anal. Commun. 34 (1997) 339–341.
17. B.G. Scarfe, I.D. Wilson, M. Spraul, M. Hofmann, U. Braumann, J.C. Lindon, J.K. Nicholson, Anal. Commun. 34 (1997) 37–39.
18. M. Tugnait, E.M. Lenz, P. Phillips, M. Hofmann, M. Spraul, J.C. Lindon, J.K. Nicholson, J. Pharm. Biomed. Anal. 28 (2002) 875–885.
19. M. Spraul, A.S. Freund, R.E. Nast, R.W. Withers, W.E. Maas, O. Corcoran, Anal. Chem. 75 (2003) 1546–1551.
20. O. Corcoran, P.S. Wilkinson, M. Godejohann, U. Braumann, M. Hofmann, M. Spraul, Am. Lab. Perspectives in Chromatography 34 (2002) 18–21.
21. C. Seger, M. Godejohann, L.H. Tseng, M. Spraul, A. Girtler, S. Sturm, H. Stuppner, Anal. Chem. 77 (2005) 878–885.
22. T. Suna, A. Salminen, P. Soininen, R. Laatikainen, P. Ingman, S. Mäkelä, M.J. Savolainen, M.L. Hannuksela, M. Jauhiainen, M. Taskinen, K. Kaski, M. Ala-Korpela M, NMR Biomed. 20 (2007) 658–672.
23. E.J. Jeyarajah, W.C. Cromwell, J.D. Otvos, Clin. Lab. Med. 26 (2006) 847–870.
24. P.S. Belton, I. Delgadillo, A.M. Gil, P. Roma, F. Casuscelli, I.J. Colquhoun, M.J. Dennis, M. Spraul, Magn. Reson. Chem. 35 (1998) 52–60.
25. P.S. Belton, I.J. Colquhoun, E.K. Kemsley, I. Delgadillo, P. Roma, M.J. Dennis, E. Holmes, J.K. Nicholson, M. Spraul, Food Chem. 61 (1998) 207–213.
26. A.M. Gil, I.F. Duarte, I. Delgadillo, I.J. Colquhoun, F. Casuscelli, E. Humpfer, M. Spraul, J. Agric. Food Chem. 48 (2000) 1524–1536.
27. I. Duarte, A. Barros, P.S. Belton, R. Righelato, M. Spraul, E. Humpfer, A.M. Gil, J. Agric. Food Chem. 50 (2002) 2475–2481.
28. F.H. Larsen, F. Van den Berg, S.B. Engelsen, J. Chemom. 20 (2006) 198–208.
29. M.A. Bescia, I.J. Košir, V. Caldarola, J. Kidrič, A. Sacco, J. Agric. Food Chem. 51 (2003) 23–26.
30. P. Rinke, S. Moitrier, E. Humpfer, S. Keller, M. Moertter, M. Godejohann, G. Hofmann, H. Schaefer, M. Spraul, Fruit Processing 1 (2007) 2–10.
31. G. Martin, J. Koziet, A. Rossmann, J. Dennis, Anal. Chim. Acta 321 (1996) 137–146.
32. J.L. Cross, T.M. Gallaher, J.J. Leary, S. Schreiner, The Chemical Educator 3 (1998) 1–9.
33. N. Ogrinc, I.J. Košir, J.E. Spangenberg, J. Kidrič, Anal. Bioanal. Chem. 376 (2003) 1618–2642.
34. P.S.C. Wu, G. Otting, J. Magn. Reson. 176 (2005) 115–119.
35. R. Wagner, S. Berger, J. Magn. Reson., Ser A 123 (1996) 119–121.
36. S. Akoka, L. Barantin, M. Trierweiler, Anal. Chem. 71 (1999) 2554–2557.
37. G.S. Remaud, V. Sylvestre, S. Akoka, J. Qual., Comp., Rel. Chem. Measurement 10 (2005) 415–420.

38. W.P. Aue, E. Bartholdy, R.R. Ernst, J. Chem. Phys. 64 (1976) 2220–2246.
39. Bruker BioSpin GmbH Germany BBIOREFCODE.
40. AMIX software by Bruker BioSpin GmbH.
41. K.D. Lawrence, J.L. Arthur, "Robust Regression", CRC, New York (December 11, 1989) ISBN-10: 0824781295.
42. J.T. Kent, J.M. Bibby, K.V. Mardia, "Multivariate Analysis", Academic Press Inc., US, 1980.
43. S.E. Eugene, "Randomization Tests", Marcel Dekker Ltd, New York 1995.
44. D.L. Massart, B.G.M. Vandeginste, L.M.C. Buydens, S. De Jong, P.J. Lewi, J. Smeyers-Verbeke, "Handbook of Chemometrics and Qualimetrics: Part A", Elsevier Science B.V, Amsterdam 1997.

CHAPTER 2

NMR Assays for Carbohydrate-Based Vaccines

C. Jones *and* N. Ravenscroft

Contents

Abstract

Antibodies against surface carbohydrates of many microbial pathogens protect against infection. Initially exploited through purified polysaccharide vaccines, glycoconjugate vaccines, in which the surface carbohydrate of a microbial pathogen is attached to a carrier protein, now prove the best means to generate protective immunity. Carbohydrate-based vaccines against several organisms are already licensed: similar products are in development. As biological assays are often unavailable, nuclear magnetic resonance (NMR) spectroscopy proves an invaluable tool for their characterisation and quality control.

Keywords: carbohydrate-based vaccines, glycoconjugate, capsular polysaccharide, identity, *O*-acetylation, meningitis, pneumonia, typhoid

1. Vaccines Based on the Cell Surface Carbohydrates of Microbial Pathogens

Bacterial pathogens cause many of the most important diseases which still lead to millions of deaths each year. Infants and children in developing countries are at particular risk, although typhoid is a major cause of mortality in older children and adults, and pneumonia remains a serious problem for the elderly in both developed and developing countries. Many of these pathogens possess an outer shell of cell surface carbohydrates, or glycocalyx. Information on the mortality from these pathogens is available.[1–9] Several of these organisms cause more than one disease, with, for example *Streptococcus pneumoniae* or *Haemophilus influenza* type b (Hib) causing both meningitis and acute respiratory infections, and otitis media. Even when cured of meningitis caused by Hib, the pneumococcus or the meningococcus, approximately 10% of patients suffer long-term neurological sequelae. In addition some fungal pathogens such as *Cryptococcus neoformans*, an important pathogen amongst immunodeficient populations, and parasites also possess cell surface glycoconjugates which are important factors in virulence.

Vaccines are potentially the most cost-effective means to prevent and control infectious diseases caused by encapsulated bacteria such as *Streptococcus pneumoniae*. An international economic analysis that projected pneumococcal vaccination to the same rate as diphtheria–tetanus–pertussis (DTP) vaccine coverage in 72 developing countries predicted it would prevent 262,000 deaths per year in young children, thus averting 834 million disability-adjusted life years (DALYs) yearly.[10] If every child could be reached, up to 407,000 deaths per year would be prevented. Vaccination at $5 per dose was projected to be highly cost effective in 68 of 72 countries when each country's per head gross domestic product per DALY averted was used as a benchmark. Although vaccination of infants in the world's poorest countries against the pneumococcus has the potential to prevent many deaths, it would need substantial funding as this is currently an expensive vaccine. The pneumococcal conjugate vaccine against seven serotypes is the world's largest selling vaccine by global net revenue. It is currently available in 84 countries and included in 16 national immunisation schedules and earned its manufacturer $1249 million net revenue during the first 6 months of 2007.[11] This is likely to increase as more countries include this vaccine in their national immunisation schedules.

One way to bring down the costs of vaccination and thereby extend coverage is through the rise of developing world manufacturers focussing on regional needs. The uptake of Hib conjugate vaccines in the developing world has been hampered by cost and supply. In 2003 the WHO estimated that current global Hib vaccine production capacity was enough for about 50 million of the 125 million children born each year,[12] thus there is an opportunity for developing world manufacturers to enter the market of conjugate vaccines.

1.1. Cell surface carbohydrates

The cell surface polysaccharides of many bacteria – capsular polysaccharides (CPSs) and the O-chains of lipopolysaccharides (LPSs) – and the CPSs of some fungi such as *Cryptococcus neoformans* have a key role in protecting the pathogen from immunological and other host defence mechanisms whilst the organism establishes an infection. The CPS hides cell surface proteins from the immune system, inhibits complement deposition or phagocytosis and protects the bacterium from killing if phagocytosis occurs.[13] However, antibodies against cell surface polysaccharides are often protective against infection, as attachment of the immunoglobulin to a cell surface molecule on the pathogen provides a signal that activates host defences to attack the invading bacterium. (For the classical study on the role of serum antibodies in protection from meningococcal disease, see Goldschneider et al.[14]). Hence CPS and CPS- or LPS-derived conjugates can be used to elicit antibodies recognising the cell envelope and stimulating a protective immune response against these pathogens – that is, they are vaccines. The major challenge in the development of polysaccharide-based vaccines has been to manufacture an immunogen which stimulates a strong, long-lasting, high-avidity antibody response, with appropriate complement-activating antibody isotypes present, and which induces immunological memory. Whilst pure polysaccharide vaccines are still in use against pneumonia, meningococcal meningitis and typhoid, better immune responses and responses in target populations who could not otherwise be protected can be achieved by covalently attaching the polysaccharide, or an oligosaccharide fragment derived from it to a suitable carrier protein. These are the glycoconjugate vaccines.[15–18] Repeated doses of some polysaccharide vaccines have been shown to induce a state of immune tolerance, or hyporesponsiveness;[19] this has added impetus to the development of the corresponding conjugate vaccines.

1.2. Carbohydrate-based vaccines

At present, purified polysaccharide vaccines are available for protection against *Salmonella enterica* serovar Typhi (formerly *S. typhi*), against two, three or four serogroups of *Neisseria meningitidis* and against 23 serotypes of *Streptococcus pneumoniae*.[20] The first licensed glycoconjugate vaccine was against Hib infections and a number of variants are now available, differing in the carrier protein, conjugation chemistry, and whether the saccharide chain is a high-molecular-weight polymer, "size-reduced polysaccharide", or an oligosaccharide. Several monovalent conjugates against *Neisseria meningitidis* Group C and a tetravalent meningococcal conjugate against serogroups A, C, W-135 and Y are now available, and a monovalent serogroup A vaccine, intended for use to prevent epidemic Group A meningitis in sub-Saharan Africa, is expected to be licensed soon. Conjugate vaccines containing 7,

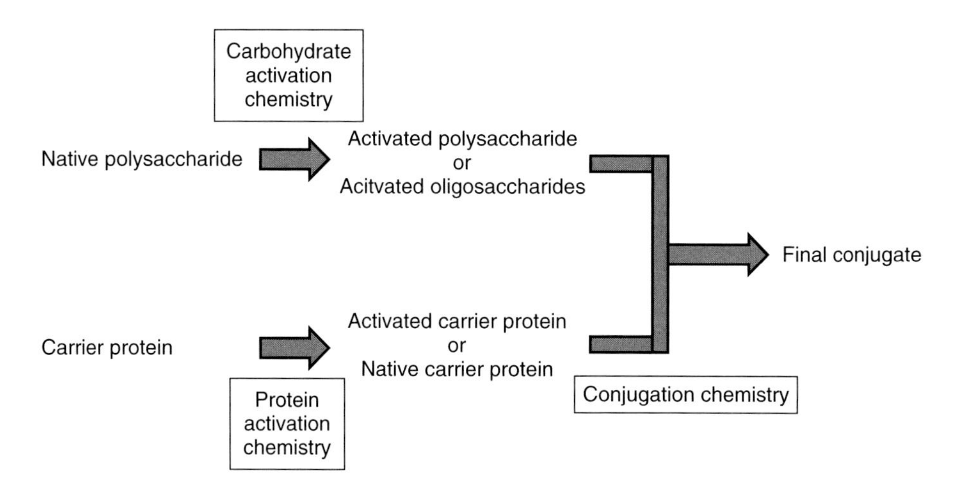

Figure 1 Diagram illustrating production of glycoconjugate vaccines from polysaccharide and carrier protein, with activation of the polysaccharide (and sometimes the protein) prior to conjugation.

10 or 13 pneumococcal serotypes have or soon will be licensed. Late stage clinical trials have been reported for an octavalent LPS-based conjugate vaccine against *Pseudomonas aeruginosa*, for protection of cystic fibrosis sufferers from colonisation of the lungs, for *S.* Typhi Vi conjugates for mass vaccination in Asia, for *Shigella* infections, and for *E. coli* O157 and *Vibrio cholerae*.[21] A novel Hib poly(ribosylribitolphosphate) (PRP) – tetanus toxoid conjugate has been licensed in which the glycan component is produced entirely by chemical synthesis,[22] an approach that allows great flexibility in the conjugation chemistry which can be applied. The coming decade promises to be an extremely exciting time in this area of science. Due to a combination of their relative simplicity and the lack of good in vivo assays that mimic responses in human patients, quality control of polysaccharide-based vaccines relies heavily on physicochemical methods to determine structural identity and other factors crucial for vaccine efficacy. Nuclear magnetic resonance (NMR) spectroscopy has been widely used to characterise the native polysaccharide, activated polysaccharide or oligosaccharide and the final conjugate (Figure 1), and provides valuable qualitative and quantitative data for each type of material.

1.3. The structures of bacterial polysaccharides

CPSs and LPS O-chains are usually assembled by polymerisation of preformed oligosaccharide units, and have a strict repeating unit of between 1 and about 10 sugar residues.[23–24] These repeat units may be linked through glycosidic bonds, phosphodiesters or alditol phosphates, and may be branched or linear.[25] Various substituents such as glycerol phosphate, pyruvate ketals or *O*-acetyl groups may be present. The *O*-acetyl groups are labile and migrate between adjacent hydroxyl groups on the same sugar ring or, less frequently, to sterically available hydroxyl groups on adjacent sugar residues, or may be lost from the polymer by hydrolysis.

This is often the only source of structural heterogeneity present and is observable in the spectrum. Bacterial polysaccharides are a rich source of unusual sugar residues.[26] Because of this repeating structure the NMR spectra of bacterial polysaccharides are relatively simple. The structures of some important bacterial CPSs used in vaccine manufacture are shown in Table 1. Although bacterial polysaccharides have high molecular weights, they give relatively simple and high-quality NMR spectra (Figure 2), especially at elevated temperature. As they have limited tertiary structure, relaxation is dominated by internal mobility, and thus spectral linewidths for resonances tend to depend on the structure of the repeat unit and the presence of a particularly flexible linkage, rather than the overall molecular weight of the CPS.[46] In a branched polysaccharide, residues in mobile side chains may have different relaxation properties than a residue forming part of a relatively rigid backbone.

2. NMR Tests for the Identity and Purity of Polysaccharides Used in Vaccine Manufacture

A key quality control test for polysaccharides used in vaccine production, whether polysaccharide or conjugate, is to ensure that they are what they are supposed to be – which is a combination of their identity and their purity. Within the context of this review, quantification and location of the *O*-acetyl substituents will be considered as an aspect of "identity", as some immunologically distinguishable CPSs differ only in the presence or not of *O*-acetyl groups (such as the pneumococcal Types 9A and 9V), and the immune responses evoked by these CPSs are not necessarily cross-protective, as is the case for pneumococcal Types 19F and 19A.

2.1. Traditional chemical approaches

Methods to establish the identity and purity of the CPSs were developed when the structures of most of the repeat units were undefined, and these remain the official pharmacopoeial methods.[47–49] These are a combination of colorimetric methods for different saccharide types (uronic acids, aminosugars and methylpentoses) or substituents (*O*-acetyl groups). It is not usually known if, for example, the response factor for a 2,6-dideoxy-2-amino sugar is the same as for a "simple" amino sugar or "simple" 6-deoxy sugar in these assays. Coverage is incomplete, with no specification for the *O*-acetyl content for some serotypes now known to be so substituted. Limit specifications on, for example, phosphorus in pneumococcal CPSs that do not contain phosphate groups are, consequently, an indirect limit on the content of C-polysaccharide (a cell wall-associated teichoic acid which is a ubiquitous contaminant of pneumococcal CPSs). Traditional "wet" chemical assays are complemented by an immunological test with specific antiserum, as several of the different pneumococcal serogroups differ only in inter-sugar linkages, not composition. Some polysaccharides, notably the *S.* Typhi Vi, are so stable that these approaches fail, and no useful information can be obtained from traditional chemical approaches.

Table 1 Structures of the repeating units of some of the polysaccharides used in vaccine production

Polysaccharide	Repeat unit	Reference
Haemophilus influenzae Type b ("PRP")	→3)-β-D-Rib*f*-(1→1)-D-Ribitol-(5→***P***→	27,28
Neisseria meningitidis Group A	→6)-α-D-Man*p*NAc(3/4OAc)-(1→***P***→	29
Neisseria meningitidis Group C	→9)-α-D-Neu5Ac(7/8OAc)-(2→	30
Neisseria meningitides Group W135	→6)-α-D-Gal*p*-(1→4)-α-D-Neu5Ac(9OAc)-(2→	31
Neisseria meningitidis Group Y	→6)-α-D-Glc*p*-(1→4)-α-D-Neu5Ac(9OAc)-(2→	31
Salmonella enterica Typhi Vi	→4)-α-D-Gal*p*NAcA(3OAc)-(1→	32
Streptococcus pneumoniae Type 1	→3)-D-AAT-α-Gal*p*-(1→4)-α-D-Gal*p*A(2/3OAc)-(1→3)-α-D-Gal*p*A-(1→	33
Streptococcus pneumoniae Type 2	→4)-β-D-Glc*p*-(1→3)-[α-D-Glc*p*A-(1→6)-α-D-Glc*p*-(1→2)]-α-L-Rha*p*-(1→3)-α-L-Rha*p*-(1→3)-β-L-Rha*p*-(1→	34
Streptococcus pneumoniae Type 3	→3)-β-D-GlcA-(1→4)-β-D-Glc*p*-(1→	35
Streptococcus pneumoniae Type 4	→3)-β-D-Man*p*NAc-(1→3)-α-L-Fuc*p*NAc-(1→3)-α-D-Gal*p*NAc-(1→4)-α-D-Gal*p*2,3(S)Py-(1→	36
Streptococcus pneumoniae Type 5	→4)-β-D-Glc*p*-(1→4)-[α-L-Pne*p*NAc-(1→2)-β-D-Glc*p*A-(1→3)]-α-L-Fuc*p*NAc-(1→3)-β-D-Sug*p*-(1→	37
Streptococcus pneumoniae Type 6B	→2)-α-D-Gal*p*-(1→3)-α-D-Glc*p*-(1→3)-α-L-Rha*p*-(1→4)-D-Rib-ol-(5→***P***→	38
Streptococcus pneumoniae Type 9N	→4)-α-D-Glc*p*A-(1→3)-α-D-Glc*p*-(1→3)-β-D-Man*p*NAc-(1→4)-β-D-Glc*p*-(1→4)-α-D-Glc*p*NAc-(1→	39
Streptococcus pneumoniae Type 14	→4)-β-D-Glc*p*-(1→6)-[β-D-Gal*p*-(1→4)]-β-D-Glc*p*NAc-(1→3)-β-D-Gal*p*-(1→	40
Streptococcus pneumoniae Type 18C	→4)-β-D-Glc*p*-(1→4)-[α-D-Glc*p*(6OAc)-(1→2)][Gro-(1→***P***→3)]-β-D-Gal*p*-(1→4)-α-D-Glc*p*-(1→3)-β-L-Rha*p*-(1→	41
Streptococcus pneumoniae Type 19A	→4)-β-D-Man*p*NAc-(1→4)-α-D-Glc*p*-(1→3)-α-L-Rha*p*-(1→***P***→	42
Streptococcus pneumoniae Type 19F	→4)-β-D-Man*p*NAc-(1→4)-α-D-Glc*p*-(1→2)-α-L-Rha*p*-(1→***P***→	43
Streptococcus pneumoniae Type 23F	→4)-β-D-Glc*p*-(1→4)-[α-L-Rha*p*-(1→2)]-[Gro-(2→***P***→3)]-β-D-Gal*p*-(1→4)-β-L-Rha*p*-(1→	44
Staphylococcus aureus Type 5	→4)-β-D-ManNAcA-(1→4)-α-L-FucNAc(3OAc)-(1→3)-β-D-FucNAc-(1→	45
Staphylococcus aureus Type 8	→3)-β-D-ManNAcA(4OAc)-(1→3)-α-L-FucNAc-(1→3)-α-D-FucNAc-(1→	45

AAT, 2-acetamido-4-amino-2,4,6-trideoxygalactose; Gro, glycerol; Pne, 2-acetamido-2,6-dideoxytalose; ***P***, phosphate in a phosphodiester linkage.

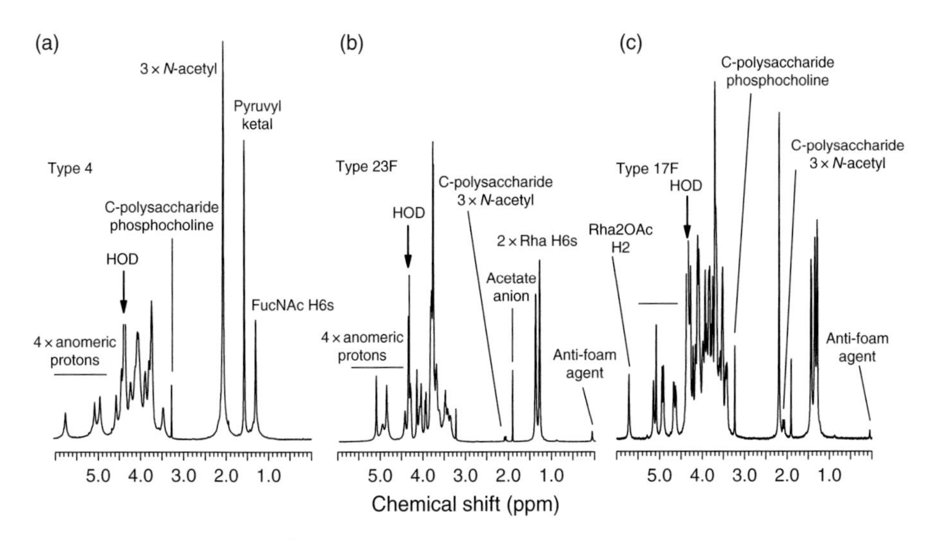

Figure 2 Partial 500 MHz ^{1}H NMR spectra of several CPSs used in vaccines, emphasising differences in line width between different samples: (a) pneumococcal Type 4, (b) pneumococcal Type 23F CPS and (c) pneumococcal Type 17F CPS. All spectra were collected at an indicated probe temperature of 70°C.

Additional tests are used to limit protein and nucleic acid contamination. However, the general policy for pharmacopoeias is that any tests which provide the same information can be used, provided they have been appropriately validated. In practice, NMR methods provide similar but not always identical information to traditional colorimetric and immunochemical assays, and a selection of complementary approaches is still required. The most recent pharmacopoeial monographs or WHO Recommendations do include NMR spectroscopy as "another method" which can be used.[50]

2.2. Polysaccharide identity determination by NMR spectroscopy

Over the past few years we and others have developed NMR spectroscopy as a means to control the identity of CPSs used in vaccine manufacture. NMR spectroscopy provides a fingerprint spectrum characteristic of the saccharide that is sensitive to the small structural differences, such as changes in a single inter-sugar linkage (Figure 3). The spectra of all of the CPSs used in vaccine manufacture that we have studied to date (more than 30) are quite distinct. The spectrum of an individual serotype is sensitive to both the degree of substitution and location of *O*-acetyl groups, and other labile substituents (Figure 4). When high-field instruments are used, NMR approaches require relatively little material (typically 2–3 mg in our laboratory, but sometimes 0.5 mg), and no calibration against reference compounds. When the spectra have been collected, there is no need for reference materials, although published data at several field strengths would be desirable. Application and validation of this approach has been published for the CPSs from Hib,[51] 4 *Neisseria meningitidis* serogroups,[52] 23 *Streptococcus pneumoniae* serotypes[53]

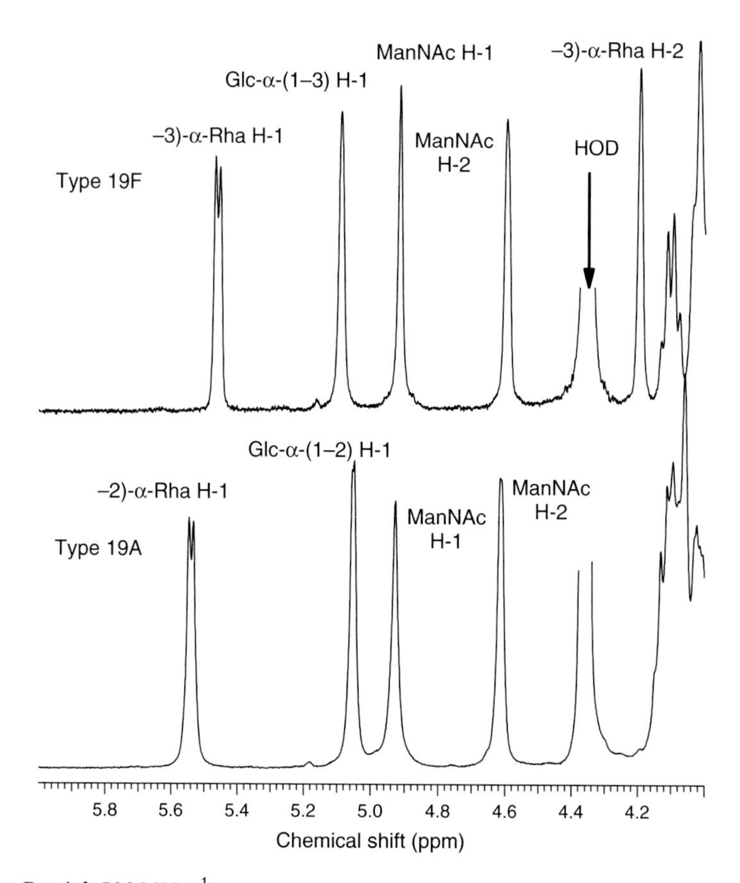

Figure 3 Partial 500 MHz ^{1}H NMR spectra of the anomeric regions of the spectra of pneumococcal Types 19F and 19A CPS. These CPSs have the same composition and differ only in the linkage between the Glc and Rha residues, which is α(1→3) in the Type 19A and α(1→2) in the Type 19F.

and the *S.* Typhi Vi CPS.[54] These approaches are being incorporated into the more recent WHO Recommendations on the production and quality control of the different glycoconjugate vaccines,[57,58] and Recommendations for the production and control of pneumococcal and meningococcal Group A conjugate vaccines.[58,59]

The "fingerprint" methodology has been validated using variable amounts of polysaccharide between 0.5 and 5 mg, by variation in temperature ±20°C from the target temperature, by the addition of salt into the solutions up to 200 mM. These variations of conditions caused, for the Hib PRP spectrum, extremely small variations in the chemical shifts of the resonances (typically 0.005 ppm between 30°C and 50°C), apart from the ribitol H-3, which moves by 0.02 ppm over the same temperature range. "Feasible" errors in shimming or setting of the 90° pulse caused no spectral variation that could compromise the assignment of the CPS identity. In most cases, comparison of the NMR spectrum of the test sample and the reference spectrum has been by visual inspection. The recommendations for a revised general EP monograph on NMR spectroscopy are

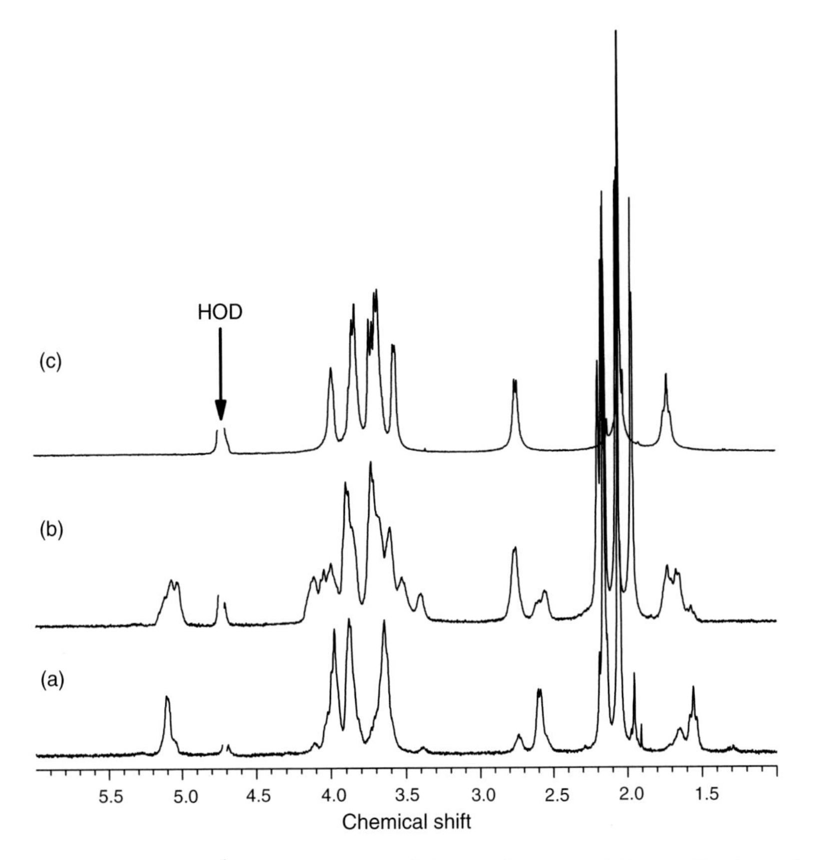

Figure 4 Partial 500 MHz ^{1}H NMR spectra of the meningococcal Group C polysaccharide: (a) the *O*-acetylated CPS with the *O*-acetyl group principally located on the Neu5Ac O-8; (b) with the *O*-acetyl group located principally on the Neu5Ac O-7, the thermodynamically favoured form and (c) chemically de-*O*-acetylated material. Spectra were collected at an indicated probe temperature of 30°C.

that "identity" requires that the peaks in the spectra of the test and reference samples, acquired using the same procedure and operating conditions, should correspond in position, intensity and multiplicity.[59] This is consistent with the EP requirements for assessing identity through the use of infrared spectroscopy.[60]

Whilst not essential, full assignment of the NMR spectrum obviously gives greater confidence, and the development of the methodology has been accompanied by an effort to gain the assignment data. In several cases, this has indicated that the published structures are incorrect, and revised structures reported. In the view of the authors, this is an area which is likely to see an increasing effort in coming years, as more complex glycoconjugate vaccines are licensed for clinical use.

The group at Merck have reported objective mathematical means to compare spectra through the calculation of correlation coefficients between test and reference spectra,[53] using only the anomeric region of the spectrum, and set a requirement that the correlation coefficient should be greater than 0.95. This was validated by calculating a matrix of the correlation coefficients from all possible combinations of the 23 serotypes. It is not clear how sensitive this approach would be, for example, to elevated levels of pneumococcal C-polysaccharide in the sample. Other mathematical approaches, such as those based on principal component analysis, are obviously possible.

A different approach, intermediate between the traditional wet chemical methods and the NMR fingerprint method, is to use characteristic resonances in the NMR spectrum, such as those from *N*- or *O*-acetyl groups or *C*-methyl groups, to calculate the proportions of different sugar residue types present in the saccharide, in these cases, *N*-acetylhexosamine residues, *O*-acetyl groups and 6-deoxysugars, respectively. In some cases, notably the pneumococcal type 1 CPS, the compositional values expected from NMR measurement will have to be adjusted from those expected from traditional chemical approaches, as the latter are skewed by the instability of some of the residues, towards values derived from the structure determined. Some polysaccharides, notably the *S.* Typhi Vi, that compositional assays using traditional chemical approaches have proven less than satisfactory – in this case the pharmacopoeial specification mentions only *O*-acetyl groups and protein and nucleic acid contaminants, and not the saccharide components.

Another recently reported approach was the application of high-resolution magic-angle spinning (HR-MAS) proton and heteronuclear one- (1D) and two-dimensional (2D) NMR approaches to identify the Hib saccharide component present in a cetavlon precipitate, which is the means by which the polysaccharide is initially isolated from the bacterial fermentation. The spectra obtained (1D ^{1}H, TOCSY and ^{1}H–^{13}C HSQC) were consistent and comparable with solution spectra of the uncomplexed polysaccharide, and data acceptable for quality control purposes could be obtained on a 300 MHz instrument.[61]

2.3. NMR analysis of blends and CPSs in the presence of excipients

Whilst control testing of the CPSs is performed at the stage of the monovalent bulk polysaccharides, high-field instruments have sufficient resolution to obtain meaningful data on the blended bulks and final formulations (no adjuvants are included in polysaccharide vaccines, but carbohydrate excipients such as lactose are added in some cases). There is a general requirement on manufacturers to demonstrate that all expected polysaccharide components are present in the expected quantities. The group at Merck reported the deconvolution of the 600 MHz proton NMR spectrum of the 23 component pneumococcal vaccine, demonstrating that all the expected CPSs are present and, potentially, providing relative quantification. This approach highlighted the presence of the pneumococcal type 17A CPS in a 23-valent pneumococcal vaccine, rather than the expected Type 17F.[62] Whilst the authors are not aware of a published report describing this, a similar approach should be possible for bi-, tri- or tetravalent meningococcal polysaccharide vaccines. The availability of pulsed field gradients allows selective detection of high-molecular-weight, slowly diffusing molecules in the presence of low-molecular-weight, rapidly diffusing molecules. This technique has been called "diffusion editing". One of the authors has applied this approach to the observation of high-mass polysaccharides in the presence of a sucrose excipient, with acceptable results and suppression of the sucrose signals by more than 90%, compared to those of the polysaccharide.

2.4. Quantitation of the *O*-acetyl contents of CPSs

Variation in the degree and position of the *O*-acetyl groups can complicate spectral comparison and validation of an NMR fingerprinting approach difficult. An extreme example of this is the meningococcal Group C CPS, where slow spontaneous migration of the *O*-acetyl group from the Neu5Ac O-8 to the Neu5Ac O-7 results in very great spectral changes (Figure 4). The solution to this problem, and of how to quantify the degree of *O*-acetylation, is to obtain two spectra on each CPS sample. Firstly the spectrum of the native CPS is collected, then NaOD in D_2O is added to the sample in the NMR tube to a final concentration of 200 mM. The spectrum of the sample is recorded again after de-*O*-acetylation. In our experience, de-*O*-acetylation occurs in the time required to re-equilibrate the sample in the magnet and to adjust the shimming and tuning, although hydrolysis of the meningococcal Group C CPS is slower. The spectrum of the de-*O*-acetylated material is characteristic of the polysaccharide backbone, and the degree of *O*-acetylation in the original sample is calculated from the integrals of the acetate anion and an appropriate resonance arising from the saccharide backbone (Figure 5). Validated methods have been published for the meningococcal and *S.* Typhi Vi CPSs.[52,54] and assays for pneumococcal and staphylococcal CPSs are under development. In our experience, it has not proven necessary to correct for the presence of trace amounts of acetate anion in the original sample. Currently, specifications for the CPSs typically quote a minimum quantity of "*O*-acetyl residues" (measured by a Hestrin assay) per gram dry weight of CPS, but the

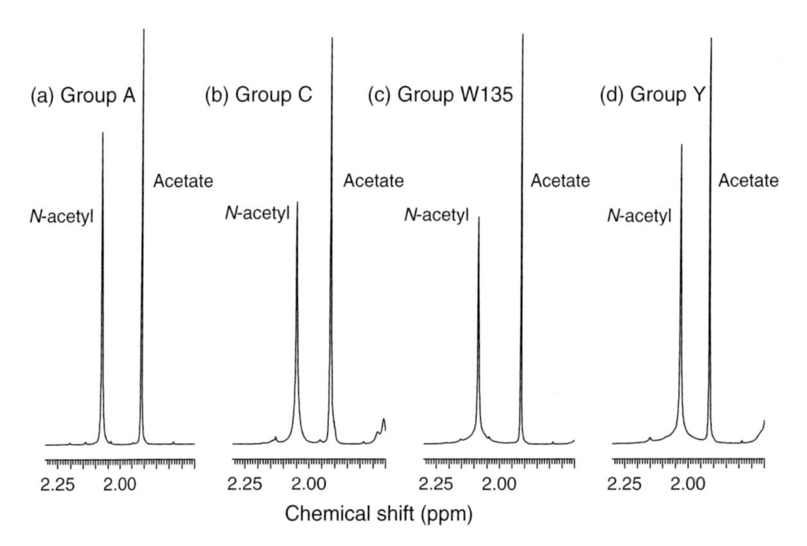

Figure 5 Partial 500 MHz ^{1}H NMR spectra of base hydrolysed meningococcal CPSs, showing resonances from the acetate anion and from the *N*-acetyl groups.

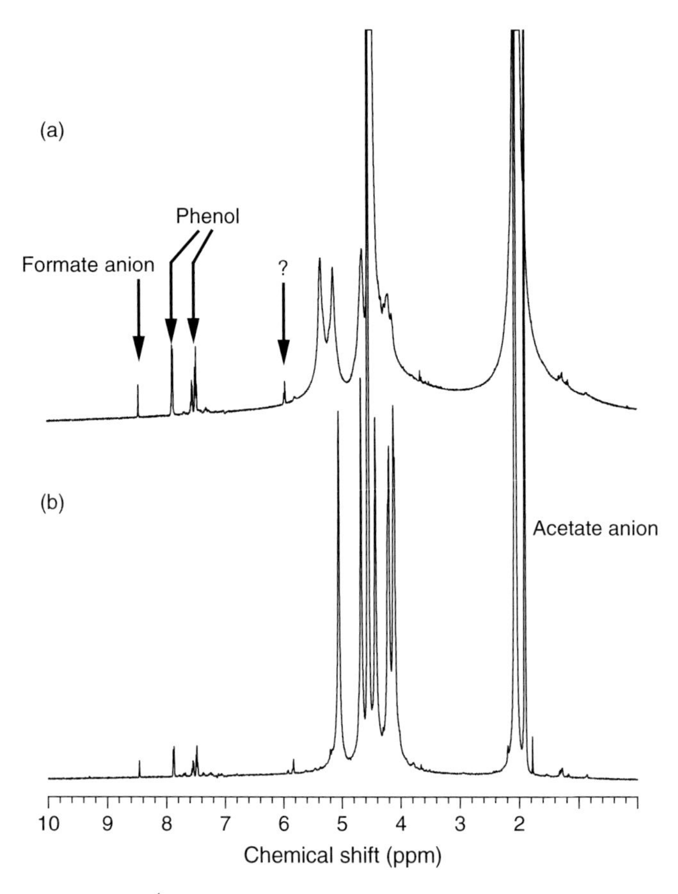

Figure 6 Partial 500 MHz ^{1}H NMR spectra of (a) native *O*-acetylated Vi CPS from *Salmonella enterica* serotype Typhi, at 50°C, and (b) the same material after a short treatment in situ with 200 mM NaOD. Resonances arising from various impurities such as formate and phenol are highlighted.

increasing use of NMR methods will lead, it is to be hoped, to adoption of a simpler specification based on the proportion of repeat units which are *O*-acetylated. The base treatment has other beneficial effects, in leading to reductions in linewidths for some of the more viscous CPSs, such as the *S.* Typhi Vi (Figure 6).[54]

2.5. Quantitation of contaminants and process-related impurities

Since NMR spectroscopy is able to detect almost all molecules in solution possessing protons (or other NMR active nuclei), it is an ideal means to identify and quantify contaminants, whether these are product- or process-related, without defining in advance what species are to be measured. The key species to consider are (i) contaminating polysaccharides produced by the bacteria, (ii) molecules carrying over from the fermentation and polysaccharide isolation steps, (iii) residual organic solvents from polysaccharide precipitation and (iv) potential contaminants arising from polysaccharide processing during activation for conjugation.

2.5.1. Product-related impurities

Although there are no formal direct pharmacopoeial limits for the content of contaminants such as C-polysaccharide[63] in the pneumococcal CPSs, they are a marker of the consistency of the production process. This information is present in the NMR spectrum. For the C-polysaccharide content, the intensity of the C-polysaccharide phosphocholine resonance with known CPS resonances can be compared. Whilst there are known differences in the number of phosphocholines per C-polysaccharide repeat unit in some serotypes (which can be defined) and loss of phosphocholine can occur during manufacture,[64] this approach has the advantages of simplicity and sensitive proton detection. The major problem is to correct for the degree of phosphocholine substitution, typically 80%, which might be by running a separate ^{31}P spectrum on the same sample,[65] or by reference to historic data. No authoritative report to validate this approach has been published. Alternatively, an internal reference compound containing both resolved proton and ^{31}P resonances could be added to quantify the CPS (through the ^{1}H spectrum) and the C-polysaccharide through the relative intensity of the in-chain phosphodiester resonance. This is a less sensitive approach, but will be unaffected by the loss of phosphocholine substituents. In some cases, where the CPS contains a phosphodiester linkage, direct quantification is feasible. The decision between these approaches is likely to depend on the required accuracy, the required sensitivity, and the time needed for validation.

In Gram-positive bacteria, the CPS is covalently attached to peptidoglycan in the cell wall, and release of the CPS may occur by digestion of the peptidoglycan framework, leaving fragments of this attached to the CPS. Relatively high levels of peptidoglycan contamination were observed in NMR studies of the capsular polysaccharides from *S. aureus* Types 5 and 8.[45] C-polysaccharide contaminating pneumococcal polysaccharides can be present in CPS-associated (presumably when both species are peptidoglycan-bound) or free forms (from lipoteichoic acid or released from peptidoglycan), with identical NMR spectra. The group at Merck quantified these two forms using two independent methods, one of which was based on measurement of self-diffusion rates by NMR spectroscopy and quantification of the two species.[66]

CPSs from Gram-negative organisms, such as Hib or *N. meningitidis*, are attached to the bacterial outer membrane through the presence of a lipid anchor, and attached to the glycan chain through a phosphodiester linkage.[67,68] In the spectra of these polysaccharides, resonances are sometimes observed which are consistent with such lipidic material, either as anchors still attached or as a fragment after hydrolysis.

2.5.2. Process-related impurities

Non-carbohydrate impurities which have been observed in these polysaccharides include trace quantities of silicone-based anti-foam agents, acetate anion, formate anion and phenol, some of which had not previously been recognised (Figure 6). It is a simple matter to obtain reference spectra of anticipated or conjectured trace impurities, which simplifies identification of the origin of minor peaks in complex spectra. Other small molecules which are observed include glycerol (arising from ultrafiltration membranes used in desalting and size fractionation), residual cetavlon from polysaccharide isolation and ethanol (from secondary polysaccharide precipitation). Some manufacturers use NMR spectroscopy to estimate the amounts of such materials, either by comparison with the intensity of the CPS resonances or by using added internal standards, such as dimethyl sulphoxide (DMSO).[53] This requires resolved identified resonances arising from the different contaminants.

However, NMR spectroscopy is not an appropriate technique to estimate contamination by high-molecular-weight contaminants such as residual protein or nucleic acid, as the broad lines reduce sensitivity. Quantification of these contaminants is pharmacopoeial requirement, with typical limit specification of 1–2%. Traditional approaches such as colorimetric protein assays or UV absorbance is required.

2.6. Polysaccharide quantification

At some stage in both polysaccharide and glycoconjugate vaccine manufacture, there is a requirement to quantify the amount of saccharide present. This has usually been undertaken using traditional chemical approaches, but for a manufacturer producing a range of products there are complications. For example, the Hib PRP can be quantified by an orcinol assay for ribose, by phosphate assay, or by specialist high-performance anion exchange chromatography (HPAEC) for ribitol[69] or for the PRP monomer.[70] The performance of each of these approaches was assessed during the development of a Hib PRP reference material,[71] and in some cases is poor. For each of the 40 or more polysaccharides present in the existing or developmental conjugate vaccines, different methods are required, or a different specification applied. In the case of the *S.* Typhi Vi CPS, colorimetric approaches fail and a dye binding assay is required.[72] If NMR spectroscopy is to be used to identify the CPS, the same assay can be used to determine the saccharide content of the sample by adding known amounts of reference compounds, providing information on sample purity.[53]

NMR spectroscopy with internal reference standards provides a simple generic approach that can be used for a much wider range of CPSs. Base removal of *O*-acetyl groups may be necessary in some cases (e.g. the Men C CPS). The Merck group report the addition of reference compounds to the test sample to allow quantification of the saccharide content of the sample, and purity can be assessed by comparison of this figure with the dry weight of the sample.

3. Identification of End Groups as Markers of Polysaccharide Degradation

For unconjugated polysaccharides, immunogenicity depends on molecular weight/size, and only high-mass CPSs are immunogenic. Practically, this has limited the number of CPS vaccines which are feasible, and the isolation of pure CPS without depolymerisation drove much of the early developmental work to establish efficacious polysaccharide vaccines. Estimation of molecular weight is a key control test, and extremely rare breaks are sufficient to reduce vaccine immunogenicity and efficacy by a clinically significant degree. Whilst chain cleavage leads to the formation of characteristic NMR resonances from newly formed end group, the sensitivity and dynamic range of NMR spectroscopy to observe small signals in the presence of intense ones means that it is not an appropriate technique. The methods used are soft gel chromatography or high-performance size exclusion chromatography (HPSEC) coupled to multi-angle laser light scattering (MALLS) detection.

The earliest glycoconjugate vaccines against Hib and *Neisseria meningitidis* Group C infections were, coincidentally, produced from some of the least stable polysaccharides used in vaccine manufacture and NMR spectroscopy can be used to investigate their degradation pathways. Hib PRP degrades by cleavage on either side of the phosphodiester linkage, with initial formation of five-membered cyclophosphate intermediates that, under basic conditions, can open to one of the two different monophosphate esters. The major cyclophosphate is on O-2 and O-3 of the ribofuranosyl residue, and can open to either the 2- or the 3-O-phosphomonoester. The minor cyclophosphate intermediate is the ribitol-4,5-cyclophosphate. The terminal β-Ribf 2,3-cyclophosphate residue has characteristic low-field H-1, H-2 and H-3 resonances which are well resolved from those of the intact repeat units (Figure 7) and which can allow quantification.[73] Full NMR assignments of the end group residues formed during degradations have been obtained (Jones and Lemercinier, unpublished data).

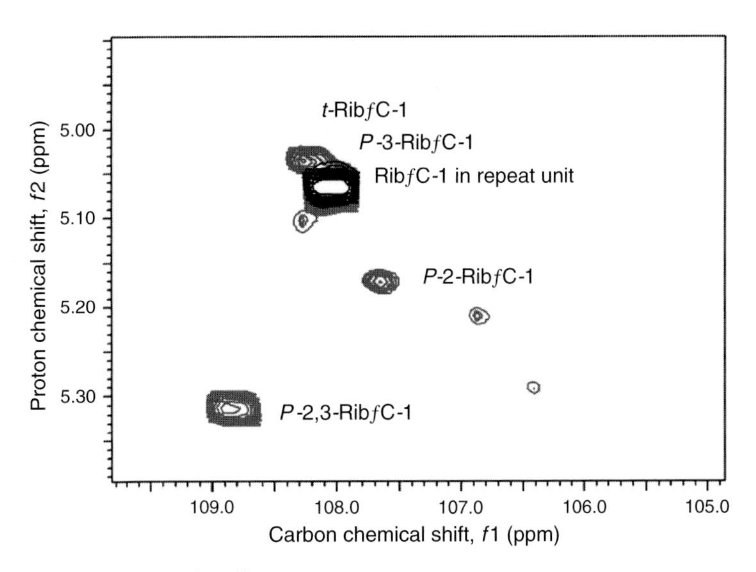

Figure 7 Partial 500 MHz $^{1}H-^{13}C$ HSQC spectra of native Hib PRP (black crosspeak) and material deliberated degraded by thermal stress (grey crosspeaks). Full analysis of the 2D spectra (including $^{1}H-^{31}P$ HSQC spectra) allows individual crosspeaks to be assigned and the mechanisms of degradation defined. Whilst formation of the Rib*f*(2,3-cyclophosphate) is preferred over the ribitol(4,5-cyclophosphate), a more detailed quantitative analysis is possible to define rates of individual reaction steps.

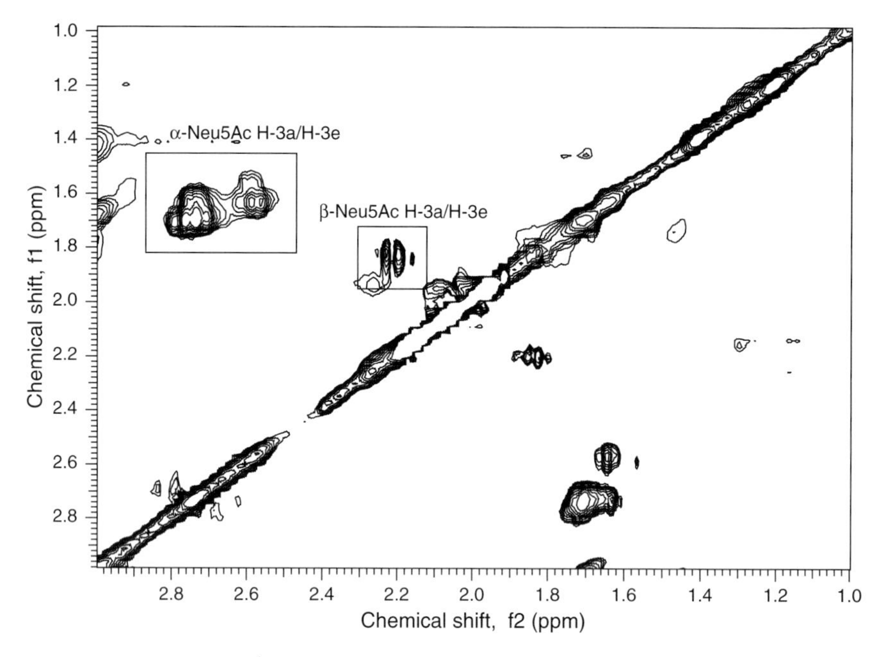

Figure 8 Partial 500 MHz ^{1}H TOCSY spectrum of a deliberately degraded Men C-CRM197 conjugate, showing Neu5Ac H-3a/H-3e crosspeaks from in-chain residues in the α-anomeric configuration, and reducing terminal Neu5Ac residues in the β-anomeric configuration. As for the Hib PRP example, identification of the crosspeaks defines the preferred mechanism of degradation, and quantitative analysis is possible, to complement measurements of the increase in free saccharide arising from saccharide depolymerisation.

On the other hand, the Group C meningococcal CPS degrades by hydrolysis of the labile ketosidic linkage, allowing migration of the *O*-acetyl group to the thermodynamically favoured O-9 position at the non-reducing terminus and anomerisation to form the favoured β-anomer at the new reducing terminus (Figure 8). Both of these groups give rise to identifiable signals in 1D and 2D NMR spectra.[74]

4. Characterisation of Activated Intermediates in Vaccine Manufacture

A common feature of the production of any glycoconjugate vaccine is the need to activate the polysaccharide for conjugation to the carrier protein (Figure 1). In some cases, the carrier protein is also activated to provide a complementary reactive group. Although it is conceivable, for example, to couple a carboxylate-containing polysaccharide to an amino group containing protein through direct amide formation, in practice, there are always too many competing reactive groups present to allow the use of such a simple strategy. Two general strategies are in use.[22,75–85] Firstly, a depolymerisation step is used (periodate oxidation, acid hydrolysis, deamidation or radiation-induced damage), which generates one or two reactive sites at the chain terminus/termini which can be used directly in the conjugation or which allow attachment of an appropriately activated linker. The resulting "neo-glycoconjugate" vaccine tends to be monomeric (although of cross-linking can occur with bifunctional saccharides from periodate oxidation), and may be similar to a typical plasma protein in size and proportion of glycan, apart from the identity of the glycan chains. The second approach is random multiple activation at sites along the "intact" polymer chain, with cyanogen bromide, periodate or amide formation at carboxylate groups. Reactive groups present on the linker, or introduced as a secondary step, react with complementary groups on the carrier protein to form complex crosslinked network vaccines.

NMR spectroscopy provides a means to characterise activated polysaccharides, to quantify the degree of activation that has occurred, to identify the most reactive sites within the polysaccharide and, if depolymerisation has occurred, to determine the mean molecular weight of the oligosaccharides formed.

4.1. Combined activation and depolymerisation using periodate oxidation or acid hydrolysis

Glycoconjugate vaccines stimulate immunity through an entirely different immunological mechanism than unconjugated polysaccharides. This pathway, even present in young infants, involves T cells and leads to avidity maturation, induction of complement-recruiting antibody isotypes and induction of immunological memory. Another consequence of exploiting a different mechanism of immunogenicity is that these vaccines can be manufactured using relatively short glycan chains: less than 10 sugar residues are necessary. For some products, therefore, oxidative depolymerisation of the Hib and Group C meningococcal CPSs to oligosaccharides with concomitant activation for conjugation is a key manufacturing step, and quality control procedures aim to ensure that successive batches of oligosaccharides are consistent in molecular weight (i.e. that the degree of activation has been consistent). Although there are several approaches to obtain these data, including high-performance liquid chromatography (HPLC)–MALLS, ion exchange chromatography and combinations of wet chemical assays, NMR can provide a mean value for the degree of polymerisation (but not a size distribution profile) and is useful as a reference method as no assumptions about response factors or reference compounds are required. Periodate oxidation leads to the formation of terminal (hydrated) aldehydes with low-field proton resonances as shown for Hib (Figure 9), In a comparative study, proton NMR analysis was used to establish the molar mass of the Hib oligosaccharides generated by periodate oxidation.[82] Good correlation was obtained with the mass determined by MALLS, whereas the colorimetric assays were shown to vary with the reference used. Oligosaccharides generated by acid hydrolysis give rise to low-field anomeric resonances, and these resonances are usually resolved and quantifiable.[86] NMR has been used together with liquid chromatography and mass spectrometry to qualify colorimetric size analysis assays for Hib, Men A and C oligosaccharides obtained by acid hydrolysis.[75] Similarly,

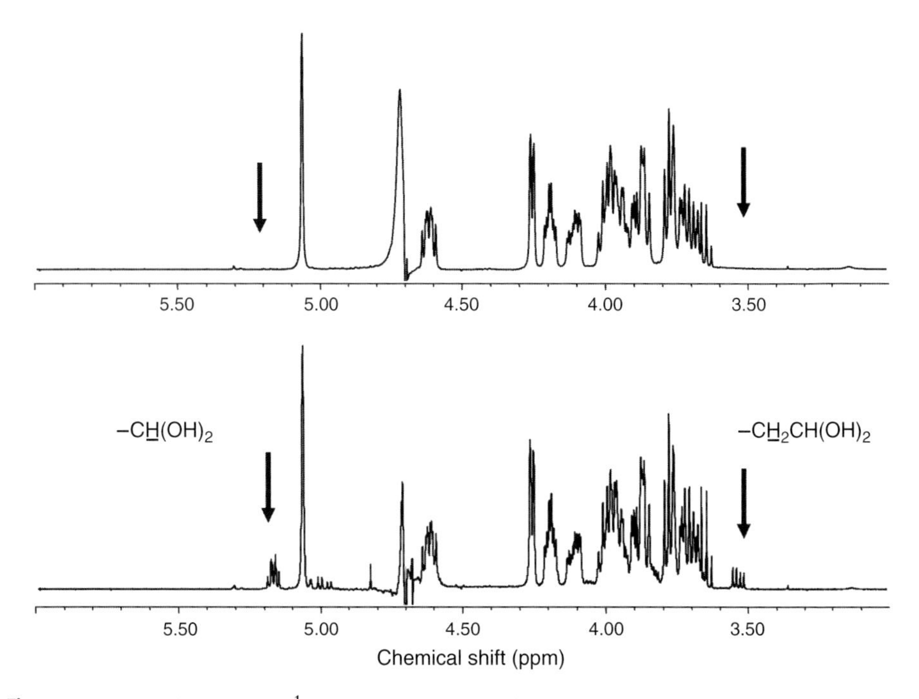

Figure 9 Partial 400 MHz ^{1}H NMR spectrum showing periodate oxidation of the Hib polysaccharide to generate end groups. The resonances due to the hydrated aldehyde (at 5.17 ppm) and an adjacent methylene proton (at 3.52 ppm) are highlighted. All spectra were collected at an indicated probe temperature of 30°C.

NMR characterisation of Men Y and W135 oligosaccharides was performed prior to size determination by use of HPAEC for obtaining the in-chain/end group ratio.[87] NMR has been used to monitor the conjugation process from polysaccharide through activated intermediates to the final conjugate.[88]

4.2. Random activation without depolymerisation

A second common approach is to randomly activate the polysaccharide for conjugation through free hydroxyl or carboxylate groups. Activation of hydroxyl groups uses cyanogen bromide, or a crystalline variant such as 1-cyano-4-(dimethylamino)pyridinium tetrafluoroborate (CDAP),[78,79] and attachment of a bifunctional linker, whilst addition of a suitable bifunctional linker, such as cystamine or adipic acid dihydrazide (ADH), to the carboxylate groups using a water-soluble carbodiimide[82,83] produces glycan suitable for selective attachment to a (suitably activated) carrier protein. Alternatively, diols present in a sugar ring may be activated by periodate oxidation to generate in-chain aldehydes in the polysaccharide. In both cases, key control tests are required to quantify the degree of activation of the polysaccharide, and to ensure that the manufacturing process is consistent. Practically, much of the work in this area has been undertaken either by academic groups who have focussed on assessing the immunogenicity of the final conjugate in animal models, and have reported minimal details of the characterisation, or by vaccine industry players who rarely publish studies of this kind. The degree of activation of polysaccharides has typically been carried out by colorimetric assays, such as trinitrobenzene sulphonate (TNBS), for free amino groups present in the linker, which may be subject to interference or may be incompatible with some polysaccharides, such as those which contain free amino groups. As the linkers contain resonances well resolved from the usual saccharide resonances, NMR provides a means to quantify the proportion of repeat units which have been modified, although wet chemical methods are more commonly applied in routine quality control. For example, periodate oxidation of Group A polysaccharide occurs between the hydroxyls at C-3 and C-4 of the unacetylated residues; this can be monitored by the disappearance of the H-3 and H-4 NMR signals (Figure 10) although quality control is performed by use of the bicinchoninic acid assay for aldehydes.[89]

In a paper which presents a model approach for the validation of a quantitative NMR procedure,[77] the group at Merck describe a method to quantify the degree of derivatisation of Hib PRP used in vaccine manufacture. Their process involves two-stage activation of the polysaccharide. The unstable product generated by the treatment of Hib PRP with carbonydiimidazole (CDI) is trapped

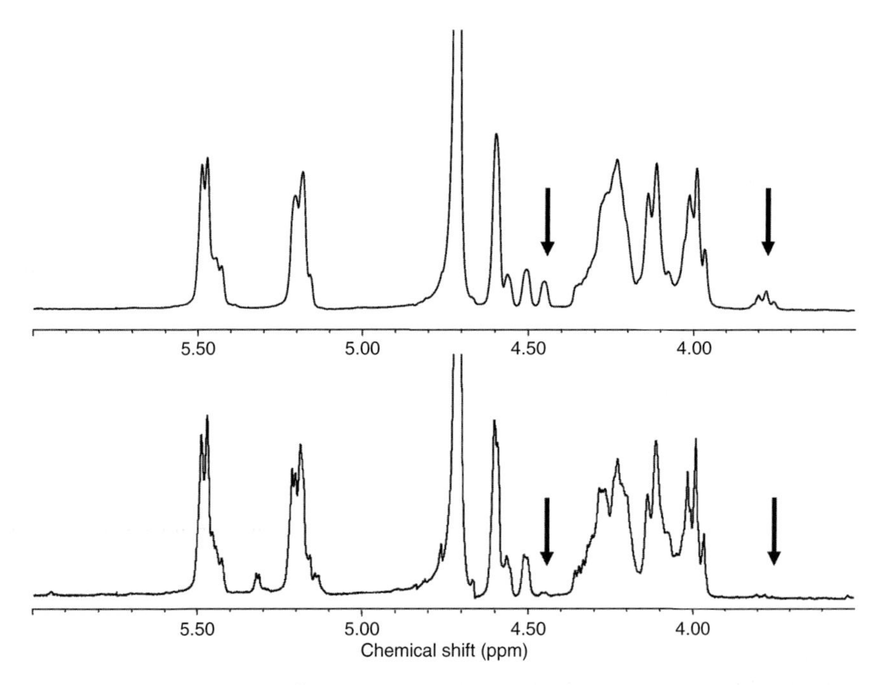

Figure 10 Partial 400 MHz ^{1}H NMR spectrum showing periodate oxidation of the meningococcal group A polysaccharide. The resonances due to H-2 (at 4.45 ppm) and H-4 (at 3.78 ppm) of the unacetylated residues are highlighted; the disappearance of these signals can be monitored during oxidation. All spectra were collected at an indicated probe temperature of 30°C.

with an excess of 1,4-diaminobutane, and this intermediate is then bromoacetylated prior to conjugation. The NMR assay of this bromoacetylated intermediate quantifies the proportion of PRP repeat units which have been derivatised, the proportion of 1,4-diaminobutane residues which have undergone bromoacetylation, and the mass of Hib PRP (measured against DMSO as an internal standard). A single 600 MHz NMR spectrum replaces several labour-intensive colorimetric and chromatographic assays. The authors report their approaches to validation of method accuracy, recovery, intermediate and inter-operator precision, specificity in the presence of small molecular contaminants, and the robustness of the assay to salt, "wet" (H_2O-containing) samples, sample quantity and temperature variation. The assay showed high precision (%RSD (relative standard deviation) of ≤1.0% for intermediate and inter-operator precision), high specificity, good linearity and to be robust to variation in the assay conditions. The same report indicated that 1,1′-carbonyldiimazole (CDI) shows no significant specificity for different hydroxyl groups within the repeat unit.[77] In the opinion of the authors, this paper highlights the potential of NMR in this area, although commercial pressures are likely to inhibit such thorough publication of similar detailed studies. An alternative approach, applied to the structurally complex pneumococcal polysaccharides, is to carry out gas chromatography-mass spectrometry (GC–MS) of derivatised fragments of the activated polysaccharide to quantify these.[90]

5. NMR Analysis of Glycoconjugate Vaccines

5.1. Identity and integrity of the saccharide component of glycoconjugate vaccines

As for the polysaccharide components, there is a requirement that the identity of the glycan chains in the glycoconjugate vaccine be determined. In some cases, this may include assessment of the degree of *O*-acetylation, as labile substituents may be lost during the manufacturing process. For many glycoconjugate vaccines, principally but not exclusively those vaccines using CRM197 as carrier protein with either oligosaccharide or polysaccharide glycans, 1D and 2D NMR spectra of sufficient quality to assess the integrity of the glycan chains can be obtained. The spectrum of the glycan moieties is essentially the same as that of the unconjugated saccharide, whilst the carrier protein gives rise to broad peaks (Figure 11). The glycan signals are, therefore, easily recognised and the glycan identity can be obtained by comparison with the spectrum with that of the parent polysaccharide.

We have applied this approach to glycoconjugate vaccines prepared from Hib PRP, meningococcal serogroups, and multiple pneumococcal serotypes. Samples which have so far failed to produce acceptable spectra in our hands have been highly crosslinked Hib conjugate vaccines with tetanus toxoid as the carrier protein, and the Hib vaccine using meningococcal outer membrane protein (OMP) vesicles

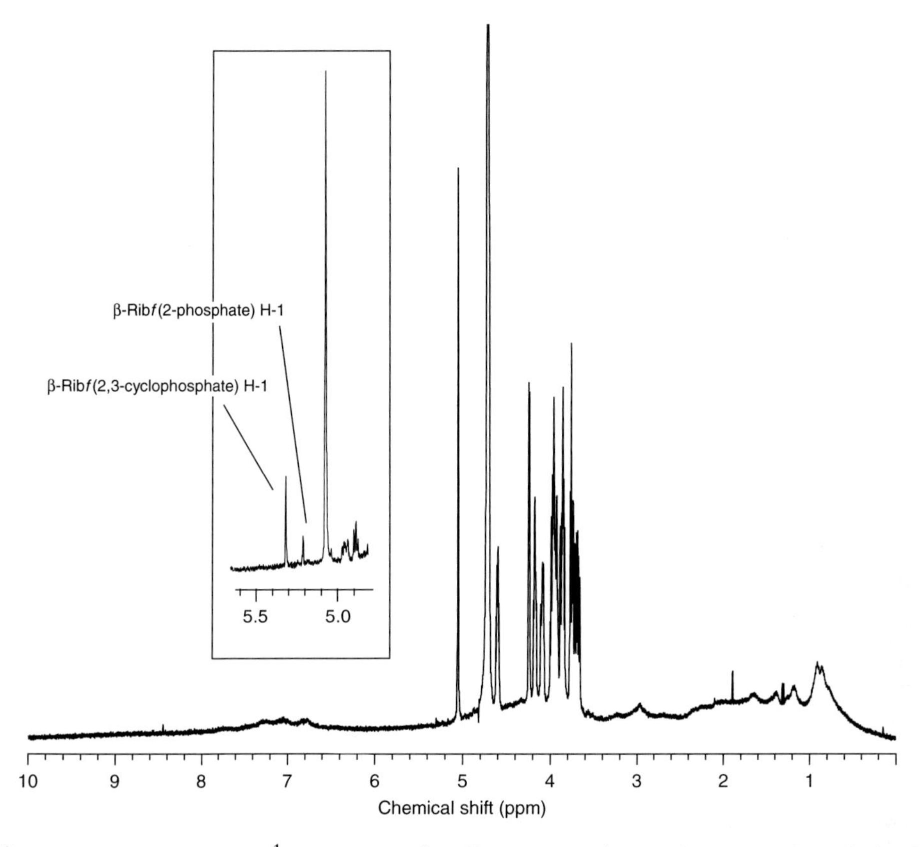

Figure 11 Partial 500 MHz ^{1}H spectrum of a Hib-CRM197 glycoconjugate vaccine, obtained at 30°C. The inset shows the "anomeric" region of the spectrum of a deliberately degraded sample, with resonances arising from the newly formed ribose-2,3-cyclophosphate end groups highlighted.

as the carrier "protein". The difference in line width between saccharide- and protein-derived resonances probably reflects the fact that relaxation in the saccharide chains is dominated by internal mobility.

Degradation/depolymerisation of the glycan chains and release of free saccharide is the most significant method by which these vaccines lose immunogenicity, and this process can be monitored either by NMR spectroscopy (as the newly formed end groups give rise to characteristic signals, see Section 3) or by quantifiable changes in the amount of free unconjugated saccharide present. The latter method is used in routine quality control. In addition to analysis of the glycoconjugate vaccines, NMR spectroscopy can also be used to monitor the identity and integrity of the saccharide component throughout the manufacturing process from starting polysaccharide, oligosaccharide intermediates to conjugate vaccine (Figure 12). These analyses are performed on bulk samples; however, recent advances in instrumentation now permit analysis of final fill samples, as shown for a Men A conjugate vaccine formulated in mannitol.

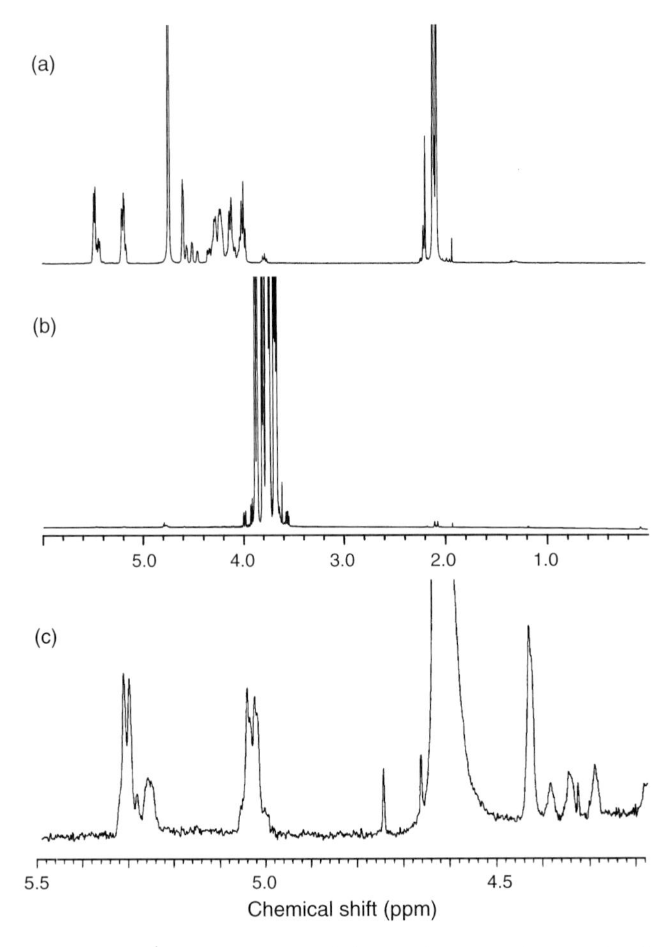

Figure 12 Partial 600 MHz ^{1}H NMR spectra of the meningococcal group A polysaccharide and conjugate vaccine: (a) group A polysaccharide; (b) a single vial of group A polysaccharide (120 μg) conjugated to tetanus toxoid and formulated with 40 mg of mannitol and (c) an expansion of the anomeric region confirming the similarity of the *O*-acetylation pattern to that of the polysaccharide. All spectra were collected at an indicated probe temperature of 30°C.

5.2. Determination of polysaccharide–protein ratio in glycoconjugate vaccines

The ratio of saccharide to carrier protein is a key test in vaccine development and quality control. If the ratio is low, excessive amounts of carrier protein are present for the required saccharide dose, whilst conjugates whose saccharide content is too high tend to be less immunogenic, possibly due to inhibition of the proteolysis which generates the peptides for display by the major histocompatibility complex class II (MHCII) to initiate T cell involvement.[91] Two basic approaches to determine this ratio have been used. Firstly, separate chemical analyses for the polysaccharide and protein components can be used. Alternatively, NMR methods are available. In the first approach, the saccharide content is measured by a colorimetric approach or HPAEC after hydrolysis, which works well for Hib and the meningococcus CPSs that are easily and quantitiatively degraded, but is problematic for pneumococcal CPSs where some serotype CPSs fail to degrade cleanly, and is unlikely to be applied to the *S.* Typhi Vi or *Staphylococcus aureus* conjugates, where the CPS is extremely resistant to hydrolysis. The methodology will be different for each individual conjugate – colorimetric and HPAEC methods have been favoured for meningococcal and Hib conjugates, whilst HPLC or GC or GC-MS methods have been proposed for pneumococcal conjugates.[92] Protein content is determined by colorimetric methods or UV absorbance. The methodology is simple, but requires careful validation of many individual steps that introduce errors. As so frequently the case with polysaccharide analysis, the difficult step is to ensure complete and reproducible hydrolysis. On the other hand, NMR methods are relatively difficult to implement and only easily applicable to conjugates prepared from pure carriers such as CRM197, but they directly provide a ratio by integration of sets of resonances from the protein and polysaccharide. Practically, the carrier protein must be denatured by the addition of guanidinium hydrochloride (GnHCl) to a final concentration of 5 M and by obtaining the spectrum at elevated temperature,[85] or by dissolution in base. This removes the sequence-specific chemical shift variation for the resonances from the side chains of the aromatic amino acids and promotes rapid deuterium exchange of the His H-2. The resonances from the aromatic side chains provide a measure of the protein content, and one of more of the sugar-derived resonances can be integrated to determine the saccharide content. This is illustrated in Figure 13. When using base denaturation, de-*O*-acetylation of the glycan occurs, simplifying the spectrum, and, since, lower salt concentrations are used, it is more compatible with the requirements of the NMR probe. On the other hand, some CPSs such as Hib PRP degrade under these basic conditions. Although, to the best of our knowledge, this approach has never been formally validated, the well-understood physical basis for NMR spectroscopy and the fact that the saccharide and protein moieties are quantified simultaneously make this approach suitable as a primary method for measurement of polysaccharide–protein ratios, against which other methods can be calibrated or validated. For each glycoconjugate vaccine analysed by NMR spectroscopy it is, of course, necessary to validate that the resonances integrated arise solely from the desired hydrogens.

When the carrier protein has been chemically toxoided or is a mixture of proteins, such as the meningococcal OMPs, it is obviously more difficult to assess the number of proteins contributing to the "protein" signal, but the approach is still viable to characterise consistency between batches. It has been our experience that the toxoids require more extreme conditions to denature them, probably as a result of the crosslinks that are introduced by the formaldehyde treatment.

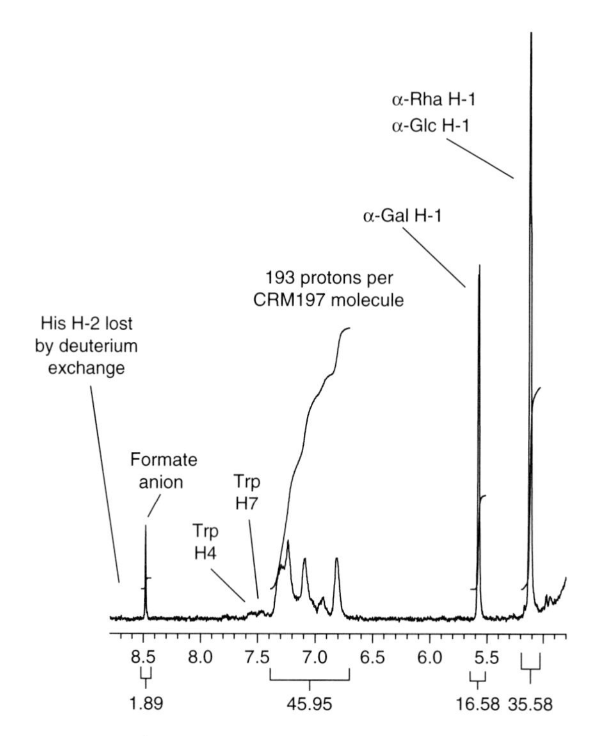

Figure 13 Partial 500 MHz ^{1}H NMR spectrum of a denatured pneumococcal Type 4 glycoconjugate vaccine with CRM197 as carrier protein, showing the region of the spectrum containing resonances from the side chains of aromatic amino acids and the anomeric protons of the sugar residues. The sample is dissolved in 5 M deuterium-exchanged guanidinium hydrochloride in deuterated water, and the spectrum collected at a nominal probe temperature of 70°C. Chemical shifts are referenced against internal acetate anion at 1.908 ppm. Key resonances are identified and the integrals of the resonances from the side chains of the aromatic amino acids and the anomeric resonances shown. From the data, the molar ratio of saccharide repeat units to protein can be calculated, and hence the weight ratio. In this sample, the molar ratio of saccharide repeat units to carrier protein is approximately 110:1, which equates to a saccharide:protein weight ratio of 1.56:1.

6. Conclusions

Conjugation of cell surface saccharides to carrier proteins is now well established as a generic technology for the production of vaccines against a wide range of bacterial pathogens. The same conjugation approaches are being developed to generate glycoconjugate immunotherapeutics to protect against cancer re-emergence after treatment, by eliciting immune responses against cell surface glycoconjugates over-expressed by the tumour cells. In addition, many parasites express unusual cell surface glycans which may be suitable targets for immune protection.[17] There are a large number of potential glycoconjugate vaccines targeted against bacterial infections in development,[16] and NMR will have a unique and valuable role in the characterisation of the saccharide component, activated saccharide and the final conjugates.

The first applications of NMR spectroscopy to polysaccharide and glycoconjugate vaccines were for structure determination and for identity testing of the glycan components, where NMR is a single generic technology able to substitute for a wide variety of disparate chemical approaches whilst providing more precise data. The NMR spectra also identify and quantify known impurities previously controlled only through indirect approaches, or not suspected to be present. It soon became apparent that in many cases the same approaches could also be applied to the characterisation of manufacturing intermediates and final conjugates, and that a single technology – capital and expertise intensive admittedly – provides precise and quantifiable data which would otherwise require the application of multiple approaches. Quantification of polysaccharide purity and, for example, saccharide–protein ratios in final conjugates are clearly possible, although there is little published data in the field so far. Even when NMR is not an appropriate option as a routine test, it offers many advantages as a means to establish reference preparations with defined properties to allow calibration of traditional chemical approaches, as these usually require multiple subsidiary measurements and assumptions about response factors. As some of the new glycoconjugate vaccines, such as those against typhoid and *Staphylococcus aureus*, utilise extremely stable polysaccharides, NMR approaches may prove the only viable methods. Progress to date has been slow because there are few laboratories working in this field.

ACKNOWLEDGEMENTS

We would like to thank all the vaccine manufacturers and the Meningitis Vaccine Project who have made available samples of polysaccharides and glycoconjugate vaccines, and with whom we have had many fruitful discussions. We would particularly like to thank Mr Xavier Lemercinier (NIBSC) for collaboration.

REFERENCES

1. Anonymous, Wkly. Epidemiol. Rec. 78 (2003) 110–119.
2. J. Eskola, M. Anttila, Pediatr. Infect. Dis. 18 (1999) 543–551.
3. Anonymous, Wkly. Epidemiol. Rec. 73 (1998) 64–68.
4. Anonymous, Wkly. Epidemiol. Rec. 77 (2002) 331–339.
5. K.L. Kotloff, J.P. Winickoff, B. Ivanoff, J.D. Clemens, D.L. Swerdlow, P.J. Sansonetti, G.K. Adak, M.M. Levine, Bull. World Health Organ. 77 (1999) 651–666.
6. J.A. Crump, S.P. Luby, E.D. Mintz, Bull. World Health Organ. 82 (2004) 346–353.
7. R. Edelman, M.M. Levine, Rev. Infect. Dis. 8 (1986) 329–349.
8. http://www.who.int/child-adolescent-health/New_Publications/IMCI/WHO_FCH_CAH_01.10/Young_Infant
9. B. Ivanoff, C.L. Chaignat, Bull Soc. Pathol. Exot. 95 (2002) 355–358.
10. A. Sinha, O. Levine, M.D. Knoll, F. Muhib, T.A. Lieu, Lancet 369 (2007) 389–396.

11. Wyeth Reports Earnings Results for the 2007 Second Quarter and First Half, http://www.wyeth.com/news/announcements, July 19, 2007 (accessed 26 August 2007).
12. http://www.who.int/vaccine_research/about/gvrf/Jadhav.pdf, accessed 26 August 2007.
13. I. Roitt, *Essential Immunology* (9th ed.), Blackwell Scientific, Oxford (1997).
14. I. Goldschneider, E.C. Gotschlich, M.S. Artenstein, J. Exp. Med. 129 (1969) 1307–1326..
15. A.A. Lindberg, Vaccine 17 (1999) S28–S36.
16. C. Jones, An. Acad. Bras. Cienc. 77 (2005) 293–324.
17. J.F.G. Vliegenthart, FEBS Lett. 580 (2006) 2945–2950.
18. C. Jones, *Comprehensive Glycoscience*, J.P. Kamerling Ed., Elsevier, Oxford, UK (2007).
19. K.L. O'Brien, M. Hochman, D. Goldblatt, Lancet Infect. Dis. 7 (2007) 597–606.
20. H.J. Jennings, Curr. Top. Microbiol. Immunol. 150 (1990) 97–127.
21. The Jordan Report on the Accelerated Development of Vaccines: see http://www.niaid.nih.gov/dmid/vaccines/jordan20.
22. V. Verez-Bencomo, V. Fernández-Santana, E. Hardy et al. Science 305 (2004) 522–525.
23. V.N. Shibaev, Adv. Carbohydr. Chem. Biochem. 44 (1986) 277–339.
24. M.M. McGowen, J. Vionnet, W.F. Vann, Glycobiology 11 (2001) 613–620.
25. J.P. Kamerling, Pneumococcal polysaccharides: a chemical view, A. Tomasz, Ed., *Streptococcus pneumoniae*, pp. 81–114, Mary Anne Liebert, Larchmont (2000).
26. B. Lindberg, Adv. Carbohydr. Chem. Biochem. 48 (1990) 279–318.
27. P. Branefors-Helander, C. Erbing, L. Kenne, B. Lindberg, Acta Chem. Scand., B 30 (1976) 276–277.
28. R.M. Crisel, R.S. Baker, D.E. Dorman, J. Biol. Chem. 250 (1975) 4926–4930.
29. D.R. Bundle, I.C.P. Smith, H.J. Jennings, J. Biol. Chem. 249 (1974) 2275–2281.
30. A.K. Bhattacharjee, H.J. Jennings, C.P. Kenny, A. Martin, D.R. Bundle, J. Biol. Chem. 250 (1975) 1926–1932.
31. A.K. Bhattacharjee, H.J. Jennings, C.P. Kenny, A. Martin, D.R. Bundle, Can. J. Biochem. 54 (1976) 1–8.
32. K. Heyns, G. Kiessling, Carbohydr. Res. 3 (1967) 340–353.
33. C.J.M. Stroop, Q. Xu, M. Retzlaff, C. Abeygunawardana, C.A. Bush, Carbohydr. Res. 337 (2002) 335–344.
34. P.E. Jansson, B. Lindberg, M. Andersson, U. Lindquist, J. Henrichsen, Carbohydr. Res. 182 (1998) 111–117.
35. R.E. Reeves, W.F. Goebel, J. Biol. Chem. 139 (1941) 511–519.
36. C. Jones, F. Currie, M.J. Forster, Carbohydr. Res. 221 (1991) 95–121.
37. P.E. Jansson, B. Lindberg, U. Lindquist, Carbohydr. Res. 140 (1985) 101–110.
38. L. Kenne, B. Lindberg, J.K. Madden, Carbohydr. Res. 73 (1979) 175–182.
39. C. Jones, B. Mulloy, A. Wilson, A. Dell, J.E. Oates, J. Chem. Soc. Perkin Trans. 1 (1985) 1665–1673.
40. B. Lindberg, J. Lonngren, D.A. Powell, Carbohydr. Res. 58 (1977) 177–186.
41. J. Lindberg, Thesis Stockholm University, 1990.
42. E. Katzenellenbogen, H.J. Jennings, Carbohydr. Res. 124 (1983) 235–245.
43. H.J. Jennings, K.-G. Rosell, D.J. Carlo, Can. J. Chem. 58 (1980) 1069–1074.
44. J.C. Richards, M.B. Perry, Biochem. Cell Biol. 66 (1988) 758–771.
45. C. Jones, Carbohydr. Res. 340 (2005) 1097–1106.
46. M. Tylianakis, A. Spyros, P. Dais, F.R. Taravel, A. Perico, Carbohydr. Res. 315 (1999) 16–34.
47. European Pharmacopoeia, 6th Ed. *European Directorate for the Quality of Medicines*, pp. 816–818, Strasbourg, France, (2008).
48. European Pharmacopoeia, 6th Ed. *European Directorate for the Quality of Medicines*, pp. 827–829, Strasbourg, France, (2008).
49. European Pharmacopoeia, 6th Ed. *European Directorate for the Quality of Medicines*, pp. 847–848, Strasbourg, France, (2008).
50. European Pharmacopoeia, 6th Ed. *European Directorate for the Quality of Medicines*, pp. 792–794, Strasbourg, France, (2008).
51. X. Lemercinier, C. Jones, Biologicals 28 (2000) 75–83.
52. X. Lemercinier, C. Jones, Carbohydr. Res. 296 (1996) 83–96.

53. C. Abeygunawardana, T.C. Williams, J.S. Sumner, J.P. Hennessey Jr, Anal. Biochem. 279 (2000) 226–240.
54. X. Lemercinier, I. Martinez-Cabrera, C. Jones, Biologicals 28 (2000) 17–24.
55. Anonymous, World Health Organ. Tech. Rep. Ser. 897 (2000) 27. Available at http://whqlibdoc.who.int/trs/WHO_TRS_897.pdf.
56. Anonymous, World Health Organ. Tech. Rep. Ser. 924 (2004) 102–128.
57. Anonymous, World Health Organ. Tech. Rep. Ser. 927 (2005) 64–98.
58. Anonymous: Available through http://www.who.int/biologicals/publications/trs/areas/vaccines/meningococcal/en/index.html
59. Anonymous. Pharmeuropa 19 (2007) 323–327.
60. Anonymous, European Pharmacopoeia 6th Ed., strasbourg, 2008, pp. 39–41.
61. J.M. Wieruszeski, P. Talaga, G. Lippens, Anal. Biochem. 338 (2005) 20–25.
62. C. Abeygunawardana, "High resolution NMR Applications in polysaccharide-based vaccines", Abstracts of the 222nd ACS Annual meeting, Chicago IL, August 26th–30th, 2001.
63. W. Fischer, T. Behr, R. Hartmann, J. Peter-Katalinic, H. Egge, Eur. J. Biochem. 215 (1993) 851–857.
64. W. Egan, Dev. Biol. 103 (2000) 3–9.
65. X.D. Wu, R.A. Kircher, P.H. McVerry, D.A. Malinzak, Dev. Biol. 103 (2000) 269–270.
66. Q. Wu, C. Abeygunawardana, A.S. Ng, A.W. Sturgess, B.J. Harmon, J.P. Hennessey, Anal. Biochem. 336 (2005) 262–272: erratum ibid., 342 (2006) 258
67. E.C. Gotschlich, B.A. Fraser, O. Nishimura, J.B. Robbins, T.Y. Liu, J. Biol. Chem. 256 (1981) 8915–8921.
68. J.S. Kuo, V.W. Doelling, J.F. Graveline, D.W. McCoy, J. Bacteriol. 163 (1985) 769–773.
69. A. Bardotti, N. Ravenscroft, S. Ricci, S. D'Ascenzi, V. Guarnieri, G. Averani, P. Constantino, Vaccine 18 (2000) 1982–1993.
70. C.M. Tsai, X.X. Gu, R.A. Byrd, Vaccine 12 (1994) 700–706.
71. F. Mawas, B. Bolgiano, P. Rigsby, D. Crane, D. Belgrave, M.J. Corbel, Biologicals 35 (2007) 235–245.
72. A.L. Stone, S.C. Szu, J. Clin. Microbiol. 26 (1988) 719–725.
73. C. Jones, X. Lemercinier, D.T. Crane, C. Gee, B. Bolgiano, S. Austin, Dev. Biol. Stand. 101 (1999) 177–183.
74. C. Jones, X. Lemercinier, D.T. Crane, C.K. Gee, S. Austin, Dev. Biol. 103 (2000) 121–136.
75. N. Ravenscroft, G. Averani, A. Bartoloni, S. Berti, M. Bigio, V. Carinci, P. Costantino, S. D'Ascenzi, A. Giannozzi, F. Norelli, C. Pennatini, D. Proietti, C. Ceccarini, P. Cescutti, Vaccine 17 (1999) 2802–2816.
76. A. Pawlowski, S.B. Svenson. FEMS Microbiol. Lett. 174 (1999) 255–263.
77. X. Cai, Q.P. Lei, D.H. Lamb, A. Shannon, J. Jacoby, J. Kruk, R.D. Kensinger, R. Ryall, E. Zablackis, P. Cash, Anal. Chem. 76 (2004) 7387–7390.
78. Q. Xu, J. Klees, J. Teyral, R. Capen, M. Huang, A.W. Sturgess, J.P. Hennessey Jr, M. Washabaugh, R. Sitrin, C. Abeygunawardana, Anal. Biochem. 337 (2005) 235–245.
79. J.Y. Tai, P.P. Vella, A.A. McLean, A.F. Woodhour, W.J. McAleer, A. Sha, C. Dennis-Sykes, M.R. Hilleman, Proc. Soc. Exp. Biol. Med. 184 (1987) 154–161.
80. O. de Weers, M. Beurret, L. van Buren, L.A. Oomen, J.T. Poolman, P. Hoogerhout, Bioconjug. Chem. 9 (1998) 309–315.
81. A. Lees, B.L. Nelson, J.J. Mond, Vaccine 14 (1996) 190–198.
82. A.J. D'Ambra, J.E. Baugher, P.E. Concannon, R.A. Pon, F. Michon, Anal. Biochem. 250 (1997) 228–236.
83. Z. Kossaczka, S. Bystricky, D.A. Bryla, J. Shiloach, J.B. Robbins, S.C. Szu, Infect. Immun. 65 (1997) 2088–2093.
84. S.C. Szu, A.L. Stone, J.D. Robbins, R. Schneerson, J.B. Robbins, J. Exp. Med. 166 (1986) 1510–1524.
85. C.A. Laferrière, R.K. Sood, J.M. de Muys, F. Michon, H.J. Jennings, Vaccine 15 (1997) 179–186.
86. N. Ravenscroft, *Pharmeuropa Special Issue: Biologicals Beyond 2000: Challenges for Quality Standards in an Evolving Field*, European Directorate for the Quality of Medicines, Strasbourg, France, 2000.

87. A. Bardotti, G. Averani, F. Berti, S. Berti, C. Galli, S. Giannini, B. Fabbri, D. Proietti, N. Ravenscroft, S. Ricci, Vaccine 23 (2005) 1887–1899.
88. N. Ravenscroft, S. D'Ascenzi, D. Proietti, F. Norelli, P. Costantino, Dev. Biol. 103 (2000) 35–47.
89. N. Ravenscroft, J. Chen, M. Hearshaw, P. Mensah, M. LaForce, J.-M. Prenaud, S. Viviani, S. Beri, J. Joshi, K. Suresh, 15th International Pathogenic Neisseria Conference, Cairns, Australia, 2006. P8.3.09.
90. J.S. Kim, E.R. Laskowich, F. Michon, R.E. Kaiser, R.G. Arumugham, Anal. Biochem. 358 (2006) 136–142.
91. G.R. Siber, Science 265 (1994) 1385–1387.
92. J.S. Kim, E.R. Laskowich, R.G. Arumugham, R.E. Kaiser, G.J. MacMichael, Anal. Biochem. 347 (2005) 262–274.

CHAPTER 3

Fluorine-19 or Phosphorus-31 NMR Spectroscopy: A Powerful Technique for Biofluid Metabolic Studies and Pharmaceutical Formulation Analysis of Fluorinated or Phosphorylated Drugs

R. Martino, V. Gilard, *and* M. Malet-Martino

Contents

Abstract

Fluorine-19 or phosphorus-31 nuclear magnetic resonance (NMR) (^{19}F NMR or ^{31}P NMR) spectroscopy provides a highly specific tool for detection, identification and quantification, in a single run, of all the fluorine- or phosphorus-containing species, including unexpected substances in the bulk solutions analysed (biofluids or pharmaceutical formulations). The first part of the chapter presents an overview of the advantages and limitations of NMR spectroscopy. The second part of the chapter deals with fluorinated drugs such as anti-bacterials, neuroleptics, analgesics, anti-inflammatory agents, anaesthetics, anti-cancer agents and anti-fungals. The ^{19}F NMR analysis of biofluids leads to metabolic data while the analysis of pharmaceutical formulations provides drug quality control. The third part of the chapter reports the ^{31}P NMR metabolic studies of three phosphorylated drugs from the analysis of biofluids.

Keywords: flucloxacin, ciprofloxacin, linezolid, flupentixol, niflumic acid, methoxyflurane, fluorouracil, capecitabine, gemcitabine, fluorocytosine, amifostine, cyclophosphamide

1. Introduction

Nuclear magnetic resonance (NMR) is the only physical method used routinely for the direct study at the molecular level of biological samples, from biofluids, cell or tissue extracts, excised tissues, packed intact cells (in vitro studies) to isolated living cells or isolated perfused organs (ex vivo studies), and finally, animal models and human subjects (in vivo studies). Consequently, NMR is unique in its ability to permit the analysis of the metabolism of both endogenous and xenobiotic compounds such as drugs. The method is non-selective and unexpected substances are not overlooked during the investigation, as all low-molecular-weight molecules in solution (provided they bear the nucleus under investigation and are present at sufficient concentrations) are detected simultaneously in a single analysis. This contrasts with chromatography that usually requires some prior knowledge of the structure of the compounds analysed to optimise sample preparation and/or detection. NMR also avoids the use of a number of different chromatographic techniques, which is sometimes necessary when analytes have different chemical structures. This is an important attribute of NMR in the search for novel compounds when, often, the analyst will have no idea of the type of molecule to look for. Under appropriate recording and processing conditions, the area of each NMR peak is directly proportional to the number of corresponding nuclei. Thus, at variance with other techniques, the response factor is not dependent on the molecular structure. Absolute concentration determinations can be assessed with calibrating on a reference substance that gives resonance signal(s) in the excited frequency range observed.[1,2] In bulk solutions (biofluids, cell or tissue extracts, pharmaceutical formulations), the concentrations of the analytes are calculated by comparing the analyte signal integrals with the reference signal integral, corrected from the number of nuclei giving rise to the resonances. NMR is thus a suitable technique for the quantitative detection of drugs and metabolites in such solutions. Nevertheless, the binding of drugs and/or metabolites to macromolecules in plasma or the presence of micellar substructures in bile can induce significant signal broadening that leads to reduced signal-to-noise (S/N) ratio or even to NMR invisibility of the signal. Some sample pre-treatment may be required in such cases.

There are several magnetically active nuclei that can be routinely used for drug metabolism studies or pharmaceutical formulation analysis (hydrogen, carbon-13, fluorine-19, phosphorus-31, lithium-7, and, to a much lower extent, deuterium, oxygen-17, nitrogen-14 and -15, platinum-195). This chapter will be devoted to metabolic studies in biofluids and pharmaceutical formulation analysis of fluorinated or phosphorylated drugs using fluorine-19 NMR (^{19}F NMR) or phosphorus-31 NMR (^{31}P NMR) as the analytical technique. The first part presents an overview of the advantages and limitations of ^{19}F and ^{31}P NMR, emphasising the quantification procedures and the sensitivity limits of the technique. The second part deals with the fluorinated drugs, their metabolic studies in biofluids as well as the quantitative analysis of their pharmaceutical formulations using ^{19}F NMR. The third part reports ^{31}P NMR studies of phosphorylated drugs and their metabolites in biofluids.

2. Advantages and Limitations of ^{19}F and ^{31}P NMR for in vitro Studies

The studies reported here focus on fluorine-19 and phosphorus-31 nuclei because of their favourable NMR characteristics: nuclear spin of 1/2, relatively narrow lines, 100% natural abundance, high sensitivity for fluorine-19 nucleus (83% that of proton) and correct sensitivity for phosphorus-31 nucleus (6.6% that of proton), large chemical shift range (about 500 and 800 ppm for fluorine-19 and phosphorus-31, respectively), which minimises signal overlap.

An important advantage, specific to ^{19}F NMR, is that the negligible level (below the limit of NMR detection) of endogenous mobile fluorinated metabolites of low molecular mass (the only ones that can be detected) eliminates interfering background signals. Concerning the phosphorus nucleus, the presence of endogenous phosphate and derivatives (e.g. phosphomonoesters, phosphodiesters) may interfere with signals from phosphorylated drugs and their metabolites. In practice, this is not a large obstacle since these endogenous compounds are relatively few, at least in biofluids, to produce detectable signals. Moreover, ^{19}F and ^{31}P NMR of biological matrices are not hampered by dynamic range problems as encountered in proton NMR due to the intense signal of water protons that must be suppressed.

Set against these advantages, however, are a number of disadvantages that need to be taken into account. The total volume of sample required for NMR analysis ranges between 0.3–0.7 and 2.0–3.0 ml with 5 and 10 mm tubes, respectively, depending if a coaxial capillary is or not employed. This can be a hindrance for pharmacokinetic studies that require numerous plasma samples or for difficult-to-obtain biofluids such as bile, cerebrospinal fluid or those from neonates. But the volume needed for an experiment may be reduced without having to redesign the radio frequency (RF) coil, using several approaches: utilisation of smaller tubes (e.g. 3 mm microtubes requiring 130–150 μl) or tubes fitted with plastic plugs matched to the susceptibility of the solvent or spherical microcells. In the two last cases, the volume sample (60–80 μl under favourable circumstances, or 18 μl, respectively) is placed within the active region of the RF coil.[3,4]

However, compared with most chromatographic and other spectroscopic techniques, NMR is relatively insensitive, which represents the principal drawback of the technique. About 20 years ago, Nicholson and Wilson[5] considered that the detection limit for ^{1}H NMR in biological matrices was theoretically about 10 μmol/l for 500 MHz (or above) spectrometers using 5 mm probe. Today, the sensitivity limit for low-molecular-weight (≤500 Da) metabolites in body fluids is the micromolar range (5–30 μmol/l) depending on the number of hydrogens that contribute to a signal, the multiplicity of the resonance and the difficulty to quantitate the metabolites of interest due to chemical noise caused by considerable signal overlaps in some regions of the spectrum especially between 3.5 and 4.1 ppm.[6–8]

The limits of sensitivity for ^{19}F NMR in biological matrices is the low micromolar range for one fluorine nucleus. Indeed, the detection thresholds reported in the literature range between 1 and 3 μmol/l depending on the proton resonance frequency of the spectrometer (300–500 MHz), the recording time (10–24 h) and the NMR diameter tube (5 or 10 mm).[9–13] The sensitivity is enhanced threefold for the trifluoromethyl (CF_3) singlet resonance where the three fluorine nuclei are equivalent. The detection limit for ^{31}P NMR in biofluids is ≈10 μmol/l.[14]

The detection sensitivity may be increased by concentration of the biofluid of interest or by extraction of a higher amount of tissue or pharmaceutical formulation. Moreover in recent years, the NMR detection limits have been lowered by a factor of 4 with the introduction of the so-called cryoprobes. In these NMR probes, the electronic components are cryogenically cooled to ≈20 K, while the sample remains at ambient temperature. This greatly reduces the electronic noise leading to a fourfold increase in sensitivity, which for a given amount of sample reduces the experiment time by a factor of 16 over that of a conventional probe.

These ^{19}F and ^{31}P NMR sensitivity limits deal with recording conditions where fully relaxed spectra are obtained with peak areas directly proportional to concentrations. The detection sensitivity may be also increased by using the Nuclear Overhauser Effect (NOE) enhancement of the signal induced by continuous proton decoupling application, but large quantitation errors (up to 50%) are expected as the signal integrals will not be strictly proportional to concentration.[13] To obtain quantitative data, a calibration curve is established, for each substance under investigation, by linear regression analysis using peak integral of the substance and its nominal concentration[15] or the ratio (peak height of the substance/peak height of the internal standard) and the concentration of the pure substance.[16] This methodology avoids the problems of total relaxation between consecutive scans and differential NOE enhancements and leads to significant reduction in acquisition time or in detection threshold. For example, for ^{31}P NMR, a concentration of 10 μmol/l is detected in 2 h and the sensitivity limit is lowered to 7.5 μmol/l[16], whereas for ^{19}F NMR, the detection sensitivity reached 0.3 μmol/l.[13] Besides the necessity to have pure reference standard of each compound to be quantified, this approach does not take advantage of the unique property of NMR to quantify, in a single analysis, all the compounds detected whatever their structure provided their concentrations reached the limit of detection of the method. In that case, the methodology of NMR quantification is similar to that used with chromatographic methods but with a lesser sensitivity.

The accuracy and precision of the ^{19}F NMR concentration determinations are generally on the order of 5–10% for concentrations >50 μmol/l and approximately ±1 μmol/l for a concentration of a few μmol/l near the limit of detection.[9] These values seem more realistic than that of ±2% reported by Rengelshausen et al.[13] as the use of a coaxial insert tube reference introduces a level of error of ≈5%[17] (between ≈3%[13] and ≈7%[18]). For the ^{31}P NMR assays, the results obtained by Joqueviel et al.[14] show that accuracy and precision are less than ±10% for concentrations $\geq 5 \times 10^{-5}$ mol/l and approximately ±20% for concentrations between 5×10^{-5} and 10^{-5} mol/l whereas those reported by Busse et al.[16] exhibit a good accuracy ($\pm 15\%$) and reproducibility ($\pm 5\%$) even at low concentrations ($\geq 7.5\,\mu$mol/l).

Because of the length of time required for quantification recording (10–24 h), the NMR data must be acquired in 2–3 h data blocks. These blocks are then compared to check the stability of the compounds detected during the period of NMR recording. Quantification is therefore carried out using spectra resulting from the sum of all blocks. If at room temperature, there is a degradation of some compounds, it is essential to perform NMR recording at 4°C.[16,19]

3. ^{19}F NMR Studies of Fluorinated Drugs

^{19}F NMR has been used since the early 1980s for in vivo and in vitro studies of many fluorine-containing compounds. This chapter will be devoted to ^{19}F NMR metabolic studies in biofluids and pharmaceutical formulation analysis of therapeutic agents in clinical use or involved in clinical or animal trials. As a consequence, the applications of ^{19}F NMR to metabolic studies of organofluorine pesticides (e.g. atrazine, flumetsulam, profluralin and fipronil), industrial agents (e.g. chlorofluorocarbons and their potential substitutes, hydrochlorofluorocarbons or hydrofluorocarbons, substituted benzoic acids or anilines) as well as tumour markers (e.g. 2-fluoro-2-deoxy-D-glucose, 3-fluoro-3-deoxy-D-glucose) are not reviewed in this paper. Therefore, the following classes of drugs will be considered: anti-bacterials, neuroleptics, analgesics, anti-inflammatory drugs, anaesthetics, anti-cancer drugs and anti-fungals.

3.1. Anti-bacterials

^{19}F NMR has been used to quantify the antibiotic *flucloxacillin* (Figure 1) and its three major metabolites (5′-hydroxymethyl-flucloxacillin and (5*R*)- and (5*S*)-flucloxacillin penicilloic acids) in rat urine. The detection of discrete resonances for these four compounds is a good indication of the sensitivity of ^{19}F chemical shifts to remote structural change even to 7–9 bonds distant.[20,21]

^{19}F NMR has also been applied to quantitate an *m*-(trifluoromethyl)-containing *penicillin V analogue* (LY 242072) (Figure 1) in rat urine collected in the bladder of sacrificed animals. The concentrations determined were nearly identical to those measured by high-performance liquid chromatography (HPLC).[22]

^{19}F and ^{1}H NMR spectra of urine samples obtained from healthy volunteers receiving a normal therapeutic dose of the fluoroquinolone anti-microbials *ciprofloxacin* or *ofloxacin* (Figure 1) exhibited very broad resonances[23] as a result of the complexation of these drugs with metal ions, the association constant being relatively insensitive to changes in fluoroquinolone structures, but significantly dependent on the nature of the metal ion, increasing in the order $Ca^{2+} < Mg^{2+} < Fe^{3+} < Al^{3+}$ (see, e.g. Ross et al.[24] and Sakai et al.[25]). Addition of increasing amounts of EDTA, a very strong metal chelator, leads to a progressive sharpening of ^{19}F and ^{1}H resonances (with an upfield shift of ^{19}F peaks) resulting in a significant improvement in spectral quality.[23] These preliminary observations indicate that it is feasible to detect and quantify

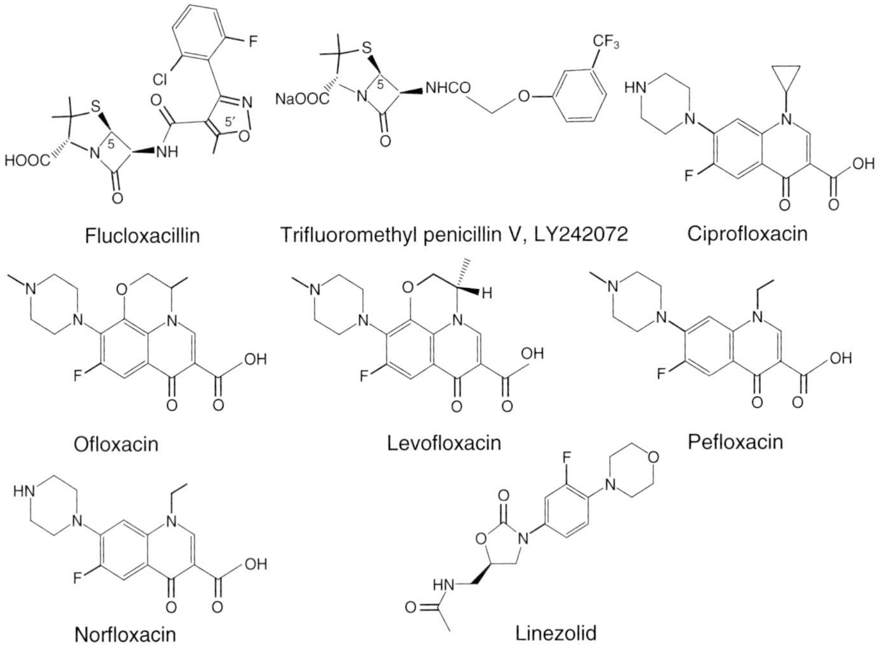

Flucloxacillin Trifluoromethyl penicillin V, LY242072 Ciprofloxacin

Ofloxacin Levofloxacin Pefloxacin

Norfloxacin Linezolid

Figure 1 Chemical structures of fluorinated anti-bacterials discussed in this chapter.

fluoroquinolones in biofluids using ^{19}F and ^{1}H NMR. To our knowledge, despite the numerous fluoroquinolones in clinical use, no other ^{19}F NMR study of these drugs in biofluids has been reported. On the contary, quantitative determinations of *levofloxacin* (Figure 1) in pharmaceutical capsules and urine samples have been performed by ^{1}H NMR using DMSO-d_6 as solvent. Insignificant difference in precisions between the NMR method and each of the fluorimetric and HPLC methods demonstrates the reliability of the NMR quantification.[26] ^{19}F and ^{1}H NMR were applied successfully to assay commercial pharmaceutical preparations of *pefloxacin*, *norfloxacin* and ofloxacin (Figure 1).[27] ^{19}F NMR was also used to demonstrate that 16 commercial pharmaceutical formulations of ciprofloxacin tablets purchased in different countries or via the Internet contain the stated amount of drug within the specification (90–110% of the labelled amount) of the US Pharmacopeia, 12 out of the 16 containing the active ingredient within $100 \pm 5\%$ of the stated concentration. The values for total impurities, determined by ^{19}F and ^{1}H NMR, ranged between 0.3% and 1%, but the formulations do not present the same impurity profile that is characteristic of the manufacturer. HPLC used to cross-validate the ^{19}F NMR data shows a good correlation between the two methods (less than 1.5% difference).[28]

Linezolid (Zyvox®) (Figure 1) is the first of a new class of antibiotics, the oxazolidinones, approved for the treatment of Gram-positive bacterial infections. The abundance of the unmetabolised drug and metabolites was quantified by ^{19}F NMR in urine of healthy volunteers receiving oral dose of linezolid labelled with ^{14}C on the carbonyl carbon of the acetamide moiety (Figure 1).[29] ^{19}F NMR-calculated mass balance for the four compounds representing more than 1% of the dose (parent drug, two carboxylic acid metabolites resulting from morpholine ring oxidation and a by-product of one of these latter compounds) ranged between 82% and 105% (mean 88 $\pm$ 7%, $n = 8$) of the results obtained by radiochemical detection. ^{19}F NMR was also used to determine the abundance of two minor metabolites coeluted by HPLC, or two non-radioactive due to the loss of the ^{14}C-acetyl group and to evidence three acid-labile metabolites arising from HPLC sample processing under acidic conditions.

3.2. Neuroleptics

The anti-psychotic drug *flupentixol* administered in an oral form (tablets or drops) is an equimolecular mixture of active *Z* and inactive *E* isomers (Figure 2). Because the CF_3 resonances of *Z* and *E* isomers are well separated, ^{19}F NMR provided a suitable and simple method for the identification and determination of *Z* and *E* flupentixol in complex biological matrices.[30] In human serum, the limits of detection (LOD) and quantification (LOQ) (defined as the analyte amount that yields an *S*/*N* ratio of 3 and 10, respectively) were ~1.7 mg/l (4 μmol/l) and 7.2 mg/l (17 μmol/l), respectively, for 6.5 min recording time on a 500 MHz spectrometer.[30] Since the S/N ratio for a peak in the NMR is known to increase with scan numbers as the $(S/N)^{1/2}$, the LOD and LOQ calculated for a recording time of 15 h, under the same experimental conditions, were ~0.3 and ~1.4 μmol/l, respectively. Indeed the presence of three

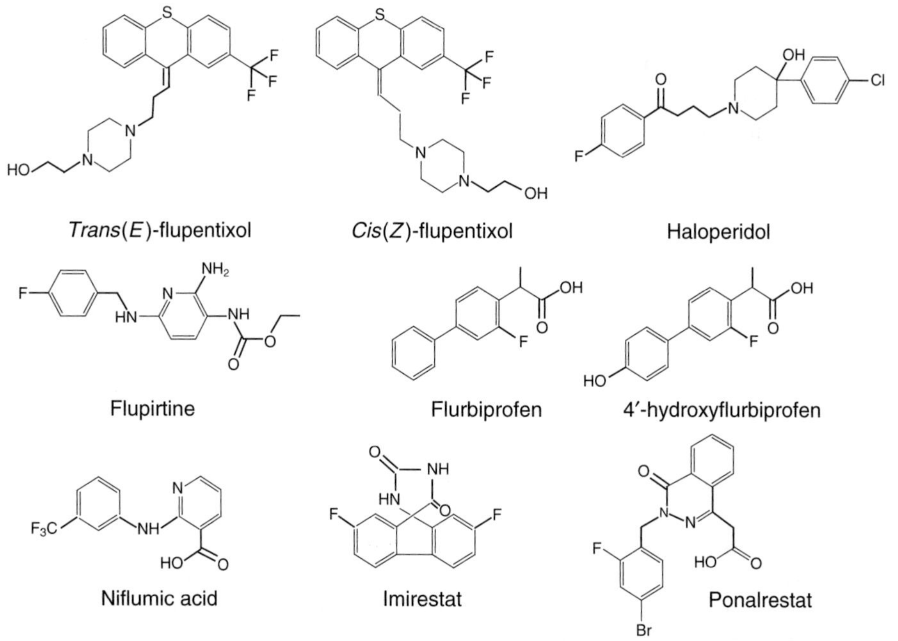

Trans(*E*)-flupentixol | *Cis*(*Z*)-flupentixol | Haloperidol

Flupirtine | Flurbiprofen | 4′-hydroxyflurbiprofen

Niflumic acid | Imirestat | Ponalrestat

Figure 2 Chemical structures of fluorinated anti-psychotics, analgesics, anti-inflammatory drugs and aldose reductase inhibitors discussed in this chapter.

equivalent fluorine nuclei in the trifluoromethyl group increases the sensitivity by a factor 3 compared to the values reported in Section 2 of this chapter. ^{19}F NMR was also used for the selective and accurate determination of another neuroleptic drug, *haloperidol* (Figure 2), in pharmaceutical formulations and human serum with a LOD of 1.4 mg/l (3.7 μmol/l) as calculated from the calibration graph at an S/N ratio of 3.[31]

3.3. Analgesics

Flupirtine (Katadolon®) (Figure 2) is a non-opiate analgesic agent in clinical use[32] and a potential drug for treatment of Creutzfeldt–Jakob disease with clinical trials under way.[33] ^{1}H and ^{19}F NMR assays of the active agent in pharmaceutical capsules have been developed. The contents measured are in excellent agreement (within ≈1%) with those obtained by HPLC.[34]

3.4. Anti-inflammatory drugs

^{1}H and ^{19}F NMR spectroscopy was applied to the study of the metabolic fate of the anti-inflammatory racemic drug *flurbiprofen* (Figure 2). ^{19}F NMR spectra of urine samples from a healthy volunteer to whom the drug was administered show the presence of 24 separate peaks among which 4 are major by far. Using a combination of one- and two-dimensional ^{1}H and ^{19}F spectroscopy, solid-phase extraction chromatography, chemical hydrolysis and HPLC coupled with ^{1}H and ^{19}F NMR, the four major ^{19}F resonances were identified. They correspond to two diastereoisomeric pairs of compounds, namely the β-D-glucuronides of the *R* and *S* isomers of flurbiprofen and its major metabolite 4′-hydroxyflurbiprofen (Figure 2).[35,36]

The metabolic fate of the trifluoromethylated anti-inflammatory drug *niflumic acid* (Figure 2) was investigated by in vivo and in vitro ^{19}F NMR in humans receiving a single oral dose of the drug.[37] The in vivo spectra of the liver during the 240 min investigation period and the in vitro spectra of protein denaturated plasma samples collected each 30 min over the same period display two resonances attributed to the parent drug (P) and one of its main metabolites (M), namely 4′-hydroxyniflumic acid, by spectroscopic comparison with niflumic acid and α,α,α-trifluoro-*o*-cresol (as a metabolite model). The signal intensities of P and M increase continuously in liver as well as in plasma. The concentration ratio of these resonances (*P/M*) is almost constant (ranged between 0.7 and 0.9) in liver. On the other hand, the resonance of the metabolite (M) emerges 90 min after drug uptake and the concentration ratio *P/M* decreases from 37 at this time to 16 at the end of the measurement period in plasma. All these findings suggest a strong first-pass metabolism and an efficient biliary excretion of the metabolite (M), which bypasses the blood circulation system. In the ^{19}F NMR spectra of urine sampled over 4 h periods during the 24 h following drug ingestion, both resonances of P and M are identified with a predominantly presence of P signal whose intensity remains nearly constant over the 24 h period, whereas the signal intensity of M increases slowly reaching an almost similar value to that of P at the end of the measurement. This observation demonstrates an effective renal elimination of niflumic acid and suggests the existence of an enterohepatic circulation pathway with a re-entry mechanism for the biliary excreted metabolite (M).

3.5. Anaesthetics

Many inhalational general anaesthetics are fluorinated, e.g. halothane (CF_3—CHBrCl), isoflurane (CF_3—CHCl—$OCHF_2$), desflurane (CF_3—CHF—$OCHF_2$), methoxyflurane ($CHCl_2$—CF_2—OCH_3), enflurane (CHFCl—CF_2—$OCHF_2$), sevoflurane ($(CF_3)_2$—$CHOCH_2F$) and fluroxene (CF_3—CH_2—OCH=CH_2). NMR studies of fluorinated anaesthetics form some of the earliest in vivo non-invasive applications of ^{19}F NMR.[38–41] Spectroscopic and imaging results regarding to the site of action and duration of residence of anaesthetics in the brain have been a source of debate and controversy.[42–47] ^{19}F NMR clinical studies of halothane and isoflurane anaesthetics in the brain have also been reported.[48–50]

If some in vivo and in vitro ^{19}F NMR studies of hepatic metabolism of halothane,[51–54] methoxyflurane[55,56] and enflurane[57,58] were performed, only the NMR metabolism and disposition studies of fluorinated anaesthetics in biofluids are detailed in this review.

Methoxyflurane (MF) is metabolised to the identified end products inorganic fluoride (F^-) and oxalate via two different pathways (Figure 3). One pathway, referred as the dechlorination pathway, begins with dechlorination and oxidation of the —$CHCl_2$ moiety of MF resulting in methoxydifluoroacetate (MDFA), which is largely excreted without further metabolism, although a small percentage can decompose, chemically and/or enzymatically, to F^- and oxalate. The second pathway, called the demethylation pathway, involves first the oxidation of the methoxy group of MF with further decomposition to F^- and dichloroacetate (DCA). DCA is then broken down to chloride (Cl^-) and glyoxalate ($^-$OOC—CHO), which is subsequently converted in vivo to glycine (^+H_3N—CH_2—COO^-) and CO_2, or oxalate.[55,56] However, the data used to develop this metabolic scheme are not conclusive as MDFA was not identified unquestionably and DCA not detected. Urine analysis of rats treated with MF and authentic MDFA and DCA was performed with a combination of ^{19}F NMR, ^1H NMR, fluoride ion-specific electrode and oxalate oxidase enzymatic assay kit.

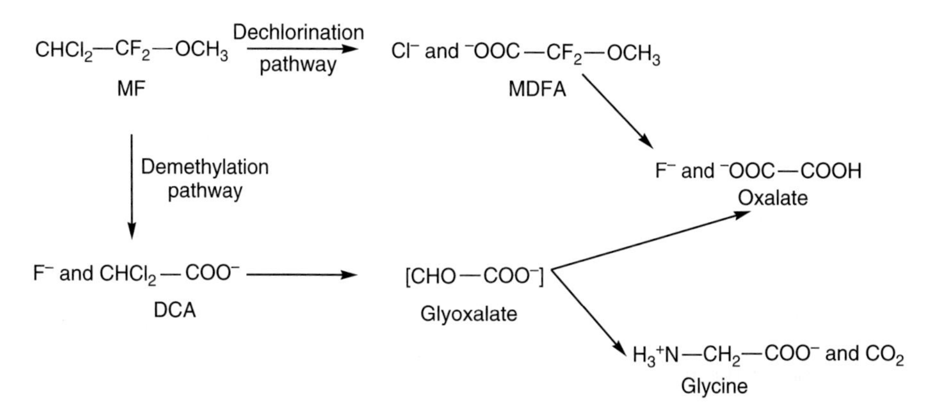

Figure 3 Abbreviated model of methoxyflurane (MF) metabolism adapted from Selinsky et al.[56]

In rat urine dosed with MF, the time course of excretion of MDFA, F^- and oxalate indicates that (i) the anaesthetic is metabolised primarily via dechlorination to yield MDFA and (ii) a significant fraction of the metabolic flux occurs via a different pathway (presumably demethylation) to yield DCA as an intermediate even if there was no evidence for excreted DCA. The absence of an observable DCA resonance in the urine reflects its rapid metabolism in vivo as demonstrated by the inability to measure unmetabolised DCA and the rapid appearance of oxalate in urine of rats dosed directly with DCA.[55,56]

^{19}F NMR spectra of urine samples collected from rats or patients exposed to *enflurane* ($CHFCl—CF_2—OCHF_2$), a fluorinated ether anaesthetic, show the presence of difluoromethoxy difluoroacetate ($^-OOC—CF_2—OCHF_2$) (DFMDFA) as major metabolite and F^-.[57] DFMDFA was also detected in rat and human plasma. Moreover, in some human urine samples, the three DFMDFA triplets were shadowed to slightly higher field by equivalent species whose concentration ranged between 6% and 12% of their DFMDFA partners. The similarity of these signals with those of DFMDFA suggests they emanate from a DFMDFA conjugate, the nature of which having not yet been elucidated. This DFMDFA conjugate was not detected by gas chromatography–mass spectrometry (GC–MS) as it is probably hydrolysed to DFMDFA during the derivatisation step required before detection.[57]. This emphasises the advantage of NMR over the more sensitive chromatographic methods for the detection of labile metabolites.

Halothane undergoes metabolism by reductive and oxidative pathways, both mediated by the cytochrome P-450 (CYP-450) system. The main end products of halothane metabolism identified in biofluids are trifluoroacetic acid (CF_3COOH; TFA) from the oxidative biotransformation, F^- and two fluorinated volatile compounds, 1,1,1-trifluoro-2-chloroethane (CF_3CH_2Cl; TFCE) and 1,1-difluoro-2-chloroethene ($CF_2{=}CHCl$), from the reductive one, as well as bromide ion which can be liberated by the two pathways.[59,60] On the contrary, *fluroxene* is metabolised by hepatic CYP-450 to 2,2,2-trifluoroethanol (CF_3CH_2OH; TFE) and CO_2 in animals, whereas the major urinary metabolite in humans is TFA.[61] No ^{19}F NMR halothane or fluroxene metabolic studies in biofluids were performed but those of two of their metabolites (TFCE and TFE) were well documented.[62–64]

The TFE metabolism was investigated using ^{19}F NMR analysis of urine of rats injected intraperitoneally with this compound.[62] Four sets of resonances (β-glucuronic acid conjugate of TFE (TFE-glu), trifluoroacetaldehyde hydrate (TFAA-hydrate), trifluoroacetaldehyde-urea conjugate (TFAA-urea) and TFA appeared in most ^{19}F NMR spectra, with a fifth set (TFE) occasionally observed. No evidence of F^- was detected in any of the ^{19}F NMR spectra or using fluoride selective electrode. TFE, TFAA-hydrate and TFA signals were assigned by spiking the urine samples with commercial authentic standards. The identification of TFE-glu resonance was established from its disappearance with an accompanying increase of that of TFE after incubation of urine samples with β-glucuronidase. The presence of TFAA-urea was indicated by the observation of an increase in its resonance and a concomitant decrease in that of TFAA-hydrate in urine samples doped with urea. It was confirmed from the formation of such an adduct after incubation of authentic TFAA-hydrate with urea. Approximately one-half of the administered TFE was excreted as TFE-glu. The remaining TFE was oxidised, primarily to TFAA-hydrate, with a small percentage oxidised further to TFA as described in the following scheme:

$$\underset{\text{TFE-glu}}{CF_3CH_2{-}O{-}glu} \longleftarrow \underset{\text{TFE}}{CF_3CH_2{-}OH} \longrightarrow \underset{\text{TFAA}}{CF_3CHO} \longrightarrow \underset{\text{TFA}}{CF_3COOH}$$

$$\underset{\text{TFAA-hydrate}}{CF_3CH(OH)_2} \rightleftharpoons \text{TFAA} \rightleftharpoons \underset{\text{TFAA-urea}}{CF_3CH(OH)NH{-}CO{-}NH_2}$$

^{19}F NMR analysis of urine from TFCE-treated rats demonstrated that this compound is totally metabolised leading to 2,2,2-trifluoromethylglucuronide (CF_3CH_2O-glucuronide) (TFE-glu), TFA, TFAA-hydrate, TFAA-urea, F^- (all identified as already described above) and a minor unidentified metabolite whose ^{19}F NMR downfield chemical shift and multiplicity indicate that it is probably an adduct of TFAA with another endogenous nucleophile.[63] The metabolites formed (Figure 4) show that TFCE undergoes oxidation of one C—H bond by CYP-450 leading to the generation of the halohydrin intermediate $CF_3CH(OH)Cl$ which may lose HCl to give TFAA. In aqueous solution, TFAA is in rapid equilibrium with its hydrate form which is predominant. TFAA may be oxidised to TFA or may react with endogenous nucleophiles such as urea or may be reduced to TFE which was not detected in urine as free alcohol but rather as glucuronide conjugate. Another metabolic pathway, possibly a CYP-450-dependent reductive dehalogenation, accounted for the production of F^-.

The same metabolites, except F^- and the minor unidentified metabolite, were also detected in urine samples of rats after TFCE inhalation.[64] The relative distribution of fluorine-containing metabolites of TFCE eliminated in urine within 24 h post-dose shows that the aldehydic compounds, TFAA-hydrate and TFAA-urea,

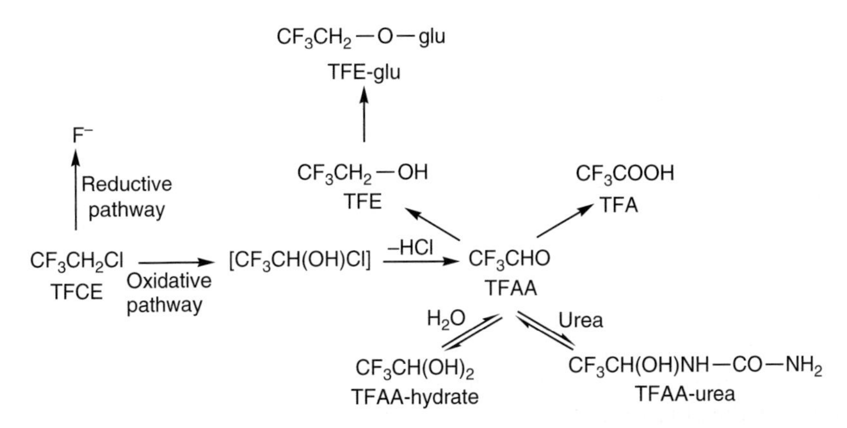

Figure 4 Proposed metabolic pathway of trifluorochloroethane (TFCE) adapted from Yin et al.[63] and Ellis et al.[64]

were the major metabolites excreted, accounting for 65%[64] or 67%[63], followed by TFA (19%[64] and 14%[63]) and TFE-glu (16% in both studies). F^- only detected by Yin et al.[63] represents 3% of the total fluorinated metabolites.

Sevoflurane is a halogenated ether inhalation anaesthetic agent that is approved for clinical use in more than 60 countries worldwide. Sevoflurane is metabolised by a CYP-450-catalysed oxidation (CYP2E1 being the major isoform involved) at the fluoromethoxy carbon to produce a transient intermediate that decomposes in equimolecular concentrations into F^- and 1,1,1,3,3,3-hexafluoro-2-propanol (HFP) that is converted by glucuronide conjugation to HFP-glucuronide (HFP-glu) (Figure 5).[65,66] Free HFP and HFP-glu were present in blood 5 min after the start of sevoflurane anaesthesia in humans,[67] but only HFP-glu was detected by ^{19}F NMR in urine of humans exposed to sevoflurane[66,68] where it represents 97 $\pm$ 3% ($n = 10$) of the total amount of organic fluorine-containing metabolites immediately after discontinuation of anaesthesia.[66] Moreover, it is well established that sevoflurane undergoes a base-catalysed elimination of hydrogen fluoride to give 2-(fluoromethoxy)-1,1,3,3,3 pentafluoro-1-propene (FPFP, also referred by the trivial name "compound A" in the official product label) as the major degradation product, which is nephrotoxic to rats. Indeed, the strong base soda lime or baralyme in the carbon dioxide absorbents of anaesthesia machines yields FPFP that is found at concentrations of 20–40 ppm in the anaesthetic circuit.[66,68–72] FPFP is mainly metabolised by conjugation with glutathione (GSH) leading to four GSH adducts corresponding to the two diastereoisomeric products of the regiospecific addition of GSH at the CF_2 carbon of the double bond (FPFP-GSH1) and the *Z* and *E* isomers of the dehydrofluorinated analogue of FPFP-GSH1 (FPFP-GSH2). These four compounds were identified by means of ion spray LC–MS–MS and ^1H and ^{19}F NMR spectroscopy from the bile of rats dosed with FPFP.[70] These GSH adducts were hydrolysed by γ-glutamyl transferase (GT) and dipeptidase (DP) enzymes to the corresponding cysteine-S conjugates, the [2-(fluoromethoxy)-1,1,3,3,3-pentafluoropropyl]-cysteine (FPFP-CYST1) and the [2-(fluoromethoxy)-1,3,3,3-tetrafluoro-1-propenyl]-cysteine (FPFP-CYST2) that were excreted in urine as mercapturate derivatives (FPFP-MER1 and FPFP-MER2) after *N*-acetylation by *N*-acetyltransferase (NAT) enzyme (Figure 5).[68,72] The cysteine-S-conjugates may also undergo β-lyase-dependent metabolism as showed by the formation of the expected products of such a reaction: F^-, pyruvate and 2-(fluoromethoxy)-3,3,3-trifluoropropanoic acid (TFPA) that is unstable and affords 3,3,3-trifluorolactic acid (TFLA) (Figure 5). These products were characterised by comparison with authentic standards or synthetic compounds using ^{19}F NMR spectroscopy (F^-), ^{19}F NMR and GC–MS (TFPA and TFLA) or after derivatisation and analysis by HPLC (pyruvate).[69]

The ^{19}F NMR examination of urine of humans anaesthetised with sevoflurane showed the resonances of HFP-glu, F^-, FPFP-MER1 and FPFP-MER2 (*Z* and *E* isomers), TFPA and TFLA.[66,68,72] The same compounds except HFP-glu were detected in rats receiving FPFP intraperitoneally.[68]

These results demonstrate that ^{19}F NMR is a very convenient technique to monitor the biotransformation of sevoflurane and that of its alkaline degradation compound formed in the anaesthesia circuit.

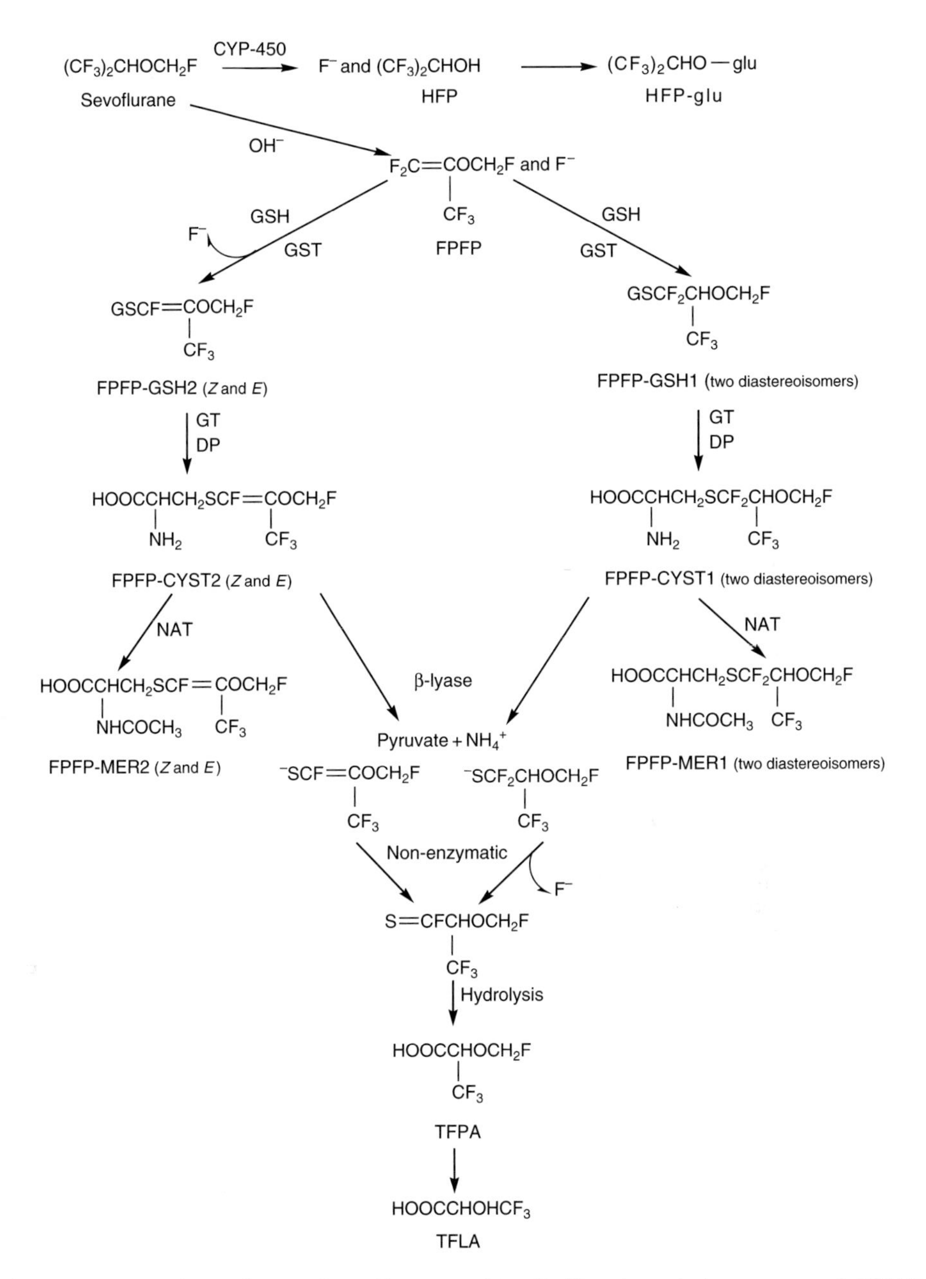

Figure 5 Metabolic pathway of sevoflurane and its alkaline degradation compound (FPFP) formed in anaesthesia circuits equipped with carbon dioxide scrubbers. Adapted from Peters and Ackland[73] and Jin et al.[70] GSH, glutathione; GST, glutathione-*S*-transferase; GT, γ-glutamyltransferase; DP, dipeptidase; NAT, *N*-acetyltransferase adapted from Orhan et al.[66] and Iyer and Anders.[69]

3.6. Anti-cancer drugs

In this part, we present the ^{19}F NMR studies dealing with three fluoropyrimidines in clinical use: 5-*fluorouracil* (FU), the mainstay of anti-metabolite treatment for solid tumours, capecitabine (CAP), a recent oral prodrug of FU, which is a fluorocytidine derivative, and gemcitabine that is a deoxycytidine analogue with two fluorine atoms attached to the carbon 2′ of the deoxyribose ring.

3.6.1. Fluorouracil

Since its introduction in clinical use 50 years ago, FU has become a component of the standard therapy for a variety of malignancies including gastrointestinal tract, head and neck, and breast cancers.[73] FU is a prodrug that requires intracellular complex metabolic conversion to fluoronucleosides (FNUCs) and then to cytotoxic fluoronucleotides (FNUCt). Besides this biochemical activation pathway, called anabolism, there is a degradative pathway called catabolism that leads to the drug elimination from the body.

FU and all its metabolites can diffuse back out of the cell, except FNUCt that are trapped within the cells due to the presence of the charged phosphate group. Biological fluid analyses are thus useful in studies of FU metabolism, especially catabolism that is the major route of elimination. Consequently, only the catabolic pathway is presented in this article and illustrated in Figure 6 (for reviews devoted to the biochemistry, mechanism of action and pharmacology of FU, see Peters and Ackland,[73] Diasio and Harris,[74] Weckbecker[75] and Grem[76]).

3.6.1.1. FU catabolism

As for natural pyrimidine bases, uracil and thymine, FU is degraded through enzymatic catabolism (Figure 6). The major fraction of the administered FU dose is converted to 5,6-dihydro-5-fluorouracil (FUH_2) by a trans-addition of two hydrogen atoms to the C5—C6 double bond of the pyrimidine ring. This reversible first step of FU catabolism requires the high-energy co-substrate nicotinamide adenine dinucleotide phosphate in its reduced form (NADPH), and is catalysed by the enzyme dihydropyrimidine dehydrogenase (DPD). FU catabolism further proceeds through reversible hydrolytic pyrimidine ring opening between C3 and C4 atoms catalysed by dihydropyrimidinase, leading to α-fluoro-β-ureidopropionic acid (FUPA). The major part of FUPA is converted via irreversible decarboxylation and deamination reactions into α-fluoro-β-alanine (FBAL) by β-alanine synthase or β-ureidopropionase. From CO_2 and NH_4^+, urea is formed by the urea cycle. FBAL, the most important product of FU catabolism, is optically active and exhibits a R configuration indicating the stereospecificity of the metabolic hydrogenation step of the C5—C6 double bond of FU.[77,78] FBAL can react non-enzymatically with bicarbonate ion (HCO_3^-) to give *N*-carboxy-α-fluoro-β-alanine (CFBAL). This is in accordance with the well-known equilibrium between compounds with an amino group and their corresponding carbamate (*N*-carboxy) derivatives in weakly alkaline aqueous carbonate solution. Owing to its acid lability, CFBAL is only detected in urine at pH $\geq$6.8–7.3 depending on the bicarbonate ion concentration, and in plasma.[9,79,80]

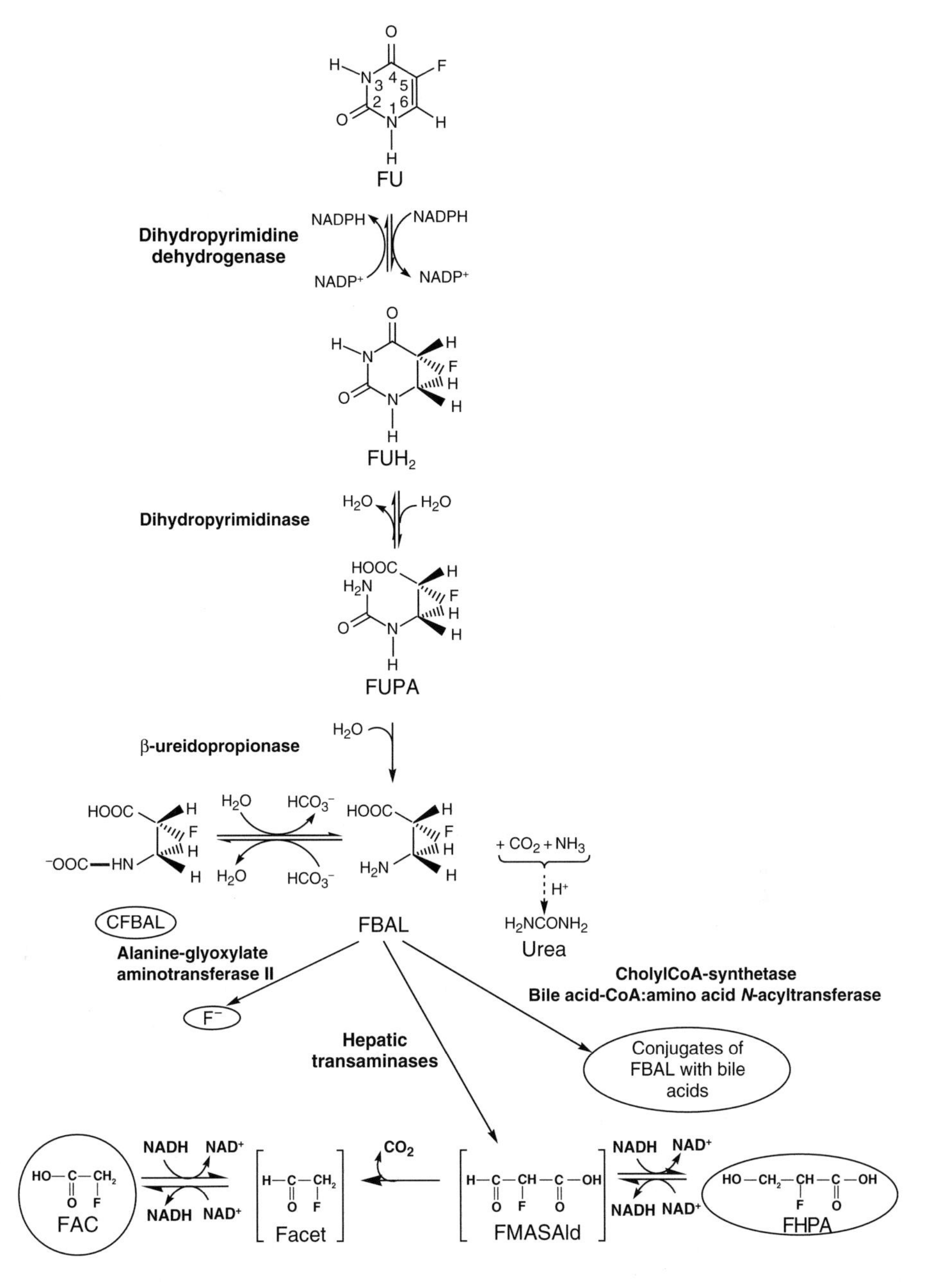

Figure 6 Catabolic pathway of 5-fluorouracil. All the compounds (except CFBAL) are represented in neutral form. The ellipses denote fluorinated metabolites identified for the first time with ^{19}F NMR. Non-detected fluorinated intermediates are represented in brackets.

FBAL can also serve as a substrate for further enzymatic reactions with the release of F^- and the formation of 2-fluoro-3-hydroxypropanoic acid (FHPA) and fluoroacetic acid (FAC) as well as that of FBAL conjugates with bile acids.

Twenty years ago, the ^{19}F NMR detection of elevated levels of F^- in acidic urine of rats treated with FBAL coupled to the fact that FBAL is defluorinated chemically in basic medium only led to the conclusion that F^- most likely results from the metabolic cleavage of the C—F bond of FBAL.[81] More recently, Porter et al.[82] demonstrated that the release of F^- from racemic (*S*,*R*)-FBAL is catalysed by the enzyme L-alanine-glyoxylate aminotransferase II, the metabolic (*R*)-FBAL being the preferred enantiomer for the defluorinating activity in rat liver homogenates.

The ^{19}F NMR detection of low amounts of two fluorinated analogues of β-alanine metabolites (FHPA and FAC) in perfusates of isolated perfused rat liver (IPRL) injected with pure FU or FBAL and in urine of rats treated with pure FU showed that the metabolism of β-alanine and its fluorinated analogue are similar.[83] The first step in β-alanine catabolism is a transamination reaction to form malonic acid semi-aldehyde (MASAld) catalysed by hepatic transaminases.[84,85] MASAld is converted either by spontaneous decarboxylation to acetaldehyde, which is further oxidised into acetate through an aldehyde deshydrogenase catalysis requiring nicotinamide adenine dinucleotide (NAD) in its oxidised form (NAD^+), or enzymatically to β-hydroxypropanoic acid ($HOCH_2$—CH_2—COOH) by propionate deshydrogenase, which needs NAD in its reduced form (NADH).[84] By analogy with the metabolism of β-alanine, we proposed that FBAL leads to the formation of FHPA and FAC according to the metabolic pathway depicted in Figure 6. The aldehydes being very reactive, FMASAld and its decarboxylated analogue, fluoroacetaldehyde (Facet), are not detected in our experiments. Similarly, during the metabolisation of fluorinated ethanes into FAC in rats, intermediate Facet was undetected in urine and kidney extracts.[86]

FHPA and FAC are detected in urine of patients treated with FU using ^{19}F NMR.[87,88] FHPA was also detected by Hull et al.[9] in urine of patients treated with FU but not identified. As FU commercial formulations are not pure (see below), FAC and FHPA can arise from the metabolism of impurities or that of FU or both. Nevertheless, Lemaire et al.[88] demonstrated that FHPA detected in urine of patients treated with FU comes, at least partly, from FU metabolism.

Biliary excretion of fluoropyrimidine drugs in humans is low: 2–3% for radiolabelled FU as measured by scintillation counting[89] and 0.8% for the FU prodrug, 5′-deoxy-5-fluorouridine (5′dFUR), as determined by ^{19}F NMR.[90] In the last case, F^- and FBAL represented ≈10% of the excreted metabolites and unknown compounds accounted for ≈90%. These metabolites were identified as conjugates of FBAL with cholic acid (choloFBAL) and chenodeoxycholic acid (chenoFBAL) in a 74/26% ratio in bile of patients with external bile drainage (Figure 7). This result is in agreement with the fact that only the "primary" bile acids (cholic and chenodeoxycholic acids in a ratio of ≈3:1) are present in case of bile derivation. The "'secondary"' bile acids (mainly deoxycholic acid) are no longer formed as the enterohepatic circulation is suppressed. In a bile sample obtained at surgery from a patient treated with intrahepatic FU, three conjugates of FBAL were detected,

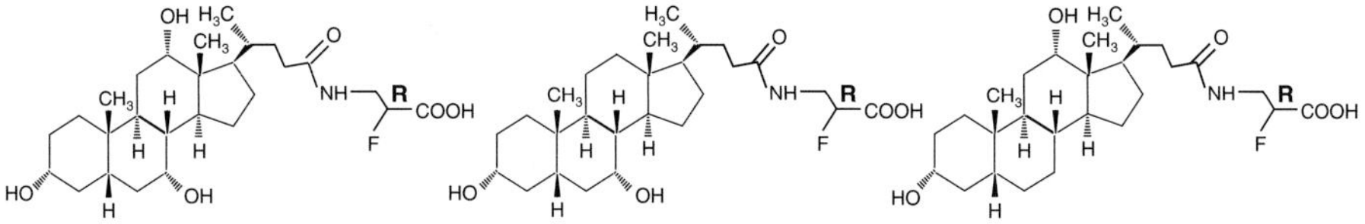

Figure 7 Chemical structures of conjugates of α-fluoro-β-alanine (FBAL) with cholic acid (choloFBAL), chenodeoxycholic acid (chenoFBAL) and deoxycholic acid (dcholoFBAL).

choloFBAL (54%), chenoFBAL (17%) and the conjugate with the third major bile acid in human bile, deoxycholic acid, dcholoFBAL (29%) (Figure 7).[90]

Independently and at the same time, Heggie et al[89] found, from an HPLC analysis of biliary excretion of radiolabelled FU administered to patients with external bile drainage, that 80–90% of the FU biliary metabolites were previously unrecognised. These metabolites were identified later, using mass spectrometry and enzymatic degradation after isolation, as conjugates of FBAL with cholic acid and chenodeoxycholic acid.[91,92] Comparing the ^{19}F NMR spectra of FBAL conjugates with the bile acids in human bile to those synthesised from racemic FBAL showed that only one of the two diastereoisomers of each FBAL conjugate forms in vivo, in accordance with the fact that only the (*R*) enantiomer of FBAL is metabolically produced from FU.[77] Two enzymes (cholyl-CoA synthetase and bile acid-CoA:amino acid *N*-acetyltransferase) are sequentially involved in the formation of bile acid–amino acid conjugates (Figure 7).[93]

In summary, ^{19}F NMR has helped to improve the knowledge of FU catabolic pathway, the present understanding of which is summarised in Figures 6 and 7. All these fluorinated metabolites were detected and quantified in a single-run analysis, the ellipses circling those identified for the first time by this method: F^-, CFBAL, conjugates of FBAL with cholic, chenodeoxycholic and deoxycholic acids, FHPA and FAC, all of them being only observed with ^{19}F NMR except FBAL conjugates with cholic and chenodeoxycholic acids.

Moreover, ^{19}F NMR confirms the importance of the urinary excretion of FU and metabolites demonstrated in the 1960s by Mukherjee and Heidelberger.[94] They reported that 90% of 6-[^{14}C]-FU administered dose (a.d.) was excreted in patients' urine within 24 h mainly as FBAL. Using ^{19}F NMR as the analytical method, a daily total urinary excretion profile of FU and metabolites following a daily intravenous (i.v.) bolus administration during a 6-day chemotherapy shows that (i) the daily excretion is nearly constant and reaches 95 $\pm$ 10% a.d., (ii) FBAL is by far the major metabolite accounting for 75 $\pm$ 10% of the injected daily dose, unchanged FU amounting to $\approx$10%, and the other metabolites $\approx$10% (FUH_2 $\approx$4%, FUPA $\approx$8%, F^- $\approx$2%) and (iii) the excretion of FU and metabolites is rapid as it occurs for 83 $\pm$ 9% within the initial 6 h.[95]

Similar data were reported by Heggie et al.[89] in patients treated i.v. bolus with 6-[^{3}H]-FU using HPLC with radioactivity quantification of each resolved peak as the analytical technique, and Hull et al.[9] in patients treated i.v. bolus with FU with or without a pre-treatment with methotrexate using ^{19}F NMR.

3.6.1.2. FU degradative pathway and FU cardiotoxicity

Because a basic medium is needed for its solubilisation, FU is dissolved for clinical use at a 50 mg/ml concentration in sodium hydroxide solution (FU-NaOH) at pH $\approx$ 9.1 and in some cases in Tris (Trometamol) buffer (FU-Tris) at pH $\approx$ 8.4.

The ^{19}F NMR analysis of commercial FU-NaOH solutions revealed the presence, besides FU, of about a hundred fluorinated signals accounting for $\approx$1.8 mol% relative to nominal FU concentration, F^- being by far the major one (Figure 8(a)).

As FU powder is pure, all the fluorinated resonances detected in the ^{19}F NMR spectra of FU clinical formulations are FU degradation compounds formed over time. Indeed, in basic conditions, FU is hydrolysed to urea and FMASAld (HOOC—CHF—CHO) that is decarboxylated with time into Facet (H_2FC—CHO) as well as to urea, F^- and non-fluorinated aldehydes.[18,96,97] The two fluorinated aldehydes that are under hydrate form are found at very low concentrations in vials (0.015% for FMASAld and 0.010% for Facet relative to nominal FU).[98] Owing to their high chemical reactivity, they are transformed by successive aldol condensations with each other and/or non-fluorinated aldehydes, and/or by reaction with urea into the numerous fluorinated compounds detected. On the contrary, the ^{19}F NMR analysis of the FU-Tris formulations revealed the presence, besides FU itself, of few fluorinated compounds accounting for $\approx$1.5 mol% relative to nominal FU concentration, with F^- representing less than 0.1 mol% of FU. The major ones were identified as adducts of Tris with FMASAld (1.0%) and Facet (0.3%) (Figure 8(b)).[87,98] Indeed, it is known that aldehydes react with nucleophilic amino and hydroxyl groups of Tris leading to the formation of oxazolidines. The oxazolidines formed with FMASAld and Facet, whose structures are presented in Figure 8(b), are stable at the pH of the FU formulation (8.4) but are in equilibrium with the starting aldehydes at physiological pH. FMASAld and Facet are thus trapped as stable depot forms in FU-Tris formulations.

FMASAld and Facet (as oxazolidine adducts or under free form) are highly cardiotoxic on the isolated perfused rabbit heart (IPRH) model. The cardiotoxicity of FU solutions on this model is correlated to the amounts of FMASAld and Facet, FU-NaOH formulations being much less cardiotoxic than FU-Tris solutions.[87,98] In the IPRH model, Facet is extensively metabolised into FAC, a highly cardiotoxic and neurotoxic poison, whereas FMASAld is converted in a very low extent to FAC and FHPA, which is also cardiotoxic on this model at high dose.[83,87,98]

In the urine of patients treated with FU-Tris formulations, ^{19}F NMR analysis revealed the presence, besides that of FU and its classical metabolites (FUH_2, FUPA, CFBAL, FBAL and F^-), of FHPA and FAC as well as that of Facet and adducts of FMASAld and Facet with urea (Figure 9).

If FMASAld and Facet resulting from FU alkaline hydrolysis are certainly the causative factor of higher cardiotoxicity and neurotoxicity of FU-Tris formulations,[99] the non-negligible frequency of cardiotoxic accidents observed after injection of the FU-NaOH formulations at pH 9.1 was too important to be explained by the very low levels of Facet and FMASAld found in these preparations. Consequently, the metabolism of FU itself, via the FBAL transamination reaction forming FMASAld that then gives FHPA, Facet and FAC (Figure 6), is also involved in the cardiotoxicity of this drug.

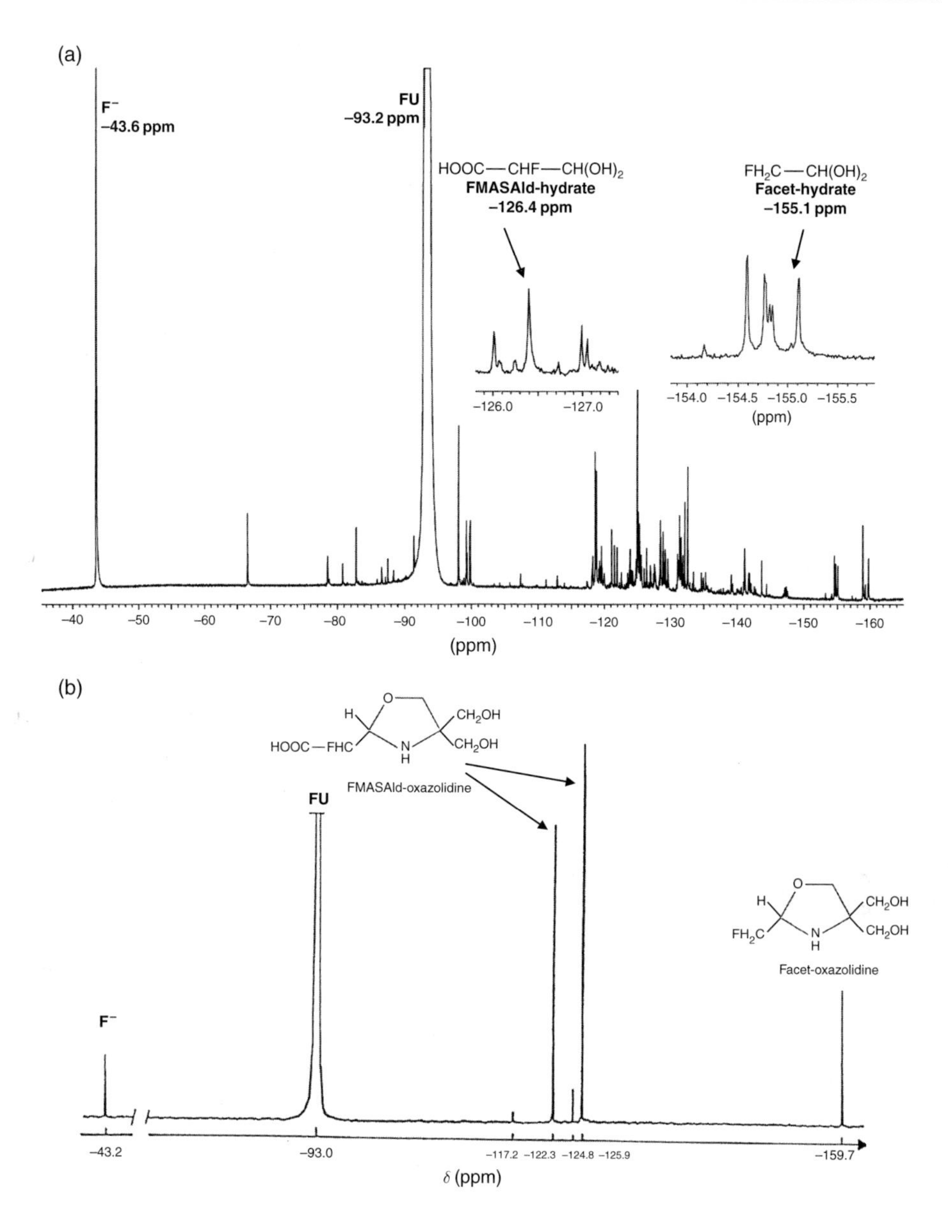

Figure 8 ^{19}F NMR spectra at 282 MHz with proton decoupling of clinical formulations of 5-fluorouracil dosed at 50 mg/ml in sodium hydroxide solutions at pH 9.1 (a) and in Tris buffer at pH 8.4 (b). The chemical shifts are expressed relative to the resonance peak of trifluoroacetic acid (TFA) (5%, w/v aqueous solution) used as an external reference. F^-, fluoride ion; FU, 5-fluorouracil; FMASAld-hydrate, fluoromalonic acid semi-aldehyde hydrate; Facet-hydrate, fluoroacetaldehyde hydrate; FMASAld-oxazolidine, two diastereoisomeric oxazolidine adducts of FMASAld with Tris; Facet-oxazolidine, oxazolidine adduct of Facet with Tris.

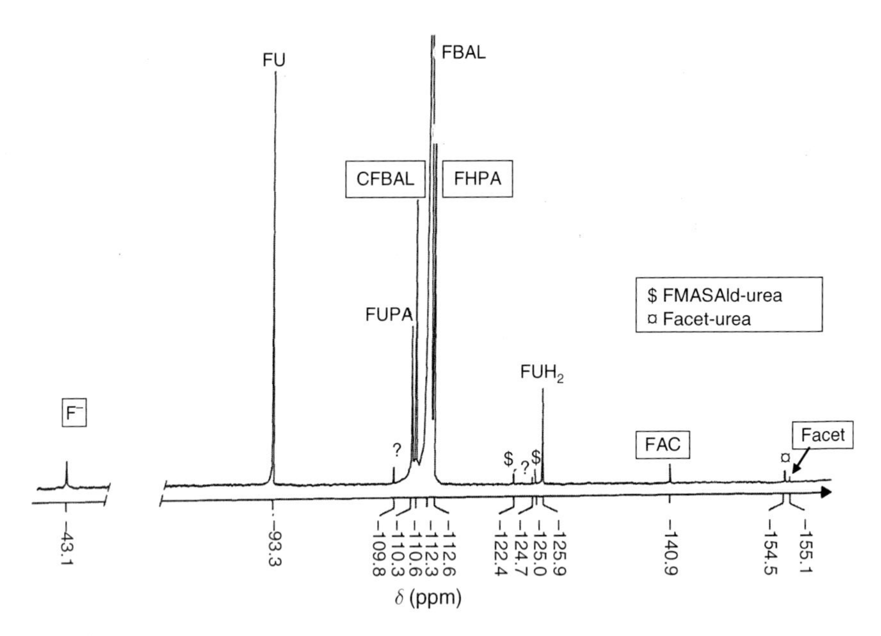

Figure 9 ^{19}F NMR spectrum at 282 MHz with proton decoupling of a urine sample from a patient treated with a continuous i.v. infusion of FU-Tris at a dose of 650 mg/m^2/day over four consecutive days. Urine fraction 48–72 h, 30-fold concentrated, pH of the sample 6.8. The chemical shifts are expressed relative to the resonance peak of TFA (5% w/v aqueous solution) used as an external reference. F^-, fluoride ion; FU, 5-fluorouracil; ?, unknown; FUPA, α-fluoro-β-ureidoproionic acid; CFBAL, *N*-carboxy-α-fluoro-β-alanine; FBAL, α-fluoro-β-alanine; FHPA, 2-fluoro-3-hydroxypropanoic acid; FMASAld-urea, two diastereoisomeric adducts of fluoromalonic acid semi-aldehyde with urea; FUH_2, 5,6-dihydro-5-fluorouracil; FAC, fluoroacetic acid; Facet-urea, adduct of fluoroacetaldehyde with urea; Facet, fluoroacetaldehyde hydrate. Metabolites identified for the first time with ^{19}F NMR are represented in boxes.

3.6.2. Capecitabine

N^4-Pentyloxycarbonyl-5′-deoxy-5-fluorocytidine, more commonly called *capecitabine* (CAP) or Xeloda®, is a recent prodrug of 5′dFUR, another FU prodrug, that is administered orally to circumvent the unacceptable toxicity of 5′dFUR without compromising its anti-tumour efficacy.[100] Because the main limitation of 5′dFUR derives from its gastrointestinal toxicity (diarrhoea) attributed to the liberation of FU in the small intestine under the action of thymidine phosphorylase (TP),[101] CAP was designed as a prodrug of 5′dFUR that could not be metabolised by TP in the intestine. Indeed, after oral administration CAP crosses intact the gastrointestinal barrier and is rapidly and almost completely absorbed;[102,103] diarrhoea should not thus occur with its use. CAP is subsequently converted into FU in a three-stage mechanism involving several enzymes (Figure 10). In a first step, it is metabolised into 5′-deoxy-5-fluorocytidine (5′dFCR) by carboxylesterase, almost exclusively located in the liver. 5′dFCR is then deaminated into 5′dFUR by cytidine deaminase mainly localised in liver and tumour tissues. Finally, 5′dFUR is transformed into FU under the action of TP, an enzyme with higher activity in malignant tissue

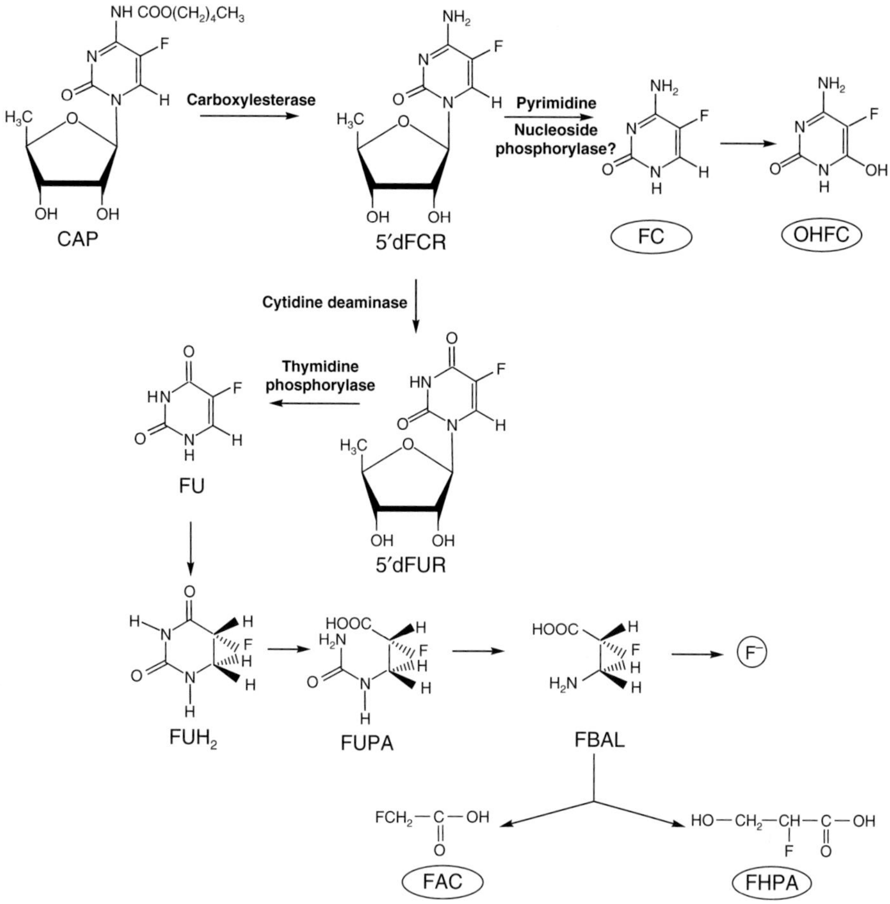

Figure 10 Catabolic pathway of capecitabine from ^{19}F NMR analysis of patients' urine. All the compounds are represented in neutral form. Metabolites identified for the first time in urine of patients with ^{19}F NMR are represented in ellipses.

than in normal tissue. Higher levels of FU are thus produced within tumours with minimal exposure of healthy tissue to FU.[104] As CAP and its two first metabolites do not show intrinsic cytotoxicity, the activation pathway is expected to combine high anti-tumour efficacy with improved clinical safety.[104]

Clinical pharmacokinetic study has demonstrated that the excretion of the intact drug and its metabolites is both rapid and almost exclusively urinary. Indeed, the percentage of radioactive dose after oral administration of a single 2 g dose of ^{14}C-radiolabelled CAP reached about 98% (95.5% in urine and 2.6% in faeces) after 6 days of cumulative collection, 84% of the urine excretion occurring in the first 12 h post-dosing.[102] The mean percentage of dose excreted in urine as parent drug and its fluorinated metabolites, 5′dFCR, 5′dFUR, FU, FUH_2, FUPA and FBAL up to 48 h post-dosing, measured by ^{19}F NMR, was 84.2% close to 92.3% of the radioactivity recovered at that time. This difference of less than 10% demonstrates that ^{19}F NMR spectroscopy is a suitable technique for quantitative studies.[102]

Moreover, in this study as well as in three other separate studies, all managed by Hoffmann-La Roche Company and involving 68 patients all in all, all the metabolites of the catabolic pathway CAP → 5′dFCR → 5′dFUR → FU → FUPA → FBAL were quantified.[102,103,105–107] As expected, the major metabolite by far is FBAL that represents 51–61% a.d., followed by 5′dFUR (6–11% a.d.) and 5′dFCR (5–7% a.d.). FUPA (3–5% a.d.) and unmetabolised CAP (2–3% a.d.) were the other significant forms, whereas FU and FUH_2 accounted for less than 1% a.d. (range 0.3–0.7% and 0.1–0.4%, respectively).[102,103,105–107]

Urine of patients receiving i.v. dose of 200–250 mg/m^2 irinotecan (CPT-11) 24 h before oral CAP treatment at a dose of 1200–1250 mg/m^2 twice daily at 12 h interval were collected over 12 h after the first CAP dose and analysed by ^{19}F NMR.[12] Pretreatment of patients with CPT-11 does not sensibly affect the pattern of CAP metabolites excreted in urine. Indeed, the total recovery of CAP and its fluorinated metabolites accounted for 71 ± 17% a.d. close to the literature values based on the same measurement technique (71–86% a.d. after 24 h urine collection). Moreover, the recovery percentages of all the pre-cited metabolites are similar in the five studies except that of FBAL.[12,102,103,105–107] The long half-life time of FBAL[89] and the limited 12 h period of urine collection can explain the lower amount of FBAL found in this study (46 ± 4% a.d.)[12] compared to that obtained after urine recovery over 48 h (57% a.d.) and 24 h (51–61% a.d.).[102,103,105–107] Three other FU metabolites were observed in small or very tiny amounts: F^- (≈0.2% a.d.), FHPA (≈0.3% a.d.) and FAC (≈0.004% a.d. only detected in 4 out of 14 samples analysed). Since CAP formulation is pure,, this demonstrates that FU can be metabolised, in humans, into FHPA and FAC via FBAL, as already shown in rats.[83] The detection of 5-fluorocytosine (FC) and 6-hydroxy-5-fluorocytosine (OHFC) in 4 out of 14 urine samples, accounting for 0.01% and 0.02% a.d., respectively, led to the identification of a novel, even minor, degradation pathway. The degradation pathway of CAP, incorporating the new fluorinated metabolites found in urine using ^{19}F NMR, is depicted in Figure 10, and a typical ^{19}F NMR spectrum of urine is presented in Figure 11.

3.6.3. Gemcitabine

Gemcitabine (2′,2′-difluorodeoxycytidine, dFdC) is a nucleoside analogue of deoxycytidine in which the two fluorine atoms are inserted into the deoxyribofuranosyl ring (Figure 12). The metabolism of gemcitabine is fairly identical to that of the deoxynucleoside analogues. Once inside the cell, gemcitabine is activated to its mononucleotide 2′,2′-difluorodeoxycytidine monophosphate (dFdCMP) by deoxycytidine kinase. dFdCMP is phosphorylated sequentially to diphosphate (dFdCDP) then triphosphate (dFdCTP). The cytotoxic action of gemcitabine is related to the incorporation of dFdCTP into DNA by DNA polymerases and the consequent inhibition of further DNA synthesis. The inhibition of ribonucleotide reductase by dFdCDP reduces the transformation of cytidine diphosphate (CDP) into 2′-deoxycytidine diphosphate (dCDP), and thus lowers 2′-deoxycytidine

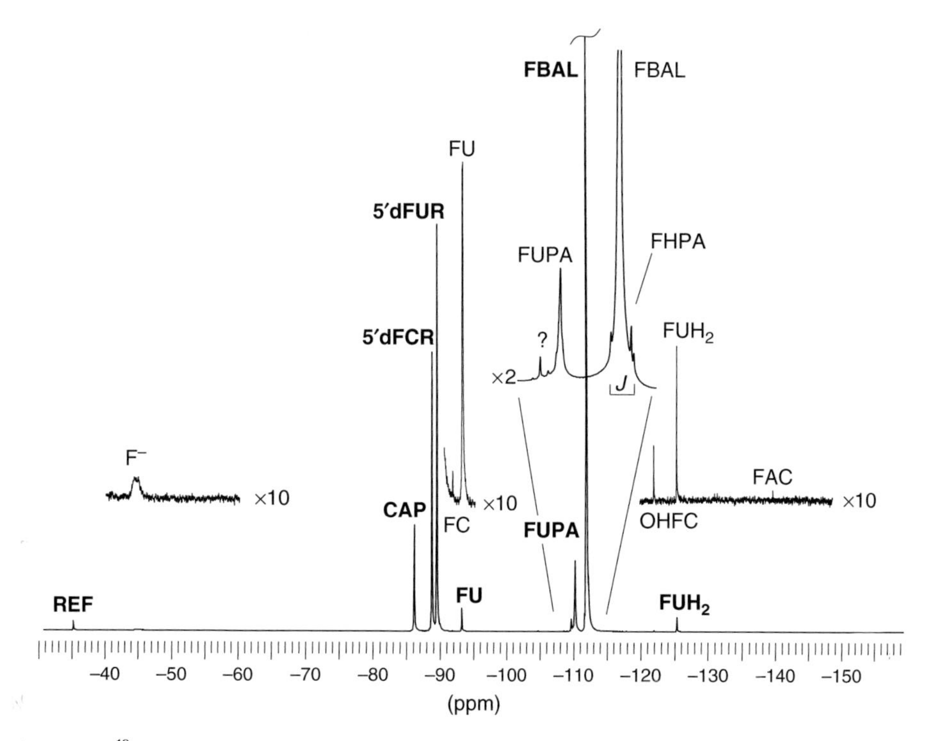

Figure 11 ^{19}F NMR spectrum at 282 MHz with proton decoupling of a urine sample from a patient receiving oral capecitabine at a dose of 3800 mg/day administered twice daily at 12 h interval, as a second treatment 3 months after the first one. Urine fraction 0–12 h collected after the first dose of 1900 mg and 10-fold concentrated, pH of the sample: 5.45. The chemical shifts are expressed relative to the resonance peak of TFA (5%, w/v aqueous solution) used as external reference. REF, external reference; F^-, fluoride ion; CAP, capecitabine; 5′dFCR, 5′-deoxy-5-fluorocytidine; 5′dFUR, 5′-deoxy-5-fluorouridine; FC, 5-fluorocytosine; FU, 5-fluorouracil; ?, unknown; FUPA, α-fluoro-β-ureidopropionic acid; FBAL, α-fluoro-β-alanine; FHPA, 2-fluoro-3-hydroxypropanoic acid; OHFC, 6-hydroxy-5-fluorocytosine; FUH_2, 5,6-dihydro-5-fluorouracil; FAC, fluoroacetic acid. J, $^1J_{13C-F}$ coupling constant.

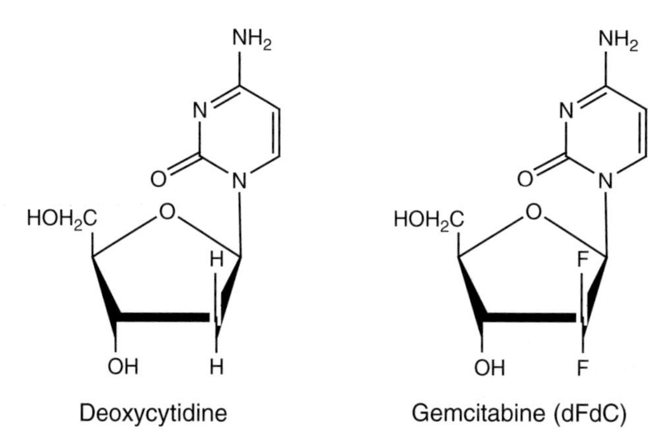

Figure 12 Chemical structures of deoxycytidine and gemcitabine.

triphosphate (dCTP) levels. Because dFdCTP and dCTP compete directly for the incorporation into DNA, this effect facilitates the DNA incorporation of dFdCTP. Gemcitabine can also be incorporated into RNA at a low extent, but neither the chemical identity of the incorporated nucleotide nor its biological consequences have been established. Gemcitabine and dFdCMP are fair substrates for deamination by cytidine deaminase and dCMP deaminase, respectively. The deamination product of gemcitabine, 2′,2′-difluorodeoxyuridine (dFdU) appears to be biologically inert and is not a substrate for phosphorylation into monophosphate. As dFdUMP, the deamination product of dFdCMP, is a substrate for phosphorylation, it can inhibit TS and thus the transformation of dUMP into dTMP.[73,108]

The ^{19}F NMR analysis of gemcitabine and dFdU was performed in plasma and urine of patients.[109] The presence of a CF_2 group at the 2′ position of the ribofuranosyl ring leads to AB-type NMR spectra consisting in four resonances. The chemical shifts of both dFdC and dFdU are pH-dependent. In the pH range 5–8, the resonances of both the compounds are distinct despite the fact that both compounds have almost identical ^{19}F chemical shifts. The accuracy of the ^{19}F NMR assay was estimated to be ≈10% and the limit of quantitation ≈10 mmol/l with a 9.4 T spectrometer equipped with a standard probe excluding proton decoupling of the spectra recorded. HPLC and ^{19}F NMR comparative assays of dFdC and dFdU in plasma samples of dFdC-treated patients showed a reasonable agreement in the results obtained with the two methods. When applied to the analysis of urine samples from patients receiving dFdC, ^{19}F NMR demonstrated that (i) ≈75% of the dFdC a.d. is excreted in 24 h, (ii) the majority of the drug is excreted as dFdU, dFdC being mainly present in the first 6 h urine samples and (iii) traces of metabolites not yet identified, probably glucuronides, are detected.

3.7. Anti-fungals

5-Fluorocytosine (FC) is an *anti-fungal* agent used for the treatment of severe fungal infections, particularly when combined to amphotericin B. The anti-fungal activity of FC results from the intra-fungal formation of FU leading to the inhibition of RNA processing and DNA synthesis via FNUCt metabolites. Susceptible fungi contain cytosine deaminase (CD), the enzyme that converts FC to FU, whereas human cells lack this enzyme thus creating a theoretical absence of toxicity for FC in humans. However, because FC and FU toxicity profiles are quite similar, it is thought that FU may account for some FC side effects.[110] Moreover, FU and FU catabolites (FUPA, FBAL) were detected in biofluids of healthy volunteers or patients receiving FC.[111–113] The evidence of FC conversion into FU by viable and non-viable *Escherichia coli* as well as by a semi-continuous culture system mimicking human intestinal microflora was clearly demonstrated.[114–116]

The ^{19}F NMR analysis of biofluids (plasma and urine) from patients treated with FC provided new information concerning FC metabolism in humans (Figure 13). More than 90% of the i.v. a.d. of FC was excreted within 24 h after the injection,

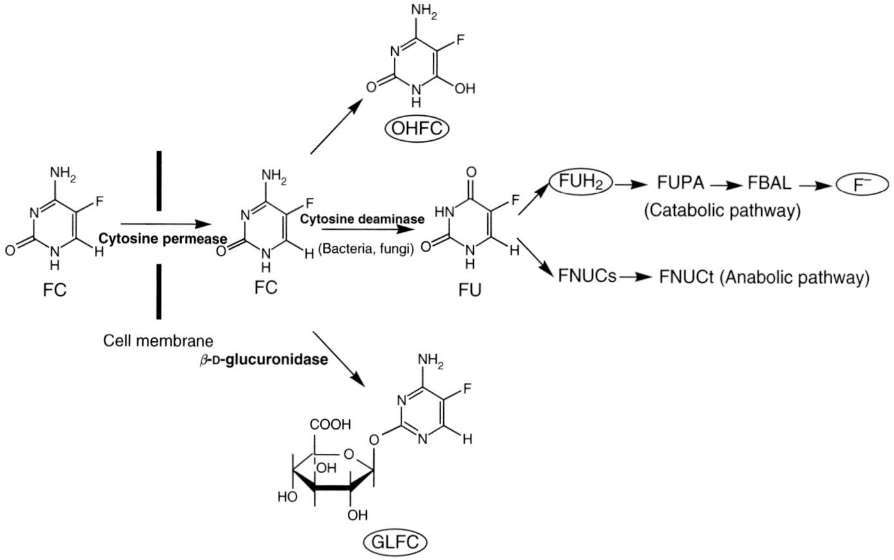

Figure 13 Metabolic pathway of 5-fluorocytosine. FC, 5-fluorocytosine; FU, 5-fluorouracil; FUH_2, 5,6-dihydro-5-fluorouracil; FUPA, α-fluoro-β-ureidopropionic acid; FBAL, α-fluoro-β-alanine; F^-, fluoride ion; OHFC, 6-hydroxy-5-fluorocytosine; GLFC, O_2-β-glucuronide of FC; FNUCs, fluoronucleosides; FNUCt, fluoronucleotides. The ellipses denote fluorinated metabolites identified for the first time in human biofluids with ^{19}F NMR.

unchanged FC representing ≈95% of the total excreted. FU was detected in plasma only, whereas all the classical FU catabolites, FUPA, FBAL, FUH_2 (only when the level of FBAL was high) and F^-, were identified in urine.[18,117] Two other compounds involving a direct metabolism of FC were found. 6-Hydroxy-5-fluorocytosine (OHFC) was detected in urine and plasma and represented less than 1.5% of the total excreted dose.[18,117] A glucuroconjugate of FC (GLFC), identified as the O2-β-glucuronide of FC,[118] was present in plasma, urine and cerebrospinal fluid (CSF). It was the major FC metabolite accounting for 1–4% of the total excreted dose.[18,117]

In conclusion, ^{19}F NMR allowed detecting and identifying four new FC metabolites in physiological fluids of FC-treated patients: FUH_2 and F^- that are well-known catabolites of FU, OHFC that was already detected in unusual circumstances in a urinary gravel excreted by a FC-treated patient,[113] and GLFC which was the first glucuronide of a fluoropyrimidine drug detected in humans.

Moreover, the ^{19}F NMR study of urine of two patients treated with FC showed a direct relationship between FU catabolites and the number of intestinal enterobacillary colonies. Indeed, the percentage of FU catabolites ($FUPA + FBAL + F^-$) was found to be extremely low (<0.6% of total fluorinated compounds excreted) when the number of enterobacillary colonies was low ($<10^3$), but considerably higher (3.5–8.8%) when such colonies were under reconstitution or within the normal range (10^5–10^8).[119] This finding demonstrated that FU-related toxicity may occur in FC-treated patients.

3.8. Miscellaneous drugs

^{19}F NMR was applied to the identification of dog urinary metabolites of *imirestat* (Figure 2), a spirohydantoin aldose reductase inhibitor, likely to alleviate certain late complications of diabetes such as cataracts, retinopathy and peripheral neuropathy.[120] The ^{19}F NMR spectrum of crude urine shows a broad signal corresponding to F^-, numerous metabolites which were predominantly glucuronide conjugates (as evidenced by ^{19}F NMR analysis of urine samples after β-glucuronidase incubation) and, as a minor component, the parent drug. Multinuclear (^{1}H, ^{13}C, proton-coupled ^{19}F as well as two-dimensional ^{19}F–^{19}F COSY (correlation spectroscopy)) NMR and GC–MS analysis of HPLC purified aglycones obtained from β-glucuronidase treatment of individual glucuronides previously isolated by semi-preparative HPLC of crude urine allowed the identification of four hydroxylated metabolites of imirestat (three containing two fluorine atoms and one containing only one).[120]

Ponalrestat (Statil®) is another potent inhibitor of the enzyme aldose reductase, which underwent clinical trials in insulin-treated diabetics (Figure 2).[121] The quantitative metabolism and excretion of this drug has been investigated by means of scintillation counting and ^{19}F NMR spectroscopy in urine and bile of rats receiving a single oral dose of ^{14}C-labelled ponalrestat. Comparison of the percentage of drug and metabolite recoveries obtained from ^{19}F NMR spectroscopy and the use of radiolabelling showed relatively good correlation ($R^2 = 0.92$), although radiotracer techniques are expected to be inherently more accurate and sensitive than NMR spectroscopy. In general, the ^{19}F NMR results are in good agreement (within 10–20%) with those obtained by scintillation counting. However, discrepancies have been observed with ^{19}F NMR signals in bile. Indeed the compartmentalisation of the compounds in micelles lead, as a consequence of restricted molecular mobility, to much broadened resonances which are difficult to quantify.[121]

4. ^{31}P NMR Studies of Phosphorylated Drugs

As they are few phosphorylated drugs in clinical use or trials, only a few ^{31}P NMR drug studies have so far been carried out. Mention can be made of WR-2721 (amifostine), a radio- and chemo-protective drug and the anti-cancer oxazaphophorine drugs, cyclophosphamide and ifosfamide.

4.1. Amifostine

S-2-[3-(aminopropylamino)] ethylphosphorothioic acid, more commonly called WR-2721 or *amifostine* (Ethyol®), is a phosphothioate that exerts a radio- and chemo-protective effect. Indeed, it protects normal tissues to a greater extent than malignant tissues from the toxicities of radiation and alkylating agent therapy of solid tumours.[122,123] Amifostine is used as a daily routine treatment to prevent neutropaenia and nephrotoxicity in patients having advanced ovarian carcinoma treated with cyclophosphamide and platinum agents (cisplatin or analogues).

WR-2721 is a prodrug that is dephosphorylated in vivo by alkaline phosphatase which is relatively lacking in tumour tissue[123] to yield the active free thiol metabolite (WR-1065) and inorganic phosphate (Pi).

$$\underset{\text{amifostine or WR-2721}}{H_2N{-}(CH_2)_3{-}NH{-}(CH_2)_2{-}S{-}PO_3H_2} \longrightarrow \underset{\text{WR-1065}}{H_2N{-}(CH_2)_3{-}NH{-}(CH_2)_2{-}SH} + \underset{\text{Pi}}{H_3PO_4}$$

The dephosphorylation of WR-2721 was monitored in abdominal and hind-quarter regions of mice by in vivo ^{31}P NMR.[124] The rate of decrease of the WR-2721 signal and the concurrent increase of urinary Pi signal constituted a direct measure of the dephosphorylation process.

Since under physiological conditions, even in the presence of alkylating agents,[125] WR-2721 is directly converted to WR-1065 and Pi, which is the ultimate phosphorylated species, ^{31}P NMR analysis of plasma or urine of patients receiving WR-2721 is without interest for the study of its metabolism and so has never been performed.

4.2. Cyclophosphamide and ifosfamide

4.2.1. Overview of the metabolism

Cyclophosphamide (CP) and its structural isomer *ifosfamide* (IF) are among the most widely therapeutically used alkylating anti-cancer agents. Both compounds are prodrugs that are bioactivated by CYP-450 enzymes to exert their toxic activity.[126–128]

The initial CYP-450-catalysed metabolic step leads to the formation of 4-hydroxymetabolites, 4-hydroxycyclophosphamide (OHCP) or 4-hydroxyifosfamide (OHIF), that equilibrate with their ring-opened aldo tautomers, aldocyclophosphamide (AldoCP) or aldoifosfamide (AldoIF) (Figures 14 and 15). These aldo intermediates are believed to play a pivotal role in anti-tumour efficacy of these drugs by partitioning between pathways giving either the cytotoxic agent or a biologically inactive compound. Indeed, the aldo derivatives undergo a non-enzymic β-elimination of urotoxic and nephrotoxic acrolein to yield the ultimate alkylating species, phosphoramide mustard (PM) or isophosphoramide mustard (IPM). Alternatively, it may be oxidised to inactive carboxycyclophosphamide (CXCP) or carboxyifosfamide (CXIF) by aldehyde dehydrogenase (ALDH) or reduced to alcocyclophosphamide (AlcoCP) or alcoifosfamide (AlcoIF) by an aldehyde reductase. OHCP and OHIF may also be partially deactivated to 4-ketocyclophosphamide (KetoCP) or 4-ketoifosfamide (KetoIF).[128]

In addition to this ring oxidation, *N*-dechloroethylation of CP or IF can occur, leading to the formation of *N*-dechloroethylcyclophosphamide (DCCP) from CP or 2-dechloroethylifosfamide (2-DECLIF) and DCCP from IF, and to the elimination of chloroacetaldehyde, a compound that may be responsible for the oxazaphosphorine-induced neurotoxicity, urotoxicity and cardiotoxicity.[129–131]

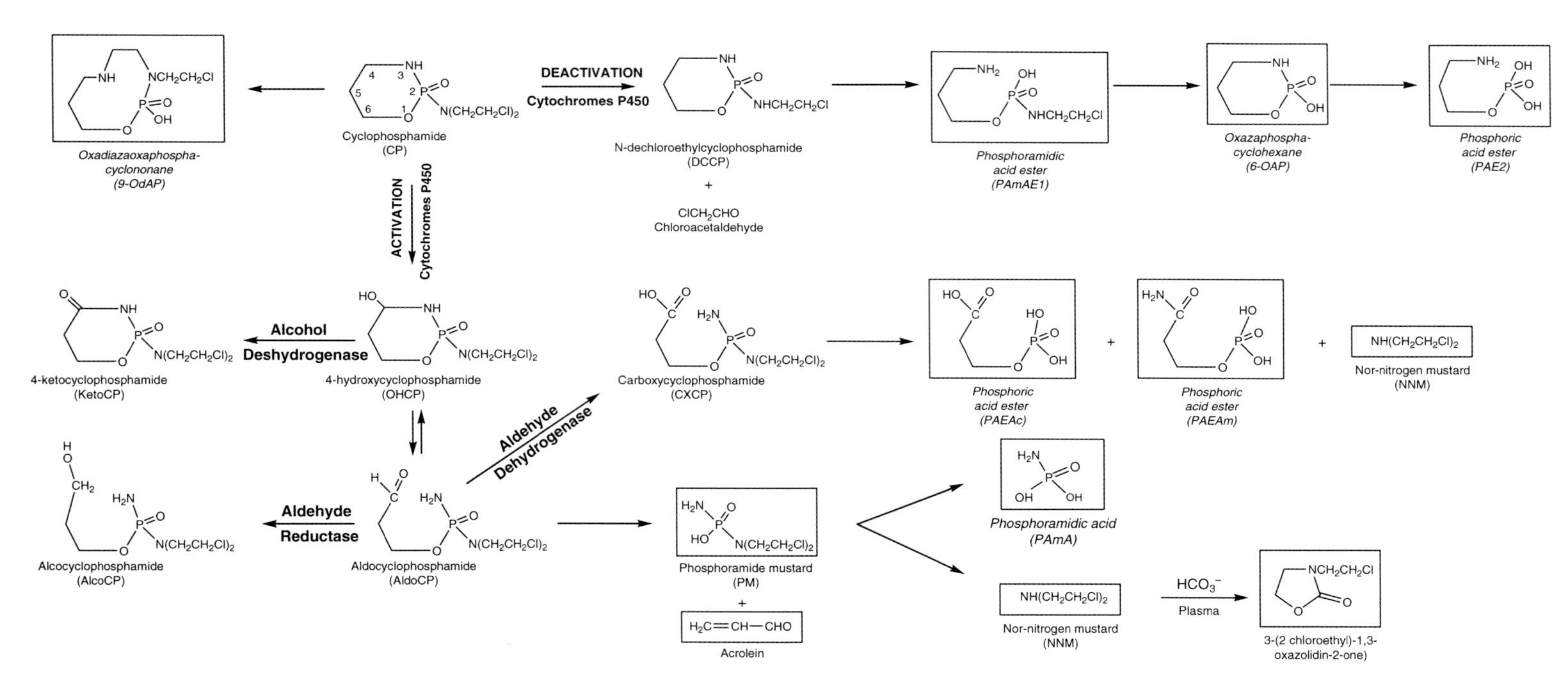

Figure 14 Metabolism of cyclophosphamide incorporating the new phosphorylated compounds found in urine using ^{31}P NMR. Metabolites that are also degradation products, i.e. spontaneously formed, are represented in boxes. The names of the new metabolites identified for the first time with NMR are in italic characters.

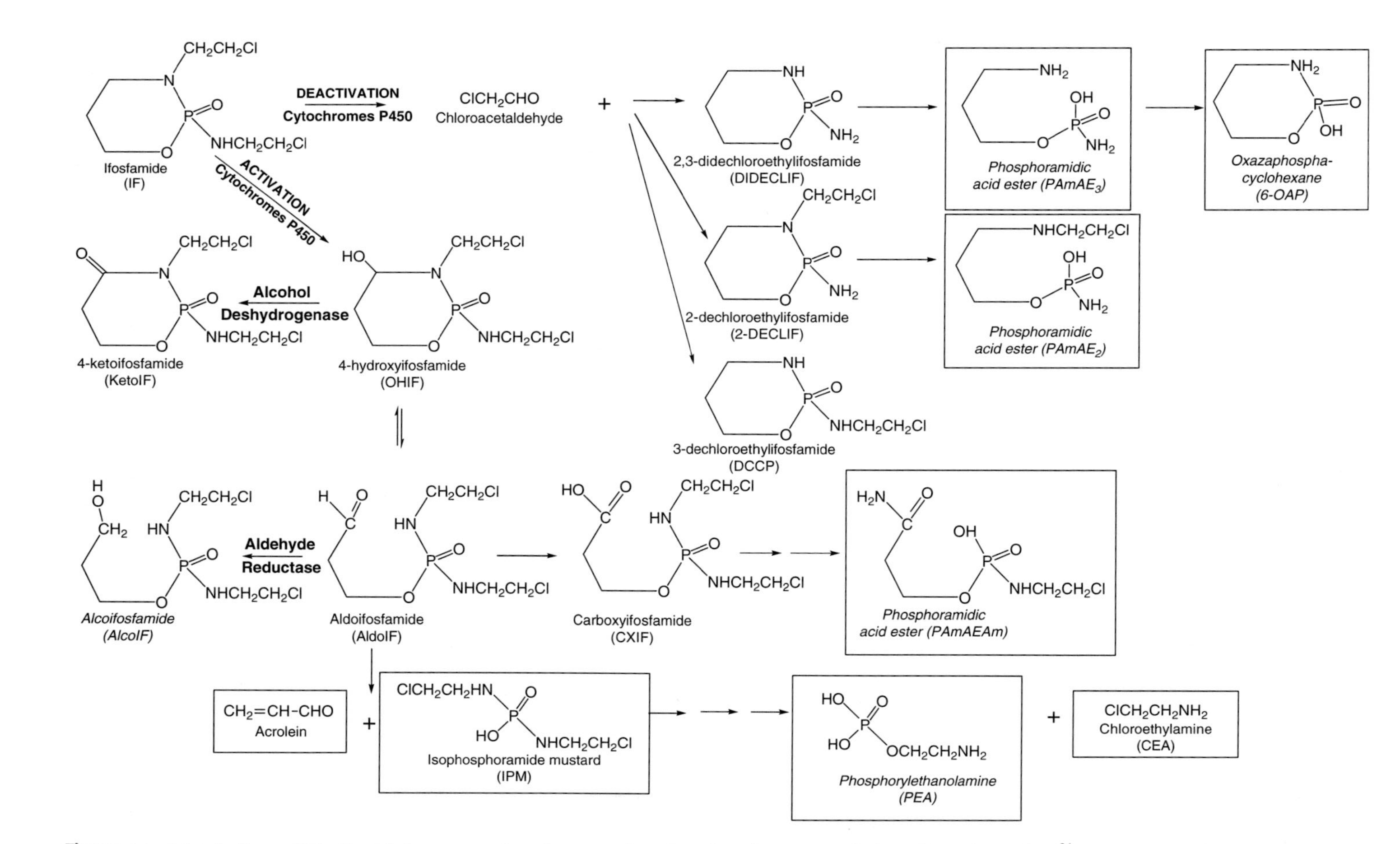

Figure 15 Metabolism of ifosfamide incorporating the new phosphorylated compounds found in urine using ^{31}P NMR. Metabolites that are also degradation products, i.e. spontaneously formed, are represented in boxes. The names of the new metabolites identified for the first time with NMR are in italic characters.

4.2.2. ^{31}P NMR studies of body fluids: a contribution to a better understanding of the metabolic pathways of CP and IF

^{31}P NMR was used to analyse urine samples from patients treated with CP[14,16,132] or IF[19,133–136] and also plasma and CSF samples from a few patients treated with IF.[133]

4.2.2.1. *Cyclophosphamide*

Busse et al.[16,132] used ^{31}P NMR to determine the 24 h urinary excretion of CP and its inactive metabolites CXCP, DCCP and KetoCP whereas plasmatic concentrations of CP were measured by HPLC. The aim of their studies was to investigate the influence of dose escalation on CP pharmacokinetics and relative contribution of activating and inactivating elimination pathways. Overall pharmacokinetics of CP is apparently not affected during eightfold dose escalation, but there is a shift in the relative contribution of different clearances to systemic CP clearance in favour of inactivation pathways. Moreover, dividing the administration of high dose of CP (100 mg/kg over 1 h) over two consecutive days (50 mg/kg over 1 h) increases the metabolism of CP,[132] demonstrating the autoinduction of CP metabolism, first reported by Bagley et al.[137] in 1973 and then confirmed by numerous authors (see, e.g. Fasola et al.[138]).

Timm et al.[139] used the same analytical methods than Busse et al.[16] (^{31}P NMR for analysis of CP, CXCP, DCCP and KetoCP in urine and HPLC for CP plasma determination) to study the effect of defined genotypes of the CYP-450 enzymes 2B6, 2C9, 2C19, 3A5 and GSTA1 on CP pharmacokinetics and metabolism. Analysis of each CYP enzyme in a group of 60 patients demonstrated that carriers deficient in CYP2C19 had lower CP elimination constants for doses below 1000 mg/m^2 (49 over 60 patients), whereas there was lack of association with other genotypes. At CP doses higher than 1000 mg/m^2, a significant increase of CP elimination was observed, possibly due to CP induction. The ^{31}P NMR analysis of urine could be performed in only 10 patients receiving a CP dose less than 1000 mg/m^2. The small sample size limited the conclusions of the study, but it was mentioned that the two samples collected from a patient carrying variant CYP2C9 alleles had lower CXCP/CP ratio than CYP2C9 wild types.

In another ^{31}P NMR study of urine samples from patients treated with CP on two consecutive days, Joqueviel et al.[14] showed that the 24 h urinary excretion of unmetabolised CP is not significantly different on the first and second days of treatment, but that of its phosphorylated metabolites is much higher after the second CP dose (37% a.d.) than after the first (20%), also suggesting autoinduction of CP metabolism. Moreover, the concentrations of CP and all its known phosphorylated metabolites, except the highly unstable tautomeric pair OHCP/AldoCP, i.e. CXCP, DCCP, AlcoCP, KetoCP and PM, were measured. Several other signals corresponding to unknown CP-related compounds were observed. Seven of them were identified. All are hydrolysis products of CP or its metabolites: one from CP (oxadiazaphosphacyclononane, 9-OdAP), two from CXCP (phosphoric acid esters called PAEAc and PAEAm), three from DCCP (phosphoramidic acid ester PAmAE1, oxazaphosphacyclohexane 6-OAP and phosphoric acid ester called PAE2) and one from PM (phosphoramidic acid $H_2N—P(O)(OH)_2$, PAmA) (Figures 14 and 16). Overall, the degradation products of CP and its metabolites

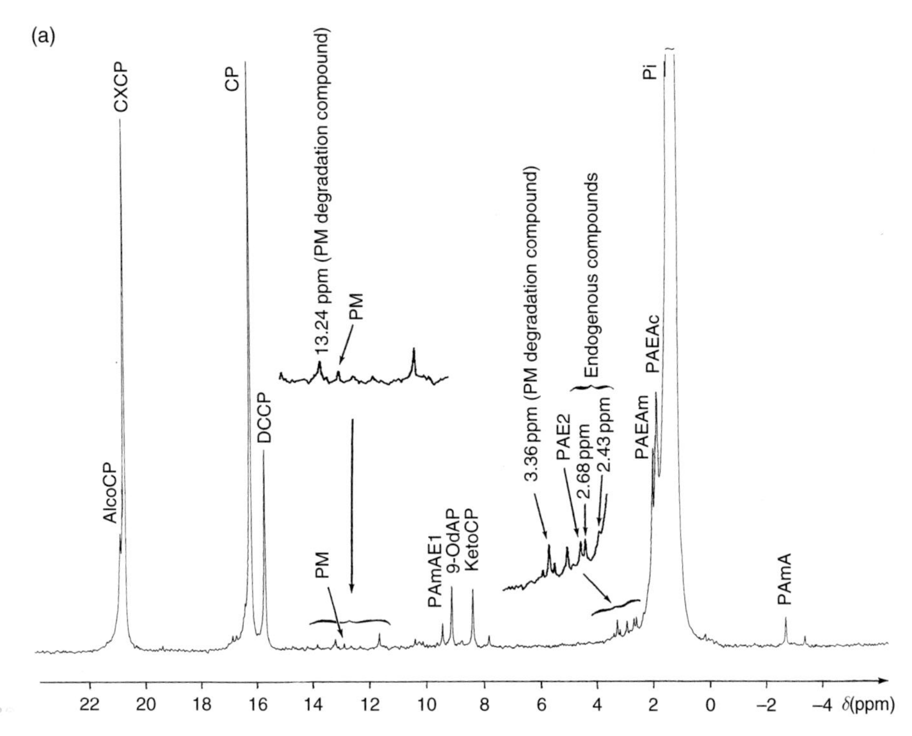

Figure 16 ^{31}P NMR spectra at 121.5 MHz with proton decoupling of urine samples from patients treated with cyclophosphamide (CP) at a dose of 60 mg/kg/day. (a) Fraction collected 18–24 h after the start of the infusion on the first day and concentrated 3.4-fold (pH 5.8). The signals at 13.24 and 3.36 ppm are derived from the degradation of phosphoramide mustard (PM) but are still unidentified. The signals at 2.68 and 2.43 ppm correspond to endogenous urinary compounds. (b) Fraction collected 0–6 h after the start of the infusion on the second day and concentrated 3.6-fold (pH 7.8). The signal at 15.26 ppm is derived from the degradation of PM. Chemical shifts (δ) are related to external 85% H_3PO_4.

newly identified in this study account for ≈3% of administered CP and make up ~15% and ~10% of the excreted metabolites of CP on days 1 and 2, respectively.

4.2.2.2. *Ifosfamide*

In a pioneering study, Misiura et al.[134] used ^{31}P NMR to quantify the urinary excretion of IF and its metabolites CXIF, DCCP, 2-DECLIF and KetoIF whose attribution is questionable with a poor-performing 60 MHz spectrometer (24.3 MHz ^{31}P resonance frequency).

Ten years later, Martino's group using a 300 MHz spectrometer analysed urine samples of nine patients treated with IF at a dose of 3 g/m^2 administered as a 3 h i.v. perfusion.[19,133] The 24 h urinary excretion of IF and its classical metabolites, CXIF, DCCP, 2-DECLIF, IPM, KetoIF, was determined. Several signals corresponding to unknown compounds were observed. Six of them were identified: AlcoIF that has not been detected at that time in human biofluids; two compounds resulting from the degradation of 2,3-didechloroethylifosfamide (DIDECLIF) whose resonance is never found, phosphoramidic acid ester called PAmAE3 and 6-OAP; one

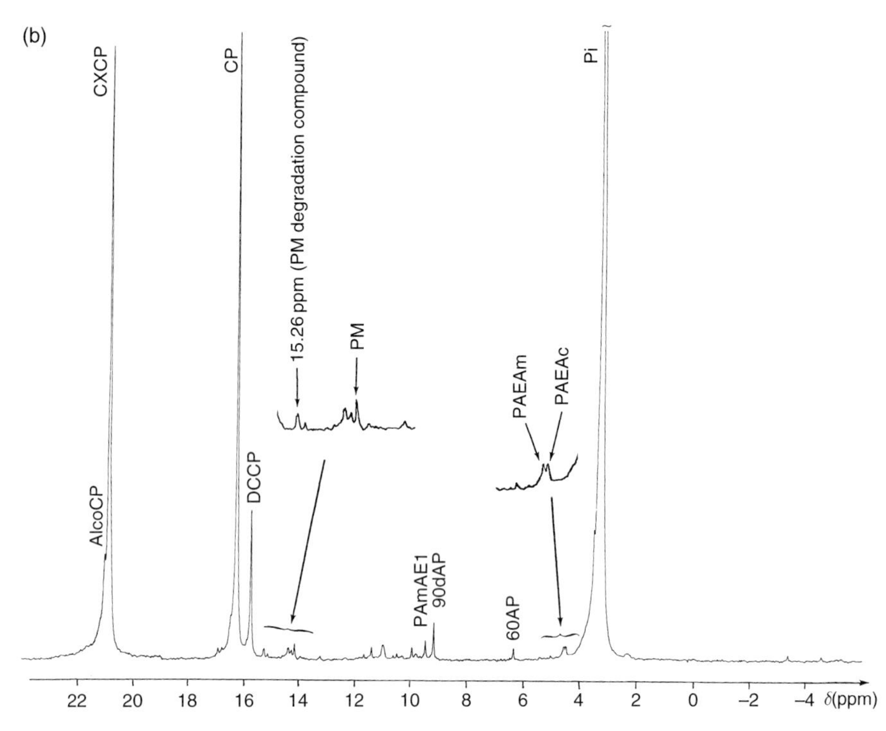

Figure 16 (*Continued*)

compound coming from the hydrolysis of the endocyclic P—N bond of 2-DECLIF, phosphoramidic acid ester called PAmAE2; one coming from the degradation of CXIF, identified as the phosphoramidic acid ester PAmAEAm from recent studies on the hydrolytic behaviour of CXIF in urine; and one as a degradation compound of IPM, phosphorylethanolamine (PEA) (Figure 15). At present, some of the other unknown compounds can be identified from recent studies on the degradative pathways of *N*-dechloroethylated IF metabolites (DIDECLIF, 2-DECLIF, DCCP),[140] IPM[141] and CXIF. The degradation compounds (identified and unknown) represent ≈14% of administered IF and ≈43% of excreted metabolites.

^{31}P NMR was used to analyse the urinary excretion of IF, DCCP and 2-DECLIF in 76 children treated with IF to evaluate the impact of GSTM1, GSTT1 and GSTP1 gene polymorphism on frequency of IF neurotoxicity and nephrotoxicity.[136] In children with polymorphic locus of GSTP1 gene, the urinary excretion of DCCP and 2-DECLIF was increased and creatinine clearance decreased. The authors' multidimensional analysis model revealed that, besides the total IF dose administered (1.5–3 g/m^2 for 3–5 days) and co-administration of other toxic drugs, polymorphic locus of this gene may be one of the factors determining a higher IF toxicity. On the other hand, no correlation was found between the GSTM1 or GSTT1 genotype and IF toxicity and the urinary excretion of *N*-dechloroethylated metabolites of IF.

Quantitative ^{31}P NMR analysis of plasma samples, even after deproteinisation, results in the detection of the sole resonance of IF, whereas that of an unique CSF sample shows the signals of IF and DCCP.[133]

^{31}P NMR was also applied to the investigation of the IF biliary metabolism in guinea pigs following i.v. infusion of 500 mg/kg of IF over 15 min.[142] Approximately 2.5 h after IF administration, a total of ≈32 μmol of phosphorus-containing compounds was measured in the bile corresponding to ≈2% of injected IF dose. The excreted metabolites are CXIF (≈25%), unmetabolised IF (≈15%), three minor metabolites for a total of ≈10% (2-DECLIF and two compounds not identified) and a mono-glutathione conjugate of IF (≈50%). This last compound was identified with liquid chromatography-mass spectrometry and ^{31}P NMR analysis of a bile sample spiked with a mixture of IF glutathione conjugates (the two monoconjugates and the double conjugate as conjugation can occur via reaction of one or/and other of the chlorine atoms of IF with the sulphur atom(s) of glutathione). Because the two monoconjugate resonances are very close ($\Delta\nu = 1.5$ Hz with a 500 MHz spectrometer), the ^{31}P NMR signal detected in bile corresponds to a glutathione moiety replacing one or other of the IF chlorine atoms.

5. Conclusion

Although optimisation procedures for a successful quantification by NMR are somewhat tedious, it is nevertheless fairly easy to use NMR routinely to obtain quantitative data. Especially, in vitro ^{19}F or ^{31}P NMR is a high potential analytical technique for absolute quantification, in a single run, of all the fluorine- or phosphorus-containing species in complex biological matrices. The limit of quantification with conventional probes is estimated at 1–3 μM and ~10 μM for ^{19}F NMR and ^{31}P NMR, respectively, in bulk solutions (e.g. biofluids, pharmaceutical formulations, tissue extracts). Comparison of assay data obtained by NMR spectroscopy and HPLC are in excellent agreement (within a few %), whereas the percentages of drug and metabolite recoveries measured by ^{19}F NMR and scintillation counting using a radiolabelled compound are in agreement within ≈10% in most studies.

Despite their limited sensitivity, ^{19}F and ^{31}P NMR can be considered as concurrent to, even in some cases, more performing than chromatographic techniques for the analysis of biofluids or pharmaceutical formulations. Indeed, as the NMR method is non-selective, all the substances in solution, even unexpected, are detected simultaneously in a single run.

References

1. W.E. Hull, Bruker Rep. 2 (1986) 15–19.
2. B.W.K. Diehl, F. Malz, U. Holzgrabe, Spectrosc. Eur. 19 (2007) 15–19.
3. M.E. Lacey, R. Subramanian, D.L. Olson, A.G. Webb, J.V. Sweedler, Chem. Rev. 99 (1999) 3133–3152.
4. W.F. Reynolds, M. Yu, R.G. Enriquez, Magn. Reson. Chem. 35 (1997) 614–618.

5. J.K. Nicholson, I.D. Wilson, Prog. Nucl. Magn. Reson. Spectrosc. 21 (1989) 449–501.
6. I.D. Wilson, J. Fromson, I.M. Ismail, J.K. Nicholson, J. Pharm. Biomed. Anal. 5 (1987) 157–163.
7. S.H. Moolenaar, G. Gohlich-Ratmann, U.F. Engelke, M. Spraul, E. Humpfer, P. Dvortsak, T. Voit, G.F. Hoffmann, C. Brautigam, A.B. van Kuilenburg, A. van Gennip, P. Vreken, R.A. Wevers, Magn. Reson. Med. 46 (2001) 1014–1017.
8. R.A. Wevers, U.F.H. Engelke, S.H. Moolenaar, C. Brautigam, J.G.N. De Jong, R. Duran, R.A. De Abreu, A.H. Van Gennip, Clin. Chem. 45 (1999) 539–548.
9. W.E. Hull, R.E. Port, R. Herrmann, B. Britsch, W. Kunz, Cancer Res. 48 (1988) 1680–1688.
10. V.J. Kamm, I.M. Rietjens, J. Vervoort, A. Heerschap, G. Rosenbusch, H.P. Hofs, D.J. Wagener, Cancer Res. 54 (1994) 4321–4326.
11. Y.J. Kamm, A. Heerschap, G. Rosenbusch, I.M.C.M. Rietjens, J. Vervoort, D.J.T. Wagener, Magn. Reson. Med. 36 (1996) 445–450.
12. F. Desmoulin, V. Gilard, M. Malet-Martino, R. Martino, Drug Metab. Dispos. 30 (2002) 1221–1229.
13. J. Rengelshausen, W.E. Hull, V. Schwenger, C. Goggelmann, I. Walter-Sack, J. Bommer, Am. J. Kidney Dis. 39 (2002)E10.
14. C. Joqueviel, R. Martino, V. Gilard, M. Malet-Martino, P. Canal, U. Niemeyer, Drug Metab. Dispos. 26 (1998) 418–428.
15. S.Y. Monte, I. Ismail, D.N. Mallett, C. Matthews, R.J.N. Tanner, J. Pharm. Biomed. Anal. 12 (1994) 1489–1493.
16. D. Busse, F.W. Busch, F. Bohnenstengel, M. Eichelbaum, P. Fischer, J. Opalinska, K. Schumacher, E. Schweizer, H.K. Kroemer, J. Clin. Oncol. 15 (1997) 1885–1896.
17. M.J. Cavaluzzi, D.J. Kerwood, P.N. Borer, Anal. Biochem. 308 (2002) 373–380.
18. Personal unpublished data.
19. V. Gilard, M.C. Malet-Martino, M. De Forni, U. Niemeyer, J.C. Ader, R. Martino, Cancer Chemother. Pharmacol. 31 (1993) 387–394.
20. J.R. Everett, K. Jennings, G. Woodnutt, J. Pharm. Pharmacol. 37 (1985) 869–873.
21. J.R. Everett, J.W. Tyler, G. Woodnutt, J. Pharm. Biomed. Anal. 7 (1989) 397–403.
22. G.D. Campbell, S. Ramaprasad, K.M. Olsen, A.F. Tryka, R.A. Komoroski, L.C. Blaszczak, T.R. Parr, Jr., J. Pharm. Sci. 82 (1993) 48–51.
23. M. Tugnait, F.Y. Ghauri, J.K. Nicholson, K. Borner, I.D. Wilson, Methodol. Surv. Biochem. Anal. 22 (1992) 291–296.
24. D.L. Ross, C.M. Riley, J. Pharm. Biomed. Anal. 12 (1994) 1325–1331 and references cited therein.
25. M. Sakai, A. Hara, S. Anjo, M. Nakamura, J. Pharm. Biomed. Anal. 18 (1999) 1057–1067 and references cited therein.
26. A.A. Salem, H.A. Mossa, B.N. Barsoum, Spectrochim. Acta A Mol. Biomol. Spectrosc. 62A (2005) 466–472.
27. G. Fardella, P. Barbetti, I. Chiappini, G. Grandolini, Int. J. Pharm. 121 (1995) 123–127.
28. S. Trefi, V. Gilard, M. Malet-Martino, R. Martino, J. Pharm. Biomed. Anal. 44 (2007) 743–754.
29. G.J. Slatter, D.J. Stalker, K.L. Feenstra, I.R. Welshman, J.B. Bruss, J.P. Sams, M.G. Johnson, P.E. Sanders, M.J. Hauer, P.E. Fagerness, R.P. Stryd, G.W. Peng, E.M. Schobe, Drug Metab. Dispos. 29 (2001) 1136–1145.
30. Z. Talebpour, S. Haghgoo, M. Shamsipur, Anal. Biochem. 323 (2003) 205–210.
31. M. Shamsipur, L. Shafiee-Dastgerdi, Z. Talebpour, S. Haghgoo, J. Pharm. Biomed. Anal. 43 (2007) 1116–1121.
32. H.A. Friedel, A. Fitton, Drugs 45 (1993) 548–569.
33. H.C. Schroder, W.E.G. Muller, Drugs Today 38 (2002) 49–58.
34. G. Fardella, P. Barbetti, I. Chiappini, G. Grandolini, Acta Technol. Legis Med. 4 (1993) 89–96.
35. K.E. Wade, I.D. Wilson, J.A. Troke, J.K. Nicholson, J. Pharm. Biomed. Anal. 8 (1990) 401–410.
36. M. Spraul, M. Hofmann, I.D. Wilson, E. Lenz, J.K. Nicholson, J.C. Lindon, J. Pharm. Biomed. Anal. 11 (1993) 1009–1015.
37. D. Bilecen, A.C. Schulte, A. Kaspar, E. Kuestermann, J. Seelig, D. Elverfeldt, K. Scheffler, NMR Biomed. 16 (2003) 144–151.

38. L. Litt, R. Gonzalez-Mendez, T.L. James, D.I. Sessler, P. Mills, W. Chew, M. Moseley, B. Pereira, J.W. Severinghaus, W.K. Hamilton, Anesthesiology 67 (1987) 161–168.
39. P. Mills, D.I. Sessler, M. Moseley, W. Chew, B. Pereira, T.L. James, L. Litt, Anesthesiology 67 (1987) 169–173.
40. A.M. Wyrwicz, M.H. Pszenny, J.C. Schofield, P.C. Tillman, R.E. Gordon, P.A. Martin, Science 222 (1983) 428–430.
41. A.M. Wyrwicz, C.B. Conboy, K.R. Ryback, B.G. Nichols, P. Eisele, Biochim. Biophys. Acta 927 (1987) 86–91.
42. L. Litt, S. Lockhart, Y. Cohen, N. Yasuda, F. Kim, B. Freire, M. Laster, N. Peterson, S. Taheri, Ann. N. Y. Acad. Sci. 625 (1991) 707–724.
43. A.S. Evers, B.W. Dubois, Ann. N. Y. Acad. Sci. 625 (1991) 725–732.
44. A.M. Wyrwicz, Ann. N. Y. Acad. Sci. 625 (1991) 733–742.
45. Y. Xu, P. Tang, W. Zhang, L. Firestone, P.M. Winter, Anesthesiology 83 (1995) 766–774.
46. P.N. Venkatasubramanian, Y.J. Shen, A.M. Wyrwicz, Biochim. Biophys. Acta 1245 (1995) 262–268.
47. P.N. Venkatasubramanian, Y.J. Shen, A.M. Wyrwicz, Magn. Reson. Med. 35 (1996) 626–630.
48. G.G. Lockwood, D.P. Dob, D.J. Bryant, J.A. Wilson, J. Sargentoni, S.M. Sapsed-Byrne, D.N.F. Harris, D.K. Menon, Br. J. Anaesth. 79 (1997) 581–585.
49. G.G. Lockwood, D.P. Dob, D.J. Bryant, J.A. Wilson, J. Sargentoni, S.M. Sapsed-Byrne, D.N. Harris, D.K. Menon, Br. J. Anaesth. 79 (1997) 586–589.
50. D.K. Menon, G.G. Lockwood, C.J. Peden, I.J. Cox, J. Sargentoni, J.D. Bell, G.A. Coutts, G.J. Whitwam, Magn. Reson. Med. 30 (1993) 680–684.
51. C.T. Burt, R.R. Moore, M.F. Roberts, T.J. Brady, Biochim. Biophys. Acta 805 (1984) 375–381.
52. B.S. Selinsky, M. Thompson, L.M. Geffreys, R.E. London, Biophys. J. 49 (1986) , 329a.
53. B.S. Selinsky, M. Thompson, R.E. London, Biochem. Pharmacol. 36 (1987) 413–416.
54. G. Urban, P. Speerschneider, W. Dekant, Chem. Res. Toxicol. 7 (1994) 170–176.
55. B.S. Selinsky, M.E. Perlman, R.E. London, Mol. Pharmacol. 33 (1988) 559–566 and references cited therein.
56. B.S. Selinsky, M.E. Perlman, R.E. London, Mol. Pharmacol. 33 (1988) 567–573 and references cited therein.
57. N.E. Preece, J. Challands, S.C.R. Williams, NMR Biomed. 5 (1992) 101–106.
58. H. Orhan, N. Vermeulen, G. Sahin, J. Commandeur, Anesthesiology 95 (2001) 165–167.
59. E.N. Cohen, J.R. Trudell, H.N. Edmunds, E. Watson, Anesthesiology 43 (1975) 392–401.
60. I.G. Sipes, A.J. Gandolfi, L.R. Pohl, G. Krishna, B.R. Brown, Jr., J. Pharmacol. Exp. Ther. 214 (1980) 716–720.
61. M.J. Murphy, D.A. Dunbar, L.S. Kaminsky, Biochem. Pharmacol. 29 (1980) 3257–3263 and references cited therein.
62. B.S. Selinsky, D.E. Rusyniak, J.O. Warsheski, A.P. Joseph, Biochem. Pharmacol. 42 (1991) 2229–2238.
63. H. Yin, J.P. Jones, M.W. Anders, Chem. Res. Toxicol. 8 (1995) 262–268.
64. M.K. Ellis, J.L. Naylor, T. Green, M.A. Collins, Drug Metab. Dispos. 23 (1995) 102–106.
65. E.D. Kharasch, A.S. Armstrong, K. Gunn, A. Artru, K. Cox, M.D. Karol, Anesthesiology 82 (1995) 1379–1388 and references cited therein.
66. H. Orhan, J.N.M. Commandeur, G. Sahin, U. Aypar, A. Sahin, N.P.E. Vermeulen, Xenobiotica 34 (2004) 301–316.
67. E.D. Kharasch, M.D. Karol, C. Lanni, R. Sawchuk, Anesthesiology 82 (1995) 1369–1378.
68. R.A. Iyer, E.J. Frink, Jr., T.J. Ebert, M.W. Anders, Anesthesiology 88 (1998) 611–618.
69. R.A. Iyer, M.W. Anders, Chem. Res. Toxicol. 10 (1997) 811–819.
70. L. Jin, M.R. Davis, E.D. Kharasch, G.A. Doss, T.A. Baillie, Chem. Res. Toxicol. 9 (1996) 555–561.
71. D.K. Spracklin, E.D. Kharasch, Chem. Res. Toxicol. 9 (1996) 696–702.
72. V. Uttamsingh, R.A. Iyer, R.B. Baggs, M.W. Anders, Anesthesiology 89 (1998) 1174–1183.
73. G.J. Peters, S.P. Ackland, Expert Opin. Investig. Drugs 5 (1996) 637–679.
74. R.B. Diasio, B.E. Harris, Clin. Pharmacokinet. 16 (1989) 215–237.

75. G. Weckbecker, Pharmacol. Ther. 50 (1991) 367–424.
76. J.L. Grem, Cancer Chemotherapy and Biotherapy: Principles and Practice, B.A. Chabner and D.L. Longo Eds, pp. 149–211, Lippincott-Raven, Philadelphia, 1996.
77. D. Gani, P.B. Hitchcock, D.W. Young, J. Chem. Soc. Perkin Trans. I (1985) 1363–1372.
78. R. Duschinsky, H. Walker, W. Wojarowski, K. Noack, H.P. Bächtold, Proc. Am. Assoc. Cancer Res. 14 (1973) , 109.
79. M.C. Malet-Martino, J.P. Armand, A. Lopez, J. Bernadou, J.P. Beteille, M. Bon, R. Martino, Cancer Res. 46 (1986) 2105–2112.
80. R. Martino, M.C. Malet-Martino, C. Vialaneix, A. Lopez, M. Bon, Drug Metab. Dispos. 15 (1987) 897–904.
81. R. Martino, A. Lopez, M.C. Malet-Martino, J. Bernadou, J.P. Armand, Drug Metab. Dispos. 13 (1985) 116–118.
82. D.J.T. Porter, J.A. Harrington, M.R. Almond, W.G. Chestnut, G. Tanoury, T. Spector, Biochem. Pharmacol. 50 (1995) 1475–1484.
83. M. Arellano, M. Malet-Martino, R. Martino, P. Gires, Br. J. Cancer 77 (1998) 79–86.
84. O.W. Griffith, Annu. Rev. Biochem. 55 (1986) 855–878 and references cited therein.
85. N. Tamaki, M. Kaneko, C. Mizota, M. Kikugawa, S. Fujimoto, Eur. J. Biochem. 189 (1990) 39–45.
86. D.A. Keller, D.C. Roe, P.H. Lieder, Fundam. Appl. Toxicol. 30 (1996) 213–219.
87. L. Lemaire, M.C. Malet-Martino, M. De Forni, R. Martino, B. Lasserre, Br. J. Cancer 66 (1992) 119–127.
88. L. Lemaire, M. Arellano, M.C. Malet-Martino, R. Martino, Proc. Int. Soc. Magn. Reson. Med. 2 (1996), 1162.
89. G.D. Heggie, J.P. Sommadossi, D.S. Cross, W.J. Huster, R.B. Diasio, Cancer Res. 47 (1987) 2203–2206.
90. M.C. Malet-Martino, J. Bernadou, R. Martino, J.P. Armand, Drug Metab. Dispos. 16 (1988) 78–84.
91. D.J. Sweeny, M. Martin, R.B. Diasio, Drug Metab. Dispos. 16 (1988) 892–894.
92. D.J. Sweeny, S. Barnes, G.D. Heggie, R.B. Diasio, Proc. Natl. Acad. Sci. U.S.A. 84 (1987) 5439–5443.
93. M.R. Johnson, S. Barnes, J.B. Kwakye, R.B. Diasio, J. Biol. Chem. 266 (1991) 10227–10233.
94. K.L. Mukherjee, C. Heidelberger, J. Biol. Chem. 235 (1960) 433–437.
95. J. Bernadou, J.P. Armand, A. Lopez, M.C. Malet-Martino, R. Martino, Clin. Chem. 31 (1985) 846–848.
96. H.A. Lozeron, M.P. Gordon, T. Gabriel, W. Tautz, R. Duschinsky, Biochemistry 3 (1964) 1844–1850.
97. B.C. Rudy, B.Z. Senkowski, Analytical Profiles of Drug Substances, K. Florey Ed, pp. 221–244, Academic Press, New York, 1977.
98. L. Lemaire, M.C. Malet-Martino, R. Martino, M. De Forni, B. Lasserre, Oncol. Rep. 1 (1994) 173–174.
99. J. Lukaschek, M. Nufer, D. Maurer, M. Asanger, H. Honegger, L. Widmer, M. Malet-Martino, R. Legay, R. Martino, J Clin. Oncol. 22 (2004) 5022–5025.
100. E.B. Lamont, R.L. Schilsky, Clin. Cancer Res. 5 (1999) 2289–2296.
101. E. Bajetta, C. Carnaghi, L. Somma, C.G. Stampino, Tumori 82 (1996) 450–452.
102. I.R. Judson, P.J. Beale, J.M. Trigo, W. Aherne, T. Crompton, D. Jones, E. Bush, B. Reigner, Invest. New Drugs 17 (1999) 49–56.
103. B. Reigner, K. Blesch, E. Weidekamm, Clin. Pharmacokinet. 40 (2001) 85–104.
104. M. Miwa, M. Ura, M. Nishida, N. Sawada, T. Ishikawa, K. Mori, N. Shimma, I. Umeda, H. Ishitsuka, Eur. J. Cancer 34 (1998) 1274–1281.
105. J. Cassidy, C. Twelves, D. Cameron, W. Steward, K. O'Byrne, D. Jodrell, L. Banken, T. Goggin, D. Jones, B. Roos, E. Bush, E. Weidekamm, B. Reigner, Cancer Chemother. Pharmacol. 44 (1999) 453–460.
106. B. Reigner, S. Clive, J. Cassidy, D. Jodrell, R. Schulz, T. Goggin, L. Banken, B. Roos, M. Utoh, T. Mulligan, E. Weidekamm, Cancer Chemother. Pharmacol. 43 (1999) 309–315.

107. C. Twelves, R. Glynne-Jones, J. Cassidy, J. Schuller, T. Goggin, B. Roos, L. Banken, M. Utoh, E. Weidekamm, B. Reigner, Clin. Cancer Res. 5 (1999) 1696–1702.
108. W. Plunkett, P. Huang, Y.Z. Xu, V. Heinemann, R. Grunewald, V. Gandhi, Semin. Oncol. 22 (1995) 3–10.
109. H.T. Edzes, G.J. Peters, P. Noordhuis, J.B. Vermorken, Anal. Biochem. 214 (1993) 25–30.
110. H.J. Scholer, Flucytosine, D.C.E. Speller Ed, Antifungal Chemotherapy, pp. 35–106, Wiley, New York, 1980.
111. R.B. Diasio, D.E. Lakings, J.E. Bennett, Antimicrob. Agents Chemother. 14 (1978) 903–908.
112. A. Polak, E. Eschenhof, M. Fernex, H.J. Scholer, Chemotherapy 22 (1976) 137–153.
113. K.M. Williams, A.M. Duffield, R.K. Christopher, P.J. Finlayson, Biomed. Mass Spectrom. 8 (1981) 179–182.
114. B.E. Harris, B.W. Manning, T.W. Federle, R.B. Diasio, Antimicrob. Agents Chemother. 29 (1986) 44–48.
115. B.W. Manning, T.W. Federle, C.E. Cerniglia, J. Microbiol. Methods 6 (1987) 81–94.
116. A. Vermes, E.J. Kuijper, H.J. Guchelaar, J. Dankert, Chemotherapy 49 (2003) 17–23.
117. J.P. Vialaneix, M.C. Malet-Martino, J.S. Hoffmann, J. Pris, R. Martino, Drug Metab. Dispos. 15 (1987) 718–724.
118. N. Chouini-Lalanne, M.C. Malet-Martino, V. Gilard, J.C. Ader, R. Martino, Drug Metab. Dispos. 23 (1995) 813–817.
119. M.C. Malet-Martino, R. Martino, M. De Forni, A. Andremont, O. Hartmann, J.P. Armand, Infection 19 (1991) 178–180.
120. P.J. Gilbert, T.E. Hartley, J.A. Troke, R.G. Turcan, C.W. Vose, K.V. Watson, Xenobiotica 22 (1992) 775–787.
121. E.M. Lenz, I.D. Wilson, B. Wright, E.A. Partridge, C.T. Rodgers, P.R. Haycock, J.C. Lindon, J.K. Nicholson, J. Pharm. Biomed. Anal. 28 (2002) 31–43.
122. D.Q. Brown, W.J. Graham, III, L.J. MacKenzie, J.W. Pittock, III, L.M. Shaw, Pharmacol. Ther. 39 (1988) 157–168.
123. L.M. Schuchter, J.H. Glick, Biologic Therapy of Cancer Updates, V.T. DeVita, S. Hellman, S.A. Rosenberg Eds, vol 3, pp. 1–10, J.B. Lippincott, New York, 1993.
124. S.A. Knizner, A.J. Jacobs, R.C. Lyon, C.E. Swenberg, J. Pharmacol. Exp. Ther. 236 (1986) 37–40.
125. D.C. Thompson, S.D. Wyrick, D.J. Holbrook, S.G. Chaney, Biochem. Pharmacol. 50 (1995) 1413–1419.
126. R.A. Fleming, J. Pharmacother. 17 (1997) 146S–154S.
127. A.V. Boddy, S.M. Yule, Clin. Pharmacokinet. 38 (2000) 291–304.
128. N.E. Sladek, Anticancer Drugs: Reactive Metabolism and Drug Interactions, G. Powis Ed, pp. 79–156, Pergamon Press, Oxford, 1994.
129. M.P. Goren, R.K. Wright, M.E. Horowitz, Clin. Chim. Acta 160 (1986) 157–161.
130. J. Pohl, J. Stekar, P. Hilgard, Arzneim. Forsch. 39 (1989) 704–705.
131. C. Joqueviel, M. Malet-Martino, R. Martino, Cell. Mol. Biol. 43 (1997) 773–782.
132. D. Busse, F.W. Busch, E. Schweizer, F. Bohnenstengel, M. Eichelbaum, P. Fischer, K. Schumacher, W.E. Aulitzky, H.K. Kroemer, Cancer Chemother. Pharmacol. 43 (1999) 263–268.
133. R. Martino, F. Crasnier, N. Chouini-Lalanne, V. Gilard, U. Niemeyer, M. De Forni, M.C. Malet-Martino, J. Pharmacol. Exp. Ther. 260 (1992) 1133–1144.
134. K. Misiura, A. Okruszek, K. Pankiewicz, W.J. Stec, Z. Czownicki, B. Utracka, J. Med. Chem. 26 (1983) 674–679.
135. K. Misiura, M. Zubowska, E. Zielinska, Arzneim. Forsch. 53 (2003) 372–377.
136. E. Zielinska, M. Zubowska, K. Misiura, J. Pediatr. Hematol. Oncol. 27 (2005) 582–589.
137. C.M. Bagley, Jr., F.W. Bostick, V.T. DeVita, Jr., Cancer Res. 33 (1973) 226–233.
138. G. Fasola, G.P. Lo, E. Calori, M. Zilli, F. Verlicchi, M.R. Motta, P. Ricci, M. Baccarani, S. Tura, Haematologica 76 (1991) 120–125.
139. R. Timm, R. Kaiser, J. Loetsch, U. Heider, O. Sezer, K. Weisz, M. Montemurro, I. Roots, I. Cascorbi, Pharmacogenomics J. 5 (2005) 365–373.

140. V. Gilard, R. Martino, M. Malet-Martino, U. Niemeyer, J. Pohl, J. Med. Chem. 42 (1999) 2542–2560.
141. S. Breil, R. Martino, V. Gilard, M. Malet-Martino, U. Niemeyer, J. Pharm. Biomed. Anal. 25 (2001) 669–678.
142. G.S. Payne, A.S.K. Dzik-Jurasz, L. Mancini, B. Nutley, F. Raynaud, M.O. Leach, Cancer Chemother. Pharmacol. 56 (2005) 409–414.

CHAPTER 4

Quantitative 2D NMR Analysis of Glycosaminoglycans

G. Torri *and* M. Guerrini

Contents

Abstract

Biological properties of heparin and structurally related glycosaminoglycans (GAGs) involve complex sequences of various sulfated uronic acid and hexosamine residues. Monodimensional nuclear magnetic resonance (NMR) spectroscopy is currently used for structural characterization of GAGs, being also reliable for quantitative measurements when signals are sufficiently separated. Heteronuclear single quantum coherence (HSQC) spectroscopy can be also exploited for quantitative purpose by proper selection of analytical signals among those with similar magnetic relaxation and one bond proton-carbon *J*-coupling. The method can be confidently used to compare different GAGs, by quantification of their various substituted monosaccharide components.

Keywords: glycosaminoglycans, HSQC, heparins, saccharides

1. Introduction

Glycosaminoglycans (GAGs) are polysaccharides formed of a repeating disaccharide units of hexuronic acid and hexosamine. As proteoglycans, where their chains are linked to the protein cores, they are ubiquitously found in plasma membranes and extracellular matrix. In contrast, the heparin is segregated in mast cells.[1,2] The GAG family includes two subfamilies, the galactosaminoglycans the most representative of which are chondroitins (ChS) and dermatan sulfate (DeS), and the glucosaminoglycans heparin and heparan sulfate (HS). ChS consists of repeating disaccharide units of (4 → 1) β-D-glucuronic acid (1 → 3) β-N-acetyl-D-galactosamine [(4 → 1)β-D-GlcA (1 → 3) β-D-GalN$_{NAc}$].[3] DeS is formed by repeating disaccharide units of (1 → 4) α-L-iduronic acid (1 → 3) β-N-acetyl-D-galactosamine [(4 → 1)α-L-IdoA (1 → 3) β-D-GalN$_{NAc}$]. Both galactosaminoglycans can be O-sulfated in the 4- or 6-position of the aminosugar, and in the case of DeS, less frequently O-sulfated in position 2 of IdoA.[4] HS and heparin (Hep) consist of repeating disaccharide units of β-D-GlcA or α-L-IdoA (1 → 4) linked to an α-N-acetyl or α-N-sulfo-D-glucosamine (GlcN$_{NAc}$, GlcN$_{NS}$, Figure 1).

Heparin is the most extensively studied glycosaminoglycan. Under physiological conditions, heparin is secreted by the Golgi apparatus of mast cells. Its exogenous version is currently used as anticoagulant and antithrombotic drug having profound effects on the coagulation system.[5] The structure of heparin consists of a carbohydrate backbone made up of alternating disaccharide unit mainly formed of α-L-2-O-sulfated iduronic acid (1 → 4) linked to a α-D-6-O-sulfated-N-sulfo-glucosamine (IdoA$_{2S}$-GlcN$_{NS,6S}$). Minor, undersulfated sequences containing nonsulfated uronic acids, such as β-D-glucuronic acid (GlcA) or α-L-iduronic acid (IdoA), together with N-acetylated glucosamine or 6-O-sulfated N-acetylated glucosamine residues (GlcN$_{NAc}$, GlcN$_{NAc,6S}$) and 6-O desulfated N-sulfo-glucosamine (GlcN$_{NS}$), contribute to the heterogeneity of heparin chains. As shown in Figure 2, the less sulfated and more N-acetylated sequences are concentrated mainly towards the reducing side of the chain (NA domain) and the most charged ones toward the nonreducing side

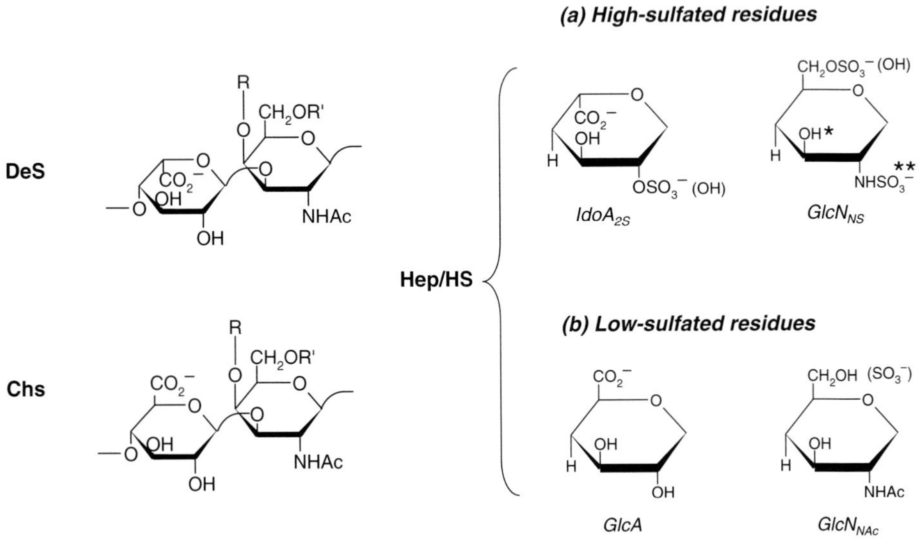

Figure 1 Disaccharide components of dermatan sulfate (DeS) and chondroitin sulfate (ChS); R=R′=H or SO_3^-. Monosaccharide components of heparin and heparin sulfate (HS). Minor GlcN residues bear a SO_3^- substituent (*) or a free amino group instead a $NHSO_3^-$ group (**).

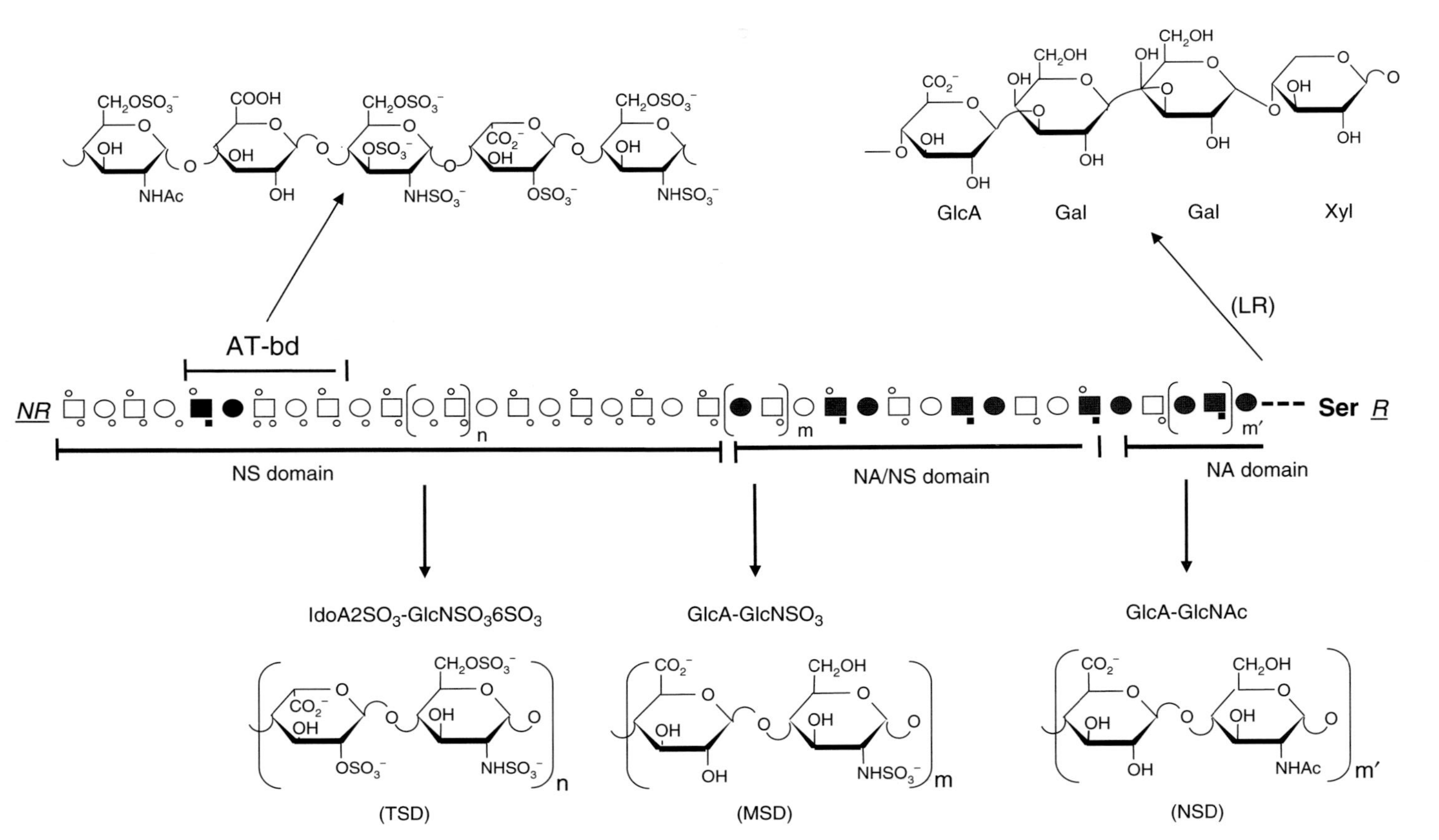

Figure 2 Schematized representation of a heparin chain constituted of N-acetylated (NA), N-sulfated (NS), and mixed NA/NS domains. Formulas of major disaccharidic sequences present in the chain, of the antithrombin-binding domain (AT-bd) and of the "linkage region" (LR) originally linked through serine to the polypeptide core, are shown.

(NS domain). Mixed domains are mostly inter-dispersed between the two regions (NA/NS domain). In addition, about one-third of the heparin chains contains residues of *N*,3,6-trisulfated glucosamine ($GlcN_{NS,3S,6S}$), mainly occurring in the pentasaccharide sequence essential for the binding to antithrombin (AT) and for the expression of the anticoagulant activity.[6] Heparin and HS structures differ from each other in the relative proportion and distribution of undersulfated sequences. Whereas in heparin minor undersulfated sequences are dispersed along the polysaccharide chain (normally more concentrated toward the "linkage region" in to the original proteoglycan), in HS GlcA-GlcNAc is the more represented disaccharide.[7] At the reducing end, some of the Hep/HS chains preserve the tetrasaccharide [β-D-GlcA-(1 → 3)-β-D-Gal-(1 → 3)-β-D-Gal-(1 → 4)-β-D-Xyl-] connection, through a serine residue, to the core protein of the original proteoglycan. This sequence is common to other GAGs, even if different sulfate and phosphate substitutions have been found within each glycosaminoglycan.[8–10] GAGs interact with a large number of protein participating in matrix organization, cell adhesion, growth, and differentiation,[11,12] and have a fundamental role in embryo development showing patterns of expression that are tightly controlled in each organ and time. Changes in proteoglycan expression frequently correlate with developmental events such as cell differentiation and mesenchymal condensation, suggesting that the molecules play a role in their regulation.[13] Elucidation of the relationships between biological function and the structure of glycosaminoglycans remains a challenge, mainly due to the difficulty in determining their fine structure.[14] The wide range of biological processes in which heparin and heparan sulfate are involved through their interaction with many different proteins, are associated with unique oligosaccharide sequences and, most often with an ensemble of different structural features including the distribution of sulfate groups along the polymer chain.[15] A limitation to structure elucidation and correlation with biological function has been the lack of sufficiently sensitive and specific analytical techniques. Most of the studies to elucidate the structure of heparin and heparan sulfate employ strategies involving enzymatic or chemical depolymerization of the polysaccharide followed by one or more separation steps and characterization of the oligosaccharide fragments. Several high-performance liquid chromatograph (HPLC) or capillary electrophoresis methods and mass spectrometric analysis (MS) for structure assignment and quantitative profiling of GAGs of animal origin, especially heparin and HS and their fragments, generated by enzymatic or chemical depolymerization, have been developed.[14] However, these methods involve degradation of the sample and an efficient fragments separation coupled with sensitive detection procedures often limit their use. For instance, heparinase does not cleave some oversulfated GAG sequences, and the method is not applicable to GAG derivatives containing unnatural sequences. Nuclear magnetic resonance (NMR) spectroscopy is one of the few techniques that provide structural information on the polymer without sample modification, avoiding possible loss of structural information as a result of breakdown of the polymer. The usefulness of NMR in understanding GAGs structural properties has been described in many reviews.[7,15,16–20] NMR spectra provide quantitative information on multiple parameters, including the monosaccharide composition, sulfation states, and glucosamine-uronic acid linkage. NMR is also the most accurate method for direct quantification of the iduronic and glucuronic acid contents. The major limitation of the technique is its low sensitivity and the consequent relatively high amount of sample needed for the analysis. However, the recent introduction of high-sensitivity cryoprobes allowed the accurate detection and quantification of the minor residues of GAGs on amounts of sample as low as tens to hundreds in the microgram scale.

1.1. GAGs characterization via 1D NMR

The 1D ^{1}H and ^{13}C spectra have been used since long time to characterize polysaccharides. Pioneer work was done on structural characterization of GAGs during the 1960s.[21] It streamed forth into the essential contribution to the definition of heparin structure,[22] while the 1970s have seen the beginning of the conformational studies on GAGs.[23,24]

For analytical purposes, ^{13}C NMR provided the most relevant contribution to the identification of GAGs. In fact, the large signal dispersions of these spectra better allows the identification of each peculiar signal generated by the different monomers present in a GAG structure. Carbon 2 (C2) signal of hexosamine can discriminate between GlcN and GalN residues, and permits as well to recognize the possible ring substituent ($GlcN_{NAc}$, $GlcN_{NS}$, $GalN_{NAc,6S}$ and $GalN_{NAc,4S}$).[25] Similarly, C1 signals, in the so called anomeric region, provide the uronic acid composition, discriminating between $IdoA_{2S}$ and $GlcA_{2OH} + IdoA_{2OH}$. From the anomeric region, other information can be related to natural reducing end terminal residues and to the sugar involved in the protein linkage sequence. (Figure 3) This region is of particular interest in the case of low-molecular-weight heparins (LMWHs), because the chemical shift of its signals permits the recognition of the depolymerization procedure, that is based on different chemical or enzymatic methods. This aspect is relevant as the LMWH biological properties can be modified/modulated by the selectivity of the depolymerization procedure. It is possible to identify the natural reducing terminal residues, the modified ones (such as anhydromannitol), and the 4,5-unsaturated uronic acid at the nonreducing end terminal residues (see Figure 9).[26] Details will be discussed later in a separate section.

It is noteworthy, and also well established, that the previously described signals can be used for quantitative evaluation only when signals of similar nature are compared. However, especially in the case of very heterogeneous samples, there is a realistic risk of interference due to masked overlapped signals. Another informative area of the ^{13}C spectrum of GAGs is around 60–70 ppm. This region corresponds to the primary alcohols signals region; by integrating the specific signals, it is possible to establish the ratio between the sulfated and the nonsulfated hexosamine residues. The quality of a quantitative analysis depends on many factors, including relaxation time values, appropriate decoupling power, sample viscosity, presence of paramagnetic cations, spectral baseline, and especially on the power of the magnetic field. In spite of these spectroscopic limitations, 1D ^{13}C NMR spectroscopy is currently used to describe the prevalent patterns of substitution, as well as to identify the presence of contaminants (Figures 3 and 4). It was recently demonstrated that the combination of NMR spectroscopy with capillary electrophoresis and computing methods considerably reduces the number of experimental constraints required for heparin/HS oligosaccharides sequence determination.[27]

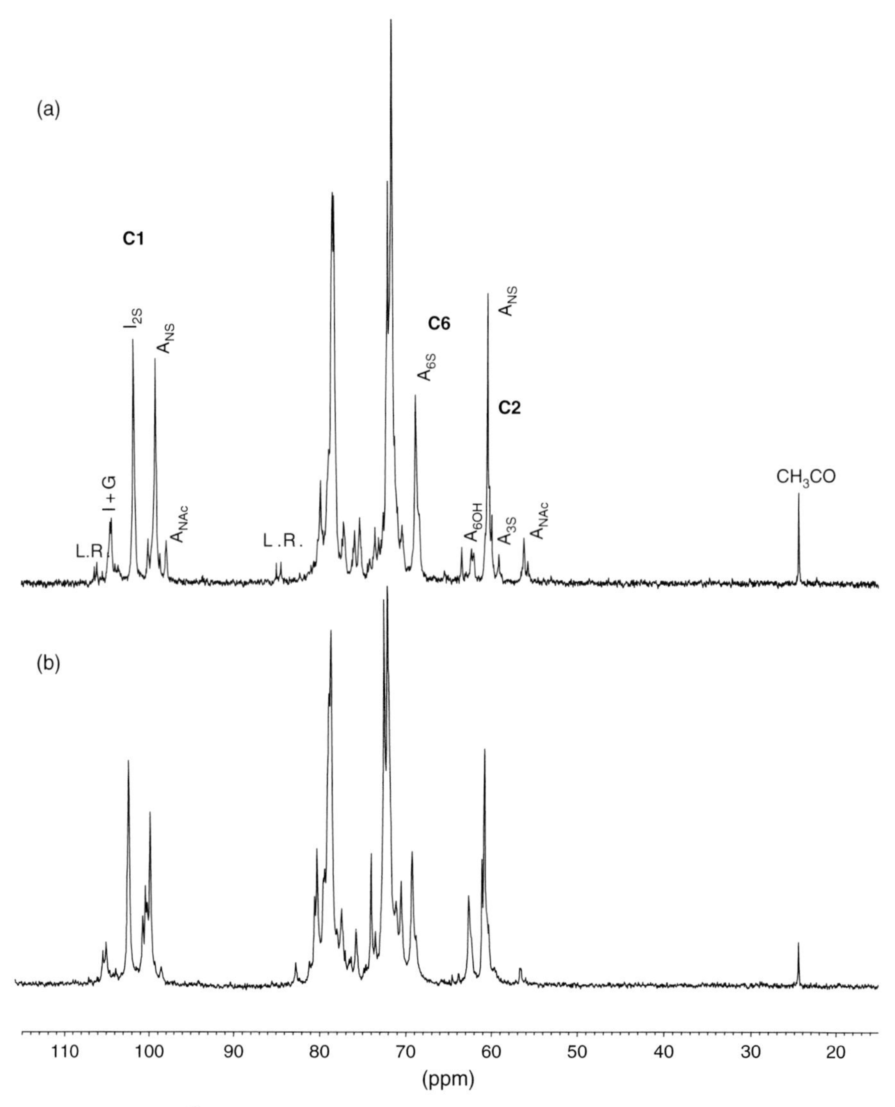

Figure 3 100 MHz ^{13}C NMR spectra of typical (a) porcine mucosal heparin and (b) bovine mucosal heparin. The spectra were recorded at 40°C. Protons were decoupled during the acquisition time. Recycle delay 4 s; number of scans 40,000. A, G, and I indicate glucosamine, glucuronic acid, and iduronic acid, respectively.

On the contrary, the most sensitive proton NMR technique is limited by its low signal dispersion and by the GAG strong signals overlapping. This limitation is particularly relevant for heparin due to its high structural complexity. In fact, earlier GAG structural studies made using the 1D ^{1}H approach were done using structurally homogeneous heparin samples extracted from beef lung, which are characterized by the prevalence of trisulfated disaccharide units.[23] The commercial heparin samples, currently extracted from intestinal mucosa of different animal species, are characterized

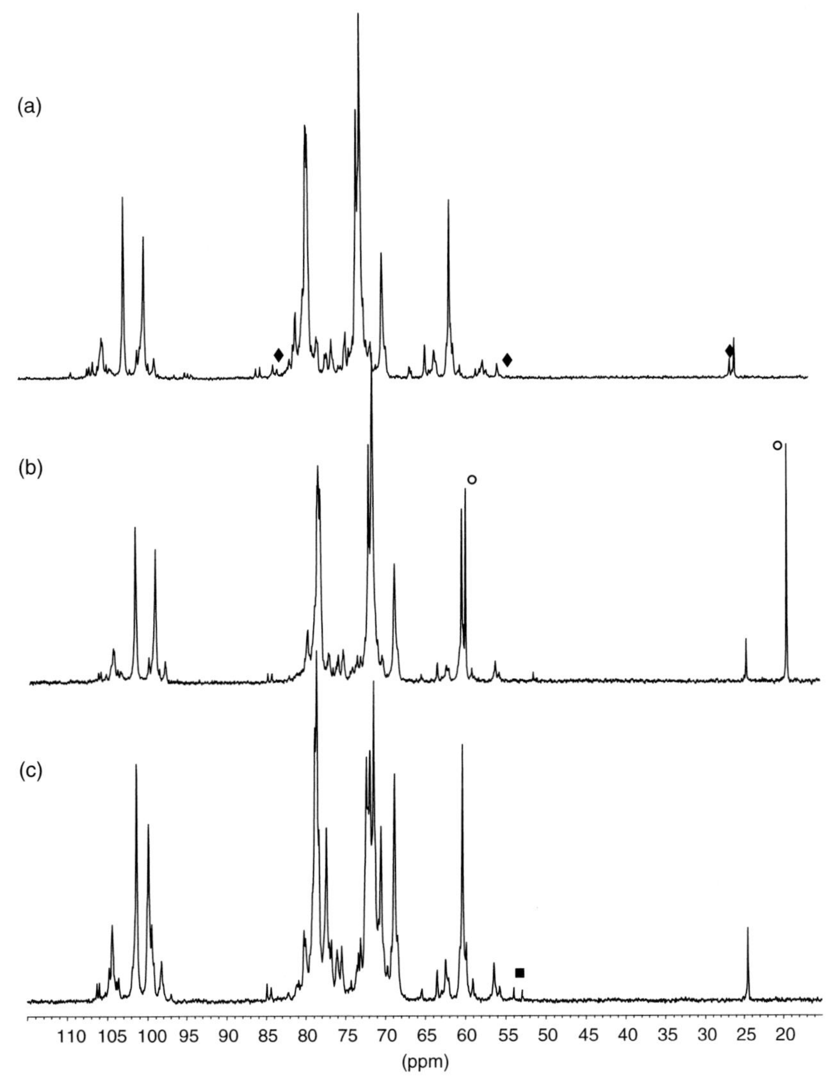

Figure 4 100 MHz ^{13}C NMR spectra of differently contaminated porcine mucosal heparin samples. Presence of (a) dermatan sulfate, ♦; (b) ethanol, °; and (c) epoxidated iduronic acid, ■.

by high structural heterogeneity. Such a heterogeneity is reflected by the fact that the heparin chains are made up of the possible combination of 48 different disaccharide building blocks, some of them, however, excluded by biosynthetic rules.

Owing to the current European regulation regarding the extraction of drugs from animal tissues, it has become important to discriminate between the two main animal species that are the conventional sources of organs in the extractive industries. The 1D NMR spectra are used to discriminate between pig and beef mucosal heparins trough quantification of specific sulfation patterns at the level of C6 of glucosamine (^{13}C NMR) and C2 of nonsulfated IdoA (^{1}H NMR) of each sample

(Figure 3).[28,29] Currently, this method is used also to discriminate the heparin origin at pharmaceutical grade, whereas the alternative DNA analysis often fails. Moreover, the technique allowed to differentiate heparins extracted from different organs, i.e., intestinal mucosa and lungs.[20]

Being the GAGs a family of similar anionic polymers, in the case of a single compound its most plausible contaminant(s) comes from other GAGs. In such a case, the ^{13}C NMR spectrum can help, though a limited extent, in the identification and, eventually, in the quantification of contaminants. The contaminant detection also depends on its relative amount with respect to the major components. However, the extraction procedure can generate some extra signals in the NMR spectra; they are due to contaminations or structural modifications caused, mainly, by residual solvent used in the process or to some extent of specific desulfation, with formation of epoxides[30] or free amines[19] (Figure 4). In the past, some of the extra signals were associated with the presence of trace EDTA (ethylenediaminotetracetic acid), compound used in the production of heparin.[31] In some case this compound was added, with mystification purpose, in an amount sufficient to increase the anticoagulant activity of heparin. Recently, NMR analysis of certain lots of heparin associated with an acute, rapid onset of serious side effects indicative of an allergic-type reaction identified the presence of a contaminant corresponding to oversulfated chondroitin sulfate.[32] 1D proton analysis, currently used to analyse heparin production batches, provides the only valid screening method able to unequivocally detect the presence of this contaminant and quantify it up to 0.05% (w/w).

2. 2D NMR Approach

As previously shown, in 1D proton spectrum of GAGs only few signals representative of a particular monosaccharide component not affected by signal overlapping can be used for quantification purpose. In addition, the difference of chemical shifts between the peaks of interest is often small. Consequently, the value of the resulting integrals may be strongly influenced by the instrument magnetic power, spectrum resolution, baseline correction and manual integration.[20] Moreover, the presence of unnatural residues in GAG preparations, generated by certain depolymerization or chemical modification procedures, causes a further signal overlapping in monodimensional NMR spectra, preventing their use for quantitative analyses.[33,34]

By resolving signals hidden in the monodimensional spectra, ^{1}H—^{13}C correlation NMR spectroscopy (HSQC) permits identification of the constituent monosaccharides of complex GAGs. The analytical signals used to characterize heparin in its monodimensional spectrum are completely resolved in the corresponding HSQC spectrum. Moreover, additional useful information on the distribution of residues in different sequences can be obtained by both the anomeric and H2/C2 cross peaks in the the 2D spectrum (Figure 5). For instance, while only the $GlcN_{NS}$ and $GlcN_{NAc}$ content can be obtained by monodimensional spectra integration, $GlcN_{NS}$ linked to IdoA can be discriminated from $GlcN_{NS}$ linked to GlcA in the HSQC spectrum. Moreover, signals at 4.61/103.7, 4.60/104.6, and 4.50/105.0 ppm indicate GlcA residue linked to $GlcN_{NS,3,6S}$, $GlcN_{NS}$, and $GlcN_{NAc}$, respectively.

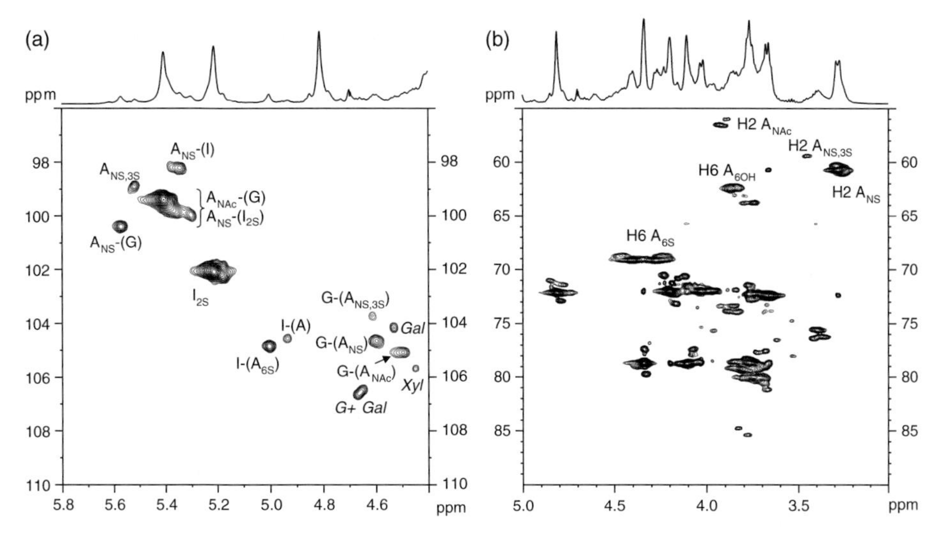

Figure 5 600 MHz HSQC spectrum of the anomeric (a) and ring signal (b) regions of the pig mucosal heparin. Labeled signals are those selected for the cross peak integration. A, G, and I indicate glucosamine, glucuronic acid, and iduronic acid, respectively. Labels in brackets indicate 1-O-linked residues. Signals corresponding to the linkage region residues are shown in italics.

The use of HSQC for molecular analysis of urinary GAG fragments to diagnose and study mucopolysaccharidoses diseases was recently described.[35] The detailed analysis of the excreted GAG fragments permits the direct diagnosis of the mucopolysaccharidoses type from urine analysis and also provides a means to monitor organ involvement, disease progression, and severity of the illness. Most peaks in the HSQC NMR spectra are sufficiently well resolved to be integrated for a quantitative analysis of the spectra, and thus for comparison of the carbohydrate compositions and sulfation patterns of GAGs from different patients. For instance, typical signals of dermatan sulfate and heparan sulfate were detected in the spectra of urines from patients with MP II, in which lack of L-iduronidase prevents cleavage of the linkage between IdoA and the aminosugar in both GAGs species, with consequent accumulation of the intact polymers in the urines.[35]

2.1. Setup of the 2D NMR method

HSQC experiments cannot be straightforwardly applied for quantitative analysis as the measured "volume" response of signals does not reflect only the atom abundance, but it is also dependent on the one-bond proton–carbon coupling constant ($^1J_{C—H}$) values and relaxation effects. The dependence of the volume of the HSQC correlation peak to the relation between the polarization transfer delay (Δ) and the actual $^1J_{C—Htrue}$ is shown in the following equation:[36]

$$V_C \propto \sin^2\left(\pi\Delta^1 J_{C-Htrue}\right)$$

This means that the optimal polarization transfer is achieved by setting $\Delta=(1/2)^1J_{C—Htrue}$ (Figure 6). Because the HSQC spectrum is normally set calculating Δ

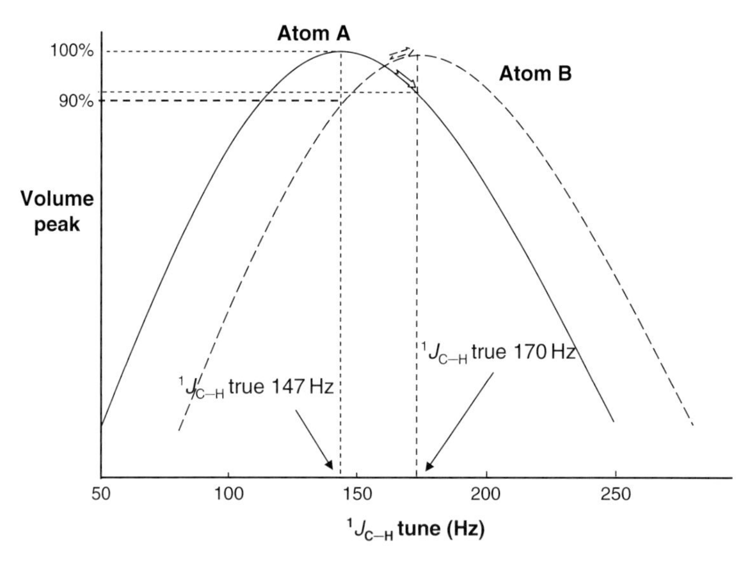

Figure 6 Schematized *J*-dependence of the cross peak volume in HSQC spectrum. When the $^1J_{C-H}$ tune is moved from 147 Hz ($^1J_{C-H}$true of atom A) to 170 Hz ($^1J_{C-H}$true of atom A) the intensity of A cross peak decreases (thin line) whereas that of B increases (dashed line).

with the average value of $^1J_{C-H}$ of the molecule atoms, the volume of each correlation peak reflects not only the amount of protons responsible for the correlation peak, but also the mismatch between the average value of $^1J_{C-H}$ and the $^1J_{C-H\,true}$. Moreover, if proton signals have many $^3J_{H-H}$ values, these couplings evolution during polarization transfer may distort the line shape of the correlation peaks compromising the accuracy of the integral value.[37]

A quantitative method for obtaining $^1H-^{13}C$ correlation NMR spectra (Q-HSQC) by suppression of $^1J_{C-H}$ dependence on the cross peak intensities has been first applied for quantitative analysis of wood lignin.[36] In that work cross peak volumes of aromatic, aliphatic, and acetate signals were compared independently from their different $^1J_{C-H}$. However, the sequence used by the authors reduces the sensitivity of HSQC spectra by about 25%, making the method useless for detection of minor peaks when using the common NMR instruments of medium magnetic field intensities.

Fortunately, for heparin and other GAGs the proper selection of analytical signals allows to minimize differences in both $^1J_{C-H}$ couplings and relaxation times. For instance, quantification of various substituted monosaccharides with similar hydrogen/carbon atoms such as H2/C2 or H1/C1 in differently substituted GlcN or uronic acid residues is weakly affected by the small differences in both $^1J_{C-H}$ and relaxation time with consequently small effects on the volume of the corresponding signal.[38]

To verify the influence of $^1J_{C-H}$ tune on the absolute volume intensities of the correlation peaks used for the compositional analysis, three HSQC spectra of a pig mucosal heparin have been measured using different $^1J_{C-H}$ tune values (Table 1). As expected, volumes slightly increase with decreasing differences between $^1J_{C-H}$ tune and $^1J_{C-H}$ true values. However, no significant variation of the percentage composition calculated for substituted glucosamine and uronic acid residues was observed for

Table 1 Pig mucosal heparin (PMH) A: cross peak volume intensities for different $^{1}J_{C-H}$ "tune" values measured for H2/C2 glucosamine signals (a) and anomeric signals of uronic acid residues (b)

(a)

$^{1}J_{C-H}$ tune (Hz)	$A2_{NS}$ ($^{1}J_{C-H}$ true 138–140 Hz)	$A2_{NAc}$ ($^{1}J_{C-H}$ true 140 Hz)	*$A2_{NS,3S}$ ($^{1}J_{C-H}$ true 140–141 Hz)*
	Volume	Volume	*Volume*
139	3058 (81%)	549 (15%)	*174 (4%)*
150	2960 (81%)	517 (14%)	*168 (4%)*
170	2811 (82%)	459 (14%)	140 (4%)

(b)

$^{1}J_{C-H}$ tune (Hz)	$I1_{2S}$ ($^{1}J_{C-H}$ true 170–173 Hz)	$I1_{2OH}$ ($^{1}J_{C-H}$ true 170–173 Hz)	*G1 ($^{1}J_{C-H}$ true 165 Hz)*
	Volume	Volume	*Volume*
139	2472 (70%)	354 (10%)	*695 (20%)*
150	2643 (71%)	367 (10%)	*729 (19%)*
170	2885 (70%)	406 (10%)	*802 (20%)*
	$I2_{2S}$ ($^{1}J_{C-H}$ true 148 Hz)	$I1_{2OH}$ ($^{1}J_{C-H}$ true 170–173 Hz)	*G1 ($^{1}J_{C-H}$ true 165 Hz)*
170	*2303 (66%)*	*406 (12%)*	*802 (22%)*

Calculated percent substitutions are given in parentheses. $^{1}J_{C-H}$ true values are from Yates et al.[38] Numbers in italic indicate the percentages of uronic acid residues calculated when the H2/C2 cross peak of the $IdoA_{2S}$ residue is used instead of the H1/C1 peak

variation of the $^{1}J_{C-H}$ "tune" values. On the contrary, comparison among volumes of cross peaks, corresponding to atoms with larger differences of $^{1}J_{C-H}$ values can generate significant errors in the compositional analysis. For instance, when the H2/C2 cross peak of the IdoA residues (4.35/78.9 ppm) are used instead of its H1/C1 peak, the calculated monosaccharide composition considerably changes because of the large differences between $^{1}J_{C1-H1}$ and $^{1}J_{C2-H2}$ (170–173 and 148 Hz, respectively).[34] However, it is important to note that the difference in the proton T_2 relaxation between low- and high-molecular-weight chains in heparin samples can interfere with the quantitative analysis. In fact, the proton T_2 relaxation during the constant-time INEPT delays affects the signal intensity according to $V_C \propto \exp(-2\Delta/T_2)$.[36] The T_2 effect is negligible for small molecules with long T_2, but it must be taken into the account for molecules with short T_2, such as polymers. According to this finding, the difference between proton T_2's of long heparin chains and small oligosaccharides (50–70 and 200–300 ms, respectively), can induce a slight overestimation of the very low-molecular-weight components content in the quantitative analysis of GAGs.

2.2. Choice of analytical signals

Since the anomeric signal of $GlcN_{NAc}$ linked to a GlcA residue overlaps to that of $GlcN_{NS}$ linked to $IdoA_{2S}$, the percentage of this residue is calculated by integration of glucosamine H2/C2 cross peaks ($GlcN_{NS}$ at 3.31/60.7 ppm, $GlcN_{NS,3S}$ at

Table 2 Monosaccharide components average contents (%), standard error (SD), and coefficient of variation (CV) for PMH A, determined by three times measurement of the 500 MHz HSQC spectrum

	A_{NS}	A_{NAc}	$A_{NS,3S}$	I_{2S}	I_{2OH}	G
Average	81%	15%	4%	70%	10%	20%
SD	1.04	0.60	0.53	0.31	0.32	0.54
CV × 100	1.29%	4.00%	11.97%	0.44%	3.17%	2.84%

Table 3 Determination of various substituted monosaccharide components (%) of PMH B

Glucosamine	A_{NS} I_{2S}	A_{NS} I	A_{NS} G	$A_{NS,3S}$	A_{NAc}	A_{NAc} α-red	A_{NS} α-red	A_{NS} β-red	A_{6S}
	64.4	10.0	8.0	4.0	11.8	1.1	0.6	–	81.8
Uronic acid	I_{2s}	I A_{6S}	I A_{6OH}	G $A_{NS,3S}$	G A_{NS}	G A_{NAc}			
	72.9	9.2	1.7	2.6	8.8	4.8			

The HSQC spectrum was measured at 600 MHz and at 308 K on a Bruker Avance 600 spectrometer equipped with 5 mm cryoprobe – italic labels indicate the 1-O-linked residues.

3.45/59.5 ppm and $GlcN_{NAc}$ at 3.97/56.2 ppm). The uronic acid composition is instead calculated from the corresponding anomeric signals ($IdoA_{2S}$ at 5.23/102.1 ppm, IdoA at 4.96–5.01/104.3–104.6 ppm, and GlcA at 4.5–4.6/103.7–105.1 ppm; Figure 6) Owing to the signals overlap, even in 2D spectra, only the overall sulfation at position 6 of GlcN can be measured. The value was evaluated by integrating the H6/C6 cross peaks of sulfated and nonsulfated residues at 4.42–4.22/69.0 and 3.84/62.4 ppm, respectively. These cross peaks correspond to the carbon signal previously used to calculate the sulfation degree of C6 by integration of monodimensional ^{13}C spectra.[29]

The reproducibility and the precision of the method were good as expressed by the standard deviation (SD) and coefficient of variation values. (Table 2) The highest coefficient of variation (CV) value (about 11%) was found for the H2/C2 of $GlcN_{NS,3S,6S}$ residue, the corresponding signal of which have a signal-to-noise ratio of about 7. To reduce CV, it is necessary to increase the signal-to-noise ratio by using longer acquisition times or higher sensitivity instruments (magnetic fields bigger than 500 MHz and/or a cryoprobe). The distribution of GlcN residues in different sequences can be better detected by integration of the anomeric region where more than one H/C correlation signal are present (Figure 5(a)). The percentage of $GlcN_{NAc}$ residues, calculated by integration of H2/C2 signals, is subtracted from that of the anomeric signal representing both $GlcN_{NAc}$ and $GlcN_{NS}$ residues (Figure 5(b)). The content of various substituted monosaccharide components for a typical pig mucosal heparin, measured at 600 MHz using a 5 mm TCI cryoprobe, is shown in Table 3.

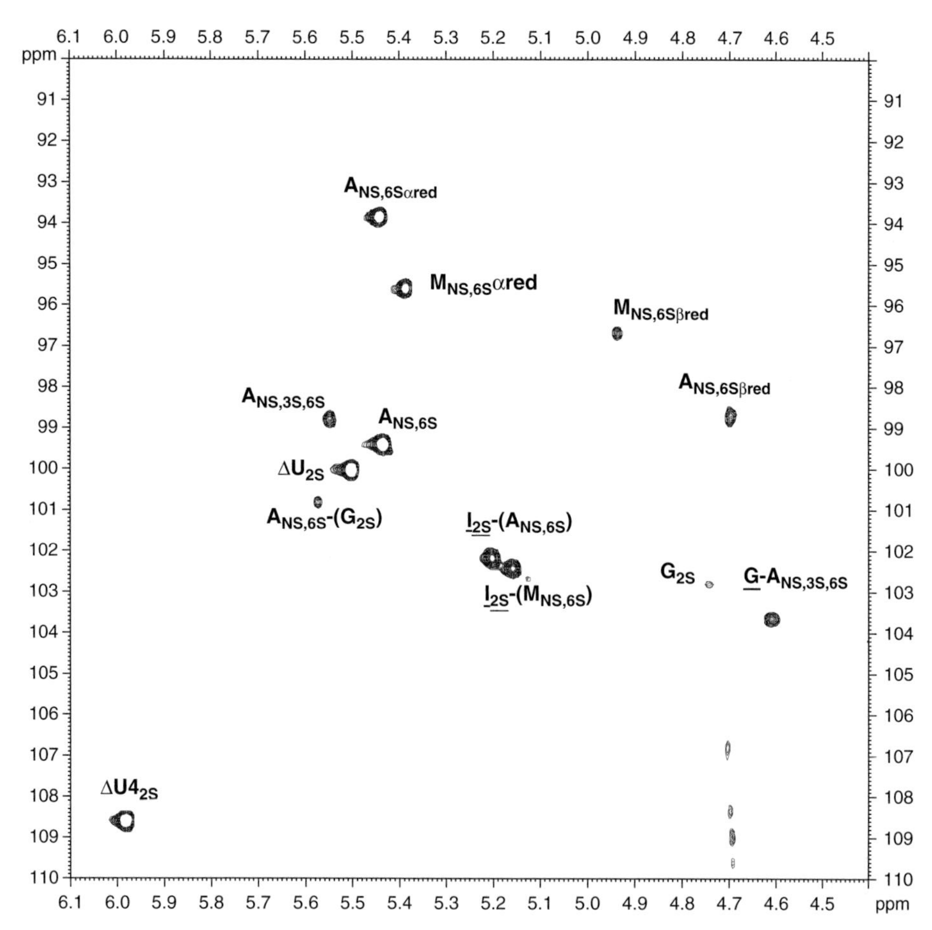

Figure 7 Anomeric region of the 600 MHz HSQC spectrum of a tetrasaccharide mixture isolated from a LMWH.

2.3. 2D analysis of a simple oligosaccharide mixture

The quantitative investigation of a series of tetrasaccharidic fractions obtained from a LMWH is a good way to verify the reliability and the limits of the 2D HSQC quantitative response. Different fractions containing only tetrasaccharide components, obtained by applying a β elimination reaction to a pig mucosal LMWH, followed by isolation by preparative size exclusion chromatography, and a further fractionation with a high-performance preparative ion pair column, have been used. To underline the high level of heterogeneity of such LMWH, and consequently the necessity of a powerful way to characterize such complex mixture, it is noteworthy that up to a total of 23 tetrasaccharide oligomers were identified by HP ion pair liquid chromatography (LC)–mass analysis. As an example, Figure 7 shows the anomeric region of HSQC spectrum of one of the fractions containing the following tetrasaccharides generated by the depolymerization process: ΔU_{2S}-$GlcN_{NS,6S}$-$IdoA_{2S}$-$GlcN_{NS,6S}$/

Table 4 Proton and carbon chemical shifts (c.s.) values of anomeric signals of a tetrasaccharide mixture (Figure 7)

Residue	NR residues		Hexosamine			Uronic acid		R residues			
	ΔU_{2S}	GlcA $A_{NS,3S}$	$GlcN_{NS}$	$GlcN_{NS,3S}$	$GlcN_{NS}$ $GlcA_{2S}$	$IdoA_{2S}$	$GlcA_{2S}$	$GlcN_{NS}$ α-red	$GlcN_{NS}$ β-red	$ManN_{NS}$ α-red	$ManN_{NS}$ β-red
^{1}H c.s. (ppm)	5.52	4.62	5.46	5.56	5.59	5.22	4.75	5.46	4.72	5.41	4.94
^{13}C c.s. (ppm)	99.6	103.9	99.7	99.0	100.9	101.7	102.9	94.2	99.0	95.8	96.8
Volume	100	18	96	18	7	100	7	68	11	48	8
recovery ratio	1		1			0.9		1.1			

Cross peak volumes are clustered as nonreducing (NR), hexosamine, uronic acid, and reducing (R) residues. The ratio between total volume of NR residues (reference) and that of other clusters is shown.

$ManN_{NS,6S}$ (tetra 1,2) and minor amounts of ΔU_{2S}-$GlcN_{NS,6S}$-$GlcA_{2S}$-$GlcN_{NS,6S}$/$ManN_{NS,6S}$ (tetra 3,4) and GlcA-$GlcN_{NS,3,6S}$-$IdoA_{2S}$-$GlcN_{NS,6S}$/$ManN_{NS,6S}$ (tetra 5,6). Each anomeric signal of the 2D map has been integrated and the corresponding values are shown in Table 4, in which nonreducing (NR), hexosamine, uronic acid, and reducing (R) residues are clustered following the tetrasaccharide sequences. The integral values of the signals permit some considerations: the intensity of NR residue signals results exactly proportional with that of hexosamine (118 vs 121), whereas a slightly difference with uronic acid and R signals has been found (107 and 135, respectively). Such difference implicates a small change in the ratio between each residue type, not far from the theoretical value of one. According to that, the composition of the mixture can be estimated by referring the volume of the tetrasaccharide "marker" residue to the volume of the reference cluster. For instance, the content of $GlcN_{NS}$ linked to $GlcA_{2S}$ has been used to quantify the amount of tetra (3,4), whereas the content of $GlcN_{NS,3,6S}$ for the quantification of the tetra (5,6). The following composition has been found: 79% of tetrasaccharides (1,2), 6% of tetrasaccharides (3,4), and 15% of tetrasaccharides (5,6), being the GlcN/ManN ratio equal to 1.4.

3. Low-Molecular-Weight Heparin Analysis

LMWHs are now replacing heparin in most of clinical applications owing to its advantages, such as improved bioavailability and decreases side effects, with respect to unfractionated heparin.[39] Interest in the development of generic version of the LMWHs requires a better understanding of their structure–activity relationships as well optimization of an analytical method suitable for comparison between products obtained with different producers, or different batches obtained by the same producer. The internal structure of LMWH should ideally match that of the parent heparin in terms of monosaccharide composition, substitution pattern, and oligosaccharide sequence. However, the depolymerization processes used in the manufacturing of LMWHs usually involves some structural modification, mostly at the level of the monosaccharidic units at the site of cleavage and typical for each de-polymerization procedure (Figure 8). Distinct pharmacological and biochemical profiles claimed for each LMWH can be associated with different structural modifications.[26] The position of the cleavage point along the heparin chain may influence the protein binding properties of the resulting fragments and the correlated biological properties. For instance, structural modification of the AT-binding pentasaccharide may reduce or cancel the affinity to AT and the corresponding anti-FXa activity.[7] Moreover, additional sequences on both sides of the pentasaccharide affect the binding properties of the oligosaccharide to AT, and consequently its activity.[40]

As shown previously for the identification of galactosaminoglycans and heparin, monodimensional NMR spectroscopy, and particularly ^{13}C NMR, is a useful tool also for the identification of each LMWH. Major signals of LMWHs correspond to

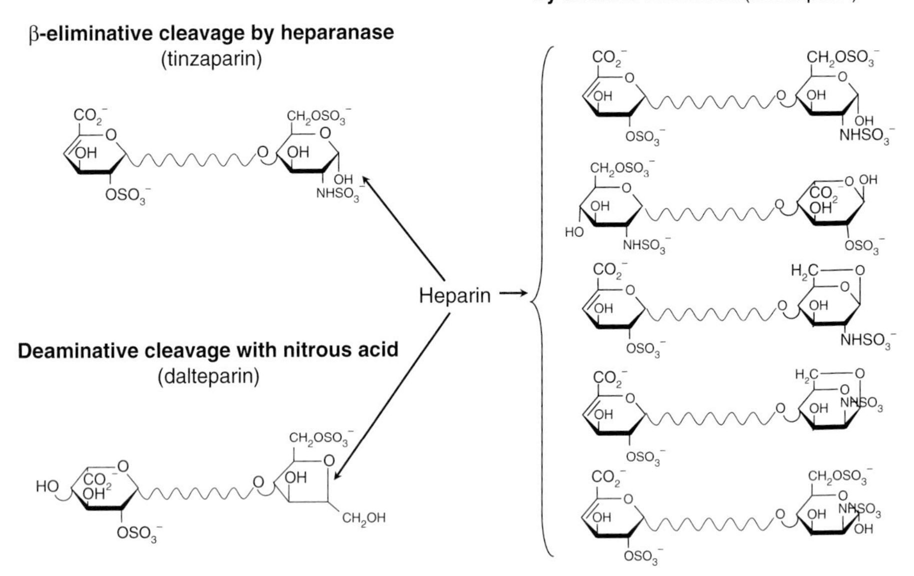

Figure 8 Scheme of depolymerization used to prepare most common LMWHs. Major reducing and nonreducing residues are shown.

the prevalent trisulfated disaccharide repeating unit; -4)-α-L-IdoA$_{2S}$-α-(1 → 4)-D-GlcN$_{NS,6S}$, typical of the parent heparin. Minor signals are mainly due to the presence of residues generated by the depolymerization process (Figure 9). For instance, dalteparin is produced using a deamination process that generate terminal 2,5-anhydromannitol (AM.ol) residue at the reducing ends. The corresponding C4, C2, and C5 signals are clearly observable in the 13C spectrum at 87.9, 85.9, and 82.3 (Figure 9(b)).[41] Enoxaparin and tinzaparin, produced by chemical β-eliminative and enzymatic methods, respectively, generate the 4–5 unsaturated 2-O-sulfated uronic acid (ΔU$_{2S}$) as the nonreducing terminal residue (C4 at 108–110 ppm), while GlcN$_{NS,6S}$ is the prevalent residue at the reducing end (93.9 ppm; Figure 9(a) and (b)). Also enoxaparin and tinzaparin can be differentiated by carbon spectra: two additional reducing anomeric signals, both of them at 95.6 ppm, and other in the C2 region of glucosamine at 55.0 and 58.6 ppm, are present in the enoxaparin spectrum. The first signal is due to the reducing N-sulfated, 6-O-sulfated mannosamine (ManN$_{NS}$), and IdoA$_{2S}$ residues, the two others to the 2-sulfo-amino-1,6-anhydro-2-deoxy-β-D-glucopyranose (1,6-an.Glc) and 2-sulfo-amino-1,6-anhydro-2-deoxy-β-D-mannopyranose (1,6-an.Man) residues, two unique bicyclic structures present at the reducing end originating from alkaline hydrolysis of the benzoyl ester of heparin.[42]

However, the structural complexity of LMWH causes a signal overlap greater than that already described in the heparin monodimensional NMR spectra,

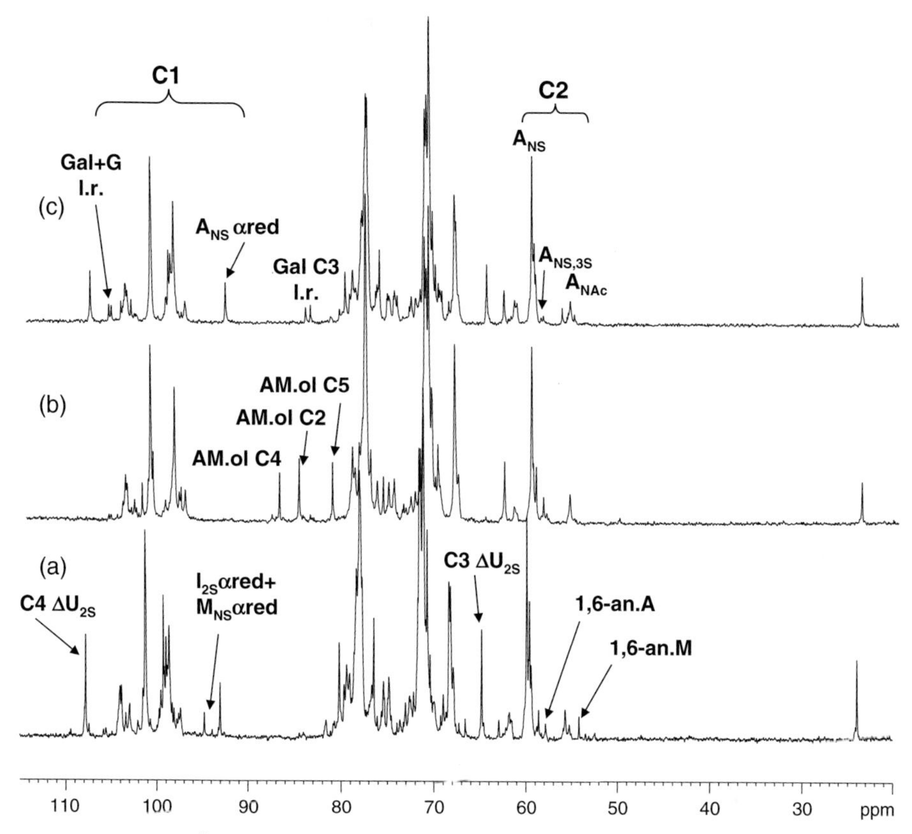

Figure 9 100 MHz ^{13}C-nuclear magnetic resonance spectra of (a) enoxaparin, (b) dalteparin, and (c) tinzaparin.

preventing complete signal assignments and the use of monodimensional spectrum for quantitative analyses.

The HSQC spectra permit to follow the structural differences among the three LMWHs not only at the level of the anomeric signals but, also of all ring signal regions (Figure 10). The spectra confirm that enoxaparin, in both the anomeric and H2/C2 regions, has the most complex structure among the current LMWHs. Besides the presence of 1,6-anhydro reducing residue, a signal corresponding to the 2-O-sulfated glucuronic acid residue ($GlcA_{2S}$), which was found in very low amounts in natural GAGs and was not detectable in spectra of nondepolymerized heparins, is clearly observable in the HSQC spectrum of enoxaparin.[43] HSQC spectrum also indicates the presence of epoxides generated by alkaline treatment, with the characteristic H2/C2 and H3/C3 signals at 3.74/54.2 and 3.82/53.3 ppm, respectively.[16,30,44]

The analytical signals of LMWHs were chosen among those with minimal overlap in the HSQC spectra.[45] Signals corresponding to the structural peculiarities of each LMWH have been selected in addition to the signals used for

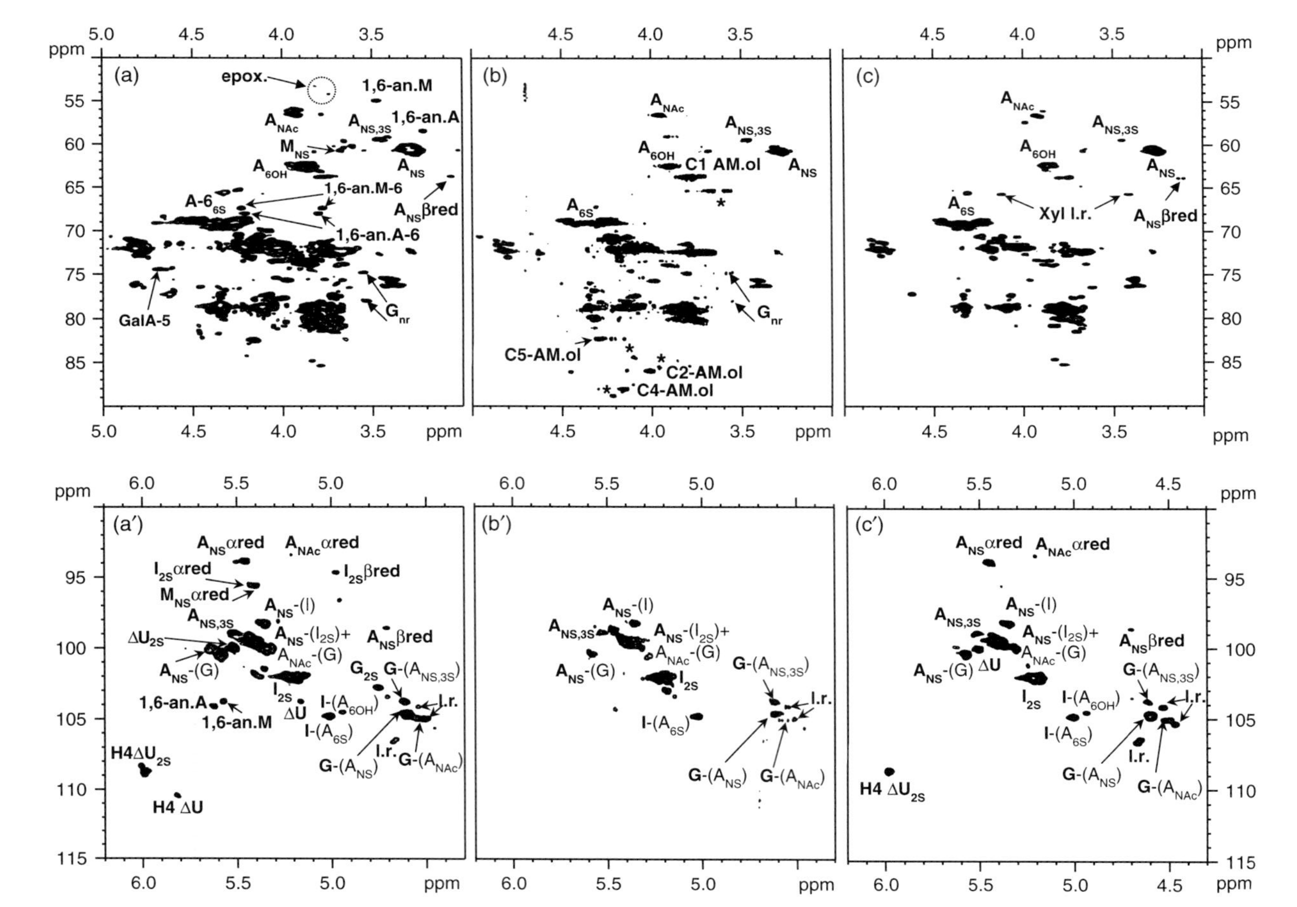

Figure 10 Ring signals of 600 MHz HSQC) spectra of (a) enoxaparin, (b) dalteparin, and (c) tinzaparin. Anomeric region of 600 MHz HSQC spectra of (a′) enoxaparin, (b′) dalteparin, and (c′) tinzaparin. Signals corresponding to H2/C2 and H3/C3 of the epoxide group are highlighted in a dotted circle. A, G, and I indicate glucosamine, glucuronic acid and iduronic acid, respectively. AM.ol, 1,6-an.A, and 1,6-an.M indicate 2,5-anhydromannitol, 2-sulfo-amino-1,6-anhydro-2-deoxy-β-D-glucopyranose, and 2-sulfo-amino-1,6-anhydro-2-deoxy-β-D-mannopyranose, respectively.

the characterization of unfractionated heparin (Figure 10). Cross peaks corresponding to the reducing anomeric residues as well as those corresponding to the nonreducing ΔU and ΔU_{2S} units were taken into account. It is necessary to consider that a LMWH chain is constituted by about 20 monomers; consequently, their reducing residues account for about 5% of total residues, 20% of which are due to beta epimers. Considering the possible distribution of the epimers in different structures, it is impossible to identify the β epimers in all of them, unless small oligomers are major components or high sensitive instruments are used.

The larger number of signals occurring in LMWHs with respect the parent heparin together with the low amount of minor sequences generated by the depolymerization reactions, require a high sensitivity and a relative high field NMR instrument to be detected and quantified. The use of cryoprobe and a magnetic field strength not lower than 500 MHz is desirable to obtain a good precision of the results in a reasonable time. In the present work, a 600 MHz instrument equipped with cryoprobe has been used with the experimental parameters indicated in Table 5. As shown from SD values, the reproducibility of the method was good. The CV, is very low (less than 3%) for almost all peaks, in any case not higher than 10% even for minor components present in amounts lower than 1–2%.

The average contents of monosaccharide components of the three batches of LMWHs are reported in Table 5. The complexity of enoxaparin with respect dalteparin and tinzaparin is immediately appreciable: eight different reducing residues are detectable compared the three of tinzaparin and the single anhydromannitol residue of dalteparin. Tinzaparin shows the higher content of $GlcN_{NAc}$ residues, whereas enoxaparin is richer in glucosamine linked to glucuronic acid ($GlcN_{NS}$-GlcA). Noteworthy, the different content of the trisulfated glucosamine, considered to be the marker of the pentasaccharidic sequence of the active site for AT, was found to be significantly less abundant in tinzaparin. As the $GlcN_{NS,3,6S}$ was found also in heparin fractions with no affinity for AT (26, 43), a direct correlation between its content and antifactor Xa (aFXa) is not straightforward. A more reliable correlation was observed between the amount of GlcA linked to $GlcN_{NS,3,6S}$ and the aFXa activity, such a better correlation is attributable to the fact that, whereas the trisulfated glucosamine $GlcN_{NS,3,6S}$ is present also in non-AT-binding sequences, this disaccharide was found prevalently in active sequences.[46]

Other parameters, such as the number average molecular weight, the sulfation pattern, and the linkage region content, can be derived from the NMR data. The total amount of reducing residues, obtained by integration of the appropriate signals in HSQC spectra, can be correlated to the average length of the chains and the Mn value, by estimation of the average molecular weight of the disaccharide. The calculated Mn are slightly lower than those calculated by low and multi-angle light scattering analysis (high-performance size exclusion chromatography/triple detector analyzer the HP-SEC/TDA Houston, TX) method,[47] (Table 5(c)). The content of linkage region of enoxaparin and dalteparin, calculated by both HSQC and 1D methods (10), is comparable with the average value of pig mucosal heparins. In contrast, the linkage region content of tinzaparin is about two times larger than that of pig mucosal heparin. The results obtained by the bidimensional method are similar to those obtained from carbon spectra integration confirming the method reliability.

Table 5 Determination of various substituted monosaccharide components (percentage) of enoxaparin, tinzaparin, and dalteparin

(a)

Sample		A_{ns} I_{2s}	A_{ns} I	A_{ns} G	$A_{ns,3s}$	A_{nac}	A_{nac} α-red	A_{ns} α-red	A_{ns} β-red	1,6 anA	1,6 anM	M_{ns} α-red	An. M.ol	A_{6s}
Enoxaparin		48.8	7.0	11.8	4.3	10.3	0.5	8.6	1.0	2.2	2.5	3.0	0.0	83.8
	SD[34]	0.49	0.21	0.23	0.16	0.02	0.05	0.05	0.06	0.05	0.05	0.09		0.10
	CV%	1.1	2.4	1.6	3.6	0.2	0.6	9.7	5.7	2.1	2.3	2.8		0.1
Tinzaparin		53.2	7.6	8.6	3.4	14.0	0.6	11.1	1.3	0.0	0.0	0.4	0.0	86.3
Dalteparin		55.3	8.9	6.0	5.4	9.6	0.0	0.0	0.0	0.0	0.0	0.0	14.8	91.5

(b)

Sample		I_{2s}	I A_{6S}	I A_{6OH}	G $A_{ns,3s}$	G A_{ns}	G A_{nac}	G_{2s}	ΔU_{2s}	ΔU	I_{2s} red	U_{2s} red	Epox
Enoxaparin		52.8	5.7	1.3	3.3	8.2	4.5	2.6	17.6	1.0	1.1	0.9	0.3
	SD[34]	0.28	0.04	0.01	0.04	0.09	0.06	0.04	0.40	0.03	0.10	0.06	0.01
	CV%	0.5	0.7	1.1	1.2	1.0	1.5	1.5	2.3	2.9	9.3	6.3	4.2
Tinzaparin		64.0	6.8	1.5	2.0	9.0	3.2	0.0	12.7	0.0	0.0	0.3	0.0
Dalteparin		76.3	8.5	0.7	4.4	6.5	3.5	0.0	0.0	0.0	0.0	0.0	0.0

(c)

Sample	Mn	Amount of linkage region calculated by HSQC method	Amount of linkage region calculated by ^{13}C method[10]	Total degree of sulfation
Enoxaparin	3200 (3600)	0.7	0.8	2.5
Tinzaparin	4700 (5500)	3.1	2.8	2.4
Dalteparin	4000 (4600)	0.6	0.7	2.6

Glucosamine (a), uronic acid (b), and Mn, linkage region content and degree of sulfation calculated from NMR data (c). Mn values have been calculated by NMR and by light scattering analysis (in parentheses).[47] For enoxaparin SD and CV are listed as previously reported.[44] The spectra were measured at 600 MHz and at 308 K on a Bruker Avance 600 spectrometer equipped with 5 mm cryoprobe. About 15 mg of samples have been dissolved in 0.6 ml deuterium oxide. HSQC spectra were recorded with 320 increments of 32 scans for each and a recycle delay of 2.5 s. The polarization transfer delay was set with a $^{1}J_{C-H}$ coupling value of 155 Hz (average between anomeric signals and H2/C2 of glucosamine).[44] Italic labels indicate the 1-O-linked residues.

4. Conclusions

The knowledge of monosaccharidic composition of natural and modified GAGs is important for deriving their structure–activity relationships. Even if monodimensional spectroscopy provides fingerprints of heparin preparations and permits characterization and quantification of specific structural features as wells, it fails when signal overlap does not allow integration of the corresponding peaks. 2D HSQC experiments can be used for the quantitative analysis of both major and minor monosaccharide constituents of heparins and other GAGs on which preliminary structural information is available. Prerequisite of the 2D NMR quantitative approach is a proper selection of analytical signals among those with similar magnetic relaxation and one bond proton–carbon *J*-coupling. The method can be applied to both high- and low-molecular GAGs even if some limitations have to be considered. The shorter proton T_2 characteristic for residues in polymeric chains, with respect its longer value in small oligomers and terminal reducing residues (50–80 and 200–300 ms respectively) causes an overestimation of the latter components. Such observation can be supported by the lower Mn of LMWHs calculated from NMR data with respect to those obtained by light scattering analysis (Table 5(c)). Besides these limitations, the method can be confidently used to compare different LMWHs as well as to control the batch to batch variability of industrial preparation of the same GAGs.

References

1. R.V. Iozzo, A.D. Murdoch, FASEB J. 10 (1996) 598–614.
2. J.E. Scott, J. Anat. 187 (1995) 259–269.
3. L.A. Fransson, Trends Biochem. Sci. 12 (1987) 406–411.
4. V. Bossennec, M. Petitou, B. Perly, Biochem. J. 267 (1990) 625–630.
5. B. Casu, Adv. Carbohydr. Chem. Biochem. 43 (1985) 51–132.
6. B. Casu, Structure and active domains of heparin, In: Chemistry and Biology of Heparin and Heparan Sulphate, H.G. Garg, R.J. Linhardt, C.A. Hales, Ed, Chapter 1, 1–28, Elsevier Inc. Oxford, 2005.
7. B. Casu, U. Lindhal, Adv. Carbohydr. Chem. Biochem. 57 (2001) 159–206.
8. M. Ueno, S. Yamada, M. Zako, M. Bernfield, K. Sugahara, J. Biol. Chem. 276 (2001) 29134–29140.
9. U. Lindahl, Biochim. Biophys. Acta 130 (1996) 368–382.
10. M. Iacomini, B. Casu, M. Guerrini, A. Naggi, A. Pirola, G. Torri, Anal. Biochem. 274 (1999) 50–58.
11. H.E. Conrad, *Heparin Binding Proteins*, Academic Press San Diego, CA, 1998.
12. I. Capila, R.J. Linhardt, J. Med. Chem. 46 (2003) 2551–2564.
13. R.V. Iozzo, Annu. Rev. Biochem. 67 (1998) 609–652.
14. I. Capila, N.S. Gunay, Z. Shriver, G. Venkataraman, Methods for structural analysis of heparin and heparan sulfate, In: Chemistry and Biology of Heparin and Heparan Sulphate, H.G. Garg, R.J. Linhardt, C.A. Hales, Ed, Chapter 3, 55–71, Elsevier Inc. Oxford, 2005.
15. M. Hricovini, M. Guerrini, A. Bisio, G. Torri, A. Naggi, B. Casu, Semin. Thromb. Hemost. 28 (2002) 325–334.
16. A.S. Perlin, B. Casu. Spectroscopic methods. In: The Polysaccharides, G.O. Aspinall, Ed, vol 1, pp. 177–185, Academic Press., 1982.
17. B. Mulloy, M.J. Forster, C. Jones, D.B. Davies, Biochem. J. 293 (1993) 849–858.

18. C.P. Dietrich, H.B. Nader, A.S. Perlin, Carbohydr. Res. 41 (1975) 334–338.
19. E.A. Yates, F. Santini, M. Guerrini, A. Naggi, G. Torri, B. Casu, Carbohydr. Res. 294 (1996) 15–27.
20. M. Guerrini, A. Bisio, G. Torri, Semin. Thromb. Hemost. 27 (2001) 473–482.
21. L.B. Jaques, L.W. Kavanagh, M. Mazurek, A.S. Perlin, Biochem. Biophys. Res. Commun. 24 (1966) 447–451.
22. A.S. Perlin, G.R. Sanderson, Carbohydr. Res. 12 (1970) 183–192.
23. G. Gatti, B. Casu, A.S. Perlin, Biochem. Biophys. Res. Commun. 85 (1978) 14–20.
24. G. Gatti, B. Casu, G. Torri, J.R. Vercellotti, Carbohydr. Res. 68 (1979) C3–C7.
25. G. Torri, Semin. Thromb. Hemost. 17 (1991) 23–28.
26. B. Casu, G. Torri, Semin. Thromb. Hemost. 25 (1999) 17–25.
27. M. Guerrini, R. Raman, G. Venkataraman, G. Torri, R. Sasisekharan, B. Casu, Glycobiology 12 (2002) 713–719.
28. B. Casu, M. Guerrini, A. Naggi, G. Torri, L. De Ambrosi, G. Boveri, S. Gonella, J. Thromb. Haemost. 74 (1995) 1205–1208.
29. B. Casu, M. Guerrini, A. Naggi, G. Torri, L. De Ambrosi, G. Boveri, S. Gonella, A. Cedro, L. Ferro, E. Lanzarotti, M. Paternò, M. Attolini, M.G. Valle, Arzneim. Forsch. Drug Res. 46 (1996) 472–477.
30. S. Piani, B. Casu, E.G. Marchi, G. Torri, F. Ungarelli, J. Carbohydr. Chem. 12 (1993) 507–521.
31. B. Casu, A. Naggi, P. Oreste, G. Torri, J. Pangrazzj, A. Maggi, M. Abbadini, M.B. Donati, Lancet (1987) 329 (1987), (8541) p. 1088–1088.
32. M. Guerrini, D. Beccati, Z. Shriver, A. Naggi, K. Viswanathan, A. Bisio, I. Capila, J. C. Lansing, S. Guglieri, B. Fraser, A. Al-Hakim, N. S. Gunay, Z. Zhang, L. Robinson, L. Buhse, M. Nasr, J. Woodcock, R. Langer, G. Venkataraman, R. J. Linhardt, B. Casu, G. Torri, R. Sasisekharan. *Nat Biotechnol.* 26 (2008) 669–675.
33. A. Naggi, B. De Cristofano, A. Bisio, G. Torri, B. Casu, Carbohydr. Res. 336 (2001) 283–290.
34. M. Guerrini, A. Naggi, S. Guglieri, R. Santarsiero, G. Torri, Anal. Biochem. 337 (2005) 35–47.
35. M. Hochuli, K. Wuthrich, B. Steinmann, NMR Biomed. 16 (2003) 224–236.
36. S. Heikkinen, M.M. Toikka, P.T. Karhunen, I.A. Kilpeläinen, J. Am. Chem. Soc. 125 (2003) 4362–4367.
37. H. Koskela, I. Kilpelainen, S. Heikkinen, J. Magn. Reson. 174 (2005) 237–244.
38. E.A. Yates, F. Santini, B. De Cristofano, N. Payre, C. Cosentino, M. Guerrini, A. Naggi, G. Torri, M. Hricovini, Carbohydr. Res. 329 (2000) 239–247.
39. J.I. Weitz, N. Engl, J. Med. 337 (1997) 688–698.
40. M. Guerrini, S. Guglieri, D. Beccati, G. Torri, C. Viskov, P. Mourier, Biochem. J. 399 (2006) 191–198.
41. T.N. Huckerby, P.N. Sanderson, A. Nieduszynski, Carbohydr. Res. 138 (1985) 199–206.
42. G. Mascellani, M. Guerrini, G. Torri, L. Liverani, F. Spelta, P. Bianchini, Carbohydr. Res. 342 (2007) 835–842.
43. S. Yamada, T. Murakami, H. Tsuda, K. Yoshida, K. Sugahara, J. Biol. Chem. 270 (1995) 8696–8705.
44. M. Hricovini, M. Guerrini, G. Torri, S. Piani, F. Ungarelli, Carbohydr. Res. 277 (1995) 11–23.
45. M. Guerrini, S. Guglieri, A. Naggi, G. Torri, Semin. Thromb. Hemost. 33 (2007) 478–487.
46. Z. Shriver, M. Sundaram, G. Venkataraman, K. Drummond, J. Turnbull, T. Toida, R.J. Linhardt, K. Biemann, R. Sasisekharan, Proc. Natl. Acad. Sci. U.S.A. 97 (2000) 10365–10370.
47. S. Bertini, A. Bisio, G. Torri, D. Bensi, B. Terbojevich, Biomacromol. 6 (2005) 168–173.

CHAPTER 5

The Use of Proton NMR as an Alternative for the Amino Acid Analysis as Identity Test for Peptides

E. Kellenbach, K. Sanders, *and* P.L.A. Overbeeke

Contents

Abstract

Proton NMR was evaluated as an alternative to amino acid analysis (AAA) as an identity test for peptides. Proton NMR can readily distinguish and identify all peptides currently described in the *European Pharmacopoeia*. A comparison with AAA as an identity test is presented.

Keywords: peptides, amino acid analyser, buserelin, goserelin, oxytocin, tetracosatide

1. Introduction

A number of monographs for synthetically produced peptides are included in the current edition 6.1 of the European Pharmacopeia.[1] To control the identity of these peptides, analytical methods such as TLC, infrared spectroscopy, electrophoresis and LC are included in the Identification section of the monographs. The monographs for buserelin and goserelin also include NMR as a test for identity.

Amino acid analysis (AAA) is included in the peptide monographs. Initially AAA was included in the Test section. However, most peptide monographs have been revised or are under revision to move the AAA from the Test section to the Identification section.

Using AAA, the relative amount (ratio) and absolute content of amino acids in a peptide can be determined. Thus, AAA is a powerful technique to be used both as an identity test and to quantify peptides and glucosylaminoglycans. Recently, the use of proton NMR was proposed as an identity test for peptides. At the request of the EDQM, the possibility of NMR as an alternative to AAA as identity test for peptides was evaluated.

2. Spectral Comparison

Solution proton NMR spectra are widely recognized, accepted and established as powerful tools for the structural elucidation and characterization of peptides and proteins.[2–4] Proton and Carbon-13 NMR spectra readily distinguish all 20 naturally occurring amino acids and unnatural amino acids present in synthetic peptides, and the random coil chemical proton chemical shifts in water have been tabled.

Quantification of the proton signals of peptides is relatively straightforward. The intensity of the proton signal is directly proportional to the relative number of protons at a particular position in a peptide.[5,6] As a result, a proton NMR spectrum of a peptide depends primarily on the relative amounts and identity of amino acids present in the sequence. This information concerning the identity of a peptide is the same as that obtained from an AAA. From a scientific point of view, NMR is a suitable alternative for AAA.

Moreover, because for an NMR measurement no hydrolysis is performed before analysis, a distinction can be made between Glu and Gln, and Asp and Asn. AAA is not able to distinguish these "critical pairs". Amino acids prone to hydrolysis or oxidation such as Trp and Cys can be detected without problems.

As a demonstration of the application of proton NMR spectra of eight different peptides, including two different salt forms of one peptide, were recorded at 400 MHz.[7] This is a moderate field strength widely available within the pharmaceutical industry. The peptides are listed in Table 1. Spectra are displayed in Figures 1 (aliphatic region) and 2 (aromatic region). All peptides

Table 1 Composition of the peptides reported in this study

Peptide	Number of AA	Sequence
Oxytocin	9	H-Cys-Tyr-Ile-Gln-Asn-Cys-Pro-Leu-Gly-NH2
Desmopressine acetate	9	Mpa-Tyr-Phe-Gln-Asn-Cys-Pro-D-Arg-Gly-NH2
Gonadorelin acetate	10	Glp-His-Trp-Ser-Tyr-Gly-Leu-Arg-Pro-Gly-NH2
Gonadorelin diacetate*	10	Glp-His-Trp-Ser-Tyr-Gly-Leu-Arg-Pro-Gly-NH2
Buserelin acetate	10	Glp-His-Trp-Ser-Tyr-D-Ser(tBu)-Leu-Arg-Pro-Gly-NHEt
Goserelin	10	Glp-His-Trp-Ser-Tyr-D-Ser(tBu)-Leu-Arg-Pro-Azagly-NH2
Protirelin	3	Glp-His-Pro-NH2
Tetracosactide	24	H-Ser-Tyr-Ser-Met-Glu-His-Phe-Arg-Trp-Gly-Lys-Pro-Val-Gly-Lys-Lys-Arg-Arg-Pro-Val-Lys-Val-Tyr-Pro-OH

*Gonadorelin diacetate contains 4 mo of water (4AQ) per mole of Gonadorelin diacetate.

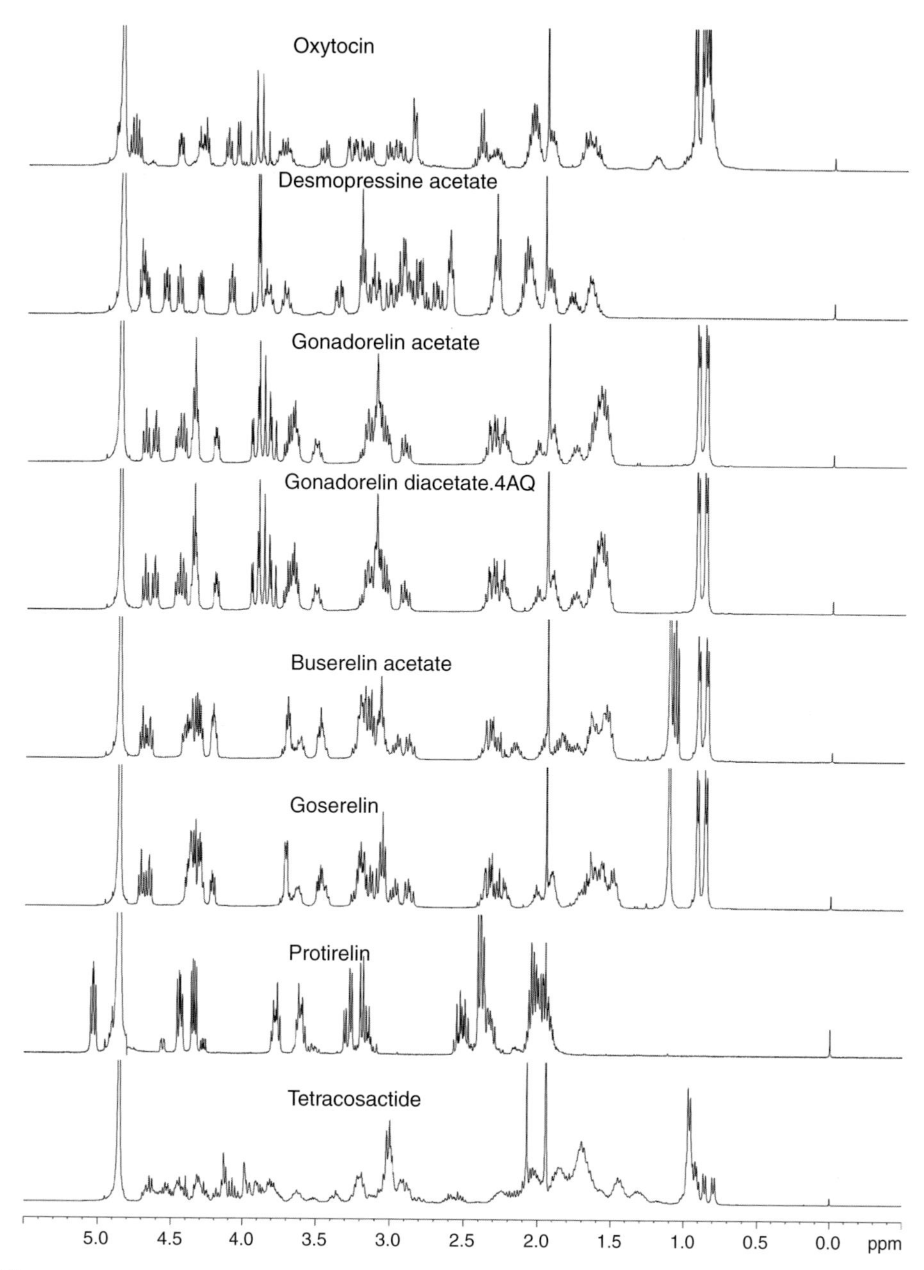

Figure 1 Aliphatic region of the 400 MHz NMR spectra of the eight different peptides.

can readily be distinguished. Even goserelin and buserelin, the peptides showing only very small differences in sequence, can be discriminated. Additionally, the protonation state as well as signals of non-exchangeable proton-bearing counter ions can be identified using proton NMR. Thus, it even proved possible to

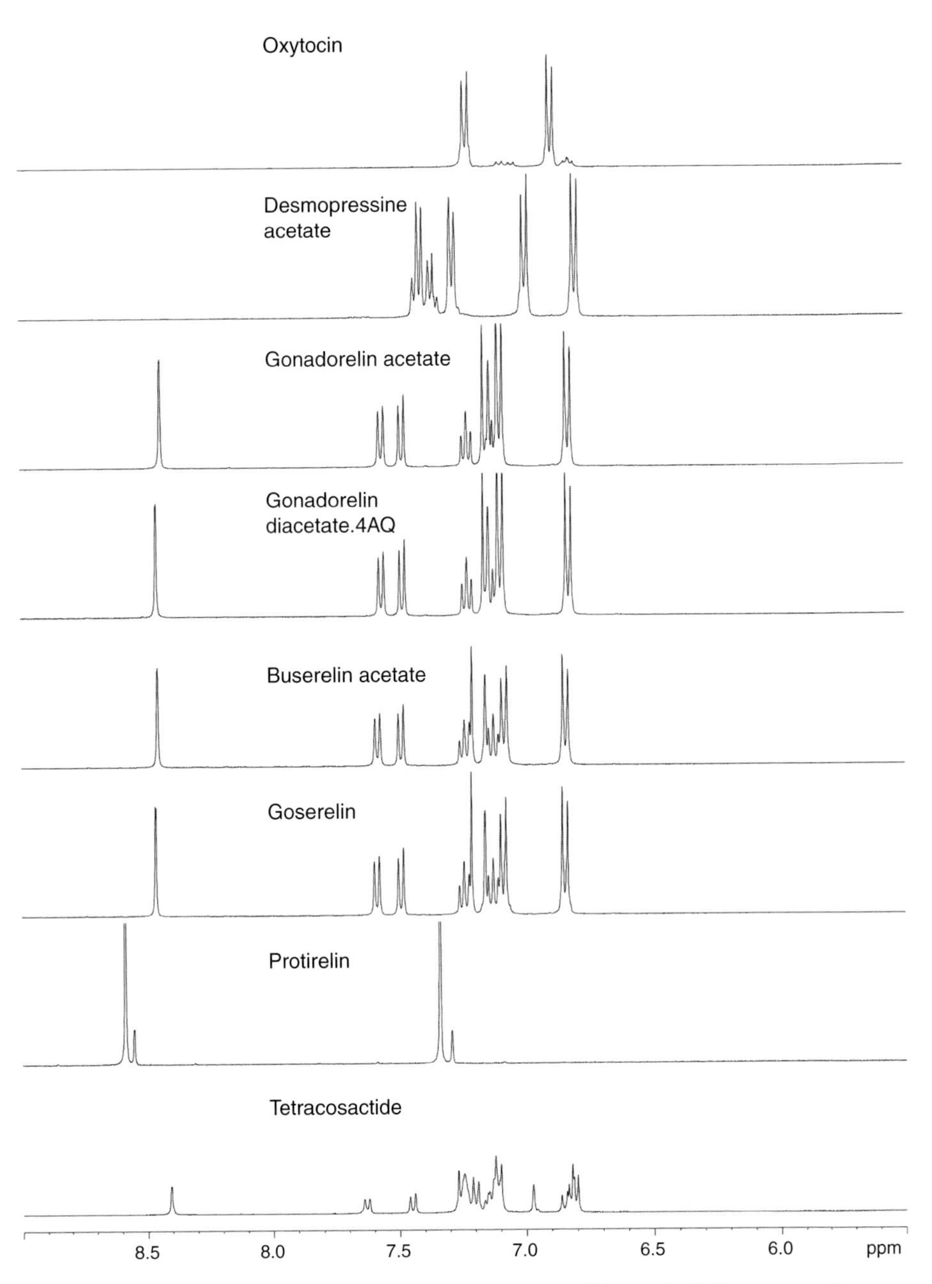

Figure 2 Aromatic region of the 400 MHz NMR spectra of the eight different peptides.

differentiate between gonadorelin acetate and gonadorelin diacetate as is demonstrated in Figure 3.

The NMR spectrum of a peptide depends on the amino acid sequence and to a lesser extent on the measuring conditions (e.g. solvent, pH, concentration,

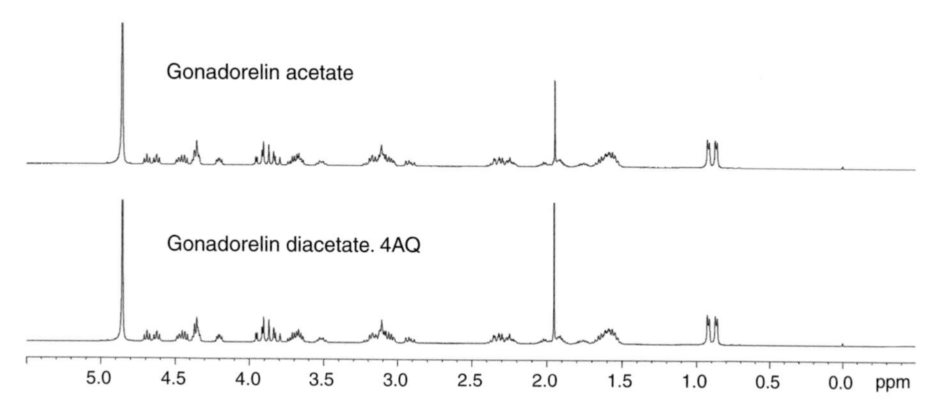

Figure 3 NMR spectra of gonadorelin acetate (top) and gonadorelin diacetate (bottom). The intensity of the acetate peak (singlet at about 1.9 ppm) allows distinction between the two peptides.

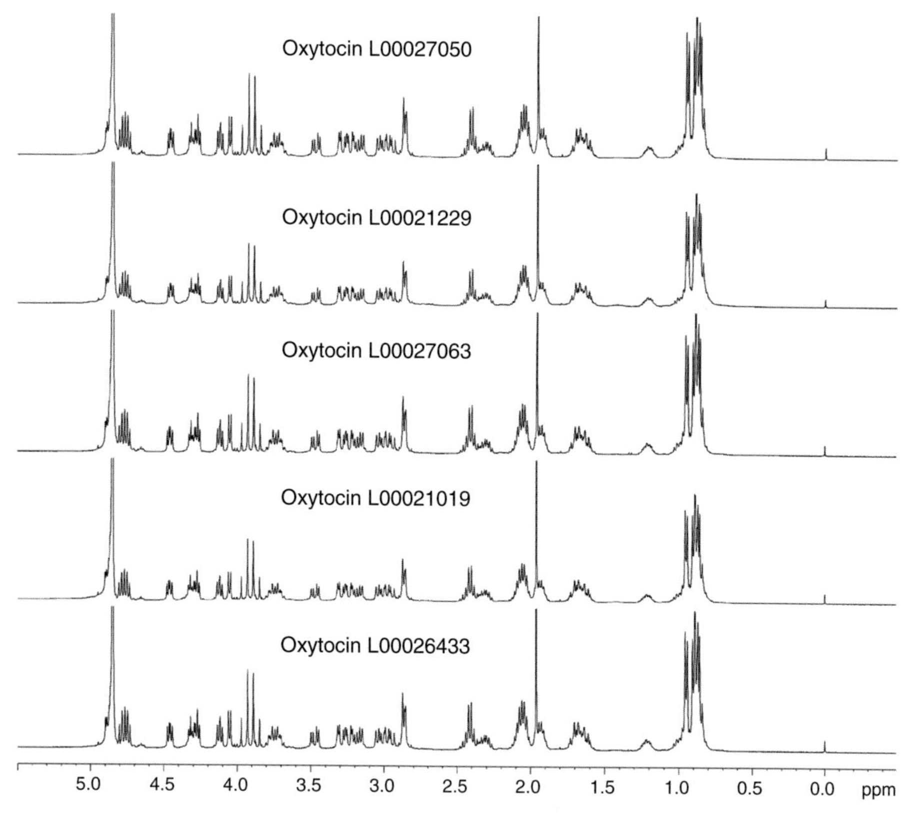

Figure 4 Aliphatic region of the 400 MHz NMR spectra of five oxytocin batches.

temperature, buffer composition, NMR acquisition and processing parameters). To allow a straightforward identification, the spectrum of the standard and the sample should be acquired under the same experimental conditions. A buffer is used to increase reproducibility of the spectra by reducing the pH effects on the chemical shifts and to prevent aggregation.

The use of NMR as an identity test for larger peptides may give problems as spectra become less well resolved for example see the tetracosatide spectrum. This is due to both the increasing number of resonances and the broader signals as a consequence of slower tumbling rates because of molecular size and, in some cases, aggregation. For each peptide, especially peptides consisting of approximately more than 15 amino acids, applicability of 1D proton NMR needs to be evaluated on an individual basis. The extension from 1D- to 2D-NMR may allow the use of NMR as an identification test to replace AAA for larger peptides. For example, it has been shown that human, porcine and bovine insulin (51 amino acids, differing 1–4 amino acids) can be readily distinguished by 2D-NMR.[8]

Whereas the proton NMR spectra of the eight peptides studied here are obviously different, different batches of the same peptide are identical as demonstrated in Figures 4 and 5, which respectively display the aliphatic and aromatic region of the of the proton NMR spectra of five different oxytocin batches.

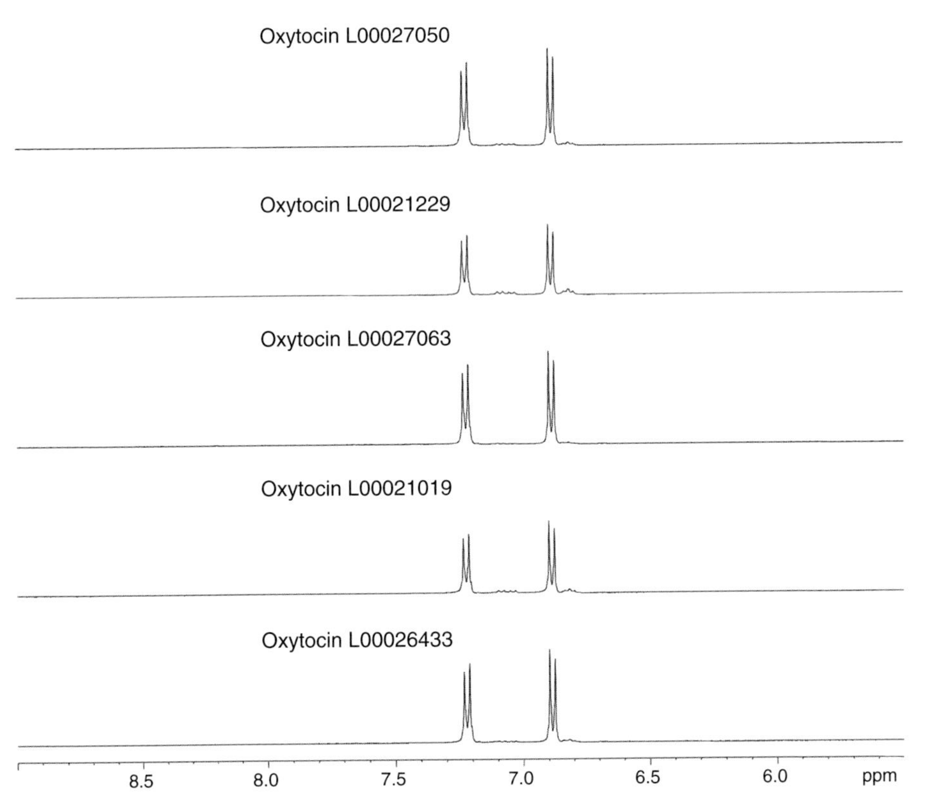

Figure 5 Aromatic region of the 400 MHz NMR spectra of five oxytocin batches.

The following procedure is proposed for recording and evaluation of the spectra:

- The spectrum of the standard and the sample should have been acquired under the same experimental conditions (concentration, solvent, NMR acquisition and processing parameter, etc.)
- For identification purposes, a visual comparison of the spectra is performed, similar to the way infrared spectra are compared.
- The comparison should be based on the position and relative intensities of the proton NMR signal of the peptides. No quantitation or calculations are performed nor required. This is routinely done for steroids within NV Organon.

The evaluation/interpretation is in line with the recent proposal for a revision of the general NMR monograph of the Ph. Eur.[9] "*qualitative analysis*: The principal use for qualitative NMR is as an identity test in which the ^{1}H or ^{13}C spectra of a test sample is compared to the spectrum of a reference sample or, less commonly, with a published reference spectrum. Spectra of reference and test samples should be acquired using the same procedure and operational conditions. The peaks in the 2 spectra, or characteristic regions of the spectra should correspond in position, intensity and multiplicity".

3. CONCLUSIONS

The eight peptides currently mentioned in the Ph. Eur. can be distinguished and identified by proton NMR spectra obtained at 400 MHz. Identification is straightforward and simple by comparison with the standard of the corresponding peptide. No calculations are performed. Even peptides differing very slightly can be readily distinguished. For each individual peptide, applicability needs to be evaluated, experimental methods may need adaptation and CRS availability should be assured. Table 2 summarizes the benefits and drawbacks of AAA and NMR. The proposed method was recently published in Pharmeuropa[10].

Table 2 A Comparison between the advantages and drawbacks of AAA and NMR

	Advantages	Drawbacks
AAA	Widely available; classical, accepted technique; also applicable for larger peptides/proteins and applicable to Gluc/Gal determination	Destruction of Asn and Gln; Cys, Met and Trp; hydrolysis conditions differ from one peptide to the other; time consuming and laborious
NMR	General, robust technique; no hydrolysis; data acquisition fast and interpretation straightforward; allows identification/quantitation of proton bearing organic counter ions	Less widely available*

*NMR analyses can however readily be outsourced to labs working under GMP for prices in the order of hundreds of euros, that is far less than the price for outsourcing of a classical AAA.

REFERENCES

1. Ph. Eur. 6.0 edited by European Directorate for Quality of Medicines, Strasbourg France, 2008.
2. K. Wüthrich, NMR in Biological Research: Peptides and Proteins, North Holland, 15–64, 1976.
3. K. Wüthrich, NMR of Proteins and Nucleic Acids (Baker Lecture Series), John Wiley and Sons, 13–19, 1986.
4. D.S. Wishart, B.D. Sykes, Methods Enzymol. 239 (1994) 363–392.
5. G. Maniara, K. Rajamoorthi, S. Rajan, G.W. Stockton, Anal. Chem. 70 (1998) 4921–4928.
6. F. Malz, H. Jancke, J. Pharm. Biomed. Anal. 38 (2005) 813–823.
7. The NMR spectra were recorded using a Bruker 400 MHz NMR spectrometer equipped with a equipped with a DPX400 QNP 5 mm probe. Concentration of the peptides was about 10 μmol/ml buffer (200 mM phosphate buffer pH 5). Samples were dissolved in this buffer, lyophilised, dissolved in D_2O and lyophilised once more to completely substitute all exchangeable protons by deuterium. Finally, the sample was dissolved in 1 ml of D_2O containing 0.5% v/v (2,2,3,3-(d4)-3-(trimethylsilyl) propionic acid sodium salt (TSP) as a chemical shift reference. Spectra were collected at ambient temperature (not controlled) acquiring 16 scans of 64 k (64×1024) data points. Sweep width was set at about 16 ppm. An exponential multiplication using a line broadening of 0.3 Hz was applied. Spectra were subsequently phased. Chemical shifts are reported relative to TSP referenced at 0 ppm.
8. C.W. Funke, J.R. Mellema, P. Salemink, G.N. Wagenaars, J. Pharm. Pharmacol. 40 (1988) 78–79.
9. Nuclear Magnetic Resonance Spectroscopy, Monograph 2.2.33. Pharmeuropa 19 (2007) 323–327.
10. E. Kellenbach, K. Sanders, G. Zomer, P.L.A. Overbeeke, Pharmeuropa 1 (2008) 1–8.

CHAPTER 6

Assessment of the Inhibitory Potency of Antibiotics by MRI

C. Faber

Contents

Abstract

Large-scale assessment of the inhibitory potency of antibiotic substances may provide a novel approach to identifying drug candidates. High-throughput screening requires parallelization of the detection method, which is limited with established procedures. Magnetic resonance imaging allows for identification of growth inhibition in bacterial cultures by measurement of the transverse relaxation time. The measurement may be extended to investigate 10,000 samples simultaneously, opening a new dimension in throughput compared with conventional methods.

Keywords: MRI, high-throughput, antimicrobials, MIC, bacterial cultures, transverse relaxation time

1. Introduction

Drug analysis includes the aspects of identifying novel drugs, assessing their affinity and specificity, characterizing their molecular structures, and testing their efficiency in vitro and in vivo. Nuclear magnetic resonance (NMR) spectroscopy is widely used in screening for novel drug molecules as well as in structural and binding studies. The high costs of NMR equipment call for acceleration strategies to increase throughput for large-scale drug screening in industrial applications. In particular, for efficiency tests, the required high-throughput has not been reached. Therefore, NMR is normally not applied in such assays. Magnetic resonance imaging (MRI) allows for in vivo investigations in animal models and human subjects. Although MRI is routinely used to monitor disease progression in patients, applications for evaluation of drugs are rare. Only few studies were reported that investigate infections by MRI,[1–5] and no systematic and direct evaluation of the efficiency of antibiotics by MRI has been reported. Recently, it was demonstrated that MRI can be used to assess the inhibitory potency of antibiotics in bacterial cultures.[6] The method allows for a high degree of parallelization, rendering it appealing for potential industrial use in large-scale tests. The background and the concept of this novel strategy are detailed in the following sections.

2. Current Status of NMR Spectroscopy and MRI

NMR spectroscopy is one of the most important tools for structural investigations of small molecules. Similarly, for proteins and nucleic acids, NMR has evolved into one of the two major techniques for elucidation of structure and dynamics of monomeric enzymes or large supramolecular complexes composed of a number of different molecules. NMR methodology is highly developed, and a large number of experiments are available to measure bond angles, relative orientations, local and global motions, and short, medium and even long range atomic distances. Despite this unique variety of obtainable structural information, the technique is mostly applied in academic research. Industrial applications are often limited to large pharmaceutical companies, which is certainly because of the high costs of the required equipment and to the very time-consuming process of data collection and analysis. Commercial use requires cost effectiveness, which is difficult to achieve if a single experiment takes hours or even days and alone provides only limited information. To improve the efficiency of NMR, a number of different strategies have been proposed. These aim at increasing the throughput either by employing standard experimental schemes in combination with newly developed instrumentation. Dedicated probes, for example, allow for data acquisition from multiple samples simultaneously or for measurements in a stopped-flow mode.[7] Alternatively, novel experiments have been devised detecting only distinct interactions, which are to be observed for example in screening tests for novel drug candidates. In particular, for the investigation of small molecules, such accelerated acquisition strategies find increasing use in pharmaceutical industry.[8]

In contrast to NMR spectroscopy, MRI is accepted as diagnostic tool in the healthcare systems of developed countries and can be applied commercially despite its still high costs per investigation. Since the first imaging implementation in the 1970s, MRI was quickly developed and today provides a large number of techniques to measure structural, functional, physiological, or metabolic parameters. Images can be acquired as fast as in a few tens of milliseconds or alternatively with spatial resolutions down to the micrometer range. Besides faster acquisition strategies, a variety of techniques is available that is able to translate functional or physiological parameters into image intensity and thus produce parameter maps with high spatial resolution. In addition, spectroscopic imaging (SI, also referred to as chemical shift imaging, CSI) techniques afford the full NMR spectroscopic information in a spatially resolved fashion. However, spatial resolution in SI is usually limited by low signal-to-noise ratio and considerable lower than in pure imaging applications. Both strategies SI and acquisition of parameter maps with MRI sequences represent potential approaches to parallelize NMR investigations.

3. High Throughput in NMR

To increase throughput in NMR measurements, more sensitive spectrometers may be used, enabling shorter data-acquisition times. Sensitivity can be enhanced for example by either using stronger magnetic fields or employing cryoprobes. Both involve very expensive equipment, which, in terms of cost efficiency, revokes the possible gain in throughput. Furthermore, the dead-time between measurements of two successive samples must be minimized. More economical strategies make use of automation robots that shorten delays because of sample changing, or use flow probes, in which different samples are measured in a stopped-flow manner. However, even with sensitivity-optimized equipment data collection time for each sample represents the major contribution to the total experimental time. Therefore, dramatic increases in throughput are only possible if several substances can be measured simultaneously.

There are two fundamentally different strategies to implement such simultaneous measurements. One approach applies to measurements that aim at detecting molecular interactions. In particular, in screening test for novel drug molecules, the interaction with a known target is probed. Observing only the influence on the target spectra or exploiting magnetization transfer effects allows adding several compounds to one sample. The measurement will detect only samples with active compounds, which then have to be investigated individually. With this strategy, several thousands of compounds can be tested per day using a single NMR spectrometer.[8,9]

For the direct observation of a compound true parallelization of NMR spectroscopy is required, for which two strategies have been proposed. First, dedicated probes are employed, which provide several detectors in the magnet and thus allow for acquisition of spectra from four or eight samples simultaneously.[10,11] In combination with design as flow probes, different samples can be measured without the usual delay for removing one sample from the magnet, bringing the next sample into the magnet and shimming the magnetic field. A major advantage of this strategy is the optimized sensitivity of the probe, since each rf-coil can be designed to reach the maximum filling factor. The high sensitivity of the probe allows for fast acquisition of the individual spectra. However, if low-concentration substances are to be detected, averaging is required and high-throughput rates cannot be achieved. Another drawback of this multi-coil strategy is the requirement of a spectrometer with multiple receive-channel capability, which is not available in standard NMR spectrometers and expensive to install.

In a second approach, several samples are placed in a single detector and spectra are acquired in a spatially resolved manner. This is possible, for example, with the use of SI techniques.[12] For an array of sample tubes, spatial encoding is required in two dimensions in the transverse plane inside the magnet. The resulting data set contains the NMR spectrum in the third dimension. Major draw back of this method is the long minimum measurement time. Increasing the number of samples requires increased resolution. Minimum measurement time scales linearly with spatial resolution in each dimension and, therefore, is directly proportional to the number of samples that are investigated simultaneously. Despite this limitation, an application for the measurement of the enantiomeric purity of samples has been reported, which reached a throughput of up to 5600 samples per day with 19 samples measured in parallel.[13]

4. Investigation of Bacterial Cultures by NMR Spectroscopy

Besides the classical application of NMR spectroscopy for the structural characterization of small molecules, the method is increasingly used in novel applications. These include for example the characterization of molecular interactions, quality control, screening for novel drug molecules and characterization of bacterial cultures. For chemotaxonomic identification of bacteria, one can take advantage of the fact that NMR spectra can be acquired from cultures in broth media with sufficiently high spectral resolution. The metabolic profile observed in the culture is characteristic for each species. Statistical classification strategies analysing the detected composition allowed distinguishing even closely related taxa.[14] Reliable identification of different species based on the NMR spectra of large numbers of isolates has been reported.[15] Metabolic profiles may also provide information on the metabolic state of the culture. However, bacterial cultures are usually investigated in a metabolically inactive state and typically not over the course of time, which would be desirable for assessment of the inhibitory potency of antimicrobial substances added to the culture. Time course measurements require more spectrometer time, and therefore, a high degree of parallelization must be achieved to reach high-throughput rates in such assays. Using the available strategies, as described above, allows only for limited throughput, rendering the method less interesting for industrial applications.

5. Assessment of the Inhibitory Potency of Antimicrobials

One aspect in analysis of novel drug molecules is the assessment of their activity. For antimicrobial agents this is the assessment of their growth inhibiting potency, which is usually characterized by the minimum inhibiting concentration

(MIC) of a drug molecule. Standard methods to determine the MIC are either the semi-quantitative disc diffusion or Kirby–Bauer method, which relates the MIC to the radius of a directly visible growth-inhibiting area in culture dishes. Alternatively, the quantitative dilution method is applied, which uses a series of sample vials containing broth cultures with decreasing concentrations of the antimicrobial substance to be probed. From the minimum concentration at which growth is inhibited, the MIC can be determined. However, using the dilution method, bacterial growth must be observed directly in the culture. A number of methods are available that allow monitoring the growth of a culture. The most commonly applied methods are based on the detection of light, typically in the visible range:

- Turbidometric methods are often used to monitor bacterial growth. Turbidity of the culture is determined by measuring light absorption (typically at 600 nm). According to the Lambert–Beer relation, the absorption depends on the concentration of the bacteria and can thus be directly related to the number of bacteria in the culture. The result is typically given as the optical density (OD), which is the conventional measure to quantify the growth state of a culture.
- Nephelometric methods measure light scattering in bacterial cultures, which also scales with the concentration and thus provides a measure of the growth state.
- Colorimetric methods detect colour changes of the medium, which occur because of the consumption of substrates in the medium or because of pH indicators added to the medium.
- Fluorometric methods measure changes in intensities of light that is emitted by fluorescent molecules, which change their concentration because of bacterial growth or consumption of the medium.
- Observation of bacterial growth without detecting light is possible with radiometric methods, which measure the amount of radioactive carbon dioxide being produced by the culture. Such methods require that a radioactive carbon source be supplied in the medium and are therefore potentially problematic because of legal restrictions. The great advantage of such methods is that they do not simply measure the cell number, but metabolic activity instead, and thus provide direct information on growth itself.
- Finally, bacterial growth may be monitored by measuring the change of conductivity of the medium, which occurs because of changes in the composition of the culture.

All these methods have originally been developed for probing single cultures at a time. However, in particular, conductivity measurements and the light-detecting methods are suitable for parallelization. Large arrays of cultures in deep well plates may be measured simultaneously (Figure 1). In particular, light-detecting methods can be implemented easily because cameras with megapixel resolution are available at low costs. For parallelization, the samples have to be placed between the light source and the camera. Detection is not possible perpendicular to the incident beam of light, which excludes nephelometric methods and imposes limitations on fluorometric methods. Measurement of the turbidity may be hampered by sedimentation of bacteria, which occurs at high concentrations in the culture. Colorimetric methods may be influenced by scattered light from neighboring wells but appears to be the

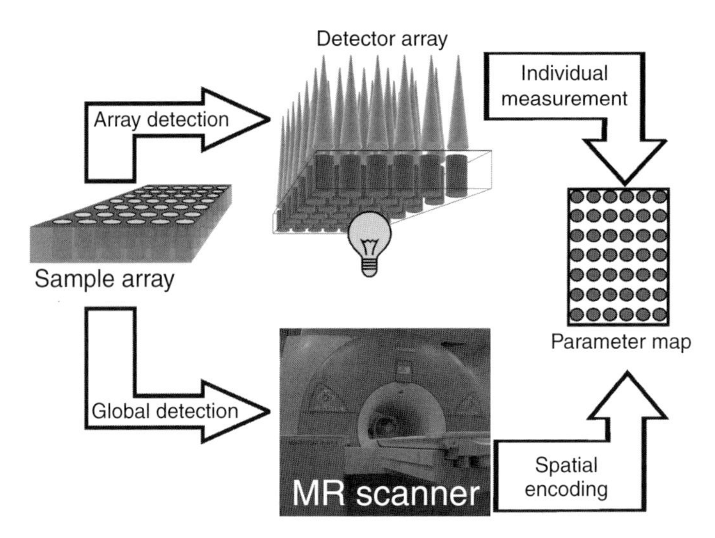

Figure 1 Parallelization strategies for test assays. Simultaneous measurement of multiple samples arranged in a sample array (left) can be achieved by two fundamentally different approaches. In particular, light-detecting methods use an array of detectors measuring each individual sample with a separate detector (top pathway). The result can be represented as a map of the probed parameter for each individual sample (right). MRI offers the possibility to perform one global measurement of the sample array, for example by using a whole body MR scanner (bottom pathway). The individual samples are resolved in a parameter (T_2) map by spatial encoding in the imaging experiment. Because only a single detector is needed, very high degrees of parallelization can be achieved.

method most straight forward to implement for a high degree of parallelization. However, any method requires a customized design to allow for high throughput.

6. Assessment of the Inhibitory Potency by MRI

A completely novel approach to parallelize bacterial growth assays is the use of MRI.[6] Because MRI is absolutely non-invasive, the measurement does not interfere with the cultures nor does it require separate excitation or detection for the individual samples (Figure 1). The number of samples that can be investigated simultaneously is only limited by the spatial resolution of the acquired image. MRI affords high spatial resolution, well below 1 mm, in a field of view of tens of centimetres. Given these geometry-constraints, several thousand samples can be imaged simultaneously. However, in order to correlate image intensity with bacterial growth and consequently with the effectiveness of antimicrobials, a suitable NMR accessible parameter, which is used as marker, has to be identified. In typical clinical MRI applications, a wealth of parameters such as the spin density, longitudinal and transverse relaxation times, magnetic field inhomogeneity, molecular diffusion, flow velocities or cross relaxation of magnetization are available and are widely applied for diagnostic purposes. For the investigation of bacterial cultures, the suitability of these parameters has to be established.

6.1. Spatial resolution and parallelization

Image intensity in MR images is normally measured in a number of different pixels, which are the smallest information-carrying units provided by the MRI measurement. To avoid errors because of fluctuations from noise in the images, data have to be averaged over a large number of pixels, providing moderate standard deviations of the mean value. Assuming that 100 pixels allow for sufficient averaging, ten by ten pixels have to be recorded for each of the samples under investigation. Feasible matrix sizes for MRI experiments are up to 2048 by 512 or 1024 by 1024 pixels. Both allow placing 10,000 ten-by-ten pixel units in the imaging matrix. If one further assumes sample dimensions of 3 by 3 mm, accordingly sample array would require dimension of 60 by 15 cm, or 30 cm^2. Both can be accommodated in clinical MRI systems without any modifications of the instrumentation. The decisive question regarding the feasibility of MRI experiments is often the achievable signal-to-noise, which again depends on the spatial resolution in an image. With the above assumption the required spatial resolution is 0.3 mm, which is easily achievable on standard clinical MRI systems. Because resolution is crucial only in two spatial dimensions, data can be acquired from a large slice along the third dimension. Using several centimetres, slice thickness grants high signal-to-noise for most imaging experiments in clinical MRI systems. Sample arrays are, therefore, best shaped as well plates that extend into the third dimension (Figure 1).

6.2. Marker of bacterial growth

Physical parameters that are typically measured by MRI and related to physiological or pathological events in humans or animals cannot be directly translated into experiments with bacterial cultures. Under reasonable growth conditions, bacteria are dilute in the medium. Therefore, the parameter spin density, which directly translates into water content, is not appropriate for cultures. Molecular diffusion, in contrast, appears at first glance to provide insight into cell concentrations in the medium. With increasing cell density, the mean free diffusion distance of molecules in the medium is expected to decrease. There is also an expected difference of diffusion between intra- and extracellular molecules, the ratio of which changes during growth of the culture. However, actual cell densities in growth experiments are too low to allow for differentiation between contributions from the different compartments. At an OD of 1, the intracellular volume for *Escherichia coli* is below 10^{-5} of the total sample volume. In a previous study, it was demonstrated that intra- and extracellular water could be separated with diffusion filters when very high cell densities were used.[16] The experiments further required very high magnetic field gradients of 0.75 T/m, which are only available in dedicated experimental MR systems. Molecular diffusion is, therefore, not suitable as marker for bacterial growth in an assessment that is easy to implement on commonly available MRI systems.

In MRI of humans and animals, the most valuable source of information are relaxation processes. These depend on the entirety of the physical and chemical parameters of the environment, which are never totally known in a living organism. Relaxation rate constants, therefore, remain at least partly phenomenological quantities. Although for humans and animals, numerous relations between relaxation time changes and physiological conditions are well established, this has to be completely reiterated for bacterial cultures. One difference is the presence of two distinct compartments, extracellular and intracellular, each of which represents a different environment and, therefore, is characterized by different rate constants of relaxation R_{extra} and R_{intra}, respectively. The experimentally observable value R_{culture} is given by both rate constants and the relative volumes V occupied by each compartment:

$$R_{\text{culture}} = V_{\text{extra}}\, R_{\text{extra}} + V_{\text{intra}}\, R_{\text{intra}}$$

Assuming that the rate constants do not change during culture growth, a change in relaxation times of the culture is expected merely from the changing relative volumes. However, even at maximum concentrations that are typically used in growth assays (up to OD 10), the relative volume occupied by the bacteria, representing the intracellular compartment, is less than 10^{-4}. On the contrary, differences in relaxation rates are less than factor 100 and, thus, the overall difference is too small to be detected in NMR relaxation time measurements.

Characterization of bacterial cultures by measuring relaxation times is, therefore, limited to observing changes in relaxation times of the medium, which represents the extracellular compartment. During growth, the bacteria consume nutrients that are supplied in the medium and in turn release metabolites. In particular, acidic compounds are produced by a number of bacteria, which lower the pH in the medium. The defined relation between pH and growth state of the culture is the basis for colorimetric growth assays that employ pH indicators. Of course, pH is not directly accessible by NMR, unless pH indicators showing defined pH-dependent chemical shift changes are used. On the contrary, pH has a major influence on chemical exchange processes, which provide an extremely efficient relaxation mechanism for transverse magnetization.[17] The impact of pH on relaxation is illustrated in Figure 2(a) for the case of a minimal medium

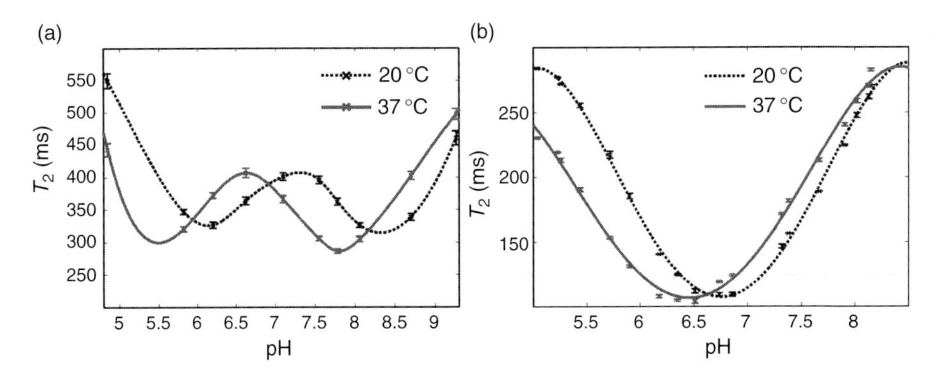

Figure 2 pH dependence of the transverse relaxation time T_2 in media for bacterial cultures. (a) Minimal medium with arginine (8 mM) as sole nitrogen source shows two minima of T_2, which correspond to the pK_a values of the side chain amino functions. (b) LB medium shows one broad minimum, which corresponds to the exchanging amide proton in the peptide backbone. Temperature shifts confirm chemical exchange as source of T_2 variations. Data were obtained at 17.6 T (from Hoffmann[18]).

containing arginine as sole nitrogen source. Transverse relaxation time T_2 shows pronounced changes over moderate pH changes with two distinct minima. The minima correspond to the exchanging side change functions of arginine. A shift of the minima with temperature indicates that chemical exchange is responsible for the variation of the relaxation time. Arginine-based media may thus serve as intrinsic pH sensors. In growth assays, the pH effect is overlaid by a concentration effect, which might cause wrong interpretations. However, this problem can be easily avoided by either calibrating a relaxation time–pH standard that takes concentration changes into account. Alternatively, the assay may be designed in a way to keep nutrient concentrations in the medium constant.

Most assays use complex instead of minimal media such as the lysogeny broth (LB) medium. Complex media typically contain a mix of all amino acids and further nutrients. Chemical exchange that is relevant for relaxation time changes is expected not only at a few distinct pH values but rather over a broader range of values. Experimentally, only one broader but still distinct minimum of T_2 is observed for LB medium [Figure 2(b)]. The pH value corresponds to the pK_a of the amide proton, common in all amino acids. The pronounced change in T_2 emphasises that the medium itself provides an intrinsic pH sensor, which may be used to characterize the growth state of a culture by measuring transverse relaxation times.

A full theoretical description for the observed relaxation time changes is not available. Although formalisms exist to treat N-site exchange processes[19] and even the starting conditions, which is the composition and concentration of the medium, may be determinable, the exact boundary conditions, which are concentration changes during growth, cannot be determined completely. The approach, therefore, relies on mainly phenomenological relations.

6.3. NMR growth curves and determination of the MIC

To characterize bacterial growth by measuring transverse relaxation time T_2, it is essential to establish the relation between T_2 and OD or pH of the culture. The relation depends critically on the conditions such as the strain of bacteria and the exact composition of the medium. For a given strain of bacteria, relaxation time, pH, and OD have to be measured during culture growth as shown in Figure 3 for *E. coli* in LB medium. The increase in OD is accompanied by a decrease in pH, which results in an initial decrease followed by a pronounced rise of T_2. After the exponential phase of growth, the pH stabilizes and also the relaxation curve rises more slowly. T_2 provides information on the growth state of the culture and can be used to distinguish growing cultures from static ones. T_2 maps of a sample array, acquired at defined time points, afford easy identification of bacterial growth (Figure 4). Changes in T_2 crucially depend on the strain under investigation and on the composition of the medium, in which the assay is performed. It is therefore a prerequisite to establish the T_2-growth relationship for the exact environment of the assay.

Figure 4 illustrates the observable difference in T_2 for a growing culture and a culture that was inhibited by presence of an antibiotic. Although the growing culture showed increasing T_2 values, the T_2 values in the inhibited culture

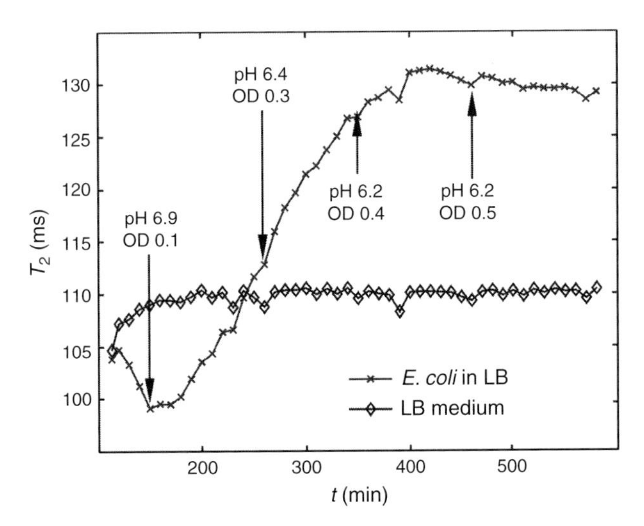

Figure 3 T_2-growth curve for *E. coli* in LB medium. Measurement at 17.6 T shows T_2 changes with bacterial growth, because of pH variations. Typical values for OD and pH of the culture are indicated.

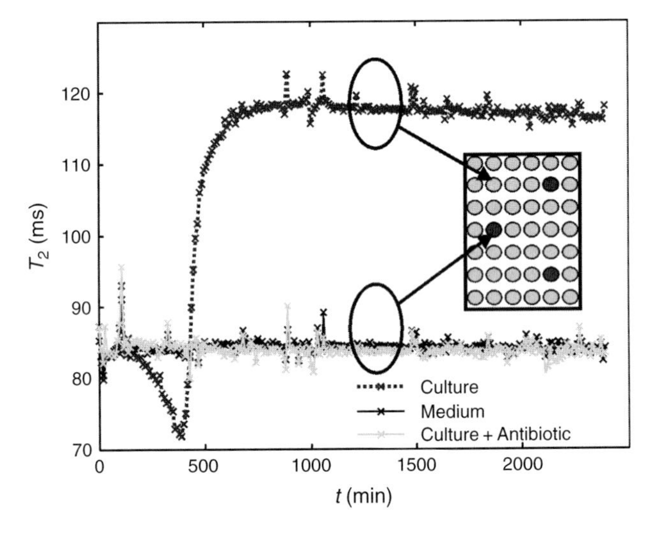

Figure 4 Generation of a parameter map in a parallel assay for inhibitory potency of antimicrobial substances. Cultures with normal growth (dashed line) can be distinguished from growth-inhibited cultures (solid line). A T_2 map of the sample array (inset), acquired at a defined time point (circles), resolves all individual samples and provides information on culture growth. The measured T_2 value discriminates growing (light gray) from inhibited (dark gray) cultures. Presented data were taken from Hoerr et al.[6] for *Streptococcus vestibularis* in BHI-medium measured at 17.6 T.

remained identical to the control medium without bacteria. A T_2 parameter map calculated from a measurement at a time point as indicated in Figure 4 discriminates between growing and growth-inhibited cultures in a sample array. From this information one can calculate the MIC of an antibiotic substance, which is the common measure for its potency. To this end, the conventional dilution method is applied and MRI used as detection method. The antibiotic to be analysed is added to the culture with decreasing concentration, according to a geometric dilution series. The lowest concentration, for which growth inhibition is observed defines the MIC, according to the DIN 58940-8 guideline. The great advantage of MRI is the potential for a very high degree of parallelization. As argued above, 10,000 samples can be measured simultaneously making throughput rates of up to 40,000 samples per hour possible. This clearly exceeds the efficiency of any established method for assessment of the inhibitory potency of antimicrobial substances. The higher throughput rates may be exploited differently. Either larger numbers of substances may be assessed simultaneously or alternatively the precision of the method may be increased by choosing smaller concentration steps or a larger number of control experiments.

Until today, only a proof of feasibility for this novel method has been given at the high magnetic field strength of 17.6 T.[6] Because of the small size of the magnet bore, only up to 20 samples were investigated in parallel. However, the method can be easily applied at lower field strength, for example at 1.5 T. It will be required to establish the relation of relaxation time and OD or pH for the strains to be tested at the lower field strength. With larger magnet bore size, the method will then offer extremely high throughput rates, which allow establishing a novel strategy in search for potential drug candidates. Large libraries can be directly screened for their influence on bacterial growth. Thus, leads for the development of novel drugs may be identified based on their antimicrobial action without knowledge of their exact mode of action and molecular interactions.

ACKNOWLEDGMENT

The author thanks Verena Hörr and Kerstin Hoffmann for providing additional data.

REFERENCES

1. P. Marzola, E. Nicolato, E. Di Modugno, C.P.,A. Lanzoni, C.H. Ladel, A. Sbarbati, Magn. Reson. Mater. Phys. 9 (1999) 21–28.
2. A.H. Kaim, T. Wischer, T. O'Reilly, G. Jundt, J. Frohlich, G.K. von Schulthess, P.R. Allegrini, Radiology 225 (2002) 808–814.
3. A.H. Kaim, G. Jundt, T. Wischer, T. O'Reilly, J. Frohlich, G.K. von Schulthess, P.R. Allegrini, Radiology 227 (2003) 169–174.
4. A.M. Lutz, D. Weishaupt, E. Persohn, K. Goepfert, J. Froehlich, B. Sasse, J. Gottschalk, B. Marincek, A.H. Kaim, Radiology 234 (2005) 765–775.
5. P. Marzola, A. Lanzoni, E. Nicolato, V. Di Modugno, P. Cristofori, F. Osculati, A. Sbarbati, J. Magn. Reson. Imaging 22 (2005) 170–174.
6. V. Hoerr, K. Hoffmann, C. Schollmayer, U. Holzgrabe, A. Haase, P. Jakob, C. Faber, Magn. Reson. Mater. Phys. 19 (2006) 247–255.

7. M.A. Macnaughtan, T. Hou, J. Xu, D. Raftery, Anal. Chem. 75 (2003) 5116–5123.
8. T. Diercks, M. Coles, H. Kessler, Curr. Opin. Chem. Biol. 5 (2001) 285–291.
9. P.J. Hajduk, R.P. Meadows, S.W. Fesik, Q. Rev. Biophys. 32 (1999) 211–240.
10. Y. Li, A.M. Wolters, P.V. Malawey, J.V. Sweedler, A.G. Webb, Anal. Chem. 71 (1999) 4815–4820.
11. H. Wang, L. Ciobanu, A.S. Edison, A.G. Webb, J. Magn. Reson. 170 (2004) 206–212.
12. A. Ross, G. Schlotterbeck, H. Senn, M. von Kienlin, Angew. Chem. Int. Ed. 40 (2001) 3243–3244.
13. M.T. Reetz, P. Tielmann, A. Eipper, A. Ross, G. Schlotterbeck, Chem. Commun. 12 (2004) 1366–1367.
14. U. Himmelreich, R.L. Somorjai, B. Dolenko, H.M. Daniel, T.C. Sorrell, FEMS Microbiol. Lett. 251 (2005) 327–332.
15. R. Bourne, U. Himmelreich, A. Sharma, C. Mountford, T. Sorrell, J. Clin. Microbiol. 39 (2001) 2916–2923.
16. K. Potter, R.L. Kleinberg, F.J. Brockman, E.W. McFarland, J. Magn. Reson. B 113 (1996) 9–15.
17. A.D. Bain, Prog. Nucl. Magn. Reson. Spectrosc. 43 (2003) 63–103.
18. K. Hoffmann, Charakterisierung des Wachstums von *Escherichia coli* mit NMR-Bildgebung, Thesis, University of Wuerzburg, Wuerzburg, 2005.
19. O. Trott, A.G. Palmer, J. Magn. Reson. 170 (2004) 104–112.

CHAPTER 7

Hypernation and Concatenation: Multiple On-Line Spectroscopic Analysis for Drug and Natural Product Characterisation

I.D. Wilson

Contents

Abstract

Linking UV, IR, MS, and nuclear magnetic resonance (NMR) spectrometers in series (concatenation) enables serial flow injection analysis to be performed resulting in comprehensive spectral data from a single injection of a pure analyte. Adding column liquid chromatography to the system (hypernation) to provide a high-resolution separation step allows comprehensive spectroscopy on the components of complex mixtures. Examples of concatenation and hypernation with NMR spectroscopy are discussed, and the practical problems and limitations of such systems are considered.

Keywords: flow injection analysis, liquid chromatography, on-line NMR spectroscopy, on-line mass spectrometry, complex mixture analysis, hypernation, concatenation

1. Introduction

The determination of structure, identity and purity is central to many fields associated with pharmaceutical research and development, and there is a continuing demand for this type of work to be performed quickly, efficiently and with the minimum amount of material. Automation and increased instrumental sensitivity are essential if these analytical aims are to be achieved, and it is fortunate that enormous technical advances in spectroscopic instrumentation have greatly reduced the amount of compound required in order to obtain diagnostic spectra. Another significant advance has been the development of liquid flow cells/interfaces for the spectroscopic and spectrometric techniques needed to enable structure determination [mass spectrometry (MS), UV/Vis spectroscopy, infra red (IR) spectroscopy, nuclear magnetic resonance (NMR) spectroscopy, circular dichrosim (CD), etc.]. The availability of such flow cells allows the routine coupling of these spectroscopies either to flow injection analysis (FIA) systems or to separations through for example liquid chromatography (LC), capillary electrochromatography, supercritical fluid chromatography (SFC) or capillary electrophoresis (CE) (see Ref. 1 and references therein as well as Refs. 2–5 for more specific examples)[1–5]. Such systems have the potential to provide relatively high-throughput analysis, and it is a logical extension of the use of these spectroscopic techniques individually to using them in combination to obtain all the information required in a single analysis thereby enabling unequivocal compound identification. As a concept, this is by no means novel having been advocated as early as 1980 by Tomas Hirschfeld.[6] Clearly, compound identification often requires the combination of diverse data from several spectroscopic techniques. So, whilst MS provides information on mass, atomic composition, and structure through fragmentation patterns, it often struggles to distinguish between positional isomers. This task can often be performed easily using NMR spectroscopy, but a definitive structure may still not be possible in the absence of MS- or IR-derived data as nitro, nitrile, thiol, phosphate or sulphate groups, etc., which do not provide an NMR-detectable signal, may be present. Information on other factors such as chirality may require further spectroscopic analysis by circular dichroism. In situations where spectroscopic information from more that one technique is required, a variety of approaches can be used, depending upon whether the sample under investigation is presented as a single pure compound or a complex mixture. A conventional approach to this type of problem would be "one at a time" analysis using the various spectroscopic techniques in turn until all the required information is obtained. This has the advantages of relative simplicity, and the analysis can be optimised for each technique individually. However, such a linear approach can be time consuming and arguably fails to exploit the current capabilities of the available instrumentation. An alternative to the conventional approach is to take advantage of the advances in flow cell/interface technology mentioned above and link the required spectroscopic techniques together in a single system. Such a flow-based analysis can be based on either FIA for pure compounds or a chromatographic separation for more complex samples. For chromatographic systems, it is a common practice to term the linking of for example LC with MS or NMR as "hyphenation".[6] To distinguish between single and multiple spectroscopic detectors linked into one chromatographic system, we have proposed the term "hypernation" (meaning more than one).[7,8] Whilst the same terminology might also be used for simple FIA systems as a separation is not involved, it might be appropriate to use "concatenation" as a description to distinguish the two. Concatenated or hypernated, the use of a single integrated system of connected spectrometers to derive the required structural information clearly represents an efficient use of time, and is intellectually attractive, but is not without obvious problems. These are associated with the cost of such an experimental set up, system control, data management and, for hypernated (separation-based) systems the, by no means trivial, problem of selecting a single eluent that is both suitable for the separation of the analytes and compatible with all the spectrometers. The cost argument is, however, more apparent than real, given that there is no absolute requirement to keep the system permanently in its hypernated/concatenated form. Thus, provided that the instruments are co-located, such systems can be rapidly configured/reconfigured as required. Potentially differences in sensitivity between for example NMR spectroscopy and MS might be expected to result in operational difficulties, but in practice, these have not proved to be a problem, as illustrated below. There is, when discussing issues of sensitivity in relation to spectroscopic methods, also the need to clearly distinguish between simple detection of an analyte (perhaps based on a single strong spectral feature) and having enough spectral information to obtain unequivocal structural data. The idea of the limit of detection (LOD) is well recognised and perhaps it is also useful when thinking about these matters to consider the minimum amount of compound that represents a reasonable *limit of spectroscopic identification* (LOSI).

2. CONCATENATED FIA SYSTEMS

If the aim is to obtain spectroscopic information on pure compounds (or to establish purity), a chromatographic separation may be unnecessary and simple FIA can serve perfectly well. The construction of a concatenated system, with spectrometers arranged in series, or in parallel, represents a simple and efficient means of acquiring this type of data, and several prototype systems have been described that produced UV, IR, ^{1}H NMR, and MS data on simple compound libraries.[9,10] In the first example of a concatenated system,[9] a mixture of acidic and basic pharmaceuticals were subjected to FIA with the first spectrometer in line providing UV spectra through a diode array detector, followed by FT-IR using a 400-μl flow cell for attenuated total reflectance (ATR) detection. Following the IR flow cell, a fixed wavelength UV detector was placed in line, which served both to allow an evaluation of the degree of band broadening introduced by the IR flow cell to be made and also to calibrate the time taken for the analyte to reach the flow cell of the NMR spectrometer. Immediately after this UV detector, the flow was split with approximately 95% directed to the flow probe (4 mm, 120 μl cell volume) of a 500 MHz NMR spectrometer and the remainder to a time-of-flight (ToF) mass spectrometer. This instrumental system employed manual injection of samples, and these were delivered to the spectrometers at a flow rate of 1 ml/min. In practice, this instrumental set-up was capable of obtaining good spectra for all four spectroscopies from ~50 μg of compound on flow. Whilst not large in terms of acquiring ^{1}H NMR spectra, this amount of material was however sufficiently high as to overload the UV-DAD to the extent that, in practice, the UV spectra were obtained from the trailing edge of the "peak" as it exited the flow cell. In general, the limiting spectrometers in this set up, with regard to sensitivity, were the FT-IR and NMR instruments, which in this case showed rather similar sensitivity. As well as on flow spectroscopy, 1D NMR spectra were also obtain in stopped-flow mode, with good data obtained with 64 scans on 50 μg of antipyrine. Obviously, using stopped-flow FIA would have allowed access to the normal NMR suite of 2D experiments as well, at the expense of the efficient use of the other spectrometers that would be left idle whilst this was done. The use of ToF MS provided accurate mass data, which allowed atomic composition to be estimated, thereby providing further useful structural data. One point to note with respect to the ToF MS used here was that, despite the undoubted sensitivity of MS, this particular instrument could only be used in either positive or negative electrospray ionisation (ESI) mode within one run. This had the practical consequence that when a mixture of acidic and basic drugs was used as a model, analytes with negative ESI for the detection the basic compounds went undetected and the MS analysis had to be repeated for them in positive ESI. The use of such a system would therefore either require careful scheduling of acidic and basic analytes in different runs to ensure that MS data were obtained on all analytes or the use of an instrument capable of rapid switching between positive and negative ionisation. Clearly, such a limitation does not apply where the MS can be rapidly switched between positive and negative ESI within the same run. A slightly more advanced system, capable of unattended automated operation through the use of an autosampler, was described in a subsequent publication.[10] Whilst aimed at obtaining essentially the same spectroscopic data as the first prototype described above, the second FIA system differed from the first in a number of details with respect to layout. Thus, the IR spectrometer was placed before the UV-DAD, and the split to the mass spectrometer (a single quadrupole instrument in this case) was immediately after the autosampler. This resulted in 60% of the flow being sent down the branch containing the IR-UV-NMR spectrometers with the remaining 40% going to the MS (total flow was 1 ml/min). In addition, the layout was simplified by leaving out the fixed wavelength UV detector. The amounts of material required for the production of all the required spectra in this second prototype system were in the region of 150–200 μg/analyte, although with further optimisation (e.g. to reduce band broadening) this quantity could no doubt have been reduced. In particular, as these experiments were not performed using actively shielded superconducting magnets, this system (and indeed those described below) suffered from the need to place the pumps, MS, IR, and UV spectrometers outside the stray magnetic field of the NMR spectrometer. The use of actively shielded magnets would have enabled the lengths of tubing required to make the connections between the component parts of this "total organic analysis device" (TOAD) to be significantly reduced and would have undoubtedly resulted in much less band broadening (thereby giving a consequent increase in sample concentration and thus instrument sensitivity). A typical set of spectra for the β-blocker propranolol are shown in Figure 1 whilst the on-flow pseudo 2D-^{1}H NMR spectra obtained for the model library on this system are shown in Figure 2. With a modest amount of optimisation, even this fairly primitive arrangement would be capable of the unattended spectroscopic characterisation of several hundred compounds per day.

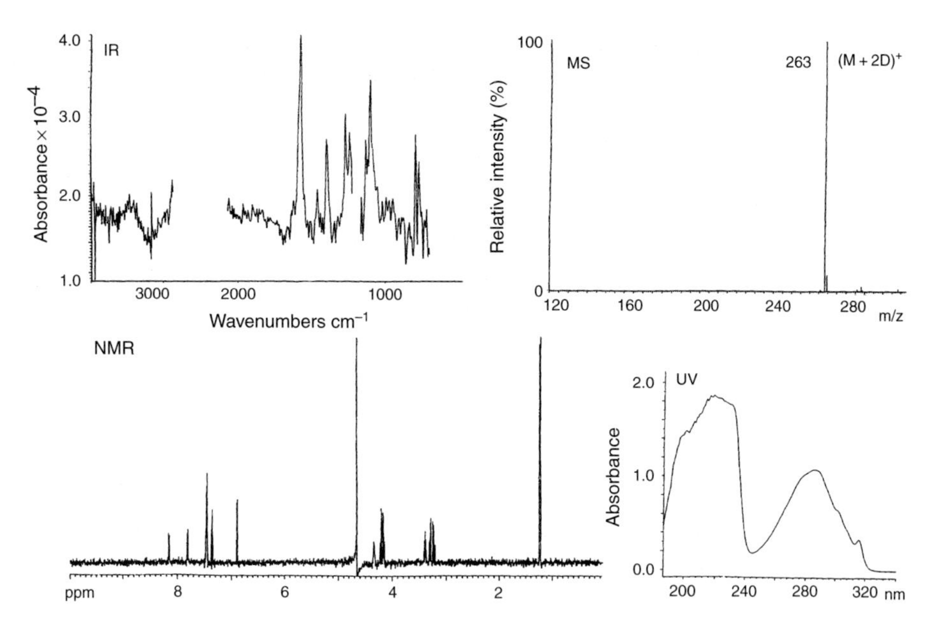

Figure 1 UV, IR, NMR and MS spectra obtained for propranolol (215 μg) using the UV–IR–NMR–MS–FIA system.[10]

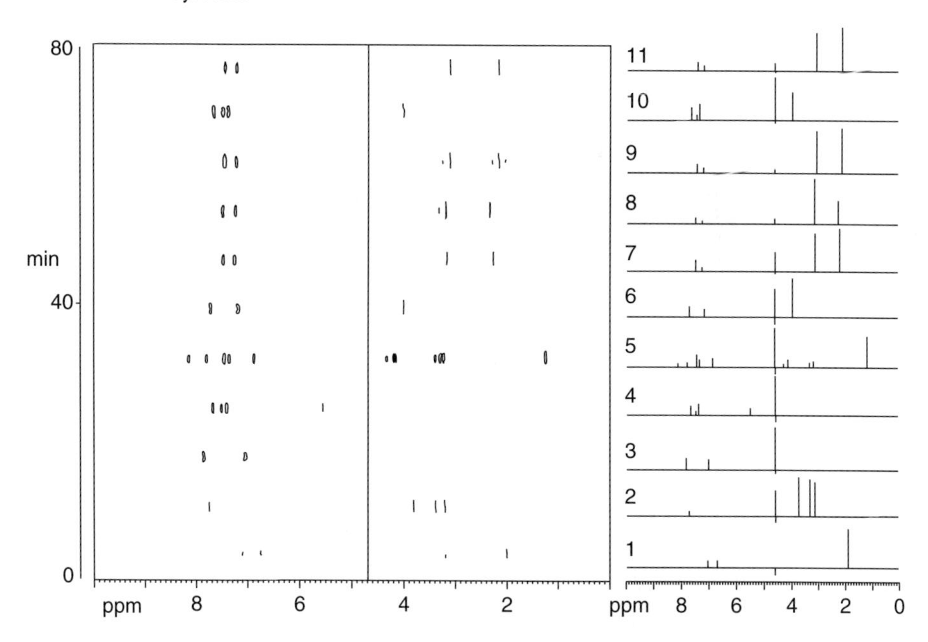

Figure 2 Pseudo 2D ^{1}H NMR spectral data for the model library of analytes obtained using the UV–IR–NMR–MS–FIA system described in ref. 10. Key: (1) acetaminophen (140 μg), (2) caffeine (165 μg), (3) 4-aminobenzoic acid (185 μg), (4) α-hydroxyhippuric acid (200 μg), (5) propranolol (215 μg), (6) 4- aminohippuric acid (270 μg), (7) 4-aminoantipyrine (295 μg), (8)4-dimethylaminoantipyrine (545 μg) (9), antipyrine (840 μg), (10) hippuric acid (365 μg) and (11) antipyrine (210 μg).[10]

3. Hybrid Concatenated/Hypernated FIA–MS/LC–NMR Systems

The instrument manufacturer Bruker has recently commercialised a system (the so-called Metabolic Profiler™) designed to perform simultaneous flow injection NMR spectroscopy and LC–MS on biological samples such as urine from neonates or fruit juices. In this system, the link between the two spectrometers is through the autosampler, where automatically prepared samples are "cloned" and then presented in two 96-well plates for analysis by the two systems individually using a double injection port. The system, which comprises a 600-MHz NMR spectrometer and a time-of-flight mass spectrometer, is shown in Figure 3. Sample preparation, for e.g. urine, which is performed as the samples are "cloned" involves adding 10% (v/v), pH 7 buffer to the aliquots destined for ^{1}H NMR spectroscopy or mixing 1:10 with water for LC–MS. Chromatography for the LC–MS is through a short (50 mm × 2.1 mm i.d. column) using an acidic (0.2% formic acid) reversed-phase gradient with acetonitrile as the organic modifier. The Metabolic Profiler™ provides automatic acquisition of NMR and LC–MS data from a large number of samples in a fully automated system followed by integration of the data. The capabilities of the system can be further extended by incorporating an on-line solid-phase extraction (SPE) step for the collection and concentration of peaks of particular interest seen through the LC–MS part of the analysis. Such an on-line SPE enables the analyte(s) to be collected for subsequent ^{1}H NMR, or indeed any other, spectroscopy (see below).

Figure 3 The metabolic profiler LC–MS and NMR system from Bruker.

4. Hypernated LC–UV–NMR–MS Combinations

Hypernation offers the advantage over simple concatenation that complex, multicomponent, samples can be analysed with (always assuming a competent and efficient separation) essentially pure components presented to the spectrometers. However, concatenated and hypernated systems are not mutually incompatible and can readily be inter-converted, one to the other, by the simple expedient of introducing or removing chromatography. Indeed, the first step in constructing a hypernated set-up should be to make the concatenated system first to check out the flow linkages between the spectrometers, and only when these have been optimised should chromatography be attempted. The most readily achieved hypernation of NMR with another spectroscopy is the addition of an LC separation and a UV/Vis photodiode array detector (DAD), as most chromatographic eluents that are compatible with NMR are also UV/Vis-transparent, enabling good-quality UV spectra to be obtained. Indeed, commercially available LC–NMR systems often come with such UV–DAD detectors as a standard accessory. This combination is reliable, easy to operate and available with relatively integrated data-handling possibilities. However, whilst easy to implement, it can be argued that addition of a capability for obtaining UV spectra, although useful, adds relatively little to the structure determination potential of the NMR technique. A more useful approach to structure determination results from the combination of LC with both MS and NMR spectroscopy (whilst retaining the UV DAD). However, the linking of NMR and MS to a single LC separation, if it is to be performed on-line with LC, does require compromises on the type of eluents that can be used for the separation as these must be compatible with both spectrometers. Thus, buffers used to control pH need to be volatile for the MS, excluding inorganic buffers such as phosphate, and ideally should not contribute interferences to the NMR spectrum. An acidic modifier suitable for use in chromatography in combination with NMR spectroscopy is trifluoroacetic acid (TFA), which, as it has no protons, does not result in interferences in the resulting ^{1}H NMR spectra. Indeed, in studies on paracetamol metabolites from urine, or propranolol, as model analytes showed that 0.1% (v/v) of TFA could be tolerated in HPLC–MS–NMR for this limited range of analytes (when these were present at high concentration).[11] However, when used for analytes such as ibuprofen and its metabolites, the suppression of ionisation was complete such that no MS data could be collected, even when the analytes were present at high concentrations. As a compromise, formic acid is generally suitable for LC–NMR–MS applications as it causes less problems for MS and with respect to NMR spectroscopy the single proton of formic acid, which has a sharp, readily suppressable ^{1}H NMR singlet near $\delta = 8.5$, gives minimal interference in the resulting NMR spectra. It also causes no problems for the acquisition of UV spectra.

Similarly, the organic modifiers used in reversed-phase separations must also be carefully considered. These are best if the chromatographic eluents are based on a single, simple, organic solvent such as acetonitrile or methanol (rather than more complex mixtures of several solvents, or the use of e.g. tetrahydrofuran) so that they contribute the minimum interference to the resulting NMR spectra. Acetonitrile and methanol are both good choices as they give rise only to a singlet resonance in the ^{1}H NMR spectrum that can be readily suppressed easily. They are also available in a deuterated form relatively cheaply further simplifying their use in LC–NMR applications. There are also the ^{13}C satellite peaks from these solvents to be considered for these solvents resulting from the one-bond ^{1}H–^{13}C spin couplings from the 1.1% of molecules that contain the ^{13}C isotope in the methyl carbon. These resonances remain following the suppression of the main peak and may cause problems of peak overlap with signals from trace analytes. These signals provide another reason for using fully deuterated solvents although other technical solutions are also available. Similarly, where gradient elution is performed, the NMR resonance frequency of solvents such as acetonitrile changes with solvent composition during the run and the solvent suppression frequency must be corrected for this, although again there are a variety of technical solutions for this, and these problems are no greater for hypernated systems than routine LC–NMR applications. Perhaps the final difficulty that needs to be recognised when MS and ^{1}H NMR are used with deuterated solvents is the fact that any exchangeable protons present on the analytes are likely to be replaced by deuterium atoms, leading to higher than expected masses. This can be corrected for by mixing the eluent to the mass spectrometer with protonated solvents in order to re-exchange these deuteriums for protons to obtain the correct masses. However, it is possible to use this phenomenon to advantage by deliberately performing MS on the peaks of interest with both deuterated and non-deuterated solvents as chromatographic eluents in order to determine the number of exchangeable protons as part of structure determination.

However, whilst it is clearly important to get these details correct, the difficulties surrounding solvent composition and instrument compatibility are more apparent than real, as demonstrated by the numerous published examples of HPLC–NMR–MS hypernations (reviewed in Refs. 12–14) used for applications as diverse as drug metabolite characterisation, pharmaceuticals and natural products (e.g. Refs. 15–46). Usually NMR and MS spectroscopy are linked in parallel with, because of the differences in sensitivity between the two, the bulk of the flow directed to the NMR spectrometer and a smaller proportion (usually ~5%) to the MS although the alternative approach with MS located in line with but after the NMR flow probe, has also been exemplified.[15]

There are a numerous examples of the use of LC–NMR–MS in drug metabolism studies, both in humans and in animals, that illustrate the practice of the technique. Thus, in the case of the human metabolism of the non-steroidal inflammatory drug ibuprofen, solid phase extracts of urine were examined by reversed-phase gradient HPLC with both NMR and MS detection (in parallel).[16] The drug was subject to extensive metabolism, giving rise to a wide range of hydroxylated, side-chain oxidised and glucuronidated products (including ibuprofen glucuronide). The resulting HPLC–NMR–MS data were able to confirm the identities of a wide range of both major and minor metabolites as well as detecting and identifying a number of additional minor components.[16] An interesting feature of this study, which nicely illustrates the complementary nature of the two techniques was that the poor ESI properties of some of these ibuprofen metabolites made their detection difficult by MS. This despite the presence of high concentrations of these components in the LC-eluent that readily allowed ^{1}H NMR spectra to be obtained.

In studies of complex, multi-component samples such as urine for drug metabolites, one of the difficulties is the detection of the analytes of interest hidden within the large number of un-related endogenous components. Here, a range of strategies can be adopted. Classically, in drug metabolism work, the use of radiolabels is employed in order to provide a specific tracer, allowing the facile detection of compound-related material. An example of the use of such a strategy, where an on-flow radioactivity detector was used to monitor the chromatographic eluent, is provided by the LC–NMR–MS study of the metabolism of the β-blocker practolol in the rat.[17] By using a [^{14}C]-radiolabel, the direct detection of the drug metabolites was possible enabling the investigators to perform stopped-flow ^{1}H NMR only on components of interest (in this case unchanged practolol, a ring-hydroxylated metabolite and the glucuronide of the ring- hydroxylated metabolite were unequivocally identified). In this investigation, the metabolic fate of an N-acetyl group that was present as a substituent on the drug was also of interest because of its potential to be metabolically labile. To investigate this, the acetyl group was stable isotope labelled with [^{13}C] in the N-acetyl group. Because of the spin–spin coupling between the ^{13}C and the ^{1}H on the CH_3 of the acetyl methyl, a characteristic doublet in the ^{1}H NMR spectrum of the drug and its metabolites was observed that further aided metabolite detection and characterisation. Through this group, it was possible to show (by both NMR and MS) that a significant portion of the [^{13}C]-labelled acetyl was removed by metabolism, only to be replaced with endogenous acetyl groups by subsequent phase II metabolism. This de- and re-acetylation has been termed futile acetylation and some 7–10% of the, apparently, "unmetabolised" drug had undergone this process.

Other options for selective detection of drug metabolites for HPLC–NMR–MS studies can employ pre-existing features of the molecules under study. So, for example the presence of fluorine in the drug enables the chromatographic eluent to be monitored by ^{19}F NMR spectroscopy, whilst the specific isotope patterns produced by the inclusion of halogens such as bromine or chlorine can be used for facile detection by MS. An example of the above is the identification of 2-bromo-4-trifluoromethyl-aniline metabolites in rat urine.[18] Here, both ^{19}F NMR detection and negative ion ESI MS (for the bromine isotope pattern) were used to track the metabolites in the LC-eluent. In this way, the major biotransformation was determined as the formation of 2-amino-3-bromo-5-trifluoromethylphenylsulfate, which accounted for some ~23% of the dose of 50 mg/kg of the aniline. More minor metabolites were identified as 2-bromo-4-trifluoromethylphenyl-hydroxylamine-*N*-glucuronide (7% of the dose) and 2-amino-3-bromo-5-trifluoromethylphenyl-glucuronide(1.4% of the dose). There were a number of additional metabolites that were only seen using MS because they were below the detection limits of the NMR under the conditions employed.

A persistent criticism of the hypernated LC–NMR–MS approach is that the relative sensitivities of the two techniques are very different such that they are poor bedfellows. Arguably the recent advances in NMR probe design in the form of "cryroprobes" have reduced the apparent disparity between NMR and MS, and examples of this experimental approach have demonstrated the utility of this combination (e.g. for the characterisation of paracetamol metabolites[39]). Another technical development involves systems where peaks eluting from the column are trapped and concentrated using an on-line solid phase extraction (SPE) system[39,47–56] from which they can then be eluted directly into the NMR flow probe using a deuterated solvent. This has the practical consequence that the composition of the chromatographic eluent is now much less critical than it was if flow into both NMR and MS instruments is being performed. Furthermore, deuterated solvents are no longer required for chromatography as trapping through SPE can be performed with protonated solvents with analytes eluted into the NMR-flow probe with the required deuterated solvents. There may also be some advantages with these systems resulting from the possibility of concentrating trace analytes at the SPE step by combining several runs to increase the amount for NMR. There have been a number of examples of the use of this approach for the determination of the metabolic fate of drugs such as acetaminophen (paracetamol)[39] and phenacetin[55] in man. In the case of the phenacetin study, the focus of the investigation was the extent of the "futile deacetylation" (described above for practolol[17]) undergone by the drug and its metabolites[55]. This type of metabolism is of interest because a potential cause of the analgesic nephropathy associated with phenacetin is the transient production of 4-aminophenol, a potent nephrotoxin. To examine the extent of deacetylation of the drug, a volunteer was dosed with a version of the drug incorporating a stable isotope labelled acetyl (phenacetin-C^2H_3). Thus, following oral administration of 150 mg of phenacetin, urine samples were collected for the periods 0–2 h and 2–4 h post-dose. ^{1}H NMR spectroscopy directly on the urine showed that the expected O-deethylation had occurred to give acetaminophen, which had subsequently been further metabolised to the respective glucuronide, sulphate and *N*-acetyl-L-cysteinyl conjugates. From measurement of the ratio of the acetyl resonance to the aromatic resonances, an estimated level of futile deacetylation of 20% was determined for the sulphate and glucuronide conjugates. However, partial overlap of the acetyl signals for the mercapturate precluded an assessment of futile deacetylation for this metabolite. To obtain more accurate results, further samples of the urine were then freeze-dried, redissolved in acidified water (pH 2) and subjected to partial purification by SPE on a C18 cartridge. The methanolic eluate from SPE was then analysed by LC–MS–SPE–NMR. The phenacetin metabolites were detected by MS and collected on the in-line SPE cartridges from which they were subsequently eluted using deuteromethanol into the flow-probe of a DRX 600 MHz NMR spectrometer for further characterisation. The percentage of futile deacetylation of the SPE-purified metabolites was found to be between 25% and 37% (by MS) depending upon the structure, which is much higher than that observed when acetaminophen is administered to humans. An interesting methodological point that came out of this study was the apparent partial chromatographic resolution of the deuterated and non-deuterated metabolites, which seems to have led to the selective enrichment of the former on the SPE cartridge (fractions were collected based on mass-selective detection of the deuterated metabolites). This resulted in the degree of futile acetylation being underestimated in the NMR-derived measurements compared with the MS data. Sensitivity was also an issue in this study with a number of minor metabolites only detected by MS and, had definitive NMR spectra been required for their identification, additional studies would have been needed to increase the quantities available for spectroscopy.

5. LC–NMR–MS WITH SUPERHEATED D_2O AS THE ELUENT

Whilst conventional eluents have been used with considerable success for LC–NMR–MS-based applications, a useful alternative for certain applications is the use of superheated D_2O (i.e. at temperatures in excess of 100°C) as the mobile phase.[42–44] Such separations are based on the fact that as the temperature of water is increased its properties as a solvent change, giving it a more non-polar character and enabling the elution of more lipophilic compounds from reversed-phase columns than would otherwise be possible. The use of superheated (or even simply hot) water as the mobile phase has advantages as the organic solvent component can be greatly reduced, or even eliminated, thereby also eliminating their signals from the resulting NMR spectra. The first example of this approach was to the investigation of the chromatography of salicylamide, salicylic acid, acetylsalicylic acid (which proved to be unstable under the conditions employed), phenacetin and caffeine.[42] Separations were performed on PS-DVB (polymer) or C18 bonded (Novapak) stationary phases at a range of temperatures up to 190°C, and including a thermal gradient from 80°C to 130°C (8°C/min) on the C18 phase, at 1 ml/min. Stopped-flow ^{1}H NMR spectra were obtained for the analytes, including a 2D-COSY spectrum of salicylamide on 100 μg of material on column. Subsequent work on this system on sulfonamides[43] used the PS-DVB column, a flow rate of 1 ml/min and a thermal gradient from 160°C to 200°C. One unexpected result of this work was that it revealed that using superheated D_2O as an eluent resulted in the rather efficient, on column, deuteration of methyl groups attached to the pyrimidine ring of certain of the analytes (sulfamerazine and sulfamethazine). This phenomenon was rapidly discovered because of the combination of both NMR and MS at the same time. Arguably, had only HPLC-^{1}H NMR been employed, a cursory examination of the data would have suggested, based on the loss of the prominent and diagnostic NMR signals for these methyl groups, that degradation of these analytes had occurred. As it was, the combination of a higher than expected mass, together with the absence of the expected signals in the NMR spectrum, immediately indicated the likely explanation.

LC–NMR and LC–NMR–MS were also used with the water-soluble vitamins pyridoxine (B6), riboflavin and thiamine at temperatures between 160°C and 200°C.[44] Whilst both pyridoxine and riboflavin (^{1}H NMR spectrum illustrated in Figure 4) were stable under these conditions, there was clear evidence from the ^{1}H NMR spectra collected for thiamine, when compared with those of a pure standard of the analyte in D_2O, that changes had occurred (possibly including deuteration of a ring methyl as seen with the sulfonamides) on chromatography. These suggestions were confirmed to some extent by the MS results, which suggested that up to eight protons might have been replaced by deuterium.

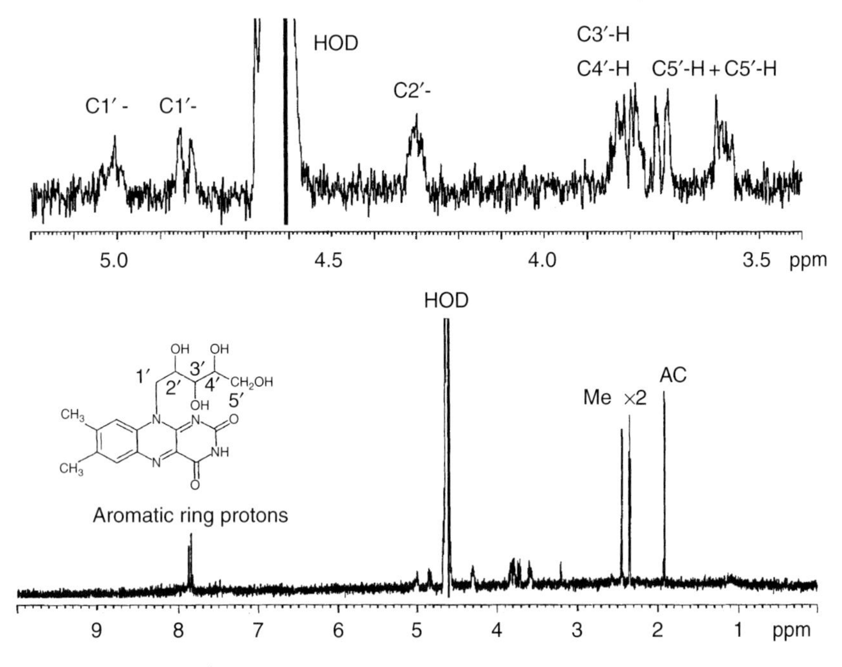

Figure 4 Stopped-flow ^{1}H NMR spectrum for riboflavin obtained following LC–NMR–MS with chromatography on a PS-DVB column at 200°C with D_2O as the mobile. The inset shows details of the small signals between 3.5 and 5.1 ppm.[44]

Decomposition products were also detected. However, these problems were circumvented when less drastic conditions were employed for chromatography using only mildly elevated temperatures (50°C). Again, the combination of NMR and MS data, acquired simultaneously on the same chromatographic peaks, was very valuable in both highlighting the problem and, by providing appropriate structural data, rapidly leading to its resolution.

6. LC–NMR AND LC–NMR–MS WITH ON-LINE COLLECTION FOR OFF-LINE IR

There have been a number of descriptions of the combination of LC–IR–NMR and LC–NMR–MS with on-line NMR (and MS) but off-line IR.[57] LC–IR–NMR was first applied to the analysis of polymer additives (2,6-di-tert-butyl-4-methylphenol, octadecl-3-(3,5-di-tert-butyl-4-hydroxyphenyl) propionate (Irganox 1076) and diisocotyl phthalate), with normal-phase LC on a size-exclusion column. To obtain IR spectra through this system, the eluent (dry deuterochloroform) emerging from the NMR flow probe was deposited onto a slowly rotating germanium disk, with concomitant solvent evaporation. The disk, on which a dry residue of the analytes remained, was then placed in a FT-IR microscope, and spectra for the separated components were obtained.[57] A benefit for the acquisition of the ^{1}H NMR spectra that derived from the use of dry deuterochloroform as the solvent was that exchangeable protons on for example phenolic hydroxyls were readily observed. In a following investigation, MS was added to the combination with the aim that on-line LC–NMR–MS spectra and off-line IR could be obtained.[58] In the event, the dry deuterochloroform that had initially been employed for LC–NMR–IR, as described above, proved to be quite unsuited for use with electrospray ionisation (ESI) as the lack of a proton donor prevented ionisation. To obtain mass spectra, it was necessary to add deuteromethanol and ammonium acetate (10 g/l in D_2O)[58] to the mobile phase. Addition of these "excipients" (which were not needed to perform the separation) to the deuterochloroform used as the eluent did enable MS data to be obtained, but the cost was additional solvent interferences in the resulting NMR spectra, highlighting again the potential for problems of solvent/spectrometer compatibility in hypernated systems. A possible solution to such difficulties is, of course, to add the deuteromethanol and ammonium acetate post-chromatography, immediately before introduction into the mass spectrometer.

7. Fully On-Line LC–UV–NMR–IR–MS Systems

The hybrid on-line/at-line LC–NMR–MS–IR system described above, although interesting, as it did enable the required spectroscopic data to be obtained, was cumbersome to operate, and the advantages of fully on-line spectral collection are clear. Such an on-line system, with acquisition of UV, IR, MS, and NMR spectra on-flow, was demonstrated in an application to a model mixture of four NSAIDs (ibuprofen, flubiprofen, naproxen and indomethacin).[59] In this application ToFMS was performed, thereby giving accurate mass data that allowed the elemental composition of each of the analytes to be determined, with obvious benefits for structure determination. This initial demonstration of LC–UV–IR–NMR–MS, with isocratic reversed-phase LC, used ~2 mg of each analyte on column and so did not really challenge the overall sensitivity of the various spectrometers. Nevertheless, the ability to acquire a complete set of on-flow spectra represented a clear indication of what might be done.

Application of this instrumental set-up to a rather more challenging and realistic application in natural product characterisation and identification was demonstrated through an example relating to the phytoecdysteroids (polyhydroxylated steroids that function as moulting hormones in insects and crustaceans) present in an extract of the plant *Lychnis flos-coculi.*[60] The LC–IR–NMR–UV–MS system was used to obtain a complete set of spectra for the major ecdysteroids present, which in this instance were identified as 20-hydroxyecdysone and polypodine B. For this work, the ^{1}H NMR spectra were obtained in the stopped-flow mode rather than on-flow.

In a later example, using a slightly different instrumental setup, reversed-phase chromatography was used to obtain spectra on-flow on a mixture of polymer additives[61] of the type discussed above. In this case, immediately after the Hypersil H5ODS HPLC column, the eluent was split with the bulk (~95%) directed first to the FT-IR spectrometer. This was fitted with an attenuated total reflectance flow cell (25 μl, with a zinc selenide ATR crystal) with which IR spectra were acquired at 8 cm^{-1} spectral resolution (with the sample ratioed against a background spectrum of the solvent to allow the automated subtraction of the solvent spectrum from the sample spectra). On leaving the FT-IR, the eluent was then sent to the UV spectrometer (188–1000 nm) and thence to the NMR flow probe. On-flow ^{1}H NMR detection was carried out in the pseudo-2D mode at 500.13 MHz. The remaining ~5% of the eluent was analysed using a single-quadrupole mass spectrometer operated with an APCI interface (in positive ion mode). Before introduction of the sample into the ion source of the mass spectrometer, the eluent was mixed with a make up flow of methanol–water (90:10, v/v) at 0.5 ml/min. Because of the relatively non-polar nature of the analytes [triethyleneglycol bis-3(3-tertiary-butyl-4-hydroxy-5-methylphenylpropionate (Irganox 245), butylated hydroxyanisole (BHT), butylated hydroxytoluene (BHA), Bisphenol A and 1,1,3-tris-(2-methyl-4-hydroxy-5-tertiarybutylphenyl)butane (Topanol CA)], the mobile phase used here was required to have a high organic content (80:20 acetonitrile–water). To minimise the interference from the organic modifier deuterated acetonitrile was used, as well as D_2O in the chromatographic eluent. Isocratic chromatography, on quantities of analyte ranging from ~230–990 μg on column with this solvent allowed the separation of four of the five analytes in 10 min, with Bisphenol A and BHA co-eluting in a peak centred at ~2 min. However, inspection of the spectroscopic information for this peak revealed that in fact there was a partial separation, with Bisphenol A eluting on the leading edge and BHA on the tailing edge, allowing "clean spectra" to be obtained for both. The quantities of material available for analysis were such that there was no difficulty in obtaining diagnostic spectra, and the system was also used to conclusively identify an unknown suspected polymer additive as BHT.

8. LC–UV–IR–NMR–MS with Hot and Superheated D_2O as the Eluent

In addition to its use for polymer additives, the LC–IR–UV–NMR–MS system described above was, with the addition of an oven to control temperature, used for the analysis of samples with water (D_2O) as the mobile phase. The use of water as an eluent at elevated temperatures brings the same advantages to LC–IR–UV–NMR–MS as described for the less-complex LC–NMR–MS systems above. Thus, addition of a column heater enabled superheated D_2O to be used for the chromatography of model drugs[62] and an ecdysteroid-containing plant extract[63] with spectroscopic analysis performed on flow.

The LC–IR–UV–NMR–MS system used in these investigations (centred around a DRX 500 NMR spectrometer with a 3-mm i.d. flow cell with a cell volume of 60 μl) is shown schematically in Figure 5.

The model drugs investigated included phenacetin, acetaminophen (paracetamol), antipyrine, norantipyrine, caffeine, phenacetin, *p*-aminobenzoic acid, propranolol, sulfacetamide and sulphanilamide.[62] Chromatographic separations were performed on either Oasis HLB or XTerra phases at either 85°C or 185°C and 0.8 or 1 ml/min

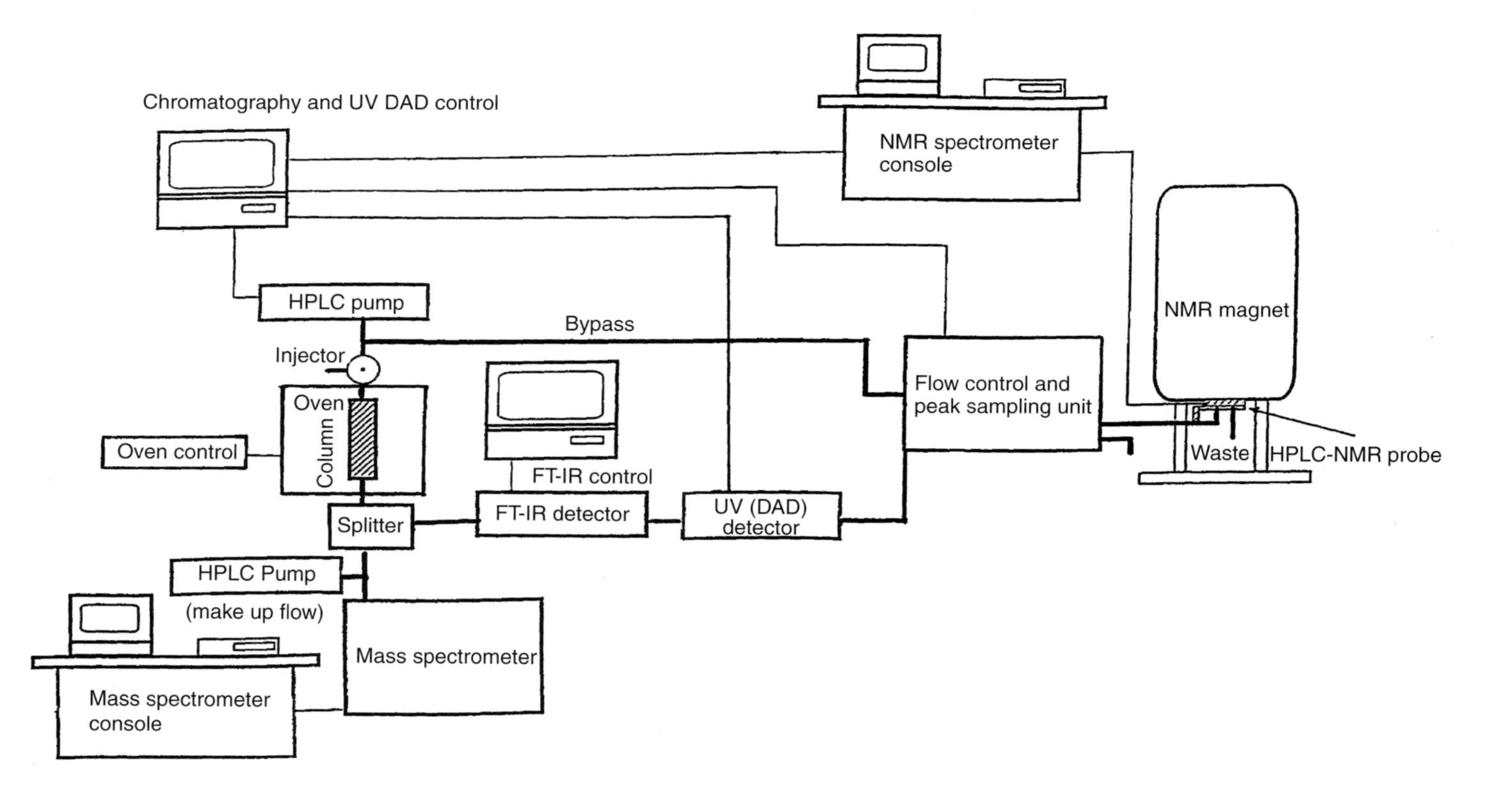

Figure 5 Schematic of the LC–UV–IR–NMR–MS system with an oven enabling superheated water chromatography to be performed.

respectively. With the XTerra phase in fact superheated conditions were not required and 85°C was sufficient to obtain resolution of acetaminophen, caffeine, antipyrine and phenacetin at 1 ml/min. Characteristic on-flow spectra for 23 μg of caffeine were obtained under these conditions. The separation of these compounds on the much more retentive Oasis material required a temperature of 185°C (at a flow rate of 0.8 ml/min), but, despite this relatively high temperature, there was no evidence of compound degradation. The separations achieved for these phases are shown in Figure 6, together with a representative set of spectra for phenacetin following chromatography on the XTerra phase (Figure 7). Clearly this example, based on pure compounds (injected either singly or as mixtures), illustrates the potential of such a combination but can be criticised as not representing a genuine application. These criticisms can be refuted to some extent by the studies undertaken on samples obtained from a number of ecdysteroid-containing plants[63] as these were undertaken on typical, crude extracts. The

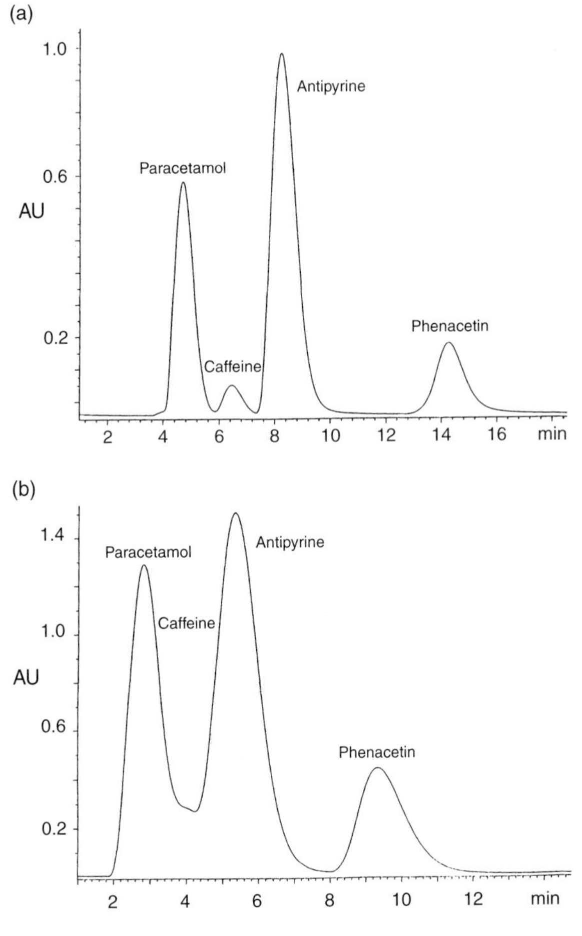

Figure 6 Chromatographic separation of a mixture of four model drugs on (A) XTerra at 85°C and 1 ml/min and (B) Oasis at 185°C and 0.8 ml/min.[62]

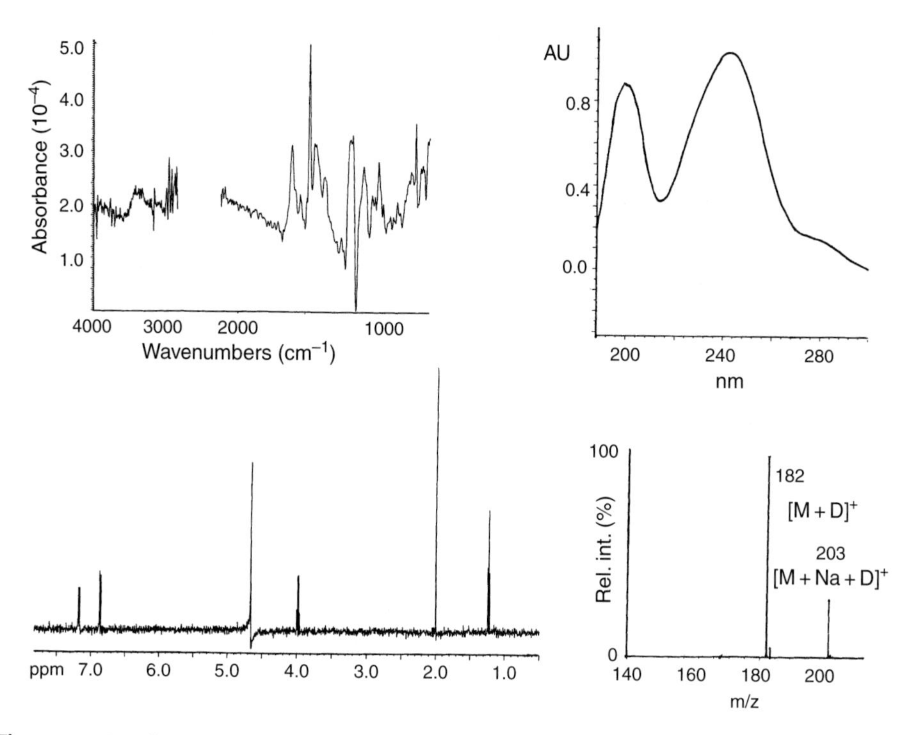

Figure 7 On-flow spectra obtained for phenacetin (144 μg) on flow (1 ml/min) following chromatography on XTerra (see Figure 6).[62]

same experimental set up was used as described for the initial studies described above, but, because of the relatively less polar nature of the ecdysteroids compared with that of the model drugs, the highly retentive Oasis phase proved unsuitable for their chromatography. So, in the case of these plant extracts (obtained from *Silene frivaldskyana, Silene nutans* and *Silene otites*), the XTerra material was used instead. For this work, D_2O heated to 160°C was delivered at 0.8 ml/min to a C8 XTerra HPLC column.[63] In the case of the *Silene frivaldskyana* extract, for example, a complete set of spectral data was obtained on the major component providing unequivocal identification of 20-hydroxyecdysone. These spectra were obtained on ~150 μg on column. The presence of more minor ecdysteroids in the extract was hinted at by the MS data, but these were below the LOSI for NMR and IR. In the same study, the *S. otites* and *S. nutans* extracts were shown to contain 20-hydroxyecdysone and 20-hydroxyecdysone and polypodine B, respectively, as major components (with integristerone A also identified in the *S. nutans* extract as a minor component). No doubt with further optimisation, and state of the art equipment such as cryoflow probes, the same results could have been obtained on much smaller samples, or alternatively good spectra could have been achieved for the minor compounds. In practice, the system described here, with all its limitations, was still able to deliver full UV, IR, MS, and NMR spectra data on 50–200 μg of ecdysteroids on-flow at 0.8 ml min.

9. The Practice of Concatenation and Hypernation

It should be clear from the examples provided above that the linking of several spectrometers into a single system, with or without a separation, is not a great technical challenge and can be done easily with commercially available instrumentation. The systems described above by no means represent the limit of what could be done, and it would be a trivial exercise to add, for example, fluorescence spectroscopy, circular dichroism and inductively coupled plasma MS (or atomic emission spectroscopy) as well if required. There is no doubt that concatenated systems, where the solvent merely acts as a "carrier" for the analytes, represent the simplest and most readily implemented multi-spectrometer solution. For the hypernated systems, the major practical difficulty that has to be overcome is the need to obtain a chromatographic eluent that is compatible with all of the spectrometers in the system. Whilst obtaining such an eluent can be testing for the separation scientist, it has not proved to be an insuperable barrier to success and the use of high-temperature, including superheated, water (or D_2O) as an eluent provides an interesting solution to some of these problems. Whilst thermal gradients were used in some of the studies described here, more recent investigations on the use of thermal gradients for LC have further illustrated the potential benefits of this approach.[64] Elevated temperatures can also be used with conventional solvent–water mixtures as a means of minimising the amount of organic modifier required to perform gradient elution, etc.[65]

Clearly, differences in sensitivity between the different types of spectrometer used in combination with NMR could be a cause of practical difficulties. Usually, in a hypernated LC–NMR–MS system, it is the NMR spectrometer that limits the sensitivity that can be achieved, and generally (although not always) MS and UV spectral data can be acquired with substantially less sample than required for NMR. Such arguments can be deployed to suggest that hypernated systems are not well suited to real world problems and single hypernations, with everything optimised for that particular system, represent a superior approach. Whilst there is some logic to such arguments these can be countered. Indeed, given the dramatic improvements in NMR spectrometer performance and sensitivity that have resulted from better probe design, cryoprobes and higher field strengths, etc., arguments centred merely on sensitivity are no longer convincing. Furthermore, in situations where *both* MS and NMR data are essential for structure elucidation, the total sample requirement will be the same whether spectroscopy is performed using the individual hyphenated systems or in a single analysis by hypernation. In such a case, the advantages in terms of efficient use of time and sample obtained by performing the separation once and acquiring the spectra simultaneously, whilst removing the variables that are immediately introduced by analysis on two different instrumental set-ups, appear compelling.

The best argument against complex hypernated systems is the practical one that many (most?) structure determination problems can be solved without resort to the complete armoury of spectroscopic techniques. There is considerable merit in this view and, given the large capital investment needed to build a dedicated

LC–UV–IR–NMR–MS system versus the limited range of problems for it to solve, it would indeed be difficult to justify. There is, however, no absolute requirement for such dedicated systems. All that is needed is for the various spectrometers that would go to make up the hypernated system to be located in the same laboratory and in such proximity that they can be rapidly linked together into whatever combination is deemed to be required for the problem in hand, and that the appropriate control and data acquisition systems are in place.

10. Conclusions

Hypernation and concatenation each represent an efficient method for obtaining comprehensive spectroscopic data on samples. It is easy to envisage the use of concatenated systems for the routine acquisition of such data on large numbers of samples resulting for example from compound libraries or synthetic chemistry programs. Hypernated systems, especially those centred on the combination of LC with both NMR and MS, are already commercially available and offer a very powerful tool for complex mixture analysis. Neither concatenation nor hypernation represent the solution to all (or even most) the problems that are encountered in compound characterisation; however, in certain circumstances, these approaches represent valuable alternative strategies to the "one spectroscopy at a time" methodologies normally employed for problems of compound identification.

References

1. On-Line LC-NMR and Related Techniques. Albert, K. Ed. Wiley: Chichester, 2002.
2. P. Gförer, L.-H. Tseng, E. Rapp, K. Albert, E. Bayer, Anal. Chem. 73 (2001) 3234–3239.
3. G. Schlotterbeck, L.-H. Tseng, H. Händel, U. Braumann, K. Albert, Anal. Chem. 69 (1997) 1421–1425.
4. U. Braumann, H. Händel, S. Strohschein, M. Spraul, G. Krack, R. Ecker, K. Albert, J. Chromatogr. A 761 (1997) 336–340.
5. K. Albert, U. Braumann, L.-H. Tseng, G. Nicholson, E. Bayer, M. Spraul, M. Hofmann, C. Dowle, M. Chippendale, Anal. Chem. 66 (1994) 3042–3046.
6. T. Hirschfeld, Anal. Chem. 52 (1980) 297A–3012A.
7. I.D. Wilson, U.A Th. Brinkman, J. Chromatogr. A 1000 (2003) 325–356.
8. I.D. Wilson, U.A. Th. Brinkman, Trends Analyt. Chem. 26 (2007) 847–854.
9. D. Louden, A. Handley, S. Taylor, E. Lenz, S. Miller, I.D. Wilson, A. Sage, Analyst 125 (2000) 927–931.
10. E. Lenz, S. Taylor, C. Collins, I.D. Wilson, D. Louden, A. Handley, J. Pharm. Biomed. Anal. 27 (2002) 191–200.
11. S.D. Taylor, B. Wright, E. Clayton, I.D. Wilson, Rapid Commun. Mass Spectrom. 12 (1998) 1732–1736.
12. J.C. Lindon, N.J.C. Bailey, J.K. Nicholson. In Bioanalytical Separations, Wilson, I.D. Ed. Elsevier Science B.V: Amsterdam, 2003, pp. 293–329.
13. J.P. Shockor. In On-Line LC-NMR and Related Techniques, Albert, K. Ed. Wiley: Chichester, 2002, pp. 89–108.

14. M. Sandvoss. In On-Line LC-NMR and Related Techniques, Albert, K. Ed. Wiley: Chichester, 2002, pp. 109–140.
15. G.J. Dear, J. Ayrton, R. Plumb, B.C. Sweatman, I.M. Ismail, I.J. Fraser, P.J. Mutch, Rapid Commun. Mass Spectrom. 12 (1998) 2023–2030.
16. E. Clayton, S. Taylor, B. Wright, I.D. Wilson, Chromatographia 47 (1998) 264–270.
17. G.B. Scarfe, J.C. Lindon, J.K. Nicholson, P. Martin, B. Wright, S. Taylor, E. Lenz, I.D. Wilson, Xenobiotica 30 (2000) 717–729.
18. G.B. Scarfe, B. Wright, E. Clayton, S. Taylor, I.D. Wilson, J.C. Lindon, J.K. Nicholson, Xenobiotica 28 (1998) 373–388.
19. F. Pullen, A.G. Swanson, M.J. Newman, D.S. Richards, Rapid Commun. Mass Spectrom. 9 (1995) 1003–1006.
20. J.P. Shockor, S.E. Unger, I.D. Wilson, P.J. Foxall, J.K. Nicholson, J.C. Lindon, Anal. Chem. 68 (1996) 4431–4435.
21. G.B. Scarfe, I.D. Wilson, M. Spraul, M. Hofmann, U. Braumann, J.C. Lindon, J.K. Nicholson, Anal. Commun. 34 (1997) 37–39.
22. R.M. Holt, M.J. Newman, F.S. Pullen, D.S. Richards, A.G. Swanson, J. Mass Spectrom. 32 (1997) 64–70.
23. I.D. Wilson, E.D. Morgan, R. Lafont, J.P. Shockor, J.C. Lindon, J.K. Nicholson, B. Wright, Chromatographia 49 (1999) 374–378.
24. G.B. Scarfe, B. Wright, E. Clayton, I.D. Wilson, J.C. Lindon, J.K. Nicholson, Xenobiotica 29 (1999) 77–91.
25. S.H. Hansen, A.G. Jensen, C. Cornett, I. Bjornsdottir, D. Taylor, B. Wright, I.D. Wilson, Anal. Chem. 71 (1999) 5235–5241.
26. J.P. Shockor, S.E. Unger, P. Savina, J.K. Nicholson, J.C. Lindon, J. Chromatogr. B 748 (2000) 267–279.
27. N.J.C. Bailey, P.J. Cooper, S.T. Hadfield, E.M. Lenz, J.C. Lindon, J.K. Nicholson, P.D. Stanley, I.D. Wilson, B. Wright, S.D. Taylor, J. Agric. Food Chem. 48 (2000) 42–46.
28. A. Lommen, M. Godejohann, D.P. Venema, P.C.H. Hollmann, M. Spraul, Anal. Chem. 72 (2000) 1793–1797.
29. M. Sandvoss, L.H. Pham, K. Levsen, A. Preiss, C. Mügge, G. Wünsch. Eur. J. Org. Chem. (2000) 1253–1262.
30. M. Sandvoss, A. Weltring, A. Priess, K. Levsen, G. Wuensch, J. Chromatogr. A 917 (2001) 75–86.
31. J. Fritsche, R. Angoelal, M. Dachtler, J. Chromatogr. A 972 (2002) 195–203.
32. G.B. Scarfe, J.K. Nicholson, J.C. Lindon, I.D. Wilson, S. Taylor, E. Clayton, B. Wright, Xenobiotica 32 (2002) 325–337.
33. J. Fritsche, R. Angoela, M. Dachtler, J. Chromatogr. A 972 (2002) 195–203.
34. D. Bao, V. Thanabal, W.F. Pool, J. Pharm. Biomed. Anal. 28 (2002) 23–30.
35. A.M. Gil, I.F. Duarte, M. Godejohan, U. Braumann, M. Maraschin, M. Spraul, Anal. Chim. Acta 488 (2003) 35–51.
36. M. Sandvoss, A. Preiss, K. Levsen, R. Weisemann, M. Spraul, Magn. Reson. Chem. 41 (2003) 949–954.
37. A.M. Gil, I.F. Duarte, M. Godejohann, U. Braumann, M. Maraschin, M. Spraul, Anal. Chim. Acta 488 (2003) 35–51.
38. I.F. Duarte, M. Godejohann, U. Braumann, M. Spraul, A.M. Gil, J. Agric. Food Chem. 51 (2003) 4847–4852.
39. M. Spraul, A.S. Freund, R.E. Nast, R.S. Withers, W.E. Maas, O. Corcoran, Anal. Chem. 75 (2003) 1536–1541.
40. B. Outtara, L. Angenot, P. Guissou, P. Fondu, J. Dubois, M. Frederich, O. Jansen, J.-C. van Heugen, J.-N. Wauters, M. Tits, Phytochemistry 65 (2004) 1145–1151.
41. C. Seger, M. Godejohan, L.H. Tyseng, M. Spraul, A. Girtler, S. Sturm, H. Stuppner, Anal. Chem. 77 (2005) 787–885.
42. R.M. Smith, O. Chienthavorn, I.D. Wilson, B. Wright, S.D. Taylor, Anal. Chem. 71 (1999) 4493–4497.
43. R.M. Smith, O. Chienthavorn, S. Saha, I.D. Wilson, B. Wright, S.D. Taylor, J. Chromatogr. A 886 (2000) 289–295.

44. O. Chienthavorn, R.M. Smith, S. Saha, I.D. Wilson, B. Wright, S.D. Taylor, E.M. Lenz, J. Pharm. Biomed. Anal. 36 (2004) 477–482.
45. M. Spraul, A.S. Fruend, R.E. Nast, R.S. Withers, W.E. Mass, O. Corcoran, Anal. Chem. 75 (2003) 1536–1541.
46. L. Griffiths, R. Horton, Magn. Reson. Chem. 36 (1998) 104–109.
47. M. Lambert, J.L. Wolfender, D. Stærk, S.B. Christensen, K. Hostettmatten, J.W. Jaroszewsk, Anal. Chem. 79 (2007) 727–735.
48. C. Clarkson, D. Stærk, S.H. Hansen, P.J. Smith, J.W. Jarowszewski, J. Nat. Prod. 69 (2006) 527–530.
49. C. Clarkson, D. Stærk, S.H. Hansen, P.J. Smith, J.W. Jarowszewski, J. Nat. Prod. 69 (2006) 1280–1288.
50. C. Clarkson, D. Stærk, S.H. Hansen, J.W. Jaroszewski, Anal. Chem. 77 (2005) 3547–3553.
51. M. Lambert, D. Stærk, S.H. Hansen, M. Sairafianpout, J. Jaroszewski, J. Nat. Prod. 68 (2005) 1500–1509.
52. S. Christophoridou, P. Dais, L.-H. Tseng, M. Spraul, J. Agric. Food Chem. 53 (2005) 4667–4679.
53. M. Godejohann, L.-H. Tseng, U. Braumann, J. Fuchser, M. Spraul, J. Chromatogr. A 1058 (2004) 191–196.
54. V. Exarchou, M. Godejohann, T.V. Beek, I.P. Gerothanassis, J. Vervoort, Anal. Chem. 75 (2003) 6288–6294.
55. A.W. Nicholls, I.D. Wilson, M. Godejohann, J.K. Nicholson, J.P. Shockcor, Xenobiotica 36 (2006) 615–629.
56. B. Kammerer, H. Scheible, G. Zurek, M. Godejohann, K.-P. Zeller, G.H. Gleiter, W. Albrecht, S. Laufer, Xenobiotica 37 (2007) 280–297.
57. M. Ludlow, D. Louden, A. Handley, S. Taylor, B. Wright, I.D. Wilson, Anal. Commun. 36 (1999) 85–87.
58. M. Ludlow, D. Louden, A. Handley, S. Taylor, B. Wright, I.D. Wilson, J. Chromatogr. A 857 (1999) 89–96.
59. D. Louden, A. Handley, S. Taylor, E. Lenz, S. Miller, A. Sage, Anal. Chem. 72 (2000) 3922–3926.
60. D. Louden, A. Handley, S. Taylor, E. Lenz, S. Miller, I.D. Wilson, A. Sage, R. Lafont, J. Chromatogr. A 910 (2001) 237–246.
61. D. Louden, A. Handley, E. Lenz, I. Sinclair, S. Taylor, I.D. Wilson, Anal. Bioanal. Chem. 373 (2002) 508–515.
62. D. Louden, A. Handley, E. Lenz, S. Taylor, I. Sinclair, I.D. Wilson, Analyst 126 (2001) 1625–1629.
63. D. Louden, A. Handley, R. Lafont, S. Taylor, I. Sinclair, E. Lenz, T. Orton, I.D. Wilson, Anal. Chem. 74 (2002) 288–294.
64. A.M. Edge, I.D. Wilson, S. Shillingford, Chromatographia 66 (2007) 831–836.
65. R.S. Plumb, P. Rainville B.W. Smith, K.A. Johnson, J. Castro-Perez, I.D. Wilson, J.K. Nicholson, Anal. Chem. 78 (2006) 7278–7283.

CHAPTER 8

Quantitative High-Resolution Online NMR Spectroscopy in Pharmaceutical Reaction and Process Monitoring

M. Maiwald, O. Steinhof, C. Sleigh, M. Bernstein, *and* H. Hasse

Contents

Abstract

Quantitative high-resolution online nuclear magnetic resonance (NMR) spectroscopy is the method of choice for investigating complex reacting mixtures. We describe the use of NMR flow cells for pharmaceutical reaction and process monitoring where reactions and processes can be covered from several hours down to minutes.

Keywords: on-line technique, PAT, hyphenation, residence time

1. Introduction

The development of online (flow) techniques has tremendously increased the value of nuclear magnetic resonance (NMR) spectroscopy for process development applications and became very attractive as noninvasive methods. Online NMR spectroscopy can meet these demands when flow probes are directly coupled to process equipment like reactors for process monitoring – from laboratory size up to industrial scale. There is a need to study complex multicomponent mixtures and gain insight into their behaviour in the real process. The use of process analytical technology (PAT) in a risk-based approach is massively recommended by pharmaceutical regulatory agencies like FDA and EMEA in their PAT initiative. PAT is considered as a system for designing, analyzing, and controlling manufacturing through timely measurements (i.e., during processing) of critical quality and performance attributes of raw and in-process materials and processes, with the goal of ensuring final product quality. The goal of PAT is to enhance understanding and thereby control the manufacturing process: quality cannot be tested into products; it should be built-in or should be by design.

Online coupling of NMR spectrometers was first developed with high-performance liquid chromatography (HPLC)[1–6] and supercritical fluid chromatography (SFC),[7] using NMR as an analytical detector with high spectral dispersion. Flow NMR probes were developed, and these have also been used for high-throughput NMR spectroscopy.[8,9]

On-line NMR spectroscopy is the method of choice for the investigation of complex fluid mixtures with analytically similar compounds, where other analytical methods (e.g., optical spectroscopy such as UV/VIS, infrared (IR), Raman, or fluorescence spectroscopy) suffer from insufficient differentiation of components. In addition, the high value of NMR in determining chemical structure and accurate quantitation, more subtle features such as speciation (e.g., protonation) are clearly indicated. Many samples are sensitive to changes in concentration, pH, temperature, or pressure, so that chromatographic methods may be ruled out. A major advantage of NMR spectroscopy is that no calibration is needed for quantification in most cases, and the method features a high linearity between absolute signal area and sample concentration. Furthermore, online NMR spectroscopy allows investigations under elevated pressures, e.g., to prevent the solutions from boiling, or for studies under process conditions. The use of self-shielding magnet technology allows compact magnets with low stray fields to be used in miniplant environments, and provides new opportunities for high-field, high-resolution experiments.

2. Quantitative Flow NMR Spectroscopy

One of the most attractive features of quantitative NMR spectroscopy is that the coefficients relating the peak area to the number of nuclei are essentially identical for all nuclei of the same species in a mixture. Provided that spin saturation is avoided, NMR peak areas can be directly used for quantification without further calibration.[10] We describe, below, how the most rigorous quantitation can be achieved with flow NMR monitoring, and Examples 1 and 2 show this. Other applications are less demanding and the apparatus can be simplified without significant compromise to the data quality. Examples 3 and 4 operate on this principle.

2.1. Technical samples on flow

Flow NMR spectroscopy allows investigations of reaction processes nearly in real time and under process conditions in a wide temperature and pressure range. Modern multipulse and pulsed field gradient (PFG) NMR experiments can be used which increase the quality of the data and reduce the experiment time.[11] Typical commercial NMR probe flow cells have an active volume of 60–120 μl and a total volume of about 120–240 μl, which is significantly smaller than that of conventional 5 mm tubes, and a small fraction of the total reactant volume. A major advantage of flow NMR probes is that they are less sensitive to changing magnetic properties of the sample because of the fixed fluid volume and the cell design. When used for online reaction monitoring, the pressure handling capacity of the flow probe is an important design consideration.

2.2. Deuterium-free samples and solvent suppression techniques

Although deuterated solvents are convenient for lock, shim, and referencing in NMR, they are not available for most process and physicochemical studies. Even a 5–10% addition of deuterated solvent to the system is not acceptable due to its cost when working with considerable reaction volumes. Furthermore, deuterium can cause unwanted isotope effects that may affect the kinetics and therefore falsify the results.[12]

As a consequence of exclusively using protonated solvents, residual solvent signals can be intense relative to solute and the NMR receiver gain would normally have to be reduced. The result is a decreased signal-to-noise ratio for the molecules of interest. In some cases solvent protons present serious problems such as analog receiver overflow or digitization faults resulting in artifacts such as baseline distortions, which can render the spectrum totally useless. To circumvent this problem, solvent suppression techniques can be applied. To prevent digital receiver overflow, pulse experiments that suppress the solvent resonance before data acquisition should be used. Further spectral improvements can be achieved as a result of signal conditioning. The former fall into two classes: some methods perturb the solvent magnetization (for instance by saturation), while others (like selective excitation techniques) leave it unchanged.

Working with fully protic samples also excludes field-frequency stabilization (deuterium lock) unless flow cells with two concentric volumes are used allowing lock solvent and reference to be placed in one chamber. Because of the excellent B_0-stability of modern NMR magnets, acquisition times of several hours unlocked can be realized without significant line-broadening or signal frequency drift. The shimming of deuterium-free samples can give excellent results when automated shim procedures based on PFG experiments are used.

Typically the shimming is optimized in less than 1 min. These processes work reliably and without operator supervision, even for concentrated samples or solvent mixtures for which solvent peaks have the same order of magnitude as analyte peaks. Furthermore, we have found that it works in our hands even under flow conditions.

2.3. Solvent suppression techniques

Whereas a variety of solvent suppression techniques yield excellent solvent-reduced spectra (Figure 1), their application may lead to considerable errors in analyte quantification. The choice of a suitable solvent suppression pulse sequences is therefore nontrivial. Generally, cautious reduction of the solvent signal in combination with data analysis (e.g., Lorentz–Gaussian curve fitting), should always be preferred to its complete suppression. Commonly used solvent suppression techniques such as presaturation are not recommended for quantitative studies because they are not sufficiently frequency-selective and can partially saturate solute signals. In order to minimize any magnetization transfer to analyte protons, selective excitation techniques such as WET can be used,[13] particularly for flow experiments. For the studies described in this chapter, WET was used with a $\pi/2$ Gaussian pulse shape, ensuring a narrow Gaussian excitation profile. The minimum frequency separation between the excitation frequency and the closest analyte signal must be determined before solvent suppression is used on this signal.[14] As an alternative to Gaussian pulses, multiple pulse decoupling schemes are described in the literature.[15]

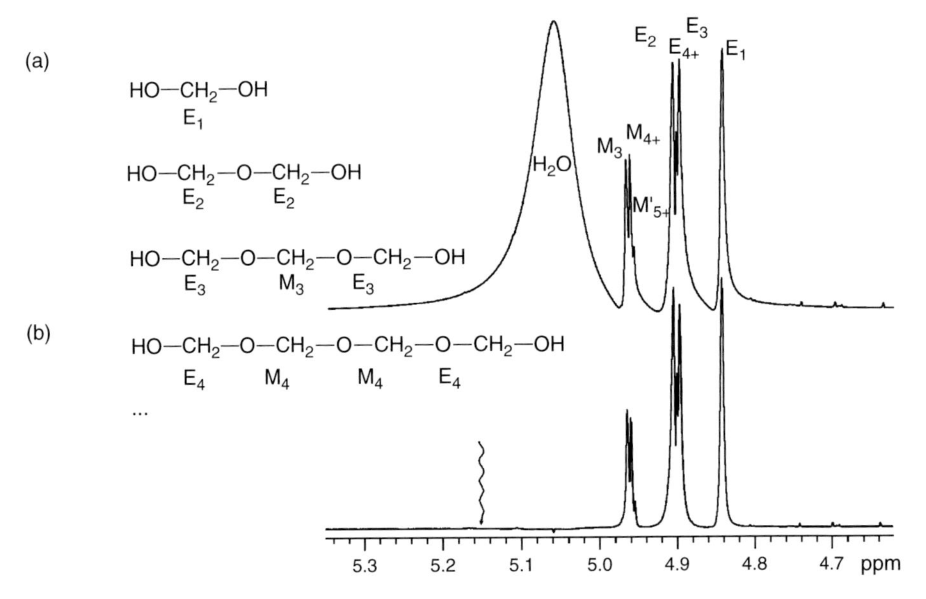

Figure 1 400 MHz ^{1}H NMR spectra of formaldehyde in water (0.340 g/g FA, pH 2, 293 K) with peak and structural assignment. (a) Section of the spectrum without solvent suppression. (b) Water suppression using WET with the transmitter frequency as indicated by an arrow ($\pi/2$ Gaussian shape, duration 47 ms, 40 Hz bandwidth).

2.4. Reacting samples

In reacting samples physicochemical properties such as magnetic susceptibility, pH, or, electrolyte concentration may change considerably through the time course of an experiment. This can lead to peak shifts of analytes, reference material, and solvents, which are related to the changing solvent structure in the molecular environment. Owing to their design that maintains a constant sample volume, NMR flow probes are quite insensitive to changes in magnetic susceptibility of the sample during the course of a reaction. Shimming or tuning of the probe during an experiment is typically not necessary.

2.5. Flowing samples

The theory of flow NMR has been well described in reviews[16–18] and papers on special topics like signal enhancement in flowing liquids.[19] Assuming complete premagnetization of all nuclei of interest, the effect of flow can be considered as a contribution to the overall magnetic relaxation in the detection volume. This is due to the flow of fully magnetized nuclei from the premagnetization volume $V_{\text{premagn.}}$, which continuously enter the active detection volume V_{active} of the probe (see Figure 2(a)). Both, the spin–lattice relaxation time, T_1, and the spin–spin relaxation time, T_2^*, (including field inhomogeneities) in flowing systems depend on the dwell time, τ, of the sample in the active volume, assuming plug flow.

$$\frac{1}{T_{i,\text{flow}}} = \frac{1}{T_{i,\text{static}}} + \frac{1}{\tau} \text{ with } \tau = \frac{V_{\text{active}}}{\dot{V}_{\text{flow}}} \tag{1}$$

where T_i indicates T_1 or T_2^*. τ is related to the flow rate, $\dot{V}_{\text{flow}}$ according to Eq. (1). To establish a complete Boltzmann distribution, which is a prerequisite for accurate quantitative measurements, the sample must reside inside the premagnetization volume of the magnetic field more than four to five times the spin–lattice relaxation

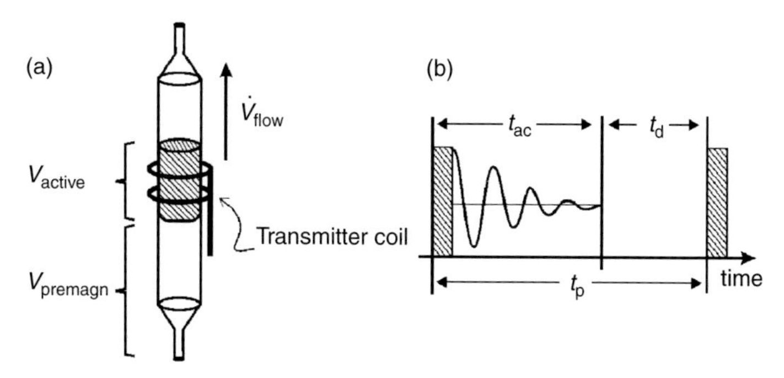

Figure 2 Flow NMR cell: (a) Flow rate $\dot{V}_{\text{flow}}$, premagnetization volume $V_{\text{premagn.}}$ and active volume V_{active} of the NMR flow cell. (b) Pulse repetition time t_p, consisting of the acquisition time t_{ac} and the relaxation delay t_d.

time of the slowest relaxing nucleus $T_{1,\max}$ prior to detection. This is accomplished by adjusting the flow rate $\dot{V}_{\text{flow}}$ to $T_{1,\max}$ and the premagnetization volume of the NMR probe to a value given by Eq. (2)

$$\dot{V}_{\text{flow, max}} = \frac{V_{\text{premagn.}}}{5T_{1,\max}}. \tag{2}$$

According to Eq. (1), the effect of flow causes a decrease of the T_2^* times resulting in a slight increase of the signal half-width $w_{1/2}$ as indicated by Eq. (3)

$$w_{1/2} = \frac{1}{\pi}\left(\frac{1}{T_{2,\text{static}}^*} + \frac{1}{\tau}\right) = \frac{1}{\pi T_{2,\text{flow}}^*} \tag{3}$$

The experimental pulse repetition time, t_p, (period for repetitive pulses, interpulse time) should be adjusted to the flow rate and the residence time of the sample in the active detection volume V_{active}, consisting of the acquisition time t_{ac} and the relaxation delay t_d (cf. Figure 2(b)) according to Eq. (4)

$$t_p = t_{ac} + t_d = \frac{V_{\text{active}}}{\dot{V}_{\text{flow,max}}} \tag{4}$$

With a typical proton T_1 of 2 s at 300 K and a 150 μl premagnetization volume, the maximum flow rate is ~0.9 ml/min.

Since narrow capillaries are undesirable because they contribute significantly to the back pressure, the resulting low flow rates would not allow fast sample transfer from the process unit under investigation to the NMR probe when a direct connection is used. By using a split valve in front of the NMR probe, as described in the experimental section, the sample can be transferred in accelerated flow to the NMR probe, while the the rest of the sample bypasses the probe at a reduced flow rate.

2.6. Acquisition

The most significant NMR parameter is T_1 under reaction conditions for the observe nucleus (cf. Eqs (1)–(4)). T_1 can be determined with the inversion-recovery NMR experiment (delay–π–τ–π/2–FID (free induction decay)).[20] The flow rate through the NMR flow probe then should be adjusted according to Eq. (3) using the measured T_1 times.

2.7. Peak deconvolution

Despite high spectral resolution that follows from use of a high-field magnet (~500 MHz or better for ^{1}H), NMR signals may overlap because the interesting chemical species are structurally related. Solvent signals can, in most cases, only be reduced but not completely eliminated, so that the remaining signal may falsify quantitative results for close solute signals. In the worst case the signal intensities can only be determined by peak deconvolution. Best results were achieved using a custom coded MATLAB^{TM} based tool[14] that uses a Lorentz–Gauss function for representing NMR peaks (Eq. (5)). The function includes four adjustable parameters: the maximum intensity of the peak, I_{max}, the chemical shift at maximum intensity of the peak, δ_{max}, and the Lorentz and Gaussian parameters a and b, respectively.

$$I(\delta) = I_{max} \frac{1}{1 + a^2(\delta - \delta_{max})^2} \exp\left[-b^2(\delta - \delta_{max})^2\right] \tag{5}$$

The parameters I_{max} and δ_{max} of each peak are roughly adjusted before starting the fit routine, whilst default values are used for a and b. The program is especially well suited for semi-automatic studies of large sets of spectra, e.g., from reaction kinetic experiments.

3. Online NMR Spectroscopy

Examples of the use of quantitative online NMR spectroscopy in physico-chemical and engineering applications which stem from recent work in our laboratories[21–24] will be used to exemplify our approach. The aim is to illustrate the applications of the practical considerations discussed in the previous section, rather than presenting in detail the chemical results of these studies.

Except for Examples 3 and 4 (see below) experiments were carried out at the University of Stuttgart using a 400 MHz NMR spectrometer (Unity Inova 400, Varian, Palo Alto, USA) equipped with a modified $^{1}H\{^{13}C,^{15}N\}$ inversely detected, triple resonance, PFG, and microflow probe with an active detection volume of 95 μl, which can be used in a pressure range up to 3.0 MPa and temperatures between 253 and 403 K.

Experiments were also performed at AstraZeneca Charnwood on a 500 MHz NMR spectrometer (DRX500, Bruker, Rheinstetten, Germany) equipped with a ^{1}H SFC flow probe with an active detection volume of 120 μl. The probe was especially designed for pressures up to 35 MPa and temperatures between 273 and 353 K. The high-pressure capacity of this probe is an advantage for this application.

3.1. Field homogeneity

Shimming is nontrivial in reaction monitoring by quantitative high-resolution NMR spectroscopy, and is normally only feasible for a stable, nonreacting mixture. Typical samples that we have considered react within several minutes, so reshimming during the kinetic experiment is not practical. The principal change in susceptibility typically occurs upon initiation of the reaction, but continues to vary through the course of the reaction as a consequence of changes in component concentrations. A variation of magnetic susceptibility leads to magnetic field inhomogeneities in the NMR probe and therefore to distorted peaks, which complicates the analysis of the spectrum. In particular, microinhomogeneities are difficult to avoid, and inevitably occur shortly after reaction initiation. Therefore, only two stable mixtures for shimming are accessible, the equilibrium mixture before the start of the kinetic experiment, and the final equilibrated mixture. In most cases shimming on the equilibrated mixture gives better results. In general, a good shim facilitates data analysis of the acquired spectra and improves its quantifiability.

3.2. Flow scheme

Figure 3 shows a typical online setup for NMR studies of reacting systems. A dosing pump P1 (0.1–20 ml/min flow rate, having a thermostated, Hastelloy pump head) was used to transport the sample from the reactor (C1) to the NMR probe. A 10 μm filter (F1) was used to prevent solid particulate matter from entering the flow system. The flow was split before entering the NMR spectrometer probe. This allowed a slow flow rate in the NMR probe (0.1–1.5 ml/min) to ensure quantitative conditions, while the flow rate in the transfer line was high to allow a rapid sample transfer and return to C1. The bypass was adjusted with a variable back pressure regulator V4. The back pressure regulator V2 (1.72 MPa) is only used to

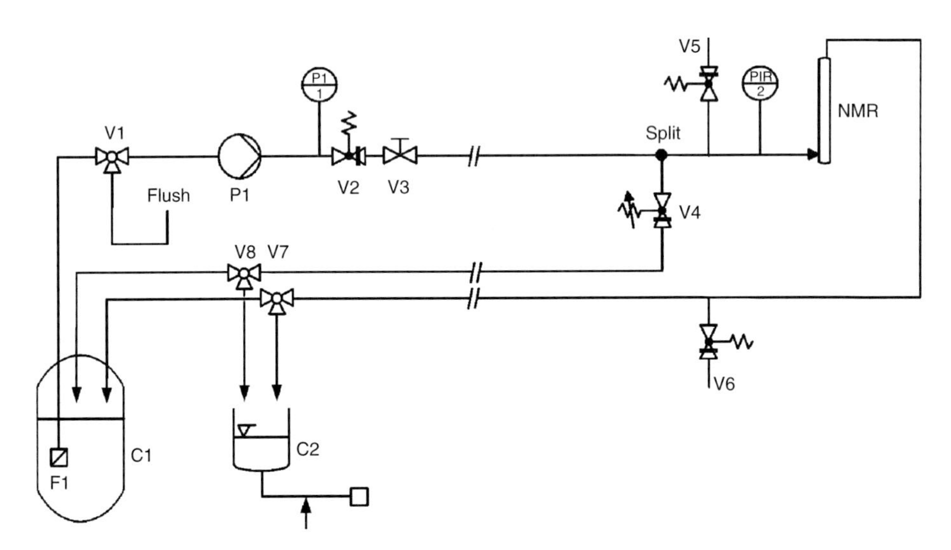

Figure 3 Typical setup for online NMR measurements. C1: laboratory reactor, F1: inlet filter, V1: (tee) purging valve, P1: thermostated dosing pump, V2: back pressure regulator, V3: shut off valve, V4: variable back pressure regulator for split adjustment, V5, V6: pressure relief valves, PI1, PIR2: pressure transducer, NMR: thermostated flow probe of NMR spectrometer, V7, V8: tee valves, C2 container on balance for mass flow control. All tubing 1/16″ OD.

give a constant resistance to the pump P1. An ultralow-volume pressure transducer was used to measure the pressure in the NMR probe.

Care was taken to avoid high pressure because this can easily damage the NMR flow probe. Every effort was made to exclude dust particles or precipitating solids as they can block the tubing. Filters, pressure relief valves (V5, 0.7 MPa and V6, 1.75 MPa in Figure 3), pressure control, and automated pump switch-off, were therefore used in all configurations. Mass flow rates were determined using a balance close to the reactor. For this purpose the flow was routed (via V7 and V8 in Figure 3) to a (pressurized) container C2 on the balance. The reflux tubing between NMR and C1 or C2, respectively, should be of equal lengths to maintain a constant flow.

3.3. Hyphenation

Inert materials such as stainless steel, Hastelloy, or polyetheretherketone (PEEK) were used for the connections between the reactor and the NMR. A choice of tubing with small inner diameter (ID) results in short delay times for sample transfer, but high pressure drops. The tubes leading to the NMR were typically 0.50 mm ID, but the return tubing back to C1 was wider (0.75 mm ID). PEEK tubing was chosen in most cases because of its good mechanical properties, chemical resistance, and biocompatibility. All lines were thermostated at the temperature of C1. For this purpose, the lines were mounted inside insulated silicon tubing filled with heat transfer fluid which was connected to the cryostat via tees. Most parts in contact to the solution were also thermostated. The total holdup of the system described (reactor–NMR–reactor) including filters, tubing, pump, valves, pressure transducer, and the NMR flow cell was kept below 5 ml and therefore small ($<1\%$) compared to the total fluid volume used. Reaction and process conditions such as dosing, thermostating, stirring, and pH control are much more realistic than in 5 mm NMR tubes.

3.4. Residence times

The delay time between a change in the reactor and the according change of the NMR signal is an important parameter in online studies. We determined the time point after the initialization of a reaction when reliable data acquisition could be started. It was shown experimentally that this time is of the order of 2–4 min for the experiments carried out in the present work, which is short compared to the reaction times of the experiments ($\sim$10 min to 12 h).

The nonideal flow of the sample from the reactor to the active region of the NMR probe is described by the residence time distribution (RTD) function which is the transfer and spreading of the sample on its way to the active detection region as a result of laminar flow in the lines (laminar for all experiments carried out here), stagnant regions (e.g., in the connection to the pressure gauge), and back mixing (e.g., in the pump head). The RTD function can be obtained from pulse tracer or step tracer experiments. In those experiments either a concentration pulse or a concentration step is produced in the reactor and the NMR signal is monitored. The desired time-dependent concentration in the reactor (analytical signal $s_{\text{reactor}}(t)$) can be reconstructed from the NMR signal $s_{\text{NMR}}(t)$, if the RTD function is known. Methods for the deconvolution of $s_{\text{NMR}}(t)$ to determine $s_{\text{reactor}}(t)$ are described in literature.[25]

When relatively slow chemical processes were monitored, it was sufficient to correct the NMR time by the mean residence time $\bar{t}_{\text{res}}$ of the sample in the system between the reactor and the NMR probe according to

$$s_{\text{reactor}}(t) = s_{\text{NMR}}\left(t + \bar{t}_{\text{res}}\right) \tag{6}$$

where $\bar{t}_{\text{res}}$ is determined by pulse tracer or step tracer experiments. t_{trans} is the transfer time, defined as the time before the first significant change of concentration is observed in the NMR experiment. The delay time t_{delay} is found as the time after which the detected tracer concentration has risen from zero to its stationary value. The time difference between t_{trans} and t_{delay} is called the dwell time t_{dwell}. When only results from step tracer experiments are available, $\bar{t}_{\text{res}}$ in Eq. (6) can be calculated as $t_{\text{trans}} + (t_{\text{dwell}}/2)$ to a first approximation as long as the monitored reaction time is long compared to any of these time segments.

For homogeneous reacting systems, the reaction continues in the system connecting the reactor and the NMR probe. In this case, the time of the NMR acquisition is directly the time for which the solutions have reacted, and, hence the time used in the evaluation of the reaction kinetics (as long as the NMR data acquisition is started upon initialization of the reaction at $t = 0$). However, reliable NMR data can only be obtained after the delay time t_{delay}, when the solution in the NMR probe is completely replaced by sample from the reactor after the perturbation (Figure 4). Effects of nonlinear flow and back mixing with starting material solution in the transfer lines during initiation of the reaction can be neglected with respect to the small volume fraction of the NMR sample lines.

As an example, Figure 4 shows original NMR data from a kinetic experiment, in which chemical processes after the dilution of an aqueous formaldehyde solution were monitored. NMR data collection was started at the time of dilution. During the transfer time t_{trans}, although the reaction has already started, NMR spectra show the composition of the initial mixture, subject to some perturbations. During t_{dwell} NMR spectra stem from poorly defined mixtures of the reacting solution and the initial solution and should not be used for quantitation. First reliable NMR spectra are only obtained after t_{delay}. It is important to minimize t_{delay} to be able to monitor reactions shortly after their initialization, which usually gives most information on the process.

As an alternative to continuous flow, stop flow techniques can also be used to monitor homogeneous reactions, i.e., the pump can be switched off after the reacting mixture was transferred to the NMR. Thus, also for those experiments the knowledge of t_{delay} is important as the flow must not be stopped before that

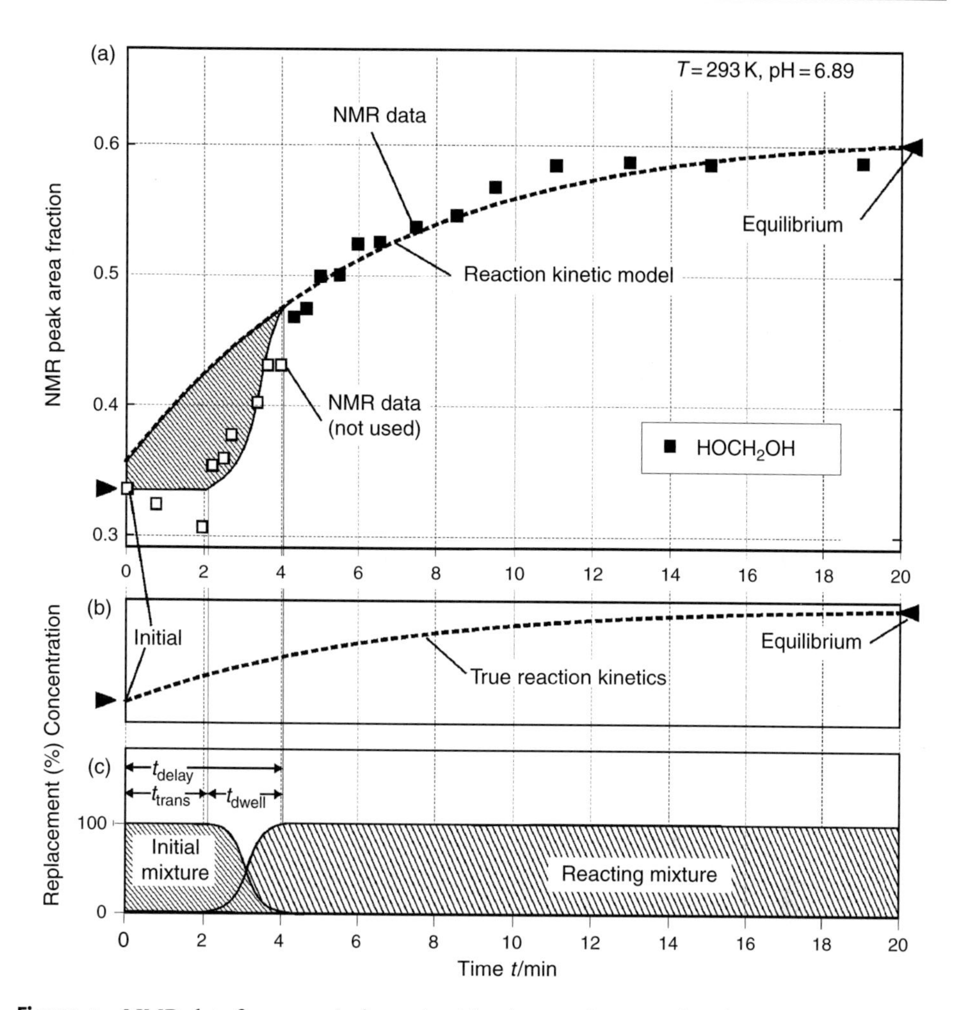

Figure 4 NMR data from a typical reaction kinetic experiment and evaluation. (a) Changes of NMR peak area fractions vs time (dilution at $t = 0$) and fit by a reaction kinetic model. Open symbols are not used for the fit. (b) Concentration changes during reaction in the reactor (qualitative). (c) Replacement of solution according to the RTD function measured independently. The delay time t_{del} is the sum of the transfer time t_{trans} and the residence time t_{dwell} in the probe.

time has elapsed. However, continuous flow has some advantages over stopped flow such as keeping a steady state of pressure and temperature in the NMR flow probe or controlling the viscosity of the studied mixture as a key advantage of online NMR spectroscopy, because a constant supply of new sample from the reactor typically provides most representative data.

It should be mentioned that t_{dwell} as defined above is not equal to the dwell time τ of nuclei in the active detection region, which is commonly used to adjust the pulse repetition time (cf. Eq. (1)). The latter is only a consequence of flow within the NMR probe flow cell, defined assuming a plug flow behavior and thus shorter than the above-mentioned t_{dwell}.

4. EXAMPLES

4.1. Ester formation kinetics

Reactive distillation can be favorable both for the reaction (e.g., increased conversion) and the separation (e.g., avoiding azeotropes). The combination of reaction and distillation in one unit, however, leads to complex process behavior that is difficult to model. Reliable reaction kinetic data are the key to success in the design and scale-up of these integrated processes. Reactive distillation is interesting for producing esters.[26,27]

4.1.1. Heterogeneously catalyzed ester formation reaction

In the present work the butyl acetate esterification according to Eq. (7) was studied.

$$C_4H_9OH + HCOOCH_3 \rightleftharpoons C_4H_9COOCH_3 + H_2O \qquad (7)$$

The common quantitative analysis with, e.g., gas chromatography suffers from long analysis runs. As a result, the monitoring of rapidly changing compositions is difficult, e.g., shortly after starting the kinetic experiment, and valuable information lost. Online ^{1}H NMR spectroscopy allows a fast and reliable analysis of all reaction compounds with sufficient data point densities for an accurate description of the reaction kinetics. It also provides reliable data despite extreme changes in composition during most of the experiments.

For the investigation of homogeneous and heterogeneous acetate kinetics a small, custom made, fully thermostated glass vessel with 30 ml total volume was used. The heterogeneous catalyst was Lewatit K2621. The catalyst was freshly rinsed six times using 15–20 g solution (*n*-butanol–acetic acid at equal molar ratio) over a period of 10–15 min for each rinsing. The reaction was started by addition of catalyst (homogeneous or heterogeneous) to a freshly prepared solution of *n*-butanol and acetic acid. The solution was sampled through a 10 μm filter. The further setup was the same as shown in Figure 3, except that narrow tubing was used (0.25 mm ID sample supply, 0.50 mm ID reflux), drastically reducing the hold-up in the lines. The total sample volume in the entire system was only 15 ml. In all, 35–40 ^{1}H NMR spectra were acquired at equal time points over a total reaction time of 6.5 h. Each spectrum used a single transient, $\pi/2$ excitation pulse (4.5 μs), and a spectral width of 4000 Hz with 64k data points.

The ^{1}H NMR spectra showed completely separated peaks of methylene groups groups in *n*-butanol (3.59 ppm) and butyl acetate (4.04 ppm), which were used to monitor the concentration of these species during the reaction.

In Figure 5 quantitative results from NMR studies of homogeneously and heterogeneously catalyzed butyl acetate kinetics are shown. At the same proton concentration homogeneously catalyzed reactions are expected to be faster than heterogeneously catalyzed reactions due to the reduced accessibility of the active catalytic centers in the solid heterogeneous catalyst and the avoidance of mass

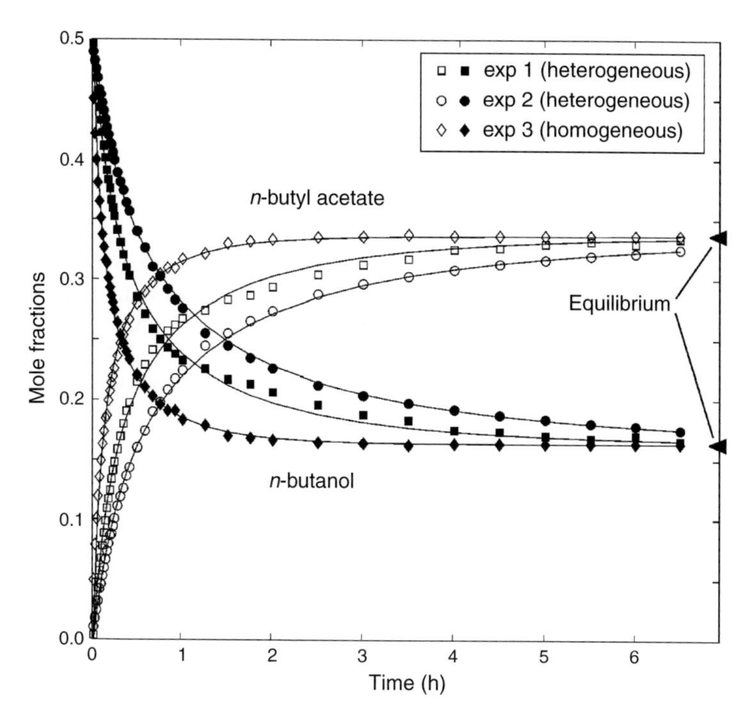

Figure 5 Homogeneously and heterogeneously catalyzed *n*-butyl acetate formation at 363 K. Experimental mole fractions from online NMR for *n*-butanol (black) and *n*-butyl acetate (open) are plotted vs time over 6.5 h for three different experiments. Feed: equimolar mixture of *n*-butanol and acetic acid. Exp. 1: heterogeneous, 14.70 g feed, 4.85 g catalyst Lewatit K2621, exp. 2: heterogeneous, 14.10 g feed, 2.38 g catalyst, exp. 3: homogeneous, 0.002 g/g conc. sulfuric acid (~10% of proton activity compared to exp. 1).

transfer effects. The effects of decreasing the catalyst amount as well as changing from heterogeneous to homogeneous catalysis are clearly seen in Figure 5 and found to be as expected. It was shown in test measurements that the accuracy of the NMR data is equal or even better than that of GC analysis (cf. low scattering of the data shown in Figure 5). In this case the major advantage over GC analysis is that NMR data can be taken 1–2 min (delay time) after initialization at short intervals of typically 3–5 s compared to about 15–20 min needed for a chromatogram. Also faster kinetics than the ones shown in Figure 5 were successfully studied, e.g., cases where equilibrium was established in less than 10 min.

A thermodynamically consistent reaction kinetic model was fitted to the experimental data in good agreement with the experimental data (cf. Figure 5). This gives useful results out of which, e.g., determination of activities of heterogeneous catalysts is possible.

4.1.2. Autocatalyzed ester formation reaction

During distillative purification of an aliphatic ester, an ester-free bottom product was expected, which would only contain starting material components.

Specifications of this process, which thoroughly based on thermodynamic calculations, could not be attained experimentally because of autocatalytic formation of the ester in the stripping section of the column. Systematic studies of this autocatalytic reaction by use of online NMR spectroscopy presented a valuable alternative to studies by GC or potentiometric titrations. The latter techniques are demanding and require pressurization of the system and a rapid analytical technique – especially at increased temperatures.

Figure 6 plots the mole fraction of the ester for the same reaction performed several times. The good reproducibility of the experimental data along the kinetic curve is represented by residues in the range of $10^{-4} < \Delta x < 10^{-3}$ including integration errors. For this experiment, commercially available 5 mm pressure Pyrex NMR tubes were used, which were pressurized to 0.05 MPa with nitrogen gas. The reaction was initiated by inserting the tube into the preheated NMR probe and followed up by NMR over 60 min. The time required to equilibrate the temperature was <3 min.

Small deviations in the repeatability can be seen, which are mainly due to insufficient thermostating of the NMR probe – in addition to small errors from adding dimethylsulfoxide (DMSO) as internal standard. The repeatability of the experiments was improved using the stopped-flow technique[28] by utilizing a flow NMR cell instead of a NMR tube, where the reaction mixture is injected into the preheated NMR flow probe and the time to heat the mixture is decreased. Using an external, precisely thermostated reactor, temperature control was improved even further, as only a small fraction of the reacting mixture passes the NMR spectrometer without precise temperature control.

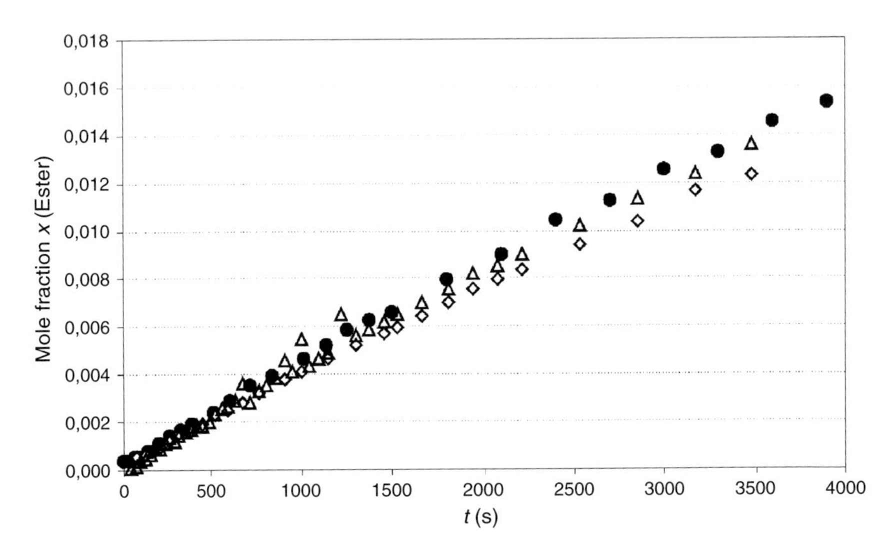

Figure 6 Observation of an autocatalyzed ester formation at $T = 354.8$ K, molar fractions of product are shown. The same experiment has been repeated three times for 3 days by the same experimentator. Quantification against DMSO as internal standard ($\sim 5 \times 10^{-3}$% of weight).

4.2. Hydroxymethylation of urea

The conversion of formaldehyde and urea to urea–formaldehyde resins is an important process in the wood industry. These resins are the most important binding agents for the production of fiber boards and are produced on a very large scale. Numerous attempts have been made on qualitative and quantitative characterization of this system during the past 80 years. For process design the kinetic parameters of the chemical reactions occurring during the two-stage production process have been of chief interest. In recent years, the reduction of formaldehyde emissions from cured resins in the final wood product has been in focus, as formaldehyde emissions received more attention as a thread to public health.

The production process consists of two steps, as shown in Figure 7. In the first step, urea reacts with an aqueous solution of formaldehyde under slightly basic conditions to form several different hydroxymethylated ureas. The degree of substitution depends on the ratio of formaldehyde to urea as well as on other reaction conditions. During the second step, these hydroxymethylated ureas condense under slightly acidic conditions to form methylene and ether bridges between the urea molecules. Long polymer chains and even cross-links can be formed this way. The degree of polymerization is usually being controlled by viscosity measurements.

The reaction network is very complex, as urea theoretically may react with up to four molecules of formaldehyde. Because of this high number of possible combinations many different intermediates can be formed. Higher intermediates can also be formed through different pathways. All reactions are equilibra, which makes isolation and characterization impossible for all but the most simple intermediates. Attempts have been made in the past to synthesize many of these intermediates.[29] Later kinetic studies have been done by titration[30] of excessive formaldehyde and formaldehyde liberating compounds. Unfortunately, these were not able to distinguish between different intermediates, so these studies gave only overall kinetics. Recent works mainly focus on the analysis of cured resins by IR[31] and NMR spectroscopy in solid form and in solution[32–34] as well as on chromatographic[35,36] and rheological investigations.[37]

To analyze in detail the kinetics of a reaction system with such a complex and permanently changing composition, it is essential to monitor formation and decomposition of as many single components as possible. Time-consuming sample preparation is not an option here, because any changes to the sample will inevitably lead to a change in composition due to the ongoing relatively fast equilibrium reactions. Online NMR spectroscopy is very well suited for analysis of this system. The sample transfer is quick and the largest part of the reaction mixture is permanently accessible for manipulations. For this reaction system, it is very important to keep the pH constant, which would be impossible if the analysis was carried out in a standard NMR tube. Figure 8 shows the reaction progress of the hydroxymethylation of urea at a formaldehyde–urea ratio of 2:1 and a reaction temperature of 343 K. The conversion of urea and formation of hydroxymethylated products can be followed quantitatively and with appropriate data workup and modeling the kinetic parameters can be derived.

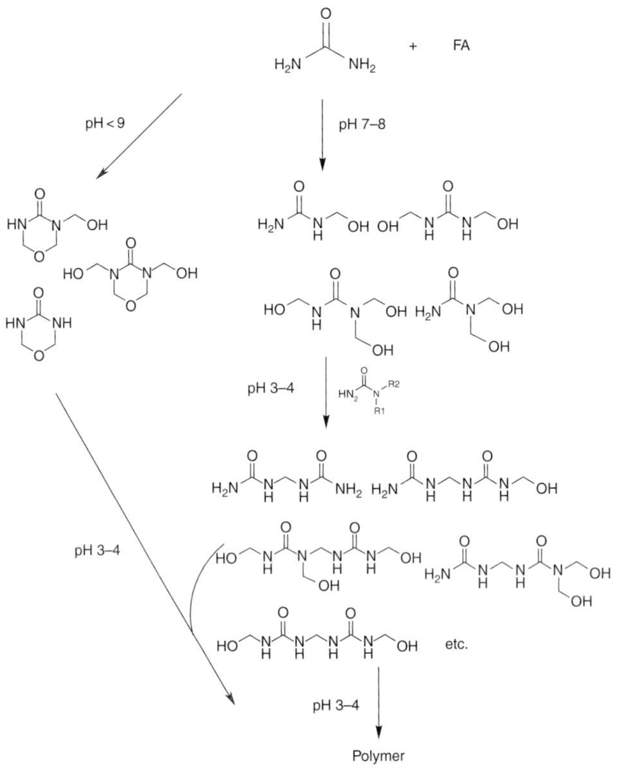

Figure 7 Simplified reaction network of the reaction of urea with formaldehyde. Hydroxymethylation during the first step, condensation during second step, and the last step including cross-linking occurs after application to the wood fiber.

4.3. Pharmaceutical R&D – chemistry

Within the pharmaceutical industry and particularly Process Chemistry laboratories, reaction monitoring by NMR is used, not necessarily to obtain precise kinetic data, but rather to obtain a greater chemical reaction understanding. This may involve identifying key intermediates or protonation states, or simply end of reaction (EOR) determination. Therefore, the apparatus used for reaction monitoring described in the next two examples is simpler and has not been subject to a critical evaluation of optimal flow rates etc to ensure absolute quantification. The apparatus is detailed in Bernstein et al.[24]

4.3.1. Solid–liquid heterogeneous reaction

There have been several reports of homogeneous reactions being studied in an NMR tube; however, the problems associated with inhomogeneous solutions – liquid–liquid or solid particles in the active volume of the NMR tube - have, on the whole, restricted NMR monitoring of in situ reactions to those done in homogeneous solution. However, within the Process Chemistry R&D arena, there are many

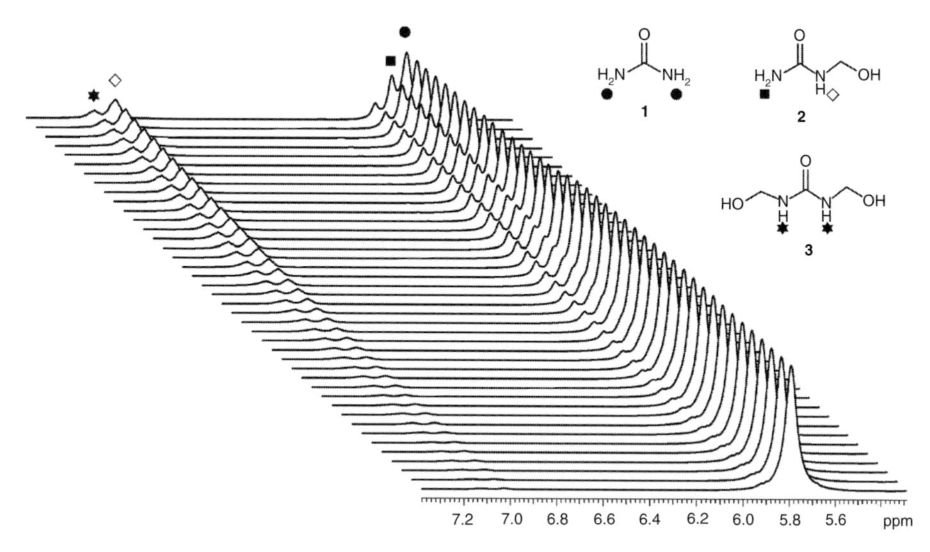

Figure 8 Observation of the reaction of urea with formaldehyde over a course of 2 h at a temperature of 343 K and pH 8.5. The molar ratio of formaldehyde to urea was 2:1. Consumption of urea (●) and formation of two different hydroxymethyl ureas (✶,◇) can be distinguished clearly making use of the amide protons. Protons of methylene groups cannot be used for quantification because of overlap with formaldehyde signals.

reactions undertaken which are heterogeneous in nature.[38] This may be as simple as a solid used as a drying agent or base (see below), or can be a liquid biphasic system having complex mass transfer issues that also affects the chemical kinetics and EOR. Flow NMR is ideally suited for such an investigation to aid chemical understanding by reaction monitoring. A filter is used at the sample point in the chemical reactor to ensure that the tubing does not become blocked, all transfer lines are temperature controlled, but this apparatus was simpler than the one described previously in that it made no special provision for residence times of the sample in the magnetic field.

A simple alkylation reaction that used solid K_2CO_3 to quench generated HBr has been studied by flow NMR.[24] In this example a 2 μm stainless-steel filter was used at the sampling point of the reactor to ensure no solid particles entered the transfer lines or the NMR flow cell, and good quality NMR data were obtained at 323 K (Figure 9).

^{1}H NMR data were acquired at regular intervals over the period of 290 min (Figure 10). There were some interesting observations here, including changes in signal intensity and chemical shifts. The NMR spectra of the alkyl region show consumption of bromoacetate, and formation of product. The reaction was followed by integrating the methyl resonances but did not follow the expected simple second order kinetics. The NMR spectra revealed an unexpected change in the chemical shifts of the starting material (2e, 2d) through the course of the reaction. This suggests protonation of the amine and that the use of solid K_2CO_3 in this reaction is therefore not ideal. This may be due to limited mass transfer to the liquid

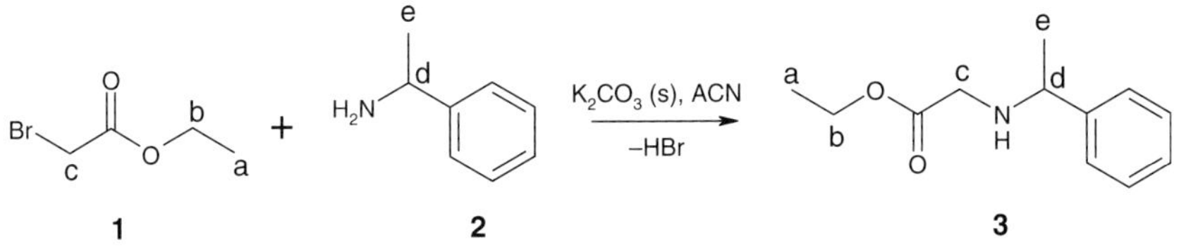

Figure 9 Reaction scheme of alkylation reaction with the presence of solid K_2CO_3 to quench generated HBr presented in Example 3.

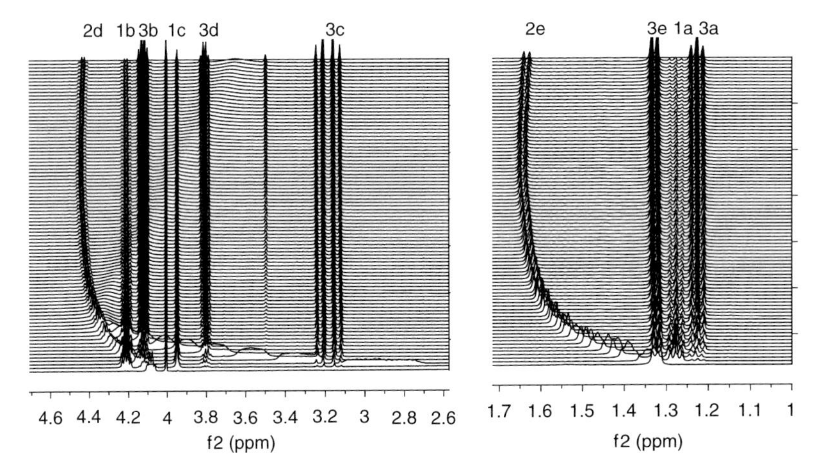

Figure 10 Expansions of ^{1}H NMR array of flow NMR spectra taken with time during the alkylation reaction.

phase or that the amine starting material is a stronger base than K_2CO_3 in this solvent system – information that other analytical techniques may fail to report. Clearly a better base would be required if this reaction was conducted under large scale.

The dramatic change in shift of the water resonance (Figure 10) also suggests a change in pH, with the medium initially becoming more acidic and then late stages of the reaction becoming less acidic. This shift change also suggests that the solid K_2CO_3 is not neutralizing the liberated HBr. This tallies with the observed protonation of the amine, and explains the observed "slowing down" of the reaction and deviation from ideal kinetics.

This example has not only highlighted the unexpected nonideal kinetic behavior of the reaction, but allowed an interpretation of the chemical shift changes of the compound and water resonances to explain this and provide a means for reaction optimization. Hence this example shows the richness of information derived from flow NMR monitoring.

4.3.2. Scaling up: simultaneous flow NMR and NIR monitoring of a demesitylation reaction

Near-infrared (NIR) monitoring of reactions is of great importance for large-scale process chemistry monitoring because it is cheap and easy to implement in large chemical reactors. Fiber-optic cables are used to transfer the response from the probe at the sample point on the reactor vessel to a distant apparatus. Unlike flow NMR there is no need to remove the sample from the reactor, making NIR ideal for use in a manufacturing environment. During chemical route development, it is therefore necessary to understand and assign the changes seen in the NIR data and create a robust monitoring method before this can be used and relied on for reactions performed at such scale. Owing to technical challenges and the complexity of NMR data interpretation, NMR monitoring within the pilot plant has not been practical or easy to perform inline. Our focus is on correlating NMR and NIR data within the laboratory prior to any large scale-up chemistry. An NIR probe was placed in the reaction vessel used for flow NMR monitoring, to allow for simultaneous monitoring by the two spectroscopic methods.

The following reaction was monitored simultaneously by NMR and NIR in a flow reactor (Figure 11).[24]

NMR spectroscopy was initially used to characterize these compounds and provide a "proof of structure." Having identified the resonances for all three compounds, it was then possible to integrate these over a time period of 420 min and their relative integrals converted to a concentration vs time plot (Figure 12). Although our NMR quantitation was not as precise as described earlier in the chapter, there was sufficient information to clearly describe the trends.

The growth of the final product NMR resonances was found to match changes in a NIR band at 5726 cm^{-1} and this band was subsequently attributed to product. We have thus been able to transfer structural understanding and concentration information from the NMR experiments to peak positions and intensity changes in the NIR spectra. Using the flow NMR data, the NIR band intensity was converted to a concentration of the product. Once the chemical understanding has been derived from the NMR experiments and transferred to the NIR data, this can then be used to independently monitor the reaction, reliably detecting species, and establishing concentrations and EOR at scale.

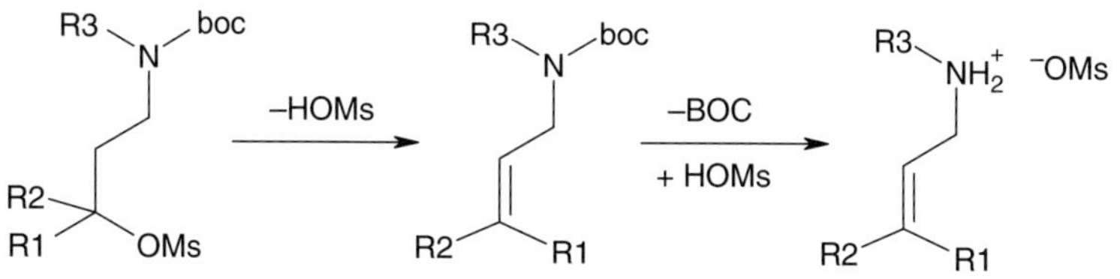

Figure 11 Reaction scheme of demesitylation reaction presented in Example 4.

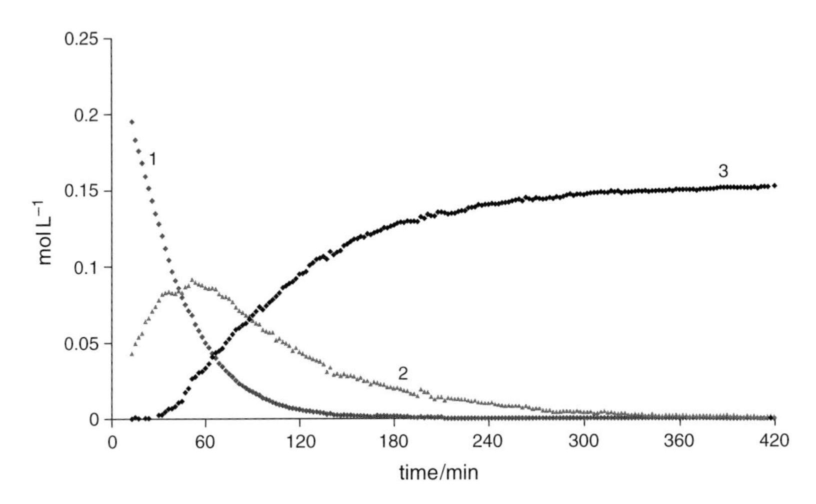

Figure 12 Plot of concentration of species 1, 2 and 3 vs experiment time during the flow NMR study. (M.A. Bernstein, M. Štefinović, C.J. Sleigh (2007) Optimising reaction performance in the pharmaceutical industry by monitoring with NMR. Copyright John Wiley & Sons Limited. Reproduced with permission).

5. Conclusions and Outlook

The use of flow NMR to monitor the concentration of chemical compartments during the course of a reaction is feasible, and provides high-quality quantitative data. The reaction is allowed to progress under conditions that closely resemble the plant, and complex conditions can be accommodated. We have discussed the considerations for the apparatus design and usage when the most accurate quantification is required and shown examples. A simpler apparatus can be used when rigorous quantification is not required. This can easily be used to allow NMR to be used to follow reaction progression and speciation. Indeed, this information can be transferred to NIR spectrum changes, and vibrational spectroscopy is better suited to in-plant monitoring.

We believe there is considerable scope for this technology to be applied to the understanding of reactions performed at large scale. This can be viewed as a simple optimization, or a broader advantage is evident when one considers the need for process understanding in the pharmaceutical sector. Future applications may also include deployment of such devices that use NMR detection in the chemical plant, sampling from the reaction itself in real time. Whilst the technical challenges for this are reduced by new magnet technology, new informatics technology will be required.

REFERENCES

1. E. Bayer, K. Albert, M. Nieder, E. Grom, T. Keller, J. Chromatogr. 186 (1979) 497–507.
2. J.F. Haw, T.E. Glass, D.W. Hausler, E. Motell, H.C. Dorn, Anal. Chem. 52 (1980) 1135–1140.
3. J.F. Haw, T.E. Glass, H.C. Dorn, Anal. Chem. 53 (1981) 2327–2332.

4. E. Bayer, K. Albert, M. Nieder, E. Grom, G. Wolff, M. Rindlisbacher, Anal. Chem. 54 (1982) 1747–1750.
5. H.C. Dorn "High performance Liquid Chromatography Nuclear Magnetic Resonance", in "Encyclopedia of Nuclear Magnetic Resonance", pp. 12070–12085, Wiley, Chichester, New York, 1996.
6. K. Albert, J. Chromatogr. A 856 (1999) 199–211.
7. K. Albert, J. Chromatogr. A 785 (1997) 65–83.
8. P.A. Keifer, Drug Discov. Today 2 (1997) 468–478.
9. B.C. Hamper, D.M. Synderman, T.J. Owen, A.M. Scates, D.C. Owsley, A.S. Kesselring, R.C. Chott, J. Comb. Chem. 1 (1999) 140–150.
10. J.N. Shoolery "Quantitative Measurement", in "Encyclopedia of Nuclear Magnetic Resonance", pp. 3907–3916, Wiley, Chichester, New York, 1996.
11. K. Albert (Ed.), "On-line LC-NMR and Related Techniques", in "On-Line LC-MR and Related Techniques", Wiley, Chichester, New York, 2002.
12. I. Hahnenstein, M. Albert, H. Hasse, C.G. Kreiter, G. Maurer, Ind. Eng. Chem. Res. 34 (1995) 440–450.
13. S.H. Smallcombe, S.L. Patt, P.A. Keifer, J. Magn. Reson. A 117 (1995) 295–303.
14. M. Maiwald, H.H. Fischer, Y.-K. Kim, K. Albert, H. Hasse, J. Magn. Reson. 166 (2004) 135–146.
15. M.A. McCoy, L. Mueller, J. Magn. Reson. A 101 (1993) 122–130.
16. A.I. Zhernovoi, G.D. Latyslev, "Nuclear Magnetic Resonance in Flowing Liquids", Consultants Bureau, New York, 1965.
17. D.W. Jones, T.F. Child, "NMR in following Systems", in "Advances in Magnetic Resonance" (J.S. Waugh, Ed.), Chapter 3, Academic Press, New York, 1976.
18. H.C. Dorn, "Flow NMR", in "Encyclopedia of Nuclear Magnetic Resonance", pp. 2026–2037, Wiley, Chichester, New York, 1996.
19. J.L. Sudmeier, U.L. Gunther, K. Albert, W.W. Bachovchin, J. Magn. Reson. A 118 (1996) 145–156.
20. T.D.W. Claridge, "High-Resolution NMR Techniques", in "Organic Chemistry, Tetrahedron Organic Chemistry Series" (J.E. Baldwin, F.R.S. Williams, R.M. Williams, Eds.), Vol. 19, Pergamon, Amsterdam, 1999.
21. M. Maiwald, H. Li, T. Schnabel, K. Braun, H. Hasse, J. Supercrit. Fluids 43 (2007) 267–275.
22. M. Maiwald, T. Grützner, E. Ströfer, H. Hasse, Anal. Bioanal. Chem. 385 (2006) 910–917.
23. M. Ott, H. Fischer, M. Maiwald, K. Albert, H. Hasse, Chem. Eng. Proc. 44 (2005) 653–660.
24. M.A. Bernstein, M. Stefinovic, C.J. Sleigh, Magn. Reson. Chem. 45 (2007) 564–571.
25. O. Levenspiel, "Chemical Reaction Engineering", pp. 255, Wiley, Chichester, New York, 1999.
26. B. Bessling, J.-M. Löning, A. Ohligschläger, G. Schembecker, K. Sundmacher, Chem. Eng. Technol. 21 (1998) 393–400.
27. M.F. Doherty, G. Buzad, Trans IChemE 70 (1992) A448–A458.
28. M. Maiwald, H.H. Fischer, M. Ott, R. Peschla, C. Kuhnert, C.G. Kreiter, G. Maurer, H. Hasse, Ind. Eng. Chem. Res. 42 (2003) 259–266.
29. H. Kadowaki, Bull. Chem. Soc. Jpn. 11 (1936) 248.
30. J. De Jong, J. De Jonge, Rec. Trav. Chim. 71 (1952) 890–898.
31. G.E. Myers, J. Appl. Polym. Sci. 26 (1981) 747–764.
32. B. Tomita, S. Hatono, J. Polym. Sci.: Polym. Chem. Ed. 16 (1978) 2509–2525.
33. G.E. Maciel, N.M. Szeverenyi, T.A. Early, G.E. Myers, Macromolecules 16 (1983) 598–604.
34. C. Soulard, C. Kamoun, A. Pizzi, J. Appl. Polym. Sci. 72 (1999) 277.
35. P.R. Ludlam, Analyst 98 (1973) 107–115.
36. A. Despres, A. Pizzi, J. Appl. Polym. Sci. 100 (2006) 1406–1412.
37. A. Suurpere, P. Christjanson, L. Siimer, Chemine Technologija 2 (2005) 16–21.
38. J.S. Carey, D. Laffan, C. Thomson, M.T. Williams, Org. Biomol. Chem. 4 (2006) 2337.

Index